Behavioral Operations in Planning and Scheduling

Jan C. Fransoo • Toni Wäfler • John R. Wilson
Editors

Behavioral Operations in Planning and Scheduling

Springer

Editors
Prof. Dr. Jan C. Fransoo
Eindhoven University of Technology
School of Industrial Engineering
P.O. Box 513
Pav F4
5600 MB
Eindhoven
The Netherlands
j.c.fransoo@tue.nl

Prof. Dr. Toni Wäfler
University of Applied Sciences
Northwestern Switzerland
School of Applied Psychology
Riggenbachstrasse 16
4600 Olten, Solothurn
Switzerland
Toni.waefler@fhnw.ch

Prof. Dr. John R. Wilson
University of Nottingham
Human Factors Group
Faculty of Engineering
University Park
Room A65 Coates
NG7 2RD Nottingham
United Kingdom
John.wilson@nottingham.ac.uk

ISBN 978-3-642-42391-8 ISBN 978-3-642-13382-4 (eBook)
DOI 10.1007/978-3-642-13382-4
Springer Heidelberg Dordrecht London New York

Cover design: WMXDesign GmbH, Heidelberg, Germany

Printed on acid-free paper

Springer is part of Springer Science+Business Media (www.springer.com)

Preface

The work presented in this book is the result of the work conducted by the research network Human and Organizational Factors in Planning and Scheduling (HOPS). The HOPS network was established in 2004 and was funded as a so-called "Action" by COST (European Cooperation in Science and Technology; cf. http://www.cost.eu) for a period of four years. COST supports new Actions when they address a relatively new and relevant field of research, in many cases interdisciplinary. COST Actions have an open nature and any researcher interested in the topic can join them.

The HOPS network has been important in the sense that researchers with a wide variety of disciplines have been working together on joint projects over the course of the Action. The joint research has been very productive, with dozens of joint papers having been published in academic journals over the past few years.

In this book, we have brought together the main results from the Action, but have furthermore complemented these by providing more extensive, expository background writing in each of the Chapters that will enable the novel and experienced researcher in this field to get quickly at grip with the various disciplinary insights and knowledge that underlie our studies. We therefore expect this book to be useful for a variety of audiences. First of all, it serves as a reference of the current state of research in the field for those interested in conducting research in this area and intending to broaden their disciplinary and methodological scope. Second, it can serve as teaching material in a graduate course on the role that humans play in planning and scheduling environments. Finally, it also serves to help practitioner to better understand the background of their experiences with planning and scheduling systems, and provides guidelines and insights in how to better manage the design and implementation of such systems.

We would like to thank COST, especially Alfredas Chmieliauskas (COST Rapporteur), David Gronbaek, Julia Stamm, Francesca Boscolo (Scientific Officers), and Jie Zhu (Administrative Officer) for their support to the Action, without which this network never would have been established. Especially the low bureaucratic burden by which financial support is provided by COST strongly stimulates new networks like ours.

Second, we would like to thank all contributors to this volume in spending their time to consolidate their multitude of findings and insights into a set of extensive and comprehensive book chapters.

Finally, we would like to thank Eindhoven University of Technology, and especially Walter Stein, for supporting us in the final editing stages of the process. Many book projects tend to fail at this important stage and without the support of Walter this book would not have been realized.

Eindhoven, The Netherlands — Jan C. Fransoo
Olten, Switzerland — Toni Wäfler
Nottingham, UK — John R. Wilson
10 August 2010

Contents

Contributors

Martina Berglund Department of Management and Engineering, Quality Technology and Management Linköping University, Linköping, Sweden, martina.berglund@liu.se

Jessica Bruch Department of Industrial Engineering and Management, Jönköping University, Jönköping, Sweden, jessica.bruch@jth.hj.se

Julien Cegarra Université de Toulouse, Toulouse, France, julien.cegarra@univ-jfc.fr

Naoufel Cheikhrouhou Ecole Polytechnique Fédérale de Lausanne, Lausanne, Switzerland, naoufel.cheikhrouhou@epfl.ch

Cees De Snoo Faculty of Economics and Business University of Groningen, Groningen, The Netherlands, c.de.snoo@rug.nl

Christos Dimopoulos European University Cyprus, Nicosia, Cyprus, c.dimopoulos@euc.ac.cy

Gael Farine Ecole Polytechnique Fédérale de Lausanne, Lausanne, Switzerland

Dieter Fischer Institute für Business Engineering, University of Applied Sciences Northwestern Switzerland, Brugg, Switzerland, dieter.fischer@fhnw.ch

Katrin Fischer School of Applied Psychology, University of Applied Sciences Northwestern Switzerland, Olten, Switzerland, katrin.fischer@fhnw.ch

Kary Främling Aalto University, Helsinki and Espoo, Finland, Kary.Framling@tkk.fi

Jan Fransoo School of Industrial Engineering, Eindhoven University of Technology, Eindhoven, The Netherlands, J.C.Fransoo@tue.nl

Roland Gasser Department of Mechanical and Industrial Engineering, University of Toronto, Toronto, ON, Canada, roland.gasser@utoronto.ca

Kathrin Gärtner School of Applied Psychology, University of Applied Sciences Northwestern Switzerland, Olten, Switzerland, kathrin.gaertner@fhnw.ch

Bernard Grabot ENIT, Université de Toulouse, Toulouse, France, bernard@enit.fr

Jane Guinery Nottingham University Business School, Nottingham, UK, jane.guinery@nottingham.ac.uk

Hannes Günter Department of Organization and Strategy, Maastricht University, Maastricht, The Netherlands, h.guenter@maastrichtuniversity.nl

Jean-Michel Hoc Centre National de la Recherche Scientifique, (CNRS: French National Research Centre), University of Technology of Compiègne, Compiègne, France, Jean-Michel.Hoc@irccyn.ec-nantes.fr

Jan Holmström Aalto University, Helsinki and Espoo, Finland, jan.holmstrom@tkk.fi

David Jentsch Department of Factory Planning and Factory Management, Chemnitz University of Technology, Chemnitz, Germany, david.jentsch@mb.tu-chemnitz.de

Johan Karltun Department of Industrial Engineering and Management, Jönköping University, Jönköping, Sweden, johan.karltun@jth.hj.se

Stefan Marsina University of Economics in Bratislava, Bratislava, Slovakia, marsina@euba.sk

Anne Mayère University of Toulouse, Toulouse, France, anne.mayere@iut-tlse3.fr

Nasser Mebarki Département Qualité, Logistique Industrielle et Organisation (QLIO), University of Nantes, Nantes, France, nasser.mebarki@univ-nantes.fr

Philip Moscoso IESE Business School, Universidad de Navarra, Pamplona, Spain, pmoscoso@iese.edu

Guillaume Pinot Centre National de la Recherche Scientifique, (CNRS: French National Research Centre), University of Technology of Compiègne, Compiègne, France, guillaume.pinot@irccyn.ec-nantes.fr

Johann Riedel Nottingham University Business School, Nottingham, UK, johann.riedel@nottingham.ac.uk

Ralph Riedel Department of Factory Planning and Factory Management, Chemnitz University of Technology, Chemnitz, Germany, ralph.riedel@mb.tu-chemnitz.de

Jan Riezebos Faculty of Economics and Business University of Groningen, Groningen, The Netherlands, j.riezebos@rug.nl

Craig Shepherd Nottingham University Business School, Nottingham, UK, craig.shepherd@nottingham.ac.uk

Ulrike Starker University of Bamberg, Bamberg, Germany, ulrike.starker@uni-bamberg.de

Wout van Wezel Faculty of Economics and Business University of Groningen, Groningen, The Netherlands, w.m.c.van.wezel@rug.nl

Rüdiger von derWeth HTW, University of Applied Sciences, Dresden, Germany, weth@wiwi.htw-dresden.de

Toni Wäfler School of Applied Psychology, University of Applied Sciences Northwestern Switzerland, Olten, Switzerland, toni.waefler@fhnw.ch

Peter Williams University of Limerick, Limerick, Ireland, peter.williams@ul.ie

Vincent Wiers School of Industrial Engineering, Eindhoven University of Technology, Eindhoven, Netherlands, v.c.s.wiers@tue.nl

John R. Wilson Professor of Occupational Ergonomics, University of Nottingham, Nottingham, UK, john.wilson@nottingham.ac.uk

Part I
Introduction

Chapter 1
Introduction

Jan Fransoo, Toni Wäfler, and John R. Wilson

Planning and scheduling play an important role in performance enhancement in any operation. This has probably been best recognized in industry, where a wide range of software applications have been developed to support decision makers with decisions such as machine scheduling, forecasting, inventory control, and sales & operations planning.

While these systems have been deployed on a large scale since the late 1990s, their actual use is a source of big debate. Symptoms abound that the actual usage is far from the intended usage. Many schedulers use their own notebooks or spreadsheets alongside (or even instead of) their advanced decision support systems. Those that do use the system may limit its functionality to that of an automated planning board and make effective use of the drag and drop facilities but do not touch the optimization engine. Also at organizational level, humans may organize processes that lead to decisions in a very different way from what the organizational chart tells us.

This book explores the role of human behavior in planning and scheduling decisions in industry. From a wide range of disciplinary perspectives, we investigate how human behavior determines performance, and how this interacts with the formal organizational processes and the decision support systems.

J. Fransoo (✉)
School of Industrial Engineering, Eindhoven University of Technology, Eindhoven, The Netherlands
e-mail: J.C.Fransoo@tue.nl

T. Wäfler
School of Applied Psychology, University of Applied Sciences Northwestern Switzerland, Olten, Switzerland
e-mail: toni.waefler@fhnw.ch

J.R. Wilson
Professor of Occupational Ergonomics, University of Nottingham, Nottingham, UK
e-mail: john.wilson@nottingham.ac.uk

J.C. Fransoo et al. (eds.), *Behavioral Operations in Planning and Scheduling*,
DOI 10.1007/978-3-642-13382-4_1, © Springer-Verlag Berlin Heidelberg 2011

We have divided this book into two main parts, supplemented by an introductory part that describes two industrial case studies to make the reader aware of the practice of production planning and scheduling, and a supportive part that concisely describes the terminology that is involved with understanding human behavior in planning and scheduling.

1.1 Part I: Introduction

The remainder of this introductory chapter consists of two case studies. In *Chapter* 2, the authors describe a case study that focuses on the information processing and decision-making that is found in Production Planning and Scheduling (PPS). Facing the growing complexity of business processes, these efforts are becoming more demanding for human planners and scheduler. Although equipped with highly developed PPS software, there are limits to the amount of complexity that can be handled by human planners and schedulers. Actually, many IT solutions create additional complexity for the user, and therefore decision-making becomes more challenging. In order to investigate decision-making as a central aspect of real-world behavior of human planners, a Naturalistic Decision-Making (NDM) approach has been used to study decision-making in PPS. NDM explicitly claims that human experts' decision-making is mostly non-algorithmic, following heuristics instead of fully rational calculations. Experts are found to be using decision strategies that safe time and reduce effort while preserving sufficient accuracy.

Chapter 3 describes a case study on the relocation of the planning department and the assignment of foremen in a manufacturing environment. Improvement of planning and scheduling processes requires integral analysis and redesign, taking into account organizational, human, technological, environmental and situational factors. This chapter presents a case study from a medium-sized scaffold manufacturer. The focus is on the activities of the humans being involved in planning and scheduling activities. The case study presented in this chapter is part of a research project of several years aimed at overall improvement of the planning performance within the company. During the project, it was decided to move the planning department to another place in the factory. Almost simultaneously, it was decided to designate production foremen who should act as communication hubs between the planners and the shop floor operators. The research team investigated the consequences of both changes by means of repeated measurements. The quasi-experimental design consisted of a pre-change and a post-change measurement regarding mutual interdependencies, interaction patterns and perceived performance. Preceding the experiment, the researchers thoroughly analyzed the planning situation and developed an integral description of it. This description as well as the setup and results of the experiment are reported in this chapter.

1.2 PART II: Organization of the Planning Process

In part II, we focus on the organization of the planning process. Planners and schedulers are assigned roles and responsibilities in a company. Moreover, they also take up specific roles and responsibilities that are not always identical to those that have been assigned. In this part, we extensively describe and study the organizational side of planning and scheduling, try to understand the role that planners and schedulers actually play in influencing the overall performance, and develop design rules on how this performance can be further enhanced.

Chapter 4 sets out to demonstrate the unsung contribution of production planners and schedulers in manufacturing businesses. In particular, it focuses on their contribution at production and sales interfaces by highlighting their activities and influence across functional, work group and organizational interfaces, and the knowledge and skills they apply to make and implement planning and scheduling decisions. To achieve this the case study addresses the following: What tasks and work activities does planning, scheduling and control consist of in relation to these interfaces? How do planners and schedulers perform their tasks? How can planners' and schedulers' activities related to production and sales interfaces be captured and modeled? How do planners and schedulers influence others in the organization? What knowledge do they contribute and how is it incorporated into decisions?

Whereas Chapter 4 focuses on the organization within a company, *Chapter* 5 addresses collaborative planning processes across different companies. While collaborative planning and relationship quality are considered key contributors to supply chain performance, their mechanisms and linkages remain unclear. In order to help address this issue this chapter introduces and unpacks the concepts of collaborative planning and relationship quality and investigates their role in supply chains. A multidisciplinary literature review was undertaken to identify conceptual and empirical work on relationship quality and collaborative planning. The chapter reveals a number of shortcomings in the literature and provides suggestions to guide future research on the links between collaborative planning, relationship quality, and supply chain performance. Implications are also provided for practitioners interested in enhancing the quality of inter-organizational relationships and collaborative planning in supply chains.

The measurement of supply chain performance is addressed in *Chapter* 6. This chapter aims to go some way towards addressing the dearth of research into performance measurement systems and metrics of supply chains by critically reviewing the contemporary literature and suggesting possible avenues for future research. The chapter provides a taxonomy of performance measures followed by a critical evaluation of measurement systems designed to evaluate the performance of supply chains. It is argued that despite considerable advances in the literature in recent years, a number of important problems have not yet received adequate attention, including: the factors influencing the successful implementation of performance measurement systems for supply chains; the forces shaping their evolution over time; and, the problem of their ongoing maintenance. The chapter provides both a taxonomy of measures and outlines specific implications for future research.

Chapter 7 addresses a particular type of collaborative planning, namely planning between large Original-Equipment-Manufactures (OEMs) and their relatively small suppliers. The increasing focus of manufacturing companies on their core business results in the development of larger and more complex supply chains, which are made up of both, large and small companies. From the perspective of large companies, the quality of coordination in supply chains is mainly dependent upon the competence of the human planners involved at each level of the supply chain in information processing methods and tools. In this interpretation, coordination problems are commonly considered the consequence of poor competence in the Small and Medium sized Enterprises (SMEs) regarding information processing, and especially in operations planning and control. Consequently, most large companies have launched programs disseminating "best practice", standardized business processes, and software tools aiming at developing efficient planning procedures throughout their supply chains, and so at a convergence of operational objectives. However, this picture does not take into account other inherent aspects of the planning problem within a supply chain, such as variety in organizational cultures and ways of gathering and interpreting information. The aim of this chapter is to illustrate the problematic situation outlined above through real industrial examples, and to suggest a framework allowing better representation and understanding of these coordination problems, to inform future system design and improvement activity. It is emphasized that critical non-technical issues have to be taken into account in the design of the planning processes across a supply chain. The authors link these to justified practice and cultural concerns of particular interest to SMEs, a voice not well represented in the literature.

In *Chapter* 8, we return to the planning problem within companies. Where Chapter 4 focuses on horizontal collaboration across different functions within a company, in this chapter the authors investigate the vertical collaboration between planners at the higher level and schedulers and dispatchers at the lower level. Instabilities in production planning and control have received considerable attention due to their negative impact on planning performance. However, extant research has been limited to theoretical (e.g. simulation) settings and has focused on specific methodologies (e.g. mathematical) to overcome instabilities. The objective of this chapter is to make two contributions to the theory development on production planning instabilities. First, it aims to make an empirical contribution through an in-depth case study, and second, it introduces a holistic framework that supports analysis of hierarchical planning systems and their potential instabilities. The in-depth case study is carried out at an industrial company that has difficulty to meet its customer deadlines and faces a significant order backlog. Planners of the company at different hierarchical levels and order chasers on the shop floor end up rescheduling open orders and updating lead times continuously when trying to meet deadlines, but eventually are not able to improve order fulfillment. Only after the introduction of an Advanced Planning System (APS) and centralization of planning decisions in a single department, on-time delivery was significantly improved and order backlog drastically reduced. This case study allows studying of the underlying mechanism of such planning instabilities, with a particular focus on the impact on stability of human and organizational factors. Based on their findings and additional conceptual research the authors have then developed a framework constituted by six key planning systems attributes. By taking into consideration these factors, a firm can address the root causes of planning instabilities, rather than merely focus on its symptoms.

Chapter 9 introduces a concept of product centric services and the use of this concept for decision-making in planning and scheduling. The development of information and communications technology has already introduced planning tools in many contexts that previously were not planned and monitored. A new emerging application area for planning and control is the after-sales supply chain. The introduction of technologies that make it possible to identify and track individual products after the sale hold the promise of improved efficiency in both the positioning of resources, such as spare parts and field engineers, as well as of improving the delivery of repair and maintenance services. However, since the number of organizations involved in the after-sales service supply chain is orders of magnitude larger than for the production and distribution of products, a major challenge is how to organize the after-sales service supply chain. Sequential process configurations, and hierarchical networks are too rigid for the constantly changing after-sales environment. A proposal for a potential solution approach is product centric services. The service concept is based on identifying and tracking product individuals (instances) independently from planning and controlling service delivery (provision). The potential benefit of the approach is a decoupling of different types of provision, and the possibility of developing new service concepts incrementally. The chapter on product centric services introduces the basic concepts and illustrative examples, and guides the reader to the emerging body of literature.

Chapter 10 focuses on the concept of control, and investigates how humans can be enabled by giving them control from a work and organizational psychology point of view. This conceptual chapter has been triggered by the experience that the implementation of new Information Technology (IT) supporting planning, scheduling, and control – although being more sophisticated than earlier systems – does not necessarily result in better control. Also, researchers experienced that the implementation of the same IT leads to different results in similar organizations. Against this background, the authors introduce a process model of control. The model proposes a set of interrelated factors determining control. At its core, it assumes that control results as a fit of control requirements and control behavior. The former is determined by operational uncertainties the latter by control opportunities, control skills and control motivation. Since the implementation of new IT can have an impact on all these factors it can lead to a misfit of control behavior and control requirements and hence to low control – even if the new IT itself is more powerful than the old IT. Furthermore, they also discuss motivational influences these changes may have on human behavior. Finally, they derive some practical dos and don'ts when implementing new IT.

Chapter 11 makes the step from the organizational and work psychology perspective to a cognitive perspective of planning and scheduling. In this chapter, the authors study the process of building decision support systems such that they are eventually accepted by the user. They take the well-known perspective of Technology Acceptance, and extend this to take into account the effect that decision support systems in planning and scheduling are usually developed together with the eventual user, and typically tailored to the specific situation of a planner or scheduler. Based on a set of case studies, they develop a number of new insights. In the second part of this chapter, they also specifically address the issue of trust in information systems, and present results of an experimental study that sheds some initial light on designing information systems such that trust is enhanced.

1.3 PART III: Design and Support of the Planning and Scheduling Task

In Part III, we study more detailed the actual scheduling tasks, and associated cognitive processes. This Part centers around the concept of task analysis and develops insights how to design decision support systems and the associated algorithms such that the task and the decision support system are well aligned.

The accomplishment of a manufacturing company's objectives is strongly connected to the efficient solution of scheduling problems that are faced in the production environment. Numerous methods for the solution of these problems have been published. However, very few of them have been adopted by manufacturing companies. *Chapter* 12 suggests that the basic reason behind this imbalance is the inadequate representation of the scheduling process when designing decision support systems. Hence, the algorithms that are designed and included in these systems might not reflect the problems that actually have to be solved. The relevance of algorithmic design can be improved by using a more complete representation of the scheduling process, which would be highly relevant for increasing the adoption rate of new support systems. The main contribution of Chapter 12 concerns the development of a theoretical framework for the design of scheduling decision support systems. This framework is based on an interdisciplinary approach that integrates insights from cognitive psychology, computer science, and operations management. The use of this framework implies that the design of a decision support system should start with an examination of the human, organizational, and technical characteristics of the scheduling situation that has to be supported. This information can be obtained and analyzed using appropriate methodologies such as hierarchical task analysis, cognitive task analysis and cognitive work analysis as well as other methodologies, such as interviews, observations, context diagrams, and data flow diagrams. The designer of the decision support system can then match the results of the analysis to the guidelines of the theoretical framework and proceed accordingly.

Chapter 13 compares various task analysis methods for planning and scheduling. Planning and scheduling experts in practice are often faced with the question of how a company can improve its planning performance. Such improvements can be related to, for example, computer support, organizational task division, performance analysis, etc. The multitude of planning and scheduling factors and their interrelatedness makes it difficult to integrally explain current performance and assess the consequences of changes. The authors analyze how different perspectives on task analysis methods complement each other for the various questions that planning and scheduling experts encounter in practice. There are two main findings. On the one hand, a combination of methods is often necessary in order to avoid myopia and biased results. On the other hand, however, the analysis shows that not all questions require a full-scale analysis of the situation.

Chapter 14 addresses the challenging question which part of a task to automate. An important part of Advanced Planning Systems (APS) are algorithms. Where algorithms are applied, the task is (partly) automated. However, the human that is supposed to use these algorithms is generally ignored when designing the system. Normally a prior investigation whether and how an algorithm can or will be used in practice is not integrated in the design process. In contrast, in the field of cognitive ergonomics, function allocation methods explicitly take into account human factors in the design of human-computer systems. The function allocation literature, however, is mainly focused on dynamic systems where humans must make decisions in situations with time pressure and important safety aspects, e.g., nuclear plants and air traffic control. The authors analyze the differences between such systems and planning and scheduling, and they propose a model for function allocation in planning and scheduling taking into account cognitive and human-machine cooperation aspects.

Chapter 15 covers three application of scheduling algorithm design. It discusses the insights developed for designing scheduling algorithms according to three design projects where algorithms have been developed. The choice of applications covers a broad spectrum. The methods used are from three different fields, namely combinatorial optimization, genetic (evolutionary) algorithms, and mathematical optimization. The application areas differ also in terms of the role of a human user of the algorithm. Some of these algorithms have been developed without detailed study of the competences of the perceived users. Others have examined humans when performing the scheduling tasks manually, but have not considered the change in cognitive load if the process of planning changes due to the implemention of the new algorithm and computerized support. Although none of the design projects fulfils all criteria developed in the framework of Chap. 12, the authors show that the framework helps to assess the design projects and the resulting algorithms, and to identify the main weaknesses in these applications. The three application areas are (1) Decision support for shunting yard scheduling using a network flow heuristic; (2) An evolutionary multi-objective decision tool for job-shop scheduling; and (3) Group sequencing: a predictive-reactive scheduling method for job-shop scheduling.

Chapter 16 presents an additional case study. The authors describe a complex planning problem, namely the train shunting scheduling for the railways in the Netherlands. The case study concerns the planning of day-to-day shunting operations at the large stations in the network, performed by 130 full-time planners. The central question is: how can these planners be supported in their task with an Advanced Planning System (APS).

1.4 PART IV: HOPSopedia

Chapter 17 describes the set-up of an online reference tool for planning terms. The tool, called Hopsopedia, enables the easy sharing of planning-related term definitions and descriptions (http://www.hops-research.org). The advanced search engine and linking features support researchers, students, and practitioners, for instance to find key references for planning terms. The tool is developed to enhance mutual understanding between people from different scientific disciplines by providing possibilities to share and discuss term descriptions. The Hopsopedia serves as an online glossary complementing this book as well as a permanent reference instrument for the further interdisciplinary shaping of the planning and scheduling sciences.

Chapter 2
Decision Making in Planning and Scheduling: A Field Study of Planning Behaviour in Manufacturing

Roland Gasser, Katrin Fischer, and Toni Wäfler

Abstract Production planning and scheduling (PPS) requires human decision making. In this chapter, we introduce two theoretical models of Naturalistic Decision Making (NDM). Their applicability to the PPS domain has not been investigated to date. A field study in a Swiss manufacturing company is described, using existing NDM methods to study 'real world' decision making. The findings indicate that planners are using substantial amounts of general production and business-related knowledge to identify and solve decision problems. In their daily work, they are very much dependent on a supportive socio-technical environment that allows efficient information provision, diagnosis and interpretation of the state of affairs, and the development of expertise. The chapter closes with a discussion of NDM-related theoretical and methodological issues, as well as some implications of our research for decision support design.

2.1 Introduction

Production planning and scheduling (PPS) involves a substantial amount of human information processing and decision making. More complex business processes create more demanding work for today's planners and schedulers. Even equipped with highly developed PPS software tools, the complexity a human decision maker can handle is limited. Furthermore, many information technology (IT) solutions *create* additional complexity for the user and increase the challenge of decision making.

R. Gasser
Department of Mechanical and Industrial Engineering, University of Toronto, Toronto, ON, Canada
e-mail: roland.gasser@utoronto.ca

K. Fischer and T. Wäfler (✉)
School of Applied Psychology, University of Applied Sciences Northwestern Switzerland, Olten, Switzerland
e-mail: katrin.fischer@fhnw.ch; toni.waefler@fhnw.ch

J.C. Fransoo et al. (eds.), *Behavioral Operations in Planning and Scheduling*,
DOI 10.1007/978-3-642-13382-4_2, © Springer-Verlag Berlin Heidelberg 2011

In order to investigate decision making as a central cognitive activity of human planners, we used a Naturalistic Decision Making (NDM) approach to study *real world* decision making in PPS. Based on field research in fire fighting, military command, and aviation domains, NDM claims that in many such environments, expert human decision making is largely non-algorithmic, following heuristics based on experience rather than analytical deliberation of expected risks and utilities. NDM suggests that experts in general tend to use decision strategies that save time and reduce effort while preserving sufficient accuracy.

After a brief introduction to the psychological background, this chapter presents findings from a field study of decision making in planning and scheduling at a Swiss manufacturing company. It concludes by discussing an integrated perspective on theoretical and practical issues, and relating findings to the design and development of decision support tools for planners.

2.2 Psychology of Decision Making

Decision making research has revealed that human decision making behaviour differs somewhat from the predictions and prescriptions of rational decision-making theories. In response to experimental observations of 'irrational' or 'biased' behaviour, behavioural models of human decision making have been formulated, most prominently by Tversky and Kahneman (Tversky and Kahneman 1974). Such models suggest that humans use contextual information in choosing between alternatives, for example weighing potential losses more heavily than potential gains or over/under-estimating probabilities. Although these models suggest that human decision making is quite fundamentally different from predictions of economically rational normative models, they retain an algorithmic perspective. Subjective biases are taken into account, and rationality is somewhat questioned, but decision making in behavioural perspectives remains a cognitive process of choosing between alternatives through rational deliberation – however *bounded*.

Other decision researchers have moved away from a strictly algorithmic view on decision making towards a non-algorithmic or heuristic understanding of human decision making. Central to the non-algorithmic stream of research is the assumption that everyday human decision-making does not entail calculating probabilities, subjective values, and expected utilities. In their overview of research on non-algorithmic decision making ('Simple Heuristics that Make us Smart'), Gigerenzer and Todd (Gigerenzer and Todd 1999) claim that such decision making *is* rational in the sense of being well-adapted to the natural human environment. Meanwhile, there is a large non-algorithmic decision making body of research based on field studies and laboratory experiments, which have led to a series of descriptive as well as predictive models (cf. Gilovich et al. 2002; Hardman and Macchi 2003; Plessner et al. 2008).

In order to develop a better understanding of *expert* or *routine* decision making in particular, researchers have described decision behaviour in real-world work

Table 2.1 Classical compared to naturalistic decision making research

Classical decision making approach	Naturalistic decision making approach
No expertise necessary	Expertise extremely important
No time pressure	Very often time pressure
Complete information is given	Often incomplete or ambiguous information
Well-defined decision goals	Ill-defined or conflicting decision goals
Risks can be calculated	Risks are difficult to calculate
Low emotional impact	High emotional impact
E.g. consumer decisions	E.g. decisions of pilots, fire-fighters
Research objective:Context-invariant decision models	Research objective:Context-related decision models

settings. Several decision making models have been developed since the 1980s as models of *naturalistic decision making* (cf. Zambok and Klein 1997). Their common claim is that expert human decision making is not algorithmic in nature but heuristic, holistic or intuitive (cf. Klein 1998; Svenson 1996). More recent work considers different *modes of thinking* and proposes a more process-oriented view of decision making (cf. Plessner et al. 2008). The differences between the classical and NDM approaches are summarized in Table 2.1.

In NDM models of decision making, deciding is considered to be a highly context-dependent process of sequentially transforming knowledge states until a decision is made. Most models also include a strategy selection of *how to decide*. Decision making as a cognitive activity is perceived as highly adaptive. Correspondingly, major assumptions for a framework of adaptive decision making (Payne et al. 1993) are:

- Strategies are characterized by different levels of accuracy, contingent on task environments.
- As a result of prior experiences and training, a decision maker is assumed to have multiple strategies available to solve a decision problem of any complexity.
- The selection of a strategy is sometimes a conscious choice and sometimes an instant (learned) association between elements of the task and the relative effort and accuracy of a specific decision strategy.

Scholars who are striving to understand the role of expert knowledge in decision making therefore need to study specific contexts, and adapt their models accordingly (Klein 1998; Klein 2009).

2.2.1 Two Models of Naturalistic Decision Making

If human planners are experts in their work domain and are making decisions in relatively uncertain and dynamic environments, under time pressure, and with incomplete information, the NDM approach is an appropriate theoretical framework for the study of decision making in PPS. Two prominent naturalistic decision making models, *recognition-primed decision making* and the *decision-ladder* will provide a theoretical background for our approach to the field study, the data analysis, and the discussion of our findings.

2.2.1.1 Recognition-Primed Decision Making

Klein's (Klein 1998) recognition primed decision making model (RPD) describes decision making processes in naturalistic environments as integrating (1) perception of the situation, (2) knowledge about the situation, and (3) knowledge about actions. In the RPD model, pattern recognition is central to decision making. Human expertise allows recognition of patterns in the information available for a certain situation. Such expert pattern recognition supports the diagnosis of the situation. Without expertise, the diagnosis is poor, the matching of situation and action is weak, and the decision maker is not able to identify adequate actions.

According to Klein (Klein 1998) experts' understanding of a situation depends on:

- Goals of the decision maker
- Critical cues
- Expectations about future developments of the situation
- Typical actions within such situations

Situational patterns are defined as sets of relevant situational cues that are interrelated to each other by conditional, causal or temporal relations, and reflect goals, constraints and expectations. Situational patterns are thus part of the planner's knowledge, and are substantial to his expertise. RPD suggests that planners use such patterns to identify critical system states and to reduce the system complexity to a number of cues sufficient to recognise prototypical situations (Klein 1998). Non-algorithmic human PPS decisions are therefore based on a pattern matching process, where actual information is compared to patterns stored in the human planner's knowledge and where a possible course of action can be retrieved immediately.

A decision is made when the preferred action path has been affirmed by what Klein calls *mental simulation*, i.e. the mental anticipation of the consequences that are to be expected when following the retrieved action path (Fig. 2.1).

2.2.1.2 Decision-Ladder

Rasmussen, Pejtersen, and Goodstein (Rasmussen et al. 1994) studied decision making and problem-solving of expert engineers and technicians. Their work revealed that these experts did not follow a linear information-processing structure, but instead took *shortcuts* using associations based on expertise when working on a familiar problem. They described these shortcuts as *shunts* from a data-processing activity directly to a knowledge state, or *associative leaps* between knowledge states. Both depend on using stereotypical knowledge stored in the experts' memory.

Their model of decision making, which is referred to as the *decision-ladder* due to its common graphical representation in the form of a ladder (cf. Fig. 2.2), can

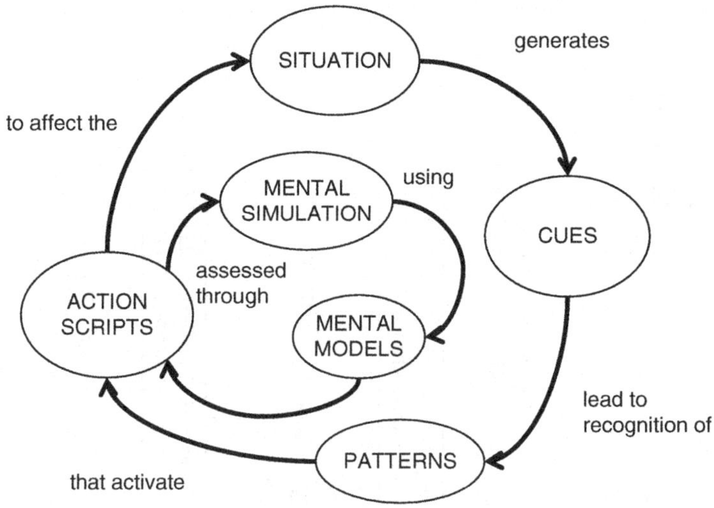

Fig. 2.1 A simplified representation of the RPD model, adapted from Klein (Klein 2009, p. 90)

nevertheless be understood as a linear model. The decision maker is step by step looking for answers to the following questions: What's going on? What lies behind? What's the effect? What goal to choose? Which is the goal state? Which is the appropriate change in operating conditions? How to do it?

According to the model, various data-processing steps are involved in decision making, each one transforming a knowledge state into another. Depending on the strategy the decision maker is using, more or less data-processing is taking place. For example, if the task is a regular routine triggered by a reminder, then the first step is activation, followed by observation. Depending on the strategy used, an observed pattern might associate directly with a procedure for execution. Or, as another example, it might lead to a knowledge state that requires interpretation, evaluation and refined task definition.

2.2.2 Summary

The recognition-primed decision making model and decision-ladder model postulate that experts use their rich experiences to reduce problem complexity, spare limited cognitive resources, and increase efficiency in decision making. Although from different backgrounds, the models clearly have parallels. Central to both models are decision making through situational pattern or stereotype recognition, and memory-driven associations. While the decision ladder allows representation of more individual variances in decision strategies, the RPD model is specific to certain contexts, usually involving high stakes and time pressure. Some of the information processing activities are quite similar in both models, such as recognition processes

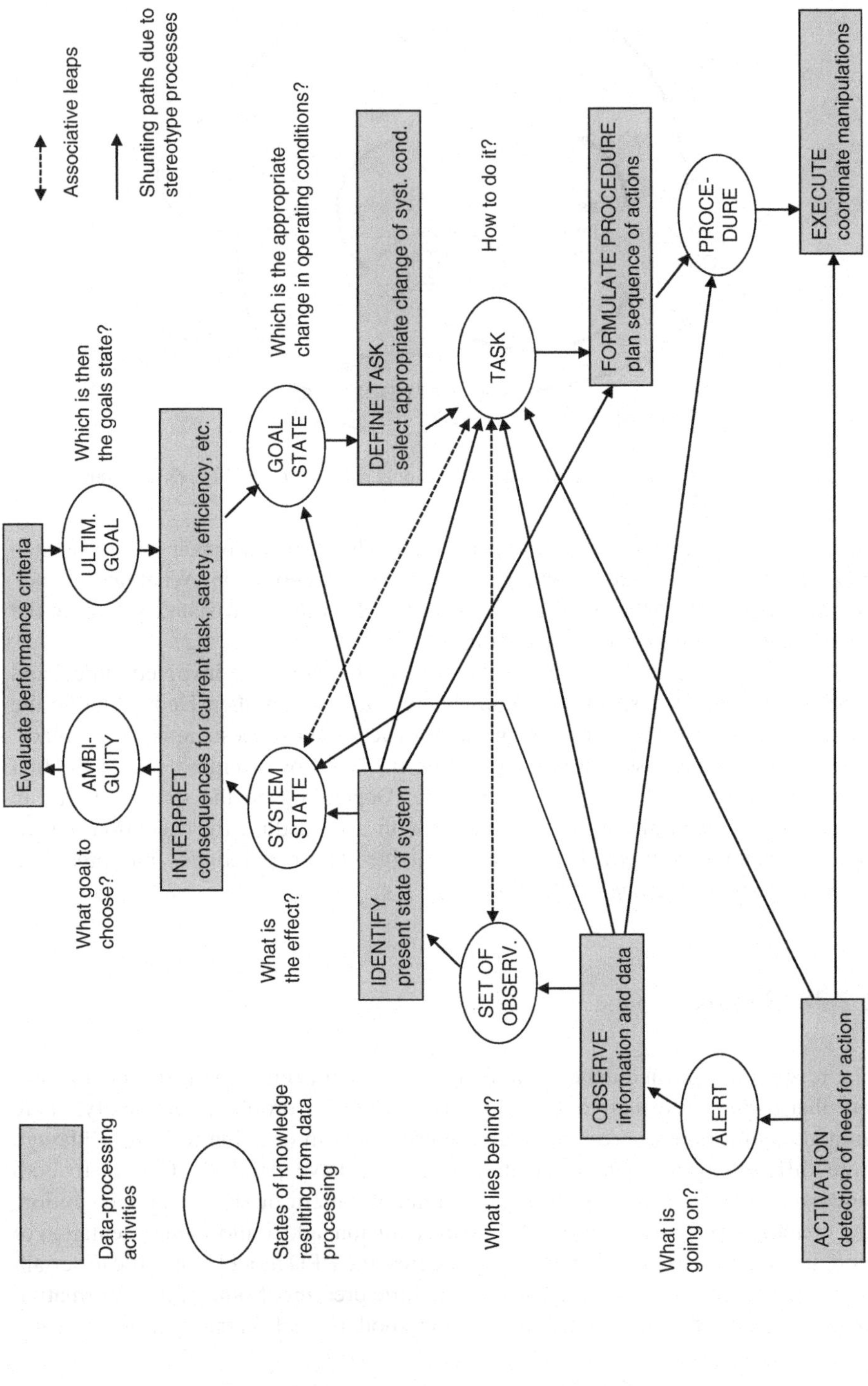

Fig. 2.2 Rasmussen's decision ladder model adapted from Vicente (Vicente 1999, p. 189)

involved in situational diagnosis, or more conscious mental processes to confirm, evaluate and redefine action paths and tasks. Both models describe expert decision making, allowing for non-algorithmic, experience-based, context-dependent strategies in a variety of complex, work-related decision situations. Hence, they both provide a useful theoretical framework for research on decision making in PPS.

From a methodological point of view, the decision ladder has mainly been used to capture think-aloud protocols in problem-solving activities, whereas there are more structured methods available for the RPD model. Some of these RPD methods were used by Crawford, MacCarthy, Wilson, and Vernon (Crawford et al. 1999) to investigate the role of industrial schedulers in organizations. The decision making and expertise involved in particular cognitive activities *within* industrial schedulers' daily routine tasks have received less scholarly attention.

2.3 Manufacturing Field Study

The aim of the field study presented below was to describe and interpret the reasoning and expert decision making behaviour of planners and schedulers as it occurs in real-world PPS rather than in artificial scheduling experiments. In order to better understand such behaviour, some preliminary questions about the organisational context were addressed: What are the roles and tasks of the persons involved? What kind of decisions do they take and how can the decision making processes be described?

Next, using the NDM approach to expert decision making, the study investigated decision situations: How do human planners and schedulers make use of information? What is the role of knowledge in their decision making? What kind of information is used to diagnose situations? How do human planners and schedulers decide on possible actions? How are information processing and expertise intertwined?

From an application-oriented perspective, we expected some insights concerning the design of support tools: How and to what extent is the human planner's decision making supported or impeded by the working conditions? What kind of support for PPS decision making is needed from a NDM-perspective? How should an information system need to be designed to support and enhance specific processes or phases in PPS decision making?

2.3.1 Company Description

The company – based in Switzerland – has approximately 500 employees and a yearly turnover of around 120 million Swiss francs[1], producing products for

[1]Approximately 80 million Euros

building and construction markets worldwide. A wide range of around 1,000 products and 4,000 versions consisting of close to 15,000 different parts are produced, assembled and shipped to hardware stores, wholesalers and building enterprises.

Manufacturing processes range from casting to surface treating. About half of the parts are produced in-house, while the rest are purchased from around 200 suppliers. Some 250 employees are working in six production units, another 50 in the assembly unit. Based on forecasts of the sales unit, PPS is performed centrally, supported by an up-to-date IT system. The parts are produced according to a production plan (make-to-stock), whereas assembling is driven by orders (assemble-to-order).

2.3.1.1 PPS Environment and Processes

The company's products are positioned in the upper price range, and delivery reliability is an important success factor on the market. A strategically defined lead-time of three days is achieved in 97% of all customer orders. The planning horizon is 12 months, with forecasts by the sales department and seasonal differences on the building market taken into account.

Production resources are generally predictable, except some raw materials like copper and steel which are affected by market cycles or scarcities due to political or other influences. External disturbances result from changes in supplier delivery reliability, raw material quality, and import/export regulations. Internal disturbances include variances in the casting processes, machine breakdowns, workflow bottle-necks, testing of new manufacturing processes, introduction of new products, and other complexities of the production processes.

A central PPS department is responsible for production planning (i.e. program planning), master scheduling, detailed production scheduling, and stock management. Key planning stages are (1) production planning, (2) scheduling of production and purchasing, (3) releasing of production and purchasing orders, and (4) central and local sequencing, dispatching and control. Goals of maximizing responsiveness and minimizing stock capital lockup are always conflicting. Scheduling hotspots include order prioritisation in casting, and the management of bottle-necks. At bottle-necks, job sequencing is centrally managed by the PPS department. Therefore, only limited amounts of scheduling and sequencing activities are done by foremen on the shop floor. Within the job shop production environment, dispatching and control of the job progress is done mainly by the foremen, who regularly report their progress to the PPS department. There is generally little distributed decision making, since most decisions are taken within the PPS department, without substantial contributions by the production units (cf. Wäfler 2002). However, there are weekly production meetings where planning-related issues and problems are discussed with the foremen. The PPS department also interacts with the sales and engineering departments to coordinate sales activities and product releases with production and purchasing activities.

Within the PPS department, there are three teams, located in the same open-space office. The planning team focuses on medium to long-term production planning, the purchasing team orders parts and raw materials, and the scheduling and dispatching team schedules and releases production orders. The third team also sequences orders for some of the production units (casting, bottle-necks) and monitors the work progress for critical (urgent) orders. The compact layout of the PPS department allows for quick cross-functional communication and coordination. This is especially important in case of major disturbances or urgent production orders.

Planners in the planning team consider setting parameters in the enterprise resource planning (ERP) software system the most difficult task. It involves many decisions, such as the method for scheduling or the optimal batch size and production lead time. Especially challenging is the introduction of new models, when the planners struggle to avoid backlogs for the new model as well as remaining stocks of the old model, while considering all interdependencies, constraints, and goals. Schedulers in the scheduling and dispatching team find the releasing of orders most difficult, e.g. when two similar orders are due within a relatively short time frame.

The company provides a good environment for a field study of decision making in PPS. Its structures and processes are comparable to many medium size manufacturing companies in Europe. When making decisions, the company's planners and schedulers are facing substantial complexity in the form of dynamics and uncertainties of the production environment.

2.3.2 *Decision Making 'in the Wild'*

2.3.2.1 Methods

Scope. A qualitative research approach was used for the field study. The retrospective *decision probe method* was developed by Crawford and her colleagues (1999) to describe decision processes in PPS based on earlier work within the RPD framework (Lipshitz and Strauss 1997). The method consists of systematic observations, structured interviews, and a structure-laying technique. The scope of our data collection and subsequent analysis extended beyond the production planning unit to other units and individuals involved in PPS activities, such as foremen and purchasing agents, to cover the whole 'secondary work system' of planning and scheduling (Wäfler 2001). Accordingly, following initial workshops and interviews on the management level, four planners were selected for workplace observations and decision probe interviews. Their roles and tasks included program planning, purchasing, scheduling, order releasing, and dispatching. They had worked for the company between 7 and 22 years (average 17 years) and all them had been in their actual positions for at least 7 years. Their average age was 46.8 (39–56).

Observations and interviews. Within a period of 6 weeks, 13 observation sessions of 1–2 h were arranged with the four participants. During the observations, the

Table 2.2 Decision probe interview questions (adapted from Crawford et al. 1999)

No.	Question
1	Describe this decision episode in your own words
2	What would an appropriate title be for this decision?
3	What caused you to have to make the decision?
4	What was it that you thought about in order to make the decision?
5	What information did you use to make this decision?
6	What knowledge did you use to make this decision?
7	Did you have to communicate the outcome of this decision to anybody else?
8	How would you rate this decision? (difficulty, time pressure, need for advice)
9	Does a documented procedure exist for this type of problem?
10	Overall, do you consider this decision to be a typical type of decision?

observer identified *decision points* without disturbing the work process if possible. In order to clarify and identify all relevant decision points within the observation period, the participant was interviewed immediately after the observation. During the first observation sessions, typical and frequent decisions (as perceived by the participants) were selected for further analysis, in order to get an overview of the variety of decisions. Subsequently, the most complex decisions were selected for further examination. All selected decisions were then examined in structured interviews consisting of a set of ten questions (see Table 2.2). In total, 90 decision points were identified during 15 h of observation. Following the observations, 30 decision probe interviews were conducted, ranging from 4 to 10 interviews per person. Out of the 30 decision probes, ten were excluded because of incomplete documentation or other shortcomings such as descriptions of workflows instead of decision making. Accordingly, the remaining sample of 20 decision probes was further analysed.

Analysis. The analysis first focused on the information 'elements' used by the decision maker in order to evaluate the need for a decision and to take the decision, i.e. to choose or to confirm an action. Potential cues for situational patterns and situation diagnosis-related processes were extracted from the decision probe interviews. The decision probes were transcribed by using an adapted notation structure proposed by Crawford and colleagues (Crawford et al. 1999). Table 2.3 shows the categories that were described within the structure. In the second step of the analysis, relations between information elements were extracted and categorised as either conditional or causal. The decision descriptions were then cross-checked by two co-researchers. Additionally, the resulting notations were validated by the interviewees.

In addition to these *structural* properties of decisions, we also considered the *process* of deciding in terms of sequentially performed cognitive activities by the decision maker and her surrounding socio-technical environment. This aspect of decision making is relevant for our study especially with regard to the design of PPS IT systems. Each sub-process can be understood as an activity that is transforming knowledge states, finally leading to a decision. From such a process-oriented perspective on decision making, a set of processes involved in decision making can be defined. According to the decision-ladder model proposed by Rasmussen and his colleagues (Rasmussen et al. 1994), up to eight information processing

Table 2.3 Categories of the decision description

Category	Specification
Context	History of the decision and trigger for actual episode
Situation	The actual decision situation/task at hand
Information	Used information during the decision process
Knowledge	Used knowledge for the identification of the decision need and the evaluation of the situation
Decision need	Result of a pattern matching process comparing the actual situation (as perceived) with knowledge about similar situations and the implications arising from deviations of the ideal state
Action options	A set of possible action paths with a preference for one of them as a result of a pattern matching process between the actual situation and the knowledge about action paths that might be feasible
Decision	Result of the decision about action paths
Communication	How and to whom was the decision communicated?
Values	Values or superordinate goals influencing the decision process

activities can be distinguished in a decision making process. These do not include psychologically-relevant processes such as learning or more perception-related processes (e.g. awareness and filtering). Therefore, we chose to use the following sub-processes in order to describe decision making in PPS:

1. Information acquisition
2. Filtering of information relevant to decision problem
3. Weighing of relevant information
4. Linking of information and activation of context-specific rules
5. Generation and selection of action paths (options)
6. Learning of decision-relevant patterns and rules (knowledge generation)

2.3.2.2 Findings

As an example, one typical routine decision concerning the planning of raw material is presented in the box below.

The planner characterised this particular decision as not very complex, not under time pressure, and without any required advice. A documented procedure for this kind of decision does not exist within the company. From this example, nine information elements and three associated relations were extracted, representing the situational pattern and the knowledge used by the planner while taking the decision. Figure 2.3 shows a schematic representation of the situational pattern that was extracted from that example. In our notation, we distinguished between (1) information elements, (2) relations between them, and (3) general rules related to the situation. Information elements consist of a variable and a related value, e.g. the variable ‘consumption’ incorporates the values ‘low’, ‘moderate’, and ‘high’. Conditional or causal relations between information elements are indicated with dotted lines, without further specification. Furthermore, situation-specific rules depending on such values and their actual values can be formulated, e.g. if copper is scarce on the market, the supplier’s reliability is affected.

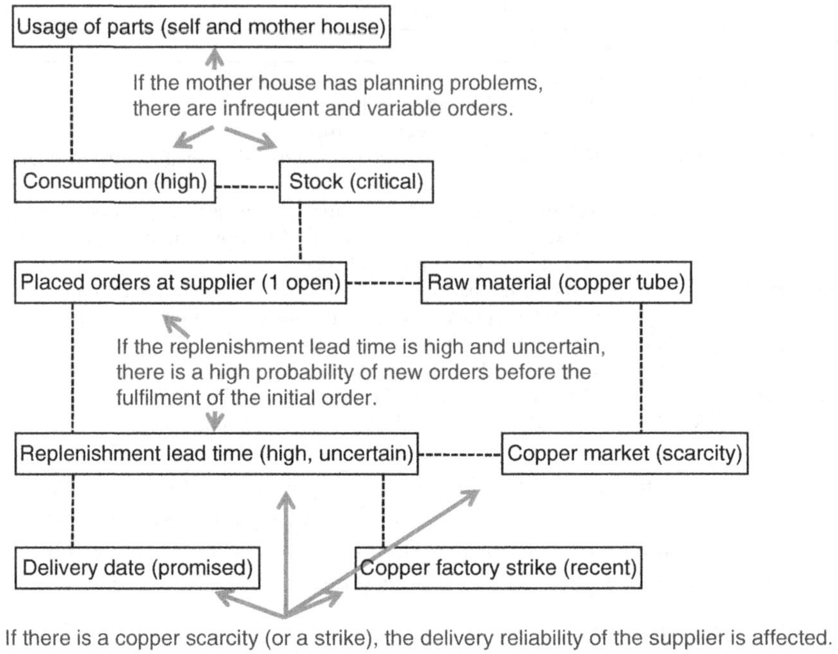

Fig. 2.3 Example of a situational pattern and relational knowledge extracted from a decision description. Information elements are displayed as *grey boxes*, *doted lines* represent relations between information elements, and relating situation-specific rules are printed in *black* with *arrows* pointing to concerned information elements

We used the above method on the sample of 20 planning and scheduling decisions and extracted 64 information elements linked by a total of 137 conditional or causal relations. The resulting situational patterns consist of between 4 and 15 information elements and they contained up to 18 relation links, with an average of 9.5. Relations were predominantly classified as conditional, with a minority of causal links (average 2.0 per pattern).

A closer look at these relations between information elements unfolds a wide body of PPS expert knowledge. The participating experts used information elements originating from local to global sources and linking them through relational knowledge adapted to the specific decision or task context. Such knowledge was applied to link seemingly independent information elements, making sense of the situation at hand. Some examples of such relational knowledge are: “If two parts require more or less the same tools on the robots, it saves cost and time to produce them right one after the other.” – “If supplier X does not have sufficient raw material, the delivery cannot be advanced.” – “If a part is novel, the production times need to be adjusted to allow for disturbances and tests.”

Classifying the information elements according to the company’s planning stages shows that almost half are related to the production order release stage. The sum of categorised information elements per planning stage is shown in Table 2.4.

Table 2.4 Classification of extracted information elements according to the planning stages (N = 64, multiple categories possible)

Planning stage	No. of information elements
Production planning	5
Material requirements planning/master prod. scheduling	14
Detailed scheduling/production order release	29
Detailed scheduling/purchasing order release	17
Sequencing/dispatching/control (centralised)	7
Sequencing/dispatching/control (local)	12

Table 2.5 Classification of relations between information elements according to knowledge domains within the company (N = 137, multiple categories possible)

Knowledge domain	Percentage of relations (%)
Marketing and sales	19
Production planning/MPS/MRP	31
Detailed scheduling and order release	62
Supply/purchasing	21
Sequencing/dispatching	19
Production processes	31
Environment	12

Classifying the relations according to correspondence with knowledge domains shows that more than 60% are linked to scheduling and order release (Table 2.5).

All of the observed planners used situation-specific rules from multiple knowledge domains. Interestingly, those involved in scheduling and order release were frequently linking rather distant knowledge domains – e.g. sales with production or scheduling with (supplier) environment.

To account for a process-oriented perspective apart from structural aspects, the decisions were described using six typical sub-processes of decision making. To illustrate that, Table 2.6 shows a *process-oriented* representation of the example decision about the procurement of raw material mentioned in the box above.

Central to the notion of decision making as a *process* is the development of action alternatives. Initially, there might be just one option to consider, i.e. to do or not to do something. After a phase of information gathering and weighing, another option might become apparent, which again requires more information and so on. As mentioned above, we did not assume a linear information processing sequence. Non-algorithmic strategies and shortcuts as well as algorithmic deliberation must be kept in mind.

To sum up the findings, we would like to point out that planners and schedulers are using extensive knowledge about their work domain, the company and its environments. Even for seemingly simple decisions like changing the batch size when releasing a production order or setting a parameter in the ERP system, they are considering many interdependences and relationships within and beyond the production system.

While not documented in detail, we observed human planners and schedulers frequently changing and modifying the outputs of the ERP system. In our view, it is

Table 2.6 Example of a process-oriented decision description

Process	Description
Trigger	The decision was made after a request by the electro-plating shop. Its foreman wanted to know if there was a possibility to start with a regular large job ahead of time in order to use underemployed resources
Information acquisition, filtering	The job that he proposed required raw material, which led the planner to check stocks and orders of the required parts
Weighing	He quickly realised that the consumption of these frequently used parts was high as normal, and that there was an open order at the supplier to get new material
Linking	He knew that the required parts are made out of copper, what in the current situation on the world market might lead to supplier-side delivery delays. In addition to general copper scarcity, a strike had recently taken place in a nearby copper factory had recently taken place, causing even more problems with copper supply
Activation of rules	Although the supplier promised to deliver on time, a decision was necessary to avoid problems due to backlogs
Linking	The risk of a backlog was high because of planning problems at the mother house of the company. The mother house, which used the parts prepared by the electro-plating shop, is known to order enormous amounts at unpredictable times
Generation of action paths	The preferred action was therefore to increase the order that was already filed to the supplier and to start working on the job (as proposed by the foreman)
Selection of action path	The decision was to charge someone else in the PPS team with inquiring at the supplier if it was possible to get more parts within the actual order

crucial to understand this decision making behaviour in order to improve the interaction between the software and the users. We next present a psychological perspective on supporting and impeding conditions for human PPS decision making that suggests potential improvements to socio-technical system design.

2.3.3 Supporting and Impeding Conditions

Based on our analysis of PPS decisions, supporting or impeding conditions for the human planners' decision making can be postulated. These conditions include management and communication processes as well as features of the technology involved:

- Supporting conditions:
  - Decision-relevant information is made more available through the coordination of formal and informal information sources and by avoiding conflicting tasks or rules.
  - Immediate feedback (e.g. about changes in customer demands, sales activities, changes in the production process, service levels) provides planners with

decision-relevant information and support the learning, correcting, and adjustment of decision-relevant situational patterns.

- Impeding conditions:
  - Relevant information is sometimes unavailable, ambiguous, and uncertain. While this is the nature of PPS, nevertheless, information is often not provided though it is potentially available. Due to poor internal or external communication processes in the company, provision of information becomes time consuming and impeding for the decision process.
  - Relations between relevant information elements and situational cues are difficult to identify by less experienced planners within the actual socio-technical environment.

In Table 2.7, we present a more detailed account of impeding and supporting conditions derived from our field study. To compensate for impeding conditions, planners and schedulers create their own tools to facilitate their reasoning and decision making using either common office software or pencil and paper. At the same time, they establish formal and informal organizational structures, practices, and routines. Where such compensating activities have been observed, they are mentioned among the supporting conditions in the right column of the table.

Anticipatory Procurement of Raw Material

The decision was made after a request by the electro-plating shop. Its foreman wanted to know if there was a possibility to start with a regular large job ahead of time in order to use underemployed resources. The job that he proposed required raw material, which triggered the planner to check stocks and orders of the required parts. He quickly realised that the consumption of these frequently used parts was high as normal, and that there was an open order at the supplier to get new material. He knew that the required parts are made out of copper, which in the current situation on the world market might lead to supplier-side delivery delays. In addition to general copper scarcity, a strike had recently taken place in a nearby copper factory, causing even more problems with copper supply. Although the supplier promised to deliver on time, a decision was necessary to avoid problems due to backlogs. The risk of a backlog was high because of planning problems at the mother house of the company. The mother house, which used the parts prepared by the electro-plating shop, is known to order enormous amounts at unpredictable times. The preferred action was therefore to increase the order that was already filed to the supplier and to start working on the job (as proposed by the foreman). The decision was to charge someone else in the PPS team with inquiring at the supplier if it was possible to get more parts within the actual order.

Table 2.7 Impeding and supporting conditions for decision making sub-processes including observed compensatory activities/tools

Process	Impeding conditions	Supporting conditions and observed compensatory activities/tools
Information acquisition	High effort required to get decision-relevant information or the information needed is not accessible, even when potentially available	Using informal communication channels; Co-location of planners and schedulers; "overhearing" discussions and communications of colleagues; Planning board; Meetings; Inspections on shop floor
Filtering of information	IT provides a wide range of information – however, in a specific situation the desired information is not "at hand", but has to be collected and filtered from different software outputs; also, a rather large amount of information provided through colleagues and other sources needs to be filtered	Individually generated spreadsheets and tables, electronically as well as printouts; Printouts of screenshots, marked with highlighter; Planning board with visual elements (colour stamps, paper clips)
Weighing of information	ERP system does not support weighing of information, e.g. for the prioritisation of production orders. Pre-decisional, preparation steps that are mostly done as part of a routine are not supported by IT	Calculations with desk-top calculator; Print-outs of spreadsheets with hand-written notes; Paper-calendars; Planning board (order queue)
Linking of information	Relevant relations between information elements are not represented because of their complex, dynamic and intransparent nature; ERP system creates additional complexity because the underlying mechanisms and algorithms are mostly hidden to the user	Individually generated spreadsheets and tables, electronically as well as printouts; Planning board with visual elements; Inspections on the shop floor (including informal exchanges with foremen and team leaders); Formal planning and scheduling meetings
Generation and selection of action paths	Pre-conditions for the execution of a specific action path are not available – or all action paths are facing risks (e.g. for backlogs) without knowing the exact probabilities of these risks	Using informal communication channels; Inspections on the shop floor (including informal exchanges with foremen and team leaders); Co-location of planners and schedulers; Formal meetings with domain experts (e.g. engineers)
(Expert) learning	Critical relations between information elements are not represented in IT – which makes it more difficult to learn about system behaviours and dynamics. Because of missing or delayed feedback incomplete or false knowledge is built up and cannot be adapted to actual developments and conditions	Feedback about production costs per part (for every production order); Informal discussions with foremen; Inspections on the shop floor

2.4 Discussion

While the theoretical foundations of NDM and some of the methods developed by its proponents have proven useful for the study of decision making in PPS, more specific questions regarding the applicability of NDM models for this domain remain to be clarified. As the findings of our study indicate, linear or quasi-linear models of decision making like the RPD model or the decision-ladder are not quite suitable for the overall description of decision making *processes* in PPS. Nevertheless, the *structural* implications of these models concerning situational patterns and expertise are highly useful to explore expert decision making in this work domain. The situational patterns described above could be understood as networks of activation, although omitting the parallel activation of motives, goals, and personal preferences involved in routine decision making (Betsch 2005).

Relational knowledge – linked to application rules for a variety of situational contexts – is considered to be essential to expertise. However, this study was not intended to elicit planners' knowledge in a systematic way. The degree to which the method described here could be used as a dedicated knowledge elicitation method would require further work employing methodological triangulation to validate the method and to systematically test its reliability.

Decisions in PPS can be understood as non-linear series of information-processing activities, whereas some are rather short and others can be very long, depending on the complexity and the criticality of the given situation. Planners and schedulers are constantly working on multiple decision problems, relying heavily on colleagues and business contacts where information provided by the IT system is insufficient. General use of NDM methods to study expert reasoning and decision making in PPS has been fruitful in terms of identifying supporting and impeding conditions for expert decision making in a complex production environment. But within this study other potentially crucial influences on PPS decision making remain unexamined. Planners are positioned as 'information brokers' within a company, embodying an important role within a network that spans across the organization (Crawford et al. 1999). Within this role, they might also be concerned about procedural issues, and therefore value and prefer some procedures more than others. This perspective would add a more inter-personal dimension to PPS decision making behaviour from information gathering to action implementation. It could further explain how seemingly irrational decisions could be rational in another perspective, i.e. through the consideration of 'procedural utility' (Benz 2007).

The 'decision-script' approach developed by Hamm (Hamm 2003) in the field of medical decision making might provide further insights into PPS decision making characteristics and mechanisms. Although this approach cannot claim to be an established model of expert decision making, it provides a valuable perspective, interlinking cognitive psychology with behavioural judgement and decision making research. For application-oriented environments in particular, the 'decision-script' approach could help to devise more appropriate interventions and design efforts to support practitioners' decision making. For example, by gaining understanding of not only the decision situation in terms of information and knowledge involved, but also in terms of the typical 'script' or learned action path followed by decision makers in such situations, sub-optimal decision-scripts could be identified and new scripts could be written and implemented. The methodological issue here is that in order to discover decision-scripts in the first place, we need to observe many decision situations. And the more similar the decision situations, the more likely it would be to discover decision-scripts.

Our field study has been able to show the activity of planners and schedulers in making sense of complex decision situations. They constantly diagnose the state of affairs, seeking to identify potential planning or scheduling conflicts and thus decision problems. These decision situations very often involve information that clearly exceeds what is available in a computerised planning system. This substantial amount of knowledge concerns not only the specialised task of planning and scheduling, but also the production process, technologies, supplier relations, marketing, and engineering. From a process-oriented perspective, decisions involve multiple phases of decision making, without a distinguishable linear order. Further complicating matters, planners and schedulers are often working on multiple decision problems in parallel. In consideration of our findings, further research should not only address methodological issues, but also try to establish a more fitting decision making model for PPS or similar work environments. From an application-oriented perspective, the interplay of socio-technical design and the acquisition of expertise in PPS is a scholarly field that remains mostly unexplored.

2.5 Outlook: Implications for Decision Support Design

When identifying a critical system state or a need for a decision, planners aggregate information coming from different information sources, i.e. the PPS software, formal and informal organizational communication, planning boards etc. Their knowledge contains expertise about *which* cue from *which* source is relevant in a particular situation, *how* an element of information is interrelated to other elements of information, and *what* conclusions can be drawn from that in the form of situation-specific rules.

Our analysis shows that the human planners' decision making is facilitated – or impeded – by the socio-technical system around them, including management, cooperation and communication, as well as IT. In order to enhance the quality of decision making, we believe at least two fundamental cognitive processes should be better supported:

1. Continuous development of expertise, i.e. planners' PPS-related skills and knowledge.
2. Situational diagnostics (i.e. pattern recognition) by facilitating the availability and presentation of decision-relevant information in a way that is compatible to the planners' knowledge structures.

Accordingly, planners should be supported in identifying and selecting relevant cues from several information sources, and in constantly learning about relevant interrelations between these cues. This could be achieved for instance by implementing existing relational knowledge into databases and interfaces. Explicit and well-structured communication paths within the organization, as well as between the organization and its suppliers, could further support decision processes by making potentially available information readily accessible for the planners. Continuous feedback about the development of market demands and about the outcomes of PPS in terms of service levels and other operating figures (e.g. production costs, capital lock-up) should finally support the adjustment and correction of expert knowledge.

Understanding naturalistic decision making by human planners and schedulers provides a base for conceptual theories of an improved socio-technical PPS system. Central to this understanding is the way expert planners and schedulers are using their IT system and other resources when making decisions. Creating more supportive conditions through a better fit of the system design to the expert's way of working will most likely lead to a higher performance in planning and scheduling. Much research and design effort is still needed to achieve this goal.

References

Benz, M. (2007). The relevance of procedural utility for economics. In B. S. Frey & A. Stutzer (Eds.), *Economics and Psychology*. Cambridge, MA: MIT.

Betsch, T. (2005). Preference theory. In T. Betsch & S. Haberstroh (Eds.), *The routines of decision making* (pp. 39–66). Mahwah: Lawrence Erlbaum.

Crawford, S., MacCarthy, B., Wilson, J. R., & Vernon, C. (1999). Investigating the work of industrial schedulers through field study. *Cognition, Technology & Work, 1*(2), 63–77.

Gigerenzer, G., & Todd, P. M. (1999). *Simple heuristics that make us smart*. New York: Oxford University Press.

Gilovich, T., Griffin, D., & Kahneman, D. (2002). *Heuristics and biases: The psychology of intuitive judgement*. Cambridge: Cambridge University Press.

Hamm, R. M. (2003). Medical decision scripts: Combining cognitive scripts and judgment strategies to account fully for medical decision making. In D. Hardman & L. Macchi (Eds.), *Thinking* (pp. 315–345). Chichester: Wiley.

Hardman, D., & Macchi, L. (2003). *Thinking: Psychological perspectives on reasoning, judgment and decision making*. Chichester: Wiley.

Klein, G. (1998). *Sources of power: How people make decisions*. Cambridge, MA: MIT.

Klein, G. (2009). *Streetlights and shadows: Searching for the keys to adaptive decision making*. Cambridge, MA: MIT.

Lipshitz, R., & Strauss, O. (1997). Coping with uncertainty: A Naturalistic decision-making analysis. *Organizational Behavior and Human Decision Processes, 69*, 149–163.

Payne, J. W., Bettman, J. R., & Johnson, E. J. (1993). *The adaptive decision maker*. Cambridge: Cambridge University Press.

Plessner, H., Betsch, C., & Betsch, T. (2008). *Intuition in judgement and decision making*. New York: Lawrence Erlbaum.

Rasmussen, J., Pejtersen, A. M., & Goodstein, L. P. (1994). *Cognitive systems engineering*. New York: Wiley.

Svenson, O. (1996). Decision making and the search for fundamental psychological regularities: What can be learned from a process perspective? *Organizational Behavior and Human Decision Processes, 65*, 225–267.

Tversky, A., & Kahneman, D. (1974). Judgment under uncertainty: Heuristics and biases. *Science, 185*, 1124–1131.

Vicente, K. (1999). *Cognitive work analysis.* London: Routledge.

Wäfler, T. (2001). Planning and scheduling in secondary work systems. In B. L. MacCarthy & J. R. Wilson (Eds.), *Human performance in planning and scheduling* (pp. 411–448). London: Taylor & Francis.

Wäfler, T. (2002). *Verteilt koordinierte Autonomie und Kontrolle.* Doctoral thesis, Universität Zürich, Zürich.

Zambok, C. E., & Klein, G. (1997). *Naturalistic decision making.* Mahwah: Lawrence Erlbaum.

Chapter 3
The Interconnectivity of Planning and Shop Floor: Case Description and Relocation Analysis

Cees De Snoo and Wout van Wezel

Abstract This chapter provides a detailed description of the planning situation within a manufacturer of office furniture. The description provides much insight into the reality of planning and rescheduling, including the dynamic relation between planning and the shop floor. A quasi-experiment is reported in which the planning department has been relocated closer to the shop floor. Expectations before and experiences after the relocation are measured and reported. Although no significant changes in communication behavior are observed, both planners and shop floor foremen perceived positive consequences of the relocation.

3.1 Introduction

This chapter presents a case description of human and organizational factors in manufacturing planning and control in a medium-sized office furniture manufacturer. The focus is on the humans performing planning and scheduling activities. The case description is based on a large, multi-year research project, in which a variety of research methods have been used, like interviews, observation studies, questionnaires, document analyses, and participative field work. The research and research design build upon several empirical studies investigating human performance in planning and scheduling (Berglund and Karltun 2007; Crawford et al. 1999; Jackson et al. 2004; MacCarthy et al. 2001; McKay et al. 1995; McKay and Wiers 2003; Van Wezel et al. 2006). After an in-depth analysis of the planning situation, a quasi-experiment has been executed to investigate the consequences of relocating the planning department. Communication behavior and human perceptions regarding changed working conditions and performance have been measured both before and after the relocation.

C. De Snoo and W. van Wezel (✉)
Faculty of Economics and Business, University of Groningen, Groningen, The Netherlands
e-mail: c.de.snoo@rug.nl; w.m.c.van.wezel@rug.nl

J.C. Fransoo et al. (eds.), *Behavioral Operations in Planning and Scheduling*,
DOI 10.1007/978-3-642-13382-4_3,

The chapter is organized as follows. Section 3.2 introduces the firm. Section 3.3 provides a detailed description of the planning situation. The quasi-experiment is introduced in Sect. 3.3. Section 3.4 presents findings based on a comparison of the results from both the pre-measurement and the post-measurement data. In Sect. 3.5, the case findings are shortly discussed.

An in-depth analysis of the rescheduling situation within the case company, including a procedure to better structure the event handling process, is described in De Snoo et al. (2010b). An in-depth analysis of the relocation is described in De Snoo et al. (2010a). It should be emphasized that we slightly changed some business characteristics to guarantee the anonymity of the company.

3.2 Description of the Firm

Desk&Storage is a furniture producer operating in a dynamic and competitive, international market. Currently, 140–200 client-specific orders are processed each day. Most clients are large firms having long-term contracts with the company. Frequently, one large order to deliver furniture for a complete building is followed by several, irregular small orders for only one or two desks or storages.

Market demands are increasing, especially regarding product mix flexibility, customization, and delivery speed. Total sales on a yearly basis are roughly predictable, but the demand per product group per month is very hard to forecast. The constantly changing product mix causes a production situation having no fixed bottleneck: each product generates a different workload per department.

Twenty-five sales agencies, both within and outside the country, are responsible for the procurement of orders and all contact with the customers. Orders are sent to the internal ordering department. This department is responsible for the right 'translation' of orders from the sales agencies into the ERP-system, which can be difficult especially in case of non-standard orders. This includes the selection of materials and suppliers, the development of production routings, and the determination of delivery date agreements.

Over 30,000 product parts are purchased from a large number of suppliers. These materials are used in three manufacturing departments: metalworking, finishing (i.e., painting), and assembly. Around 250 shop floor operators are working in the manufacturing departments in multiple shifts. Each department has a production manager and several shop floor foremen. The production manager is responsible for the long- and mid-term issues, including the development of staff schedules. The shop floor foremen act as information hubs between the planners and the operators; they work as operators at the various workstations, but have an extra responsibility in communicating problems to the planners and plan adaptations to the operators.

The standard lead time of work-in-progress is 5 days (Fig. 3.1): 1 day for each of the manufacturing departments, 1 day for testing, and 1 day for loading and transport. The sixth day is used for delivery.

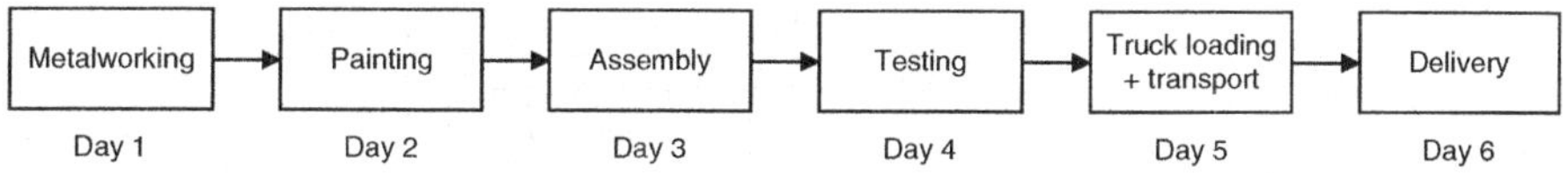

Fig. 3.1 Production process Desk&Storage

The metalworking department consists of 40 work centers in which steel operations are performed. Production orders have different routings to visit a number of these work centers. Both batch and lot-for-lot production rules are applied, depending on the number of orders and available capacity.

Within the painting department, the half-products from the metalworking department are painted and powder coated. Two powder-coating lines are available, one that is highly flexible but not so efficient, and one that requests long cleaning and setup times but is suitable for large batches.

The assembly department consists of seven production lines. Purchased and self-produced materials are assembled and packaged.

Delivery reliability is the most important business performance objective according to Desk&Storage's management. A special job function has been created, the so-called '*troubleshooter*', fulfilling the role of 'order chaser', to urge planners and operators to plan and produce products with the earliest due date. In the end, the firm realizes, on average, a delivery reliability rate of more than 90%. However, the manufacturing departments consider production efficiency to be the most important performance objective. Machine equipment and materials should be used as efficient as possible, requiring little and short setup times and clustering of similar orders.

Desk&Storage uses the ERP-system Baan, complemented with several firm-specific applications. Each night, the system calculates production and purchasing demands based on MRP-logic. The output of the system forms the basis for all further planning and scheduling activities. Recently, a newer version of the software had been implemented to improve planning flexibility. However, the implementation resulted into a failure: planners as well as operators lost control and delivery reliability rates decreased dramatically. Therefore, the old version is used again and other packages are implemented incrementally.

3.3 Description of the Planning Situation

Each production department has its own planner who is responsible for both the mid-term and short-term production plans. Planning of material supply is done by the material supply and purchasing department. Distribution of end products is outsourced, but Desk&Storage is responsible for the distribution planning. The department 'distribution planning' deals with the planning of efficient domestic and international transportation routes. Delivery of products often includes some construction or final assembly activities. Employees from the department 'service

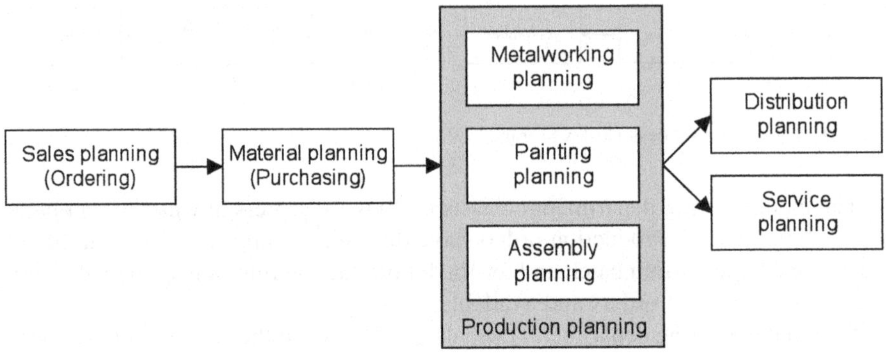

Fig. 3.2 Planning at Desk&Storage

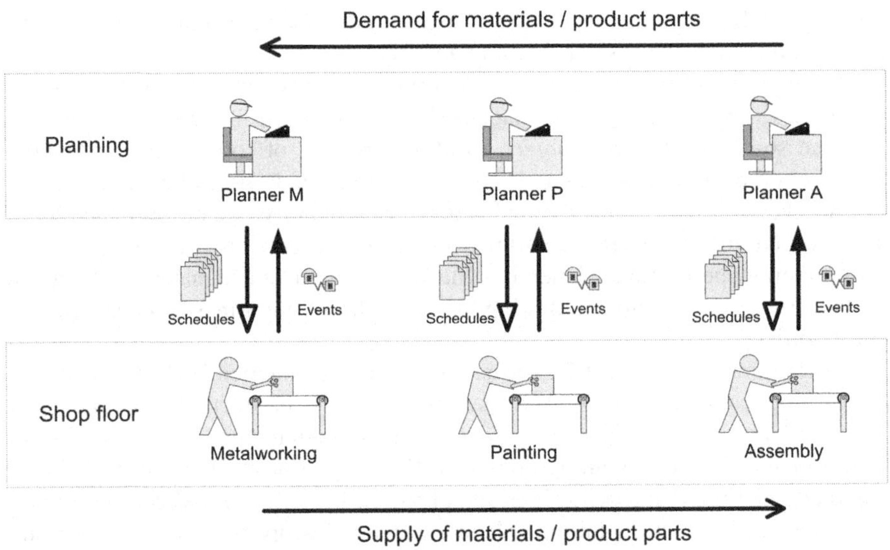

Fig. 3.3 Relation between production planning and manufacturing

planning' are responsible for scheduling the fitters' activities. Figure 3.2 provides an overview of the planning departments.

Figure 3.3 shows the basic structure of the production planning and production execution activities at Desk&Storage. Demand for products is fulfilled by means of manufacturing; manufacturing activities are scheduled by the planners. Therefore, planners provide the operators in production with schedules and schedule updates. Operators inform the schedulers about events that possibly invalidate the schedules requiring rescheduling. Planners inform each other about what has to be planned in a fixed order following MRP-logic (denoted by the arrow 'demand'). Products are produced in the various departments conform manufacturing logic (denoted by the arrow 'supply').

3.3.1 Planning and Scheduling Activities

Each night, the ERP-system calculates all material needs resulting in a basic plan for each production department. The assembly planner uses the information about accepted orders to start with the assembly plan. This plan consists of the assignment of assembly work orders to specific days. The assembly planner equalizes the assembly workload over the days of the week. Sometimes, delivery constraints like international transports are also taken into account. When the workload is equalized, the assembly plan is released. After the nightly run of the ERP-system, the painting planner starts the planning of the work orders for the painting department. The following day, the metalworking planner plans the metalworking jobs.

The main task for the production department planners is to cluster orders on certain dates and to assign them to work cells or production lines. As shown in Fig. 3.3, the planning process is serially organized: in a chain like a train with compartments, the three production planners subsequently 'optimize' the MRP-output for their department. Due to the fixed agreement about lead times per production department (cf., Fig. 3.1), planners are allowed to change the sequence of production within that day.

Alongside the task of creating plans following the backward planning logic (from assembly to purchasing), the planners face a forward moving reality as well.

3.3.2 Rescheduling Activities

All kinds of events influence both the feasibility of the plans as well as (priority of) the activities of the planners. Four types of events are distinguished:

1. Rush orders. Rush orders are all orders that are entered by a sales centre with a shorter delivery time than the standard (4–8 weeks). Rush orders can be divided into three categories:
   (a) Mock-ups. A mock-up is a pre-delivery of products that is set-up at the customer, so the customer can test the product and decide whether to order this product from Desk & Storage. These mock-ups usually need to be delivered on a very short term (within ten days). If an order has the label 'mock-up' everyone within Desk & Storage will do whatever possible in order to make sure the mock-up will be delivered in time, because everyone knows that based on the delivery of the mock-up, the customer will decide whether Desk & Storage or a competitor will get the large order that follows the mock-up.
   (b) Complaint orders. Customers complaining about deliveries have to be satisfied within ten days with repaired or new products.
   (c) 'Normal' rush orders. Some orders have to be delivered with a shorter than standard delivery time for one reason or another.

2. Order changes. Sales centers and customers regularly change orders. Changes can be divided into six types: addition of order line, cancellation of order line, change of article, change of quantity, change of delivery address, change of delivery date/time.
3. Internal disruptions. Production errors, rework, delays, et cetera cause interruptions in the production process often requiring rescheduling.
4. Supplier problems. If a supplier is not able to deliver on time, scheduled production jobs have to be postponed. Sometimes, suppliers are not able to deliver a certain material or half-product; then, order specifications have to be renegotiated with the client.

For each of these four situations, the point in time of event notification determines the number of planners and departments to be involved in the problem-solving process. If a (potential) problem-causing event is solved early, fewer planners have to be involved and fewer plans have to be adapted.

Infeasibilities in the plans and schedules are communicated in different ways. First, the planners are each morning confronted with a list of rescheduling messages from the ERP-system. The messages require manual adaptation of the production plans, for instance by adapting capacity restrictions or by using safety stock. Second, all shop floor operators have access to a MS Access-based Deficit Announcement System (DAS) to report working orders that cannot be processed due to missing materials, products, or tools. The planners handle these announcements during the day. The troubleshooter is concerned with all shortages regarding orders that have to be loaded and transported on the actual day. Third, orders that could not be loaded on the loading day, for instance because parts are still missing, are listed on the so-called 'Deficit List'. Two times a day, this Deficit List is sent to the planners. For each order, a new due date is set. The list is used by the planners to reschedule the tardy orders. Finally, each order that needs to be rescheduled due to supplier problems is listed on the so-called 'Re-plan List'. On this list, all orders affected by the material supply delay are noted with the original delivery date and the new, earliest possible delivery date. Based on this list, the production planners adapt the production plans and inform the shop floor. The ordering department informs the sales agencies and clients in case the original delivery dates could no longer be met. However, it often happens that these sales agencies do not accept the delivery postponement. In such cases, managers are involved to decide about the adaptation of capacity constraints (e.g., by allowing overtime).

Desk&Storage has two days reserved to solve problems due to possible internal tardiness problems (Fig. 3.1): 1 day is reserved for a final check for quality and completeness of the products; after this day, a full day is available for loading the products into the trucks. At the end of that day, the trucks start to transport the products to the clients. However, on average, 45% of the orders is not proceeding according to plan 2 days before this delivery! Without any action, delivery reliability would be extremely low. However, the 2 final days before delivery are often used for production activities as well. By means of rescheduling, overtime, and the troubleshooter's pressure, the firm could realize, on average, delivery reliability

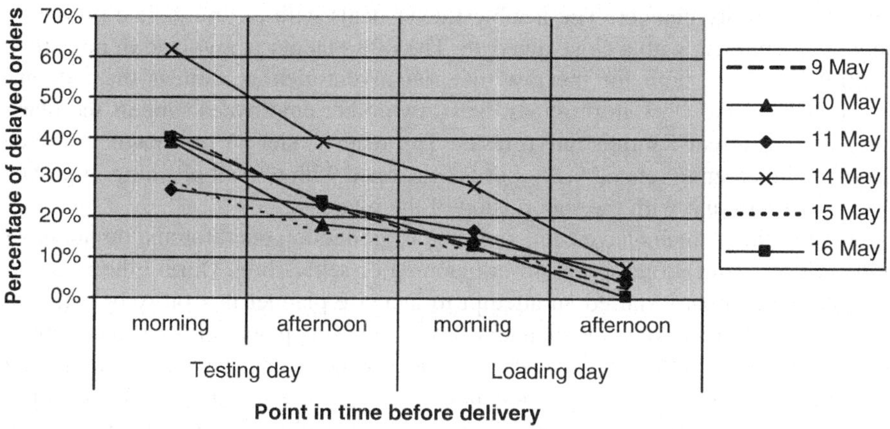

Fig. 3.4 Illustration of internal tardiness problems affecting rescheduling

over 90%. Consequently, planning consists for a large part of rescheduling. Figure 3.4 shows an illustration of the internal tardiness problems. For six normal working days, the percentage of orders that is still in progress and that is, therefore, not on schedule is shown. Two days before the due date, between 27 and 62% of the orders is not proceeding according to plan: these orders should be ready for testing and loading, but they are not. A high amount of rushing and chasing by the troubleshooter realizes a delivery performance between 92 and 99%. Sometimes, the testing activities are omitted; in other cases, the distribution plans are adapted to enable loading in the afternoon instead of during morning hours.

The so-called 'troubleshooter' plays a central role in the rescheduling process, continuously pushing the planners and production foremen to realize high delivery reliability. Rescheduling decisions have to be reconsidered, adapted, and communicated. Because a plan change often influences the plans of others, planners have to cooperate to find feasible solutions. Coordination between the planners is not formally organized, i.e. there are no formal rules which planners should be involved in what way at which moment to solve rescheduling problems. Because of the large physical distance between the planning and production department, communication between planners and shop floor operators occurs mostly via phone or using the computer system.

3.3.3 Planners

In Desk&Storage, five employees are involved in the development and 'maintenance' of capacity and production plans for the three production departments. These five employees are located together in the planning office. Each production department has its own planner: the metalworking planner, the painting planner, and the assembly planner. The troubleshooter deals with all immediate problems regarding products with a close due date. The fifth planner is involved in inventory control activities, both for the raw material inventories as well as the pipeline inventory, i.e. the work-in-progress. Next, two other employees support the planners with different administrative tasks. The team leader of the planners is not involved in planning related tasks, but is concerned with overall planning improvement programs and with the supervision of the planners.

Most of the planners have been working as production operator and do not have followed any educational program in planning or scheduling. During the research project, the company started an attempt to increase planner flexibility by introducing task rotation between the planners. The assembling planner for instance fulfills the troubleshooter-role 1 day a week. The metalworking department planner is also capable to do the planning for the painting department. To increase planning performance, the company wanted to recruit and deploy the best people for the most suitable planning tasks. However, the assessment of planner candidates appeared to be quite difficult, because of the weak predictability of a candidate's abilities to act in the dynamic planning environment. Appropriate measures to assess the required and present skills and competences for the planner were lacking.

3.3.4 Planning Environment

The planning environment consists of the elements that influence planning complexity and planning performance. McKay and Wiers (2006) show how planning and scheduling are interconnected with many functions and processes within a firm. We observed similar interconnectivity with Desk&Storage.

To align sales and production, every Monday morning a so-called 'capacity meeting' takes place with the operations manager, the head of the departments production planning and warehousing, the assembly planner, the material purchasing planner, the customer support manager, the head of the ordering department, and the account coordinator. During this meeting, backlogs in production, supplier delivery problems, special orders, and performance realization are discussed. The main goals are to discuss critical issues for the upcoming 4–6 weeks and to determine the capacity for the weeks to release (i.e. the main constraints for order acceptance and production plans).

Desk&Storage aims at a rapid delivery of a wide variety of products. Clients however are allowed to change their orders after order acceptance and suppliers sometimes delay the supply of raw material. This results in a high amount of rescheduling events complicating the planning puzzle for the planners and stability at the shop floor.

Next to this, different performance criteria are used: production is judged on the criterion of efficiency and machine utilization, whereas planning is judged on the criterion of delivery reliability. This leads to a goal conflict in the interactions between planners and operators regarding the prioritizing of orders. The operators may determine the sequence of orders within the boundaries of 1 day (conform the fixed 1-day lead time per department). Therefore, the operators often choose a sequence taking machine setups, material availability, and ease of production into account. However, because of changing orders and rescheduling, the production schedules change. Moreover, to be able to realize overall high delivery reliability, planners impose even more constraints to the shop floor or strictly prescribe the production sequence (as discussed in Sect. 3.3.1). A lot of explanation and persuasion between planning and shop floor is the consequence.

3.4 Relocation of the Planning Department

One of the interesting outcomes from the planning description above concerns the importance of efficient coordination between the planners and the shop floor. As in many manufacturing firms, Desk&Storage has separated planning and execution activities and has assigned these activities to different employees. Planners and schedulers are responsible for the development and maintenance of feasible and efficient plans, whereas foremen and operators are responsible for the execution and realization of these plans. Plan infeasibilities encountered by the operators are communicated to the planners who subsequently provide a plan update.

There are many reasons for such task splitting, for instance the higher levels of specialization possible. However, task splitting between work preparation (scheduling) and work execution (operation) will only perform well if interdependencies between both activities are managed well. Within Desk&Storage, operators frequently complain about the 'incompetence' of the planners, because they do not sufficiently take into account the actual manufacturing situation and the (im)possibilities of the shop floor. On the other hand, planners are often frustrated about self-willed operators not following the schedules, or 'lazy' operators not communicating plan disturbances quickly.

To improve the relation between the planning and shop floor departments, it was decided to move the planning department to another location in the factory. In the old situation, the planners regularly took a bicycle to meet shop floor operators because of the large distance. After the relocation, the planners were located in the center of the shop floor. It was expected that the relocation would improve coordination efficiency and result into better scheduling *and* manufacturing performance.

3.4.1 Research Methods

To investigate the consequences of the relocation of the planning department, it was decided to use a quasi-experimental research design containing a pre-change (March 2007; M1) and a post-change measurement (June 2007; M2) regarding mutual interdependencies and interaction patterns between both departments and to relate this with (perceived) changes in performance. Data collection was done by using two measurement instruments.

First, planners and production foremen were asked to respond to a number of statements about the relocation. In M1, all statements were formulated as expectations, whereas in M2, the statements were formulated as experiences. A seven-point Likert-scale was used with 1 = totally disagree and 7 = totally agree. The following statements were used, all starting with the phrase: "Because of the relocation of the planning department:

1. … my work will/has become easier.
2. … I (will) have a better overview of what happens at the planning department.
3. … I (will) have a better overview of what happens at the manufacturing departments.
4. … I (will) have less contact with the troubleshooter.
5. … the feasibility of the schedules (will/has) improve(d).
6. … the manufacturing departments (will) perform more effectively.

Second, the planners were asked to fill out a small questionnaire each time they had been involved in an interaction. In this way, detailed information about a high number of 'coordination occurrences' could be collected. Two regular working days had been chosen in close cooperation with the management during which these employees registered all their work-related interactions.

3.4.2 Results

Results from the questionnaire are reported first, followed by a short overview of findings from the interaction surveys. Table 3.1 shows the average scores on the statements. Numbers above four indicate that, on average, the respondents from a particular department agreed with the statement, whereas numbers below four indicate disagreement with the statement. We shortly explain the findings. The planners and shop floor foremen expected and experienced a change in work difficulty: due to the relocation, their work had become easier to perform. This could probably be explained by a better insight into each other's work. Indeed, the planners have more insight into what happens at the manufacturing departments, in

Table 3.1 Average scores on statements per department

Statement	Planning		Metal working		Finishing		Assembly		Total	
	M1	M2	M1	M2	M1	M2	M1	M2	M1	M2
Work more easy	6.6	6.2	4.3	5.7	4.5	4.7	5.0	6.2	5.2	5.6
More insight scheduling	–	–	2.8	5.3	4.4	3.3	2.5	6.4	3.5	4.9
More insight manufacturing	6.3	6.4	–	–	–	–	–	–	6.3	6.4
Fewer contact troubleshooter	1.7	2.0	4.0	5.3	5.0	4.2	2.5	4.4	3.4	3.9
Higher schedule feasibility	3.8	4.8	4.0	5.3	3.6	4.3	2.5	5.0	3.6	4.8
Higher effectiveness manufacturing	4.5	5.0	4.5	4.7	3.2	4.2	4.0	4.4	4.0	4.5

line with their expectations. However, the consequences of the relocation for the level of insight for the shop floor a foreman into what happens at the scheduling department are less clear. The metalworking and assembly foremen did expect no more insight, but a few months after the relocation, they indicated the reverse. The finishing foremen were not expecting much change, and their experience was almost similar to their expectation.

Whereas the planners did not expect and not experienced a decrease in the frequency of contact with the troubleshooter, the foremen indicated that they have less contact with him. Further, schedule feasibility had improved slightly according to the respondents, although expectations were not high. The effectiveness of manufacturing had not really changed due to the relocation of the planners.

From these findings, it can be concluded that the relocation has led to a better working situation for both the planners and the shop floor workers: their work has become easier and now they have more insight into the other's work. However, the respondents did not perceive significant performance improvements due to the relocation.

Collecting detailed data about all interactions of the planners appeared an intensive research method. Table 3.2 shows the number of reported interactions per planner. The planner of the assembly department was performing the tasks for the troubleshooter as well, due to absence of the troubleshooter on Wednesdays. Table 3.3 shows the departments the planners were interacting with; during M2, a significant higher number of interactions between the planners have been reported.

Relocating the planning department has decreased the physical distance between the planners and the shop floor foremen and operators enormously. However,

Table 3.2 Number of interactions per respondent during M1 and M2

Function	M1	M2	Total
Inventory controller	15	30	45
Planner assembly department / troubleshooter	37	60	97
Planner finishing department	22	14	36
Planner metalworking department	38	50	89
Total	112	154	266

Table 3.3 Number of interactions per department

Department	M1	M2	Total
Planning	18	41	59
Metal	37	42	79
Finishing	18	7	25
Assembly	3	8	11
Production management	3	6	9
Sales Agencies and Ordering	13	16	29
Warehouse	4	4	8
Purchasing	5	7	12
Distribution	4	11	15
Other	7	12	19
Total	112	154	266

Table 3.4 Medium used for communication during M1 (N = 61) and M2 (N = 63)

Channel	M1	M2	Total
Phone	91.8%	85.7%	88.7%
Face-to-face	8.2%	14.3%	11.3%
Total	100.0%	100.0%	100.0%

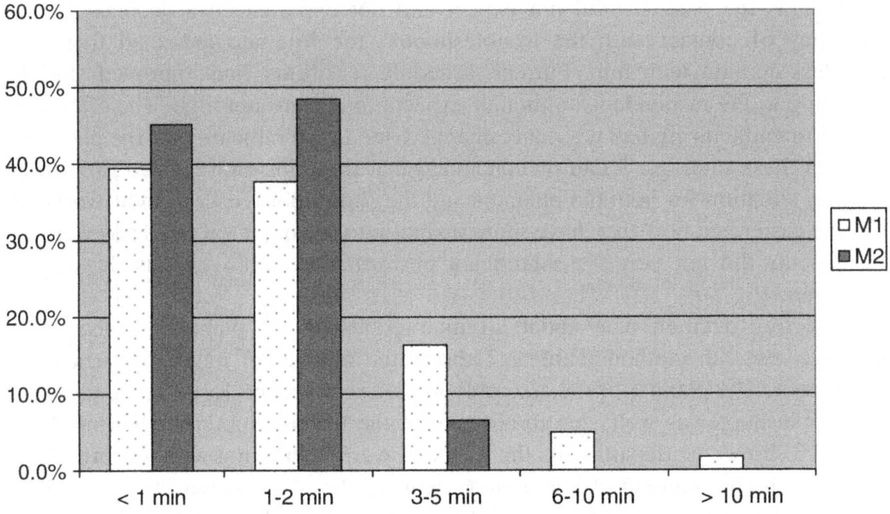

Fig. 3.5 Length of the planner-shop floor interactions (M1: N = 61), M2: N = 62)

Table 3.4 shows the communication channel used for the interactions between planning and shop floor. No significant changes appeared; the phone is still used most frequently to communicate.

Finally, Fig. 3.5 shows the duration of the interactions between the planners and the shop floor foremen and operators. It appears that, on average, the interactions during M2 day took somewhat less time that during M1. This can probably be explained by the increased level of mutual insight discussed before.

3.5 Conclusion

The case description in this chapter has shown the diversity of activities and information sources planners are dealing with. The exploratory research has resulted in several insights regarding the dynamic and often unstructured ways of event processing and problem solving in production planning and scheduling. A good relation between planning and shop floor appears to be necessary to handle the many events and schedule adaptations efficiently. Physical vicinity between the departments is one of the variables influencing this relation. The relocation of the planning department to decrease the physical distance between planning and shop floor was both expected and experienced to have positive consequences for the level of mutual understanding and coordination. Direct effects on performance could not be addressed. Further research is needed to study the effects of relocating a department and of physical vicinity between departments over a longer time period. Furthermore, the interconnectivity of planning and shop floor and the importance of the human planners interacting with many different stakeholders call for further empirical investigation and the development of theory and support tools to facilitate the coordination activities between planning and shop floor.

Acknowledgement The authors thank Desk&Storage for its willingness to participate in this research and Stephan Keuper for contributing to the data collection and analyses.

References

Berglund, M., & Karltun, J. (2007). Human, technological and organizational aspects influencing the production scheduling process. *International Journal of Production Economics, 110*(1–2), 160–174.

Crawford, S., MacCarthy, B. L., Vernon, C., & Wilson, J. R. (1999). Investigating the work of industrial schedulers through field study. *Cognition, Technology and Work, 1*(2), 63–77.

De Snoo, C., Van Wezel, W., & Wortmann, J. C. (2010a). *Integration of scheduling and manufacturing: consequences of relocating the scheduling department. Working report.* Groningen: University of Groningen.

De Snoo, C., Van Wezel, W., Wortmann, J. C., & Gaalman, G. J. C. (2010b). Coordination activities of human planners during rescheduling: Case analysis and event handling procedure. *International Journal of Production Research*, DOI: 10.1080/00207541003639626.

Jackson, S., MacCarthy, B. L., & Wilson, J. R. (2004). A new model of scheduling in manufacturing: tasks, roles, and monitoring. *Human Factors, 46*(3), 533–550.

MacCarthy, B. L., Wilson, J. R., & Crawford, S. (2001). Human performance in industrial scheduling: a framework for understanding. *Human Factors and Ergonomics in Manufacturing, 11*(4), 299–320.

McKay, K. N., Safayeni, F. R., & Buzacott, J. A. (1995). 'Common sense' realities of planning and scheduling in printed circuit board production. *International Journal of Production Research, 33*(6), 1587–1604.

McKay, K. N., & Wiers, V. C. S. (2003). Planning, scheduling and dispatching tasks in production control. *Cognition, Technology & Work, 5*(2), 82–93.

McKay, K. N., & Wiers, V. C. S. (2006). The organizational interconnectivity of planning and scheduling. In W. Van Wezel, R. J. Jorna, & A. M. Meystel (Eds.), *Planning in intelligent systems: Aspects, motivations, and methods.* New York: Wiley.

Van Wezel, W., Van Donk, D. P., & Gaalman, G. J. C. (2006). The planning flexibility bottleneck in food processing industries. *Journal of Operations Management, 24*(3), 287–300.

Part II
Organization of the Planning Process

Chapter 4
The Unsung Contribution of Production Planners and Schedulers at Production and Sales Interfaces

Martina Berglund, Jane Guinery, and Johan Karltun

Abstract This chapter sets out to demonstrate the unsung contribution of production planners and schedulers in manufacturing businesses. In particular it focuses on their contribution at production and sales interfaces by highlighting their activities and influence across functional interfaces, and the knowledge and skills they apply to make and implement planning and scheduling decisions. To achieve this it addresses the following questions in relation to these interfaces: What tasks and work activities does planning, scheduling and control consist of in relation to these interfaces? How do planners and schedulers perform their tasks? How can planners' and schedulers' activities related to production and sales interfaces be captured and modelled? How do planners and schedulers influence others in the organization? What knowledge do they contribute and how is it incorporated into decisions?

4.1 Setting the Scene

For many years the coordination of production and sales/marketing functions in manufacturing businesses has been considered to be problematic, raising significant concerns. Extensive research has highlighted conflicts between production and

M. Berglund (✉)
Department of Management and Engineering, Quality Technology and Management, Linköping University, Linköping, Sweden
e-mail: martina.berglund@liu.se

J. Guinery
Nottingham University Business School, Nottingham, UK
e-mail: jane.guinery@nottingham.ac.uk

J. Karltun
Department of Industrial Engineering and Management, Jönköping University, Jönköping, Sweden
e-mail: johan.karltun@jth.hj.se

J.C. Fransoo et al. (eds.), *Behavioral Operations in Planning and Scheduling*,
DOI 10.1007/978-3-642-13382-4_4, © Springer-Verlag Berlin Heidelberg 2011

sales/marketing (e.g. Lawrence and Lorsch 1967; Clare and Sanford 1984; Crittenden et al. 1993; Spencer and Cox 1994; Gunasekaran et al. 2002). Further to this, research has identified the challenges (e.g. Konijnendijk 1993; Mukhopadhyay and Gupta 1998; Gonzalez et al. 2004) and the crucial importance to business success (e.g. Skinner 1969; Hill 1997) of effectively coordinating production and sales/marketing. Although there may be direct interactions between production and sales/marketing functions, production planning has frequently been cited as contributing to the interface (Parente 1998) with production planning decisions influencing performance (O'Leary-Kelly and Flores 2002). Taking this perspective, production planning and scheduling may be viewed as providing a bridge between production and marketing/sales. Also, Shapiro (1977) identifies that in some cases, where production scheduling is problematic, production and sales/marketing need to cooperate more fully to resolve problems. Although there is extensive research on the production and sales/marketing interface, there is little investigating the contribution of production planning and scheduling to the interface at an operational level (Whybark 1994; Parente 1998; Swamidass et al. 2001).

Before discussing the research further the distinction between planning and scheduling needs to be considered. One definition is that scheduling is associated with decisions made when the production process is running, while planning is associated with decisions made before the production process starts (Nakamura and Salvendy 1994). However, as the precise interaction between production planning and scheduling is not easily defined it may be justified to assume that planning and scheduling in fact represent a continuum of activities across space and time (Crawford et al. 1999). For this reason, these terms are used relatively interchangeable throughout this chapter.

In manufacturing businesses, the role of humans in production planning and scheduling is often considerably underestimated. Whilst software developers and operations management academics devise complex planning and scheduling systems and managers seek to automate planning and scheduling activities they often trivialize the role of planners and schedulers seeing them purely as compensating for inadequacies of system. However, systems implementations are often highly problematic and do not achieve to expectations (MacCarthy and Wilson 2001; Davenport 1998). Empirical research has found that in many production contexts planners and schedulers are essential. This is particularly the case where planning environments are 'messy' and complex and there are multiple players (MacCarthy and Liu 1993; McKay and Wiers 2001). Clearly, the broader contributions made by planners and schedulers, including bridging between other functions, may be substantial and need to be more fully examined.

As limited research has been undertaken on planning and scheduling in relation to the sales and production interfaces, this research sets out to contribute to the body of knowledge in this area by responding to questions such as: What does the production planning, scheduling and control (PSC) process consist of in relation to the production and sales interfaces? How do planners and schedulers perform their tasks in relation to these interfaces? How can the process by which planners' and schedulers' bridge between production and sales be captured and modelled?

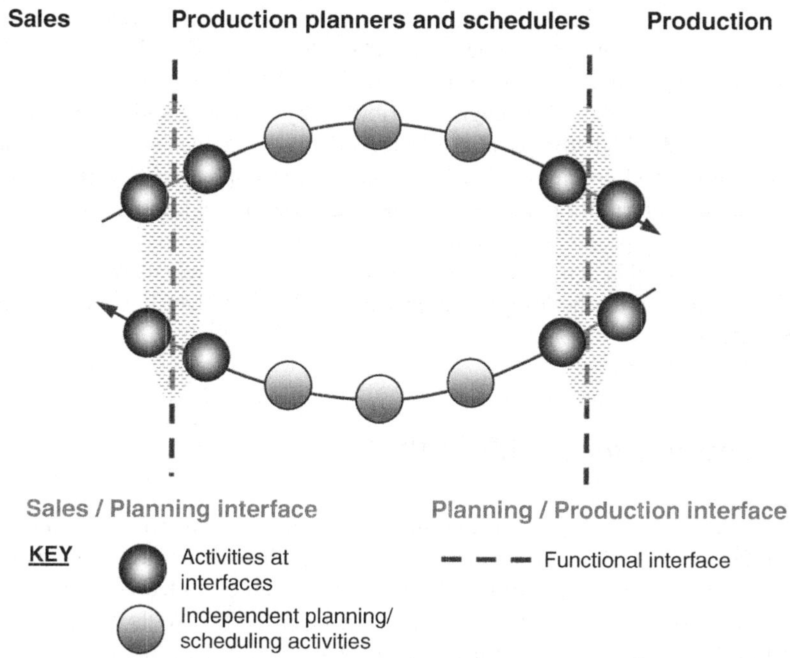

Fig. 4.1 Initial conceptual model of the PSC process

How do they influence others in the organization? What knowledge do planners and schedulers contribute to and how is it incorporated into decisions? An initial conceptual model of the PSC bridging process is used to describe the process in relation to its key features such as the functional interfaces and activities within it, see Fig. 4.1. This is provided to aid the reader's visualisation of the process under discussion.

We will be investigating the PSC process in relation to sales and production interfaces taking a number of different perspectives. These will be drawn from various theoretical domains such as operations management, organizational behaviour and work science. To tackle the questions posited above, the following themes are examined in the chapter subsections:

Capturing the reality of PSC. Examines the relevance of empirical research in general and describes the cases used in subsequent analysis;

Complexities in everyday PSC practice. Illustrates what planners and schedulers encounter in everyday work and how they respond, starting with general PSC practice and then moving on to illustrate the types of issues that are dealt with in relation to sales and production interfaces;

Planners and schedulers aligning customer demands with the production capability (Slack and Lewis 2002) – describes a model that has been developed to represent the part of the PSC process that aligns customer requests to the capability of the business through the effective handling of production enquiries.

Planners' and schedulers' influence at the sales and production interfaces. Identifies the ways in which planners and schedulers influence others;

Planners' and schedulers' knowledge use and decision making at key PSC interfaces. Looks at the knowledge contributed by planners and schedulers and how their and other's distributed knowledge is drawn together and used to enable decision making and problem solving at key interfaces and across work groups.

The primary contribution that this chapter makes is to provide empirical findings on the contribution production planners and schedulers make at sales and production interfaces. This is achieved through a combinational analysis of existing case study research. The methodological approach used to achieve this is described in the next section.

4.2 Capturing the Reality of PSC

To establish the actualities of PSC, empirical research was undertaken in a number of manufacturing businesses. A case study approach enabled an examination of phenomena in their natural setting (Meredith 1998) and provided a rich set of data on real world practice (Eisenhardt 1989; Voss et al. 2002). With the intention of supporting both exploratory and theory building research on PSC, the richness of data obtainable through detailed case study work was needed to address the research questions posited earlier (Voss et al. 2002). Whilst this empirical approach is highly fruitful the limitations of case studies have to be recognised. Unlike statistically sampled surveys, case studies cannot claim to be representative of entire populations. Therefore there are limits to the extent to which findings can be generalised. However, they still have firm validity when represented in relation to their context.

As described previously, a number of different topics are addressed in this chapter. Findings in relation to each topic are described and illustrated using a number of case studies. The case studies are drawn from six Swedish and British manufacturing businesses. All the case studies were undertaken to achieve thorough understanding of how PSC is carried out in practice; in the Swedish cases to develop a detailed understanding of work activities and in the British cases to identify best practice for responsive performance. Data collection in each case study was undertaken by a number of researchers and consisted mainly of interviews with management and staff in planning and scheduling, sales and marketing, and production. Information on these interviews is summarised in Table 4.1.

Further data was collected through observation of the PSC activities performed by individual planners and schedulers (this included observations of meetings, activities and interactions with other people and systems). The interviews concerned planning, scheduling and control processes, the interviewees' involvement, and their interactions with others. Documents, decision support packages and planning systems, characteristics of the technical production system, and market and product demand information were also studied.

Table 4.1 Number and position of interviewees

Company	Number of interviewees	Position of interviewees
Sawmill	15	Scheduler, purchaser, marketing manager, CEO, production manager, delivery officer, operators, forklift truck drivers
Parquet floor manufacturer	23	Scheduler, scheduling manager, sales manager, customer helpdesk officers, operators, supervisors, production group leaders and foreman, forklift truck drivers, purchaser, quality coordinator
Furniture manufacturer	10	Scheduler, quality manager, delivery officer, foremen, purchaser, production manager, supervisor, customer purchaser
House manufacturer	19	Scheduler, foreman, supervisors, industrial engineer, production manager, purchaser, material administrator, delivery officers sales support officers, export officer, project coordinator, construction manager, construction engineers
DIY product manufacturer	11	Planner, supply chain manager, production team leader, customer services personnel, demand manager, product manager, manufacturing manager, engineering, managing director
Steel manufacturer	9	Manager of load control, load controller, commercial business planners, manager of planning and scheduling, schedulers

An overview of each case study is provided in Table 4.2. The information includes: the case name, products, location and number of employees; the type of production undertaken; the primary objectives of planning and scheduling; contextual factors that might affect how the objectives are achieved, the key PSC interfaces that are analysed and the main PSC activities performed across them. It should be noted that in the sawmill, parquet floor, furniture and house manufacturers PSC activities are handled by production schedulers, whilst in the steel and DIY product manufacturers, people in similar roles are referred to as planners. However, for the purposes of this research these should be considered to be equivalent. These individuals are involved in both tactical and operational levels of decision making. More specific aspects of each case are presented in subsequent sections of this chapter according to their relevance to the topic under discussion. It should be noted that these cases are used for a variety of purposes and in different combinations. Whilst individual cases are used to illustrate specific issues, multiple cases are compared to develop and explore concepts. Two of the cases are examined more thoroughly in this chapter, as they provide the most informative illustrations of the pertinent issues and practice. First, the parquet floor manufacturer is used as a basis from which to model the production enquiry process, and second, the steel manufacturer is used to illustrate knowledge use and decision making across functional boundaries. For further detailed descriptions of the case studies, including

Table 4.2 Case studies used in empirical studies

Case study	Products and type of production	Planning objectives and timescales	Contextual factors impacting on PSC	Key PSC interfaces – *associated activities*
Sawmill –42 employees (Sweden)	*Beams, boards, planks* Make to order regarding high-value goods (beams and planks) from the core of the logs. Divergent production flow. By-products produced a long with high-value goods	Optimise the high value outcome of the sawing process regarding quality and volume Achieve the optimal fit between available timber and customer demand	Long-term contracts, with customer calling off required volumes Special demands from prioritized major customer Uncertainties in production process outcome	*Sales* – discuss with sales manager regarding feasibility of specific customer orders *Production* – distribute work orders; control production outcome and status
Parquet floor manufacturer – 1,400 employees (Sweden)	*Parquet floor* Predominantly make to order (regarding wearing surface of the parquet floor)	Find best possible fit between production capability and delivery demands	Uncertainties in production process outcome Large number of product variants	*Sales* – respond to enquiries for customer order; ensure that customer orders are met or alternative products suggested for delivery *Production* – schedule machinery; ensure and follow up that scheduled products are produced
Furniture manufacturer – 220 employees (Sweden)	*Book shelves* Make to stock based on prognoses from customer	Develop master schedule and optimise the relation between customer forecasts, order situation, schedule and actual deliveries	Highly automated production system Long production set-up time One dominating customer Strong dependence on forecasts	*Sales* – analyse the order situation and delivery forecasts together with main customer *Production* – provide master schedule; ensure supply of right raw material

House manufacturer – 370 employees (Sweden)	*Wooden houses* Design and make to order according to customer specification. Assembled on site.	Ensure that each house is produced as planned Order and plan material flow to production unit and building site	Project based scheduling Delivery includes installations at the building site Private customers or external building contractor (for larger houses)	*Sales* – control of design and planning stages of the project, ensure that scheduling steps are met throughout; contacts with contractor on building site regarding material *Production* – schedule production; ensure that right material is produced or delivered on time, ensure product delivered to building site
Steel product manufacturer (Britain)	*Beams, plates, rod, sections* Predominantly make to order. Batch production with process de-coupling between liquid steel production and rolling mills.	Manage demand based on profitable work selection Share liquid steel supply across mills to ensure plant utilisation and deliveries	Highly volatile demand Commodity type market Unpredictable processes Liquid steel production is the constraint	*Sales* – select orders and manage demand *Production* – know status of plan to respond to changes Order management team – share liquid steel Executive - agree plan
DIY product manufacturer – (Britain)	*DIY household adhesives* Predominantly make to stock for off the shelf delivery Batch production on flow lines. Product focused manufacturing.	Manage stock vs. service trade-offs to provide off the shelf delivery at low cost Fast track urgent orders	Customer service/ delivery performance is key Product focused plant Goals are shared throughout organisation	*Sales* – provide forecasts and product promotion information *Production* – agree and implement schedule

methodologies applied in data collections and analysis, see Guinery (2006), Coates et al. (2005), Hamlin et al. (2006), and Berglund and Karltun (2007).

Section 4.3 looks in detail at PSC in practice. It presents specific illustrations of complex issues in PSC to provide a contextual understanding of the domain.

4.3 Complexity in PSC Everyday Practice

An increasing number of studies describe the role and activities of production planners and schedulers. They can be viewed as people whose function it is to plan and schedule production according to demand and within the limitations of the operation. To achieve this they need to deal with constraints regarding operational issues including physical constraints, process issues, work force characteristics, and administrative constraints. As well as encompassing these tasks, planning and scheduling requires substantial levels of social interaction where aspects such as trust, respect, and recognised problem solving capabilities are important, as is possession of a thorough knowledge of the domain across which decisions are made (McKay et al. 1992).

In extensive field studies of production schedulers in "actual" scheduling activities, Jackson et al. (2004) identified three categories of task and three categories of role behaviour. The tasks consisted of *formal tasks* recognised by the business as scheduling tasks to be carried out, *house-keeping tasks*, in which the scheduler organised data according to the way the scheduler carried out work, and *compensation tasks*, in which the scheduler compensated for some level of problem or failure in the overall system. The research also identified work activities that were not formally recognized, but essential and in themselves required knowledge. These were described as the following three role behaviours:

- *Link and net* which includes the development of interpersonal networks, informal bargaining, friendship and favour networks, and mediating
- *Hub and filter* which consists of the scheduler acting as an implicit and explicit information hub, ensuring that information is accessible and visible, and where necessary identifying the relevance of information and filtering it
- *Balance and valve* which includes problem predicting and problem solving, changing plans and instructions, and allocating resources

Other detailed field studies investigating how knowledge is used in PSC decision making identified different role categories associated with how production planners and schedulers use and enable the sharing of knowledge (Guinery 2006; Guinery and MacCarthy 2009). These findings are described later in this chapter.

Research on schedulers' real work activities in four different businesses; the sawmill, the parquet floor manufacturer, the house manufacturer and the furniture manufacturer (see Table 4.2) revealed that scheduling in practice includes a great deal of problem solving (Berglund and Karltun 2007). Whilst some problem solving was carried out as part of the schedulers' planned activities, often the need for it was a consequence of an unplanned occurrence in production or new information from the market. (In fact scheduling work was strongly characterised by abrupt changes from one work task to another initiated by interruptions such as telephone calls, someone entering the office, or the arrival of urgent e-mails). Often several issues needed to be dealt with at the same time, and problem solving activities could extend into areas beyond the scheduler's formal remit.

The expectations held by other employees and management on the schedulers in the sawmill and parquet floor manufacturer were analysed to establish the schedulers' roles (Berglund and Karltun 2001). The analysis assumed a definition of role provided by Katz and Kahn (1978). This is, that the role an individual takes in an organisation may be defined as the combined demands with which the system is confronting the individual member (which includes the expectations held by others in the organisation). From the analysis several categories of expectations were established:

- *Information supply.* To actively supply immediate, accurate and up to date information on what should be produced, what was being produced at the moment and what had been produced (although they often lacked reliable data). Also they were expected to give and receive feedback on progress to plan, despite the information systems not supporting this process.
- *Continuous good communication.* To keep up a continuous dialogue with others to foster and support collaboration.
- *Efficient scheduling.* To produce efficient schedules to optimise production in terms of outcome and/or profitability. However in this case many interviewees viewed efficient schedules differently, relating their expectations to the impact schedules had on their own work. These included making sure that production ran smoothly and that there was sufficient material for production and few disturbances. Furthermore, the schedulers were expected to adapt to specific situations within or outside the company and show consideration of the consequences of their scheduling on other departments or group of employees (e.g. on working hours).
- *Flexibility whilst upholding agreed schedules.* To stick to agreed schedules or solutions whilst at the same time being flexible enough to discuss changes when initiated by other employees or the market.
- *Balance incompatible goals through compromise and negotiation.* To find compromises and negotiate between groups in the company and to balance the demands of different groups of employees, although they may have incompatible goals.
- *Problem solving.* To solve problems over and above those directly related to the schedule. In the parquet floor manufacturer, production personnel, production supervisors, market representatives and the parquet manufacturer's quality manager strongly maintained that they expected the production schedulers to solve problems associated with handling unexpected events, coordinating different production units, participating in developing production and products etc. Because of this, the schedulers were expected to have sufficient understanding

of problems in these areas and be able to suggest and discuss different solutions as well as carry through the ones chosen. The schedulers themselves confirmed that they believed that others in the company considered them to be problem solvers, i.e. people to turn to with all kinds of questions and problems.

- *Knowledge based on reality.* All activities carried out by the schedulers should be based in reality. The schedulers were expected to fully understand what actually occurred in the workshop and other production premises, as well as in the computer systems. As a consequence, the schedulers were expected to check frequently what was actually occurring in production.

Clearly expectations on schedulers were highly diverse and sometimes in conflict with each other (e.g. flexibility vs. schedule adherence). This is in line with Katz and Kahn (1978) who reported that roles become more complex when they require an individual to be involved in two or more parts of the organisation, as each part is likely to have its own subculture and priorities. This aspect requires investigation, particularly as it is known that production and sales and marketing functions have different priorities and their goals can conflict. For instance the sales side may strive for flexibility, short delivery times and responsive customer service while production strives to smooth production levels and stabilise manufacturing conditions. In effect each part of the organisation could be working to different objectives, in different contexts, with different agendas and paradigms and therefore ways of understanding and solving problems; in this chapter defined as 'functional logics'. These logics are influenced by aspects of the customer demand, production processes, human resources and social relations (Carballeda 1999). As these aspects differ between functions so do practical and perceived objectives and constraints and associated behaviours. In these conditions social actors strongly maintain their values, some of which may conflict.

The everyday work practice of schedulers clearly needs to be looked at *in relation to situations in which planners and schedulers have to handle the different logics of production and sales departments.* This aspect of their work was investigated in the same four case studies; the sawmill, the parquet floor manufacturer, the furniture manufacturer and the house manufacturer. These provided a number of illuminating examples of what actually occurs when planners and schedulers have to respond to the different logics of production and sales (Karltun and Berglund 2002), a few of these are presented below:

- *Negotiations concerning conflicting goals.* At the sawmill, a scheduling meeting was held every two weeks with the sawmill scheduler, the planing scheduler, the stowing leader, the marketing manager, and sometimes the timber supply manager. In these meetings, the delivery situation for approximately one month was discussed, and a delivery schedule for the main customers' orders was produced. These meetings consisted of negotiation between participants and aimed to establish the best compromise between their objectives. The marketing manager's aim was to maximise the customer service, the sawmill scheduler tried to optimise sawmill production, the planing scheduler advocated solutions that fitted into the planing schedule, the stowing leader's aim was the completion of work to complete container shipments, and the timber supply manager tried to make the sawmill operation fit with the timber supply situation.

- *The scheduler introducing informal rules.* The parquet manufacturer was very successful; this resulted in increased sales volumes during the period of the study. As these volumes exceeded the productive capacity of their production operation, the delivery lead times increased. With this change in conditions, the scheduling system's rigid rule-based handling of data quickly became obsolete or ineffective and pressure to manage the situation was placed on the scheduler. To cope with this, the scheduler introduced informal rules on how he wished the customer help-desk officers to inquire and inform him about orders before they were put into the system. He aimed to influence order agreements to improve the chances of delivering on time. This was accomplished by suggesting changes e.g. offering a different quality of product or a later delivery week. The customer help-desk officers could not see all these possibilities in the computer system, and the scheduler was mentally working with a modified schedule based on his own knowledge. Furthermore, due to the scheduler's close contacts with the production units he was aware of alternative approaches and products that could be used to resolve upcoming problems. Some customer help-desk officers did not accept this mode of working and continued to work to the information in the system. This caused significant conflict between help-desk officers and the scheduler.
- *Gaming with the scheduling software tool.* At the parquet manufacturer, the company introduced a computerised scheduling system, in which sales offices across the whole of Europe could place their orders. This system allowed each salesman to see the products and the quantities available in stock as well as what was scheduled to be produced in the next months. A salesman could then place orders on the system, reserving the products to be delivered to his special customer. Depending on the local culture in different countries and the local sales manager's policy, various orders were entered that reserved a production volume, although the customers for the orders were only prospective. When approaching the date of delivery, the orders were withdrawn which meant the schedule became inappropriate as the mix of scheduled products was not satisfying real demand. The accessibility of the scheduling system to marketing meant personnel could load the entire production system with their requested orders leaving no spare capacity for unexpected events or schedule deficiencies.

When forecasting demand for product, the scheduler kept in mind the volumes of 'withdrawn' orders that individual agents had previously requested; from this information he determined the trustworthiness of subsequent information from each agent. He then placed a schedule for next month's production that took into account the probability of the orders actually being withdrawn. By scheduling only to his predicted volume, he forced the sales people to reveal which orders were realistic. He was also very conservative in his forecasts and waited as long as possible before he submitted the schedule to production. This resulted in the available production outputs exceeding the forecast volume, which allowed the scheduler to handle minor breakdowns or other unexpected events without disrupting deliveries of orders. The scheduler was very aware of the way he used the system and described his and the other users' actions as "a little bit of gaming".

These examples demonstrate the complexity of the tasks and roles that planners and schedulers perform. In Sect. 4.4 a subset of their work activities, those associated with handling production enquiries, will be examined and captured in a model.

4.4 Planners and Schedulers Aligning Customer Demands with the Production Capability of the Enterprise

As described earlier, one subset of production and scheduling work is handling production enquiries. We will now look at that process as it is one that can link production and sales decision making. What role do planners and schedulers perform in this process? One way to describe the role is through a role activity diagram (Ould 1995). It focuses on the role as the primary unit of analysis and provides a useful way to collate and represent the PSC activities captured in field study. Activities are here defined as what is actually performed to carry out a work task (Leplat 1990; Guérin et al. 1997). Looking at one of our case studies, the parquet floor manufacturer, observations of the scheduler's work showed that a significant part of his time was devoted to handling customer requests (Berglund and Karltun 2006); these requests arrived on an ad hoc basis and were treated individually. This case is therefore used as an illustration of work activities performed within PSC to establish whether the business should accept an individual customer request based on its value to the business and whether the operation has the capability to provide products in the required quantities at the required times.

Here production capability is defined as the temporal ability of production to fulfil orders. According to Slack et al. (1998) production planning and control activities are those that reconcile supply and demand, 'in effect they coordinate and match production capability to customer requests for product and service'.

At the parquet floor manufacturer customer enquiries were normally received by telephone or by e-mail and the scheduler initially needed to assess them in relation to the operation's production capability. The assessment was sometimes made directly without consulting anyone else or collecting further data. This was possible in more straight forward situations as the scheduler continuously and proactively updated his knowledge of the current and future status of production. At other times he needed to check whether and when the specific product type was planned and scheduled using the company software scheduling tool or his own developed Excel sheets (set up to provide an overview of production). Here, the scheduler was able to directly respond to the customer enquiry with an acceptance, rejection or suggestion of an alternative delivery date and/or an alternative but similar product. In this way the scheduler represented production providing a direct response to the marketing/sales department. In effect he filtered out a large proportion of enquiries that would otherwise have to be forwarded to the production department. Where enquiries needed to be made to production, the scheduler interpreted and transformed the enquiries into the language and logics used by production, so that production could more easily assess the enquiry. The enquiry particulars were expressed, for instance, in terms of production volume, choices on how to run production including changes to routings, switching work between production areas, and the anticipated impact on overtime. The interpreted customer enquiry was directed by phone or e-mail to the production team leader or production supervisor in the relevant production area. After receiving a response from production, the scheduler either directly transferred the information to sales and marketing or transformed the enquiry to provide a positive response with a possible delivery date.

The scheduler's work activities and interactions with people in production and sales are described in a role activity diagram, see Fig. 4.2, and the activities in the production enquiry process are summarized and classified in Table 4.3.

Having modelled this process in five other cases, (Guinery and Berglund 2010) it was established that in every case this process consisted of certain common

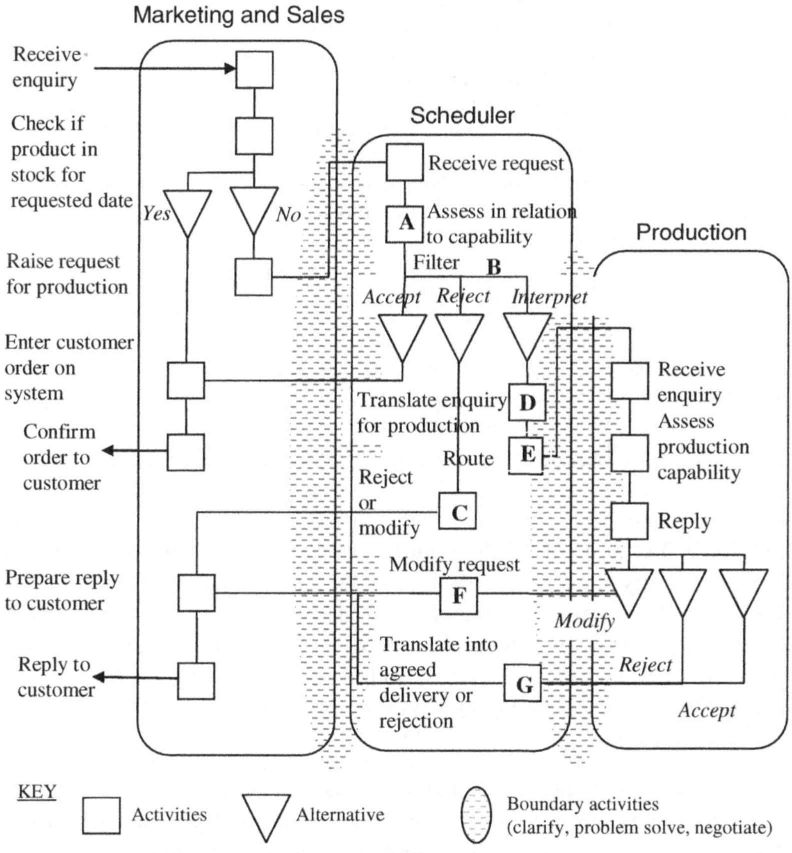

Fig. 4.2 Role activity diagram for scheduler at the parquet floor manufacturer

Table 4.3 Key activities and functional roles in the production enquiry process

Activity	Description of activity	Category	Functional role
Check customer enquiry	Check if customer enquiry fits into expected volume of products in stock for the desired delivery date		Customer service representative
Assess production *capability*	Based on own thorough knowledge about production status, PSC-tool and authority make a first assessment of production capability that guides decision-making about how to handle the customer enquiry	A Assess	Scheduler
Filter production enquiries	Filter out enquiries that the scheduler can deal with himself and direct the remaining to specific production leaders depending on type of customer enquiry	B Filter	Scheduler
Problem solve without referral	Reject the enquiry or modify it without reference to production	C Assess	Scheduler
Translate enquiry	Interpret customer enquiry and translate it into a form relevant to production	D Translate	Scheduler
Route enquiry	Route interpreted enquiry to relevant production representative	E Route	Scheduler
Assess production *capability*	Assess production capability related to issues presented by scheduler such as overtime, alternative running of production etc.		Production leaders
Suggest alternatives to customer order	Find or schedule for alternative delivery date or suggest alternative product with requested delivery date to customer	F Transform	Scheduler
Translate production reply	Translate production response to acceptance, rejection or alternative solution to customer enquiry	G Translate	Scheduler
Negotiate between marketing/sales and production	Negotiate with both marketing/sales and production regarding the possibility to fulfil production inquiries or regarding the scheduler's suggested alternatives	Negotiate	Scheduler

elements. In each case it served the common purpose of aligning sales requests to production capability and was therefore described as an alignment process between these functions. The common elements were described and subdivided into those undertaken solely by those in the alignment function (planners and/or schedulers), described here as autonomous activities, and those that occur as interactions across functional interfaces, described here as boundary activities.

The autonomous activities were found to be:

- *Assess*. assessing the enquiry in relation to its feasibility
- *Filter*. Filtering enquiries to ensure only feasible requests are taken further. This includes seeking revisions to the enquiry to support its acceptance
- *Route*. Where the enquiry is directed to relevant personnel and processes

- *Transfer*. Where a response is purely transferred from the capability function to customer interface function
- *Translate*. Translating the enquiry information into a form that makes sense to people in the other functions – this occurs in both directions; converting customer information into a form relevant to the capability function and vice versa
- *Transform*. Where a response from the capability function needs to be further developed and the alignment function applies additional information and/or its own expertise to modify the response. In effect the request is transformed into an alternative practical solution.

The boundary activities were:

- *Clarify*. Clarifying where additional information may be sought and anomalies dealt with
- *Problem solve*. Jointly problem solving where adjustments to the product or plan need to be analysed and 'best solutions' determined
- *Negotiate*. Negotiating with others, to fulfil the request where accommodations and/or trade-offs are necessary.

These autonomous and boundary activities were then used to construct a common model of the alignment process. See Fig. 4.3.

This model provides a simple representation of the elements of the process and positions them in relation to the customer facing (sales), alignment (planning or scheduling) and capability (production) functions. The elements consist of autonomous and boundary activities represented by circles. In the handling of requests

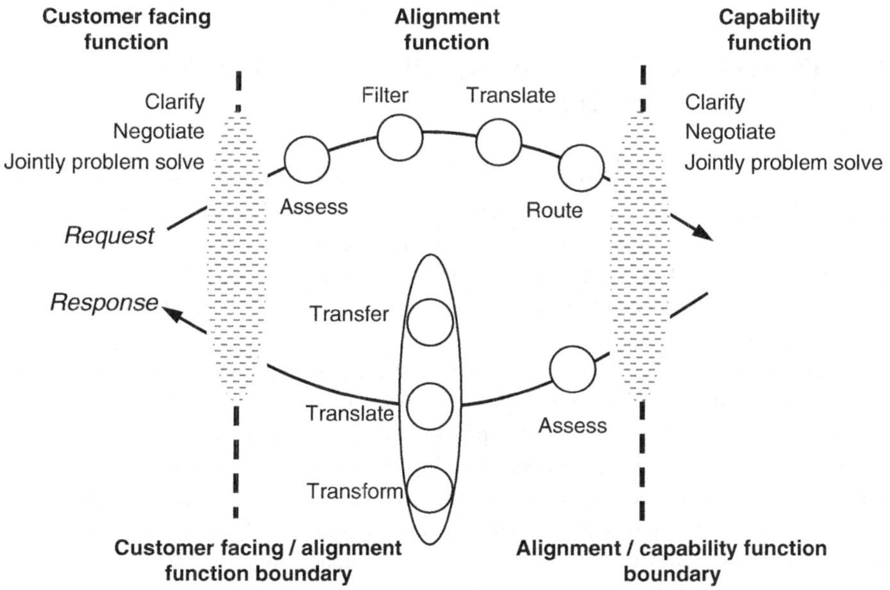

Fig. 4.3 The alignment process performed by planners and schedulers

from sales to production the autonomous activities are placed in a sequence. In reality the sequence may differ, but each element occurs to some degree. In the handling of the response from production to sales three elements are represented as occurring in parallel, this is because transfer, translation or transformation of the response are the alternative activities that may be employed prior to the response being forwarded to sales. Viewing the organisation as a system of roles (Guinery, 2006), the handling of production enquiries is an illustration of a proposed alignment role which is performed by planners and schedulers to facilitate work processes across the functional logics between sales and production (Berglund, 2009).

Although this model is purely a representation of the process associated with production enquiry handling, which constitutes a subset of activities schedulers and planners undertake to bridge between sales and production, it provides a useful 'research framework' with which to analyse other aspects of the bridging processes. It is clear from the model, that for schedulers and planners to perform effectively at boundaries they need to negotiate and jointly problem solve with sales and production. Knowing this it becomes apparent that it is important to ascertain how this is achieved. This can be achieved from closer scrutiny of activities in relation to interpersonal influence, decision making and knowledge use. The model also identifies autonomous process elements that merit further investigation e.g. what knowledge is required to successfully translate and transform the enquiry across sales and production logics; or to assess, route and filter an enquiry?

Drawing from this, Sect. 4.5 looks at how planners and schedulers influence others to perform effectively, and the subsequent one looks at decision making and knowledge use.

4.5 Planners' and Schedulers' Influence at the Sales and Production Interfaces

As described earlier, in everyday work activities planners and schedulers need to cope with and balance the sometimes incompatible agendas and perspectives of sales and production. Incompatible goals and/or means to goals provide conditions that lead to the use of power to influence others in organisations (Pfeffer 1981). It is therefore interesting to identify whether and how planners and schedulers influence sales and production employees.

There are several definitions of power and influence. Power may generally be defined as "the ability of those who possess power to bring about the outcomes they desire" (Salancik and Pfeffer 1977, p3) or as "the capacity to effect (or affect) organizational outcomes" (Mintzberg 1983, p4). Social interaction is put forward as a necessity, while interpersonal power is defined as "the ability to get one's way in a social situation" (French and Bell 1999, p283). Handy (1993) makes a distinction between influence and power by stating that influence implies the use of power, while power is the resource behind it. Handy further states that having influence is the same as having the power to influence.

Extensive literature identifies many sources of power in organizations. French and Raven (1959) put forward five different bases of what they defined as social power (1) Reward power, where the power holder has the ability to reward another person; (2) Coercive power, where the power holder can punish another person; (3) Legitimate power, where the power holder has a legitimate right to exert influence; (4) Referent power, where a person identifies him/herself with the power holder; and (5) Expert power, where the power holder possesses expertise that is needed by another person. This includes informational power, where the power holder possesses important information. Later, informational power was considered as a separate power base (Raven 1965). Mintzberg (1983) made a similar identification of five power bases (1) Control of resource; (2) Control of technical skill; (3) Control of a body of knowledge; (4) Legal prerogatives, where a person has the exclusive right to impose choices; and (5) Access to those who have power based on the first four power bases.

These different sources of power have been compiled into categories in relation to interpersonal influence. These are mainly based on Handy's (1993) power categories that in turn are derived from French and Raven (1959). See Table 4.4.

An examination of interpersonal influence (Berglund and Guinery 2008) used in four of the cases, the sawmill, the parquet manufacturer, the DIY product manufacturer, and the steel product manufacturer, revealed that a number of the above

Table 4.4 Categories of power. Derived from Berglund and Guinery (2008)

Categories of power
Resource power derives from possession of valued resources. To be an effective power base an individual must have control of the resources and these resources must be desired by others. Resources can be material or non-material as for instance in the case of grants of status. Resource power can be compared with Etzioni's (1966) control over resources and rewards
Position power is legal or legitimate power related to an individual's formal position in the company (Bolman and Deal 1991). This can also be described as the legal prerogatives to have the exclusive rights to impose choices (Mintzberg 1983). It gives the occupant of a role in the organization all the rights of that role. It includes access to information and right of access to networks. In the latter case it is possible to form alliances to gain more power. It will also give access to those who have power (Mintzberg ibid). Position power may also include the right to organize for instance the physical and social environments, the flow of communication and the right to decide. Furthermore, it includes control over agendas
Expert power is based on individuals possessing expert knowledge that is needed and acknowledged by others. It is a power base that doesn't require any sanctions and is the socially most accepted. As stated above, French and Raven (1959) define Informational power as a form of expert power; it includes control over information, which may be the result of a certain position
Personal power emanates from an individual's personality and is sometimes referred to as charisma, popularity (Handy 1993) or referent power (French and Raven 1959). It can be enhanced by an individual's position. Studies show that individuals who efficiently execute power are characterized by eloquence, ability to listen and quick comprehension (Pfeffer 1981)
Physical or coercive power is the power of superior force and rarely used in modern organizations
Negative power is when power is used contrary to accepted practice and includes the ability to interfere with things that happen. It is a latent power base and may be more evident in stressful situations

Table 4.5 Sources of influence at production and sales interfaces. Derived from Berglund and Guinery (2008)

Source of influence	Sawmill		Parquet floor manufacturer		DIY product manufacturer		Steel manufacturer	
	Prod	Com	Prod	Com	Prod	Com	Prod	Com
Resource power	–	–	–	–	–	–	–	X
Position power:	–	–	–	–	–	–	–	–
• Formal (line)	X	X	X	X	X	–	X	X
authority	X	X	X	X	–	–	X	X
• Access to	X	–	X	–	X	–	X	(X)
information								
• Access to networks								
• Right to organize								
Expert power	X	X	X	X	X	–	X	X
Personal power	X	X	X	X	X	–	X	X
Physical power	–	–	–	–	–	–	–	–
Negative power	–	–	–	–	–	–	–	–

described sources of power were applied at sales and production interfaces, see Table 4.5.

A cross case comparison of the four cases showed that:

- The right to organize occurred at the production interface in all cases as the planners and schedulers were responsible for instructions and plans.
- Position power based on access to information and to networks was in most cases prevalent at *both* sales and production interfaces.
- Expert power was identified at all interfaces.
- Resource power was utilized in one case, though in an indirect form.
- Physical power and negative power were not observed in any of the cases.

Looking at the individual cases we see that the individual scheduler at *the sawmill* used information power and expert power to influence sales to focus more on selling lower value by-products that were automatically produced from leftover wood. Furthermore, the scheduler showed great social skill in "smoothing the system" (as further described by Karltun and Berglund 2002) by making it possible to deliver products when personnel in sales had not followed accepted internal procedures for authorization from the scheduler. The interaction with production included the use of position power, expert power, information power and personal power. These were based on the scheduler's authority to instruct through work orders, his thorough knowledge and personal skills to propose overtime to meet the schedule, and a capability to solve a variety of problems with production management and personnel.

At *the parquet floor manufacturer*, the scheduler did not have any formal authority. However, position power was gained in other ways, such as through participation in several forums for decision making with representatives from both production and the sales side. He thereby gained access to information, participated in significant decision making processes, and further developed his knowledge.

This was even further developed through his informal and continuous search for information. The scheduler was highly respected for his knowledge on both production issues and customer order status, and there was no one that could directly replace him. Furthermore, the scheduler showed excellent social skills in dealing with people from both production and sales and in understanding their varied agendas. These traits are associated with personal power.

In the *DIY product manufacturer*, the organization was very clearly delineated with a hierarchical structure and clear demarcation of responsibilities between individuals and functions. The planner influenced production by having access to sales information not held by production, and having the right to organize. However, in relation to sales the planner had little influence. Despite having extensive knowledge of demand patterns at the item level (gained from planning a limited range of products on a daily basis), he had to base his plans on less accurate forecasts made by sales and marketing. The customer service ethos that pervaded the business legitimized the dominance of sales over planning. This often resulted in sales over-ruling the planner's expertise. The planner explicitly admitted that he developed compromise plans to cover both the demand he predicted and the eventuality of sales and marketing's projections being correct. This did not generate optimum plans or improve the relationship between them.

In the *steel manufacturer* planning was particularly complex and is therefore, described more fully. The Load Control (LC) manager was observed to have and maintain a pivotal role in the cross functional Order Management Team (OMT) that planned liquid steel production. The liquid steel making facility was part of a vertically integrated production facility in which it supplied a number of rolling mills. These rolling mills were effectively managed as separate businesses. The load planner strongly relied on, being the only individual with a complete overview of the production operations, having knowledge of production processes, alternative process routes and being in possession of up to date information on production order status across the liquid steel facility and all the mills it supplied. At the time of the study the company was effectively competing in a commodity market where demand had to be managed very carefully to ensure the most profitable work was selected. The steel was made to a vast range of specifications depending on the functional requirements of the finished products, and liquid steel supply had to be shared between the mills to optimise production efficiencies over the whole production system. This resulted in different mills and their associated sales functions vying for steel making production. Additionally, as liquid steel manufacture is the constraint in the overall facility, so demand had to be carefully managed to ensure there was no over commitment and the orders taken by the mill businesses were those with good margins. Clearly effective interaction with sales functions was essential for good order management.

To manage complex scenarios and associated options the LC manager and the load controller (who assisted him) developed a highly complex spreadsheet to gather accurate data and generate 'what if' analysis. Possession of this vital information enabled them to perform 'what if' analysis and present a limited range of options to senior management. His plan could be presented preferentially both to the OMT and with the business directors in the Business Management Committee (BMC). Once agreed, objectives to support the plan were set by the business directors and directed down the organization. In this way the load control planner effectively used the business director's position power to reinforce his decisions.

As negotiations between businesses on shares of production and current inventories of steel could be highly complex, a range of individuals from different functions and business areas were involved in the process. The load control planner presented background information and opinions at the weekly formal meeting to provide situation awareness to all concerned prior to these negotiations. Here the load control planner's access to the directors' agendas and the objectives set by the directors were used to legitimize his view on 'best' solutions.

Clearly, he had expert power and information power. Importantly he also had personal power developed and maintained through a network at all levels of the business that he had established over many years.

From the analysis of the above cases it is possible to conclude that planners and schedulers do have a significant influencing role at sales and production interfaces. A number of propositions were also developed. Their derivation and implications are described fully in Berglund and Guinery (2008). They are:

1. The scheduler is a social actor performing a coordinating role between sales and marketing and production.
2. A combination of expert power and personal power provides the scheduler with perceived personal integrity that helps him smooth out difficulties and conflicts.
3. Long work experience gives legitimacy to the scheduler's suggestions and contributes to both expert power and personal power.
4. The scheduler does not have formal authority but aspects of position power such as access to networks, people with power, and information as well as the right to organize (in relation to production) are evident.
5. Based on the scheduler's access to key arenas, valuable information is gained and used to influence others.
6. Information power is gained not just from formal participation in meetings but informal interaction with employees from sales and production.
7. Personal power supports negotiations with production and sales.
8. Position power based on the right to organise is effective in this organisation as it has a hierarchical structure where autonomous roles have been clearly established.
9. The need to influence can be minimised where the company ethos and goals are shared.
10. A customer focused ethos may legitimize the dominance of sales over planning; resulting in the over-ruling of planning expertise.
11. Ownership and control of information and expert knowledge create the need for a judge-adviser relationship and dependency of senior managers on their adviser.

12. By influencing the agendas and communicated objectives of senior managers the planner can legitimise his plans throughout the organisation.
13. Where multiple players vie for production the planner may be able to influence others, including senior managers, on the basis of perceived resource power.
14. Planners can consciously actively develop and maintain their sources of power; these in turn can be used to reinforce each other.

Expert and informational power was consistently seen as a strong basis for influencing others, particularly at sales and production interfaces. Knowledge and information possession and sharing was very important and therefore needs to be considered further. Section 4.6 looks at knowledge use in joint decision making.

4.6 Planning and Scheduling Knowledge Use and Decision Making at Interfaces

As identified previously, PSC needs to be understood in relation to how information and knowledge are used at interfaces. This section initially explores relevant research on knowledge and decision making. With reference to identified theory it then moves on to examine how knowledge is used and managed by planners and schedulers to support decision making at sales and production interfaces through the use of a case study.

Although much research has considered information management and the decision support systems that assist PSC, little research has focused on knowledge and how it is applied. Whilst knowledge management is a burgeoning research area it tends to focus on strategic decision making or large scale collaborative projects associated with innovation, design and development; alternatively at operational levels it tends to focus on knowledge transfer. The research described here deals with the application of knowledge in PSC at the operational level. In particular it looks at the knowledge use of planners and schedulers at interfaces with other work groups involved with PSC, where other work groups represent sales or production, sometimes at tactical or strategic levels of planning. To achieve this, four complex businesses were studied (Guinery 2006). Just one, the steel manufacturer, is described in this chapter to illustrate the overall findings. The business was selected as due to its inherent complexity. Many insights on knowledge and how it was incorporated into PSC decision making were obtained from this single case. Before presenting the case study it is important to discuss the nature of knowledge and decision making. Knowledge needs to be defined and carefully differentiated from information. It is also important to understand the environment in which knowledge is utilised as environmental factors may have significant impact on decision making and on knowledge use.

Shin et al. (2001) describe knowledge as "a high value form of information that is ready to be applied to decisions and actions". Bolisani and Scarso (1999), who consider knowledge in relation to decision making, define it as "a combination of information, ideas, procedures and perceptions that guide actions and decisions.

Knowledge differentiates itself (from information) by being relevant to context and formed by experience."

Knowledge can take different forms. Polanyi (1967) first made the distinction between tacit and explicit knowledge and its implications. Shin et al. (2001) provide eloquent definitions to differentiate them: "Tacit knowledge resides in the individual's experience and action. Explicit knowledge is codified and communicated in symbolic form or language." This view, of tacit knowledge residing solely in the individual, has since been challenged (Griffiths et al. 1998). Research has found that knowledge can be shared tacitly by members of social groups, so can be associated with the group or organization (Brown and Gray 1995; Augier and Vendelo 1999; Grant 1996). Brown and Gray (1995) argue that "with individuals, tacit knowledge means intuition, judgement, common sense – the capacity to do something without necessarily being able to explain it. With groups, tacit knowledge exists in the distinct practices and relationships that emerge from working together over time – the social fabric that connects communities of knowledge workers."

Decisions based on tacit knowledge may necessarily require decision making authority to be co-located with those who possess the knowledge, while conversely decision making is more easily decentralised where knowledge can be made explicit (Grant 1997, p453). Knowledge can therefore be a defining factor when establishing the appropriate design for the organisation of PSC personnel and processes.

Knowledge content, and explicitly the knowledge required for a decision, is also difficult to establish as knowledge is multi dimensional. It can be defined in relation to its content or source (e.g. whether sales, process, engineering etc.) or according to its contribution to decision making (e.g. whether it consists of an understanding of: what information is required; who has the required knowledge; how problems are solved or solutions implemented; why choices are important in relation to 'knock on' effects, trade-offs, and work group and individual agendas). The latter is more akin to 'sticky' knowledge.

Naturalistic Decision Making (NDM) that emanates from real-world research is one school of thought. It may be described as: "The study of NDM asks how experienced people, working as individuals or groups in dynamic, uncertain, and often fast-paced environments, identify and assess their situation, make decisions and take actions whose consequences are meaningful to them and to the larger organization in which they operate" (Klein Associates 1989). Orasanu and Connolly (1993) identified the key contextual factors that affect real world decision making and lead to NDM as:

- Ill-structured problems (not artificial, well structured problems)
- Uncertain, dynamic environments (not static, simulated situations)
- Shifting, ill-defined, or competing goals (not clear and stable goals)
- Action/feedback loops (not one shot decisions)
- Time stress (as opposed to ample time for tasks)
- High stakes (not situations devoid of true consequences for the decision maker)
- Multiple players (as opposed to individual decision making)
- Organisational goals and norms (as opposed to decision making in a vacuum)

It is worth considering whether these conditions align closely with those in which PSC decisions are made. This has implications as decisions are made based on situation assessment including the recognition of cues indicating key aspects of the situation, the sequential evaluation of alternatives, and selection of a satisfactory one, so to be effective, decision makers need to possess good situation awareness and realistic mental models to aid their decision making.

McKay et al. (1995) have observed in studies of schedulers that their 'immediate' situation awareness is not necessarily the product of experience gained over time but may be based on information proactively gained to support future decision making by an expert who knows what he will need. Similarly Klein (1993) identified that information seeking by experts is not events driven as some believe, but instead has already occurred and exists in situation awareness.

Choo (1998) postulated that the 'mode' of decision making affects the ways in which decision makers attempt to acquire and use information. Decision making modes are dependent on two properties of the organisational environment; (1) goal ambiguity about what organisational goals to pursue, and (2) technical uncertainty in relation to how goals can be achieved, see modes presented in Fig. 4.1. What is significant about these modes is the different ways in which decision makers acquire and use information and, by implication, knowledge. Where technical uncertainty is low, which is normally the case in PSC, rational and political modes of decision making are most commonly employed; the former where goal ambiguity is low, the latter where it is high. In the *rational mode* (Cyert and March 1963) decision makers generally use procedures, routines and/or rules of thumb to solve problems. This process is goal directed and problem driven. If the goals set cannot be achieved a search for a solution is triggered that first proceeds at a local level and if this fails more remote sources of information on alternatives are sought. Generally, however, solutions are identified locally that 'satisfice'. The *political mode* of decision making (Allison 1971), is one in which more game-playing is undertaken and the relevant influence of players is critical as it provides bargaining advantage. Here the focus is on dealing with conflicting goals. This will be particularly prevalent where players have different agendas and where resource sharing is an issue. Information seeking can be quite intensive and broad as players seek to justify their preferred options and draw on information (and knowledge) to do this from a broad range of experts. In this way its presentation and application is selective in supporting the players' interests.

To investigate differences between the perspectives of sales, production and engineering and their implications on knowledge at interfaces, Carlile (2002) undertook a year long ethnographic study. He used 'objects and ends analysis' to explore differences between functions. In this analysis 'objects' are items through which people communicate (such as in £s of sales for sales, units of product for production, drawings for engineering) and 'ends' are the objectives that they are working towards (such as 'close the deal' in sales and marketing and 'ship product' in production) that can also be described as their agendas. He found that different 'objects' and 'ends' complicated knowledge sharing and impacted on effective decision making. Further to this he established that 'boundary objects' (which may be simple or complex representations such as drawings) enable decision makers to share knowledge at a number of different levels. In particular, an effective boundary object establishes a shared syntax or language for individuals to represent their knowledge and provides a concrete means for individuals to specify and learn about their differences and dependencies, facilitating a process where individuals can jointly transform their knowledge (Carlile 2002).

As stated earlier, the steel manufacturer case was selected to examine knowledge use and decision making at PSC interfaces. Liquid steel planning has been described previously in the section on *Planners' and schedulers' influence at the sales and production interfaces*. In this case the environment in which decisions were made was particularly complex and dynamic, and many people needed to represent different parts of the business, in different functions and at different planning levels, for effective PSC decision making. By necessity decision making required both joint problem solving and negotiation as PSC decisions had multiple and often conflicting objectives that had implications on each mill business as well as on the performance of the liquid steel production facility that supplied them.

Three groups were identified as contributors to decision making on liquid steel production. The first was the Load Control (LC) team that planned work through the liquid steel production unit. This consisted of the LC Manager and the load controller. The second was the Order Management Team (OMT), which was a cross functional group specifically established to negotiate and firm plans. The team consisted of the LC team members, the business planners, who represented the sales interests of the individual mills that liquid steel supplied, and the Scheduling manager, who represented liquid steel production. Individuals in this group met weekly but also continuously networked to resolve problems. The third key planning group was the Business Management Committee (BMC) that planned at a more strategic level. It consisted of Business Directors, some of whom were responsible for the mill businesses, so their commercial interests were represented there. The LC team presented proposals to the BMC through a weekly meeting. Figure 4.4 shows relationships between these planning entities.

In the research, the knowledge used by different members of the cross functional OMT was identified and characterized through transcript analysis. (A small sample

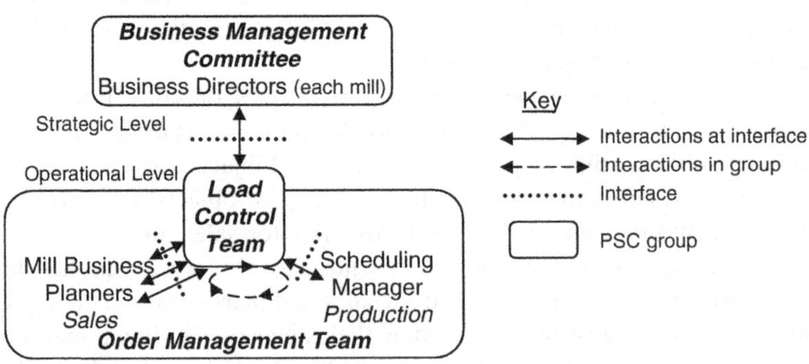

Fig. 4.4 Relationship between planning groups in steel manufacturer

of the transcript analysis is shown in Table 4.6). It was classified in relation to a range of attributes including: its content (i.e. whether on sales, production or supply issues), whether it was tacit or explicit, and according to its form and source.

It was observed that *knowledge of status* of customer orders and production supported decision-making as it ensured up-to-date information was processed. This was used in solving a range of routine and rule-based problems as well as in more complex decision making. Rule-based knowledge was incorporated into the spreadsheet that determined aggregate values on operations performance based on information on orders, production and consumption and hence inventory status. Incorporated into this was information on available capacities and efficiencies/yield, based on the load controller's interpretation of the information he received formally and informally through his networks. In effect this knowledge was codified into the spreadsheet. The inputs were constantly modified to represent the LC team's current interpretation of production and order status. Over and above this, where human judgment was required to make decisions, the tacit knowledge applied consisted of *knowledge of planning rules*, *knowledge of consequences* or impact of decisions made based on knowledge of trade-off implications, overall *situation awareness* on what was currently happening and priorities, and experience of consequences based on *knowledge of what occurred in previous events* (that related previous decisions in specific contexts to outcomes).

The LC team between them had significant understanding of the consequences of decisions, right across the operation. Use of social networks was considerable. They had extensive *knowledge of who knows what*, *who has good judgment*, and *of others' agendas*. Other people in the organization were found to use LC as a source of knowledge and LC also tapped into others' knowledge of different areas of the business. In effect the network operated as a transactive memory system (Brandon and Hollingshead 2006), in which the LC manager was accepted as a valuable knowledge source who held and provided knowledge to others. Colleagues also relied on them to orchestrate decision making, recognizing their broad knowledge and experience.

The cross functional OMT served the purpose of drawing on cross functional knowledge to make informed decisions. The content of this knowledge was also examined with members of the OMT being analysed in relation the knowledge they contribute (see Table 4.7). There was considerable coverage of 'what' knowledge on all aspects of the business operations. However, some knowledge was localized to particular areas. For example a Business Planner was needed to represent each mill as they had each type of knowledge but only in relation to their own rolling mill business (their customers, products and production).

The distribution of tacit knowledge associated with *knowledge of consequences* and *knowledge of past events* was also investigated. Here knowledge distribution was considered in relation to the workspaces that different OMT members inhabited. The OMT members covered all key business activities affected by or contributing to PSC decisions encompassing the market, customers, pricing, liquid steel and mill production. However, it was the LC team alone that had an overview.

Table 4.6 Knowledge node analysis for load control at steel manufacturer. Derived from Guinery (2006)

Quote	Knowledge Content	Form	Tacit/ explicit	Source
"...this is what I have got in for sales, do you want to change what you've got?... and this is the impact on stocks'"	Impact of decision on other areas	Rule based knowledge	Tacit to explicit presented as 'what-if'	Load control team
"I know from the time I've been here what sales I couldn't preclude in a sales plan - In the pecking order they can move up and down slightly but in terms of are they in or out, no"	Sales	Situation awareness	Tacit	Load control manager
"I sit there and I hear all this lot talking and every now and again I'll pick something up that is out with what I have been told by these sales business planners, either at OMT or at informal chit chats. And I will say 'is that right? I thought something else was happening'."	Sales	Situation awareness	Tacit	Network and meetings
"... because of my personal knowledge that I know that's not going to happen it's going to be something else and we can do something else."	N/R	Experience	Tacit	Load control manager
"Now within each of these silos there are other people looking at what sections we make and all the rest of it; but we are the only group that pull all of that together."	Operation overview and objectives	Situation awareness	Tacit	Load control manager
"They've come up from their little silos and they are very good at their little bit but there is nobody that can see and have the experience and more importantly, it is most important this, and what people are forgetting, it's the respect of this lot"	Know who knows	Knowledge of decision maker's judgement	Tacit	Social network and pivotal (political) role in it

Table 4.7 Knowledge distribution and sharing across the OMT members. Derived from Guinery and MacCarthy (2009)

Input to decision	Business planners (by mill)	Load control (all areas)	Technical manager (general)	Scheduling manager (liquid steel)	Finance manager (general)
Knowledge:					
Market/Demand	√√	–	–	–	–
Product specification	√√	√√	√√	√√	–
Production capabilities	√	√√	–	√√	–
Process alternatives	–	√	√√	√√	–
Supply capabilities	–	–	–	–	–
Trade-off implications	√	√√	√	√	√
Cost margins and contributions	–	–	–	–	√√

Double ticks indicate a high contribution, a single tick a lesser one and a dash indicates no contribution

Others were aware only of the impact of PSC decisions on those activities that related to their own specific business area, not that of others.

The analysis of both *knowledge content* and of the more 'sticky' *knowledge of consequences* shows the LC team to be the only people with a complete overview of the operation and explains why they were so relied upon in orchestrating problem solving and negotiations.

As discussed previously the type and mode of decision making affects the form of knowledge required and how it was sought and used. Characteristics of the environment in which load control decisions were made were therefore analysed (see Table 4.8).

By cross referring the observations made with the environmental characteristic associated with naturalistic decision making (NDM) (Orasanu and Connolly 1993) their similarity was immediately ascertained. This indicated that NDM predominated. The following observations of actual decision making behaviours validated this:

- The aim of joint decision making was to achieve solutions that were satisfactory to all key players, that is, they were satisficing (Simon 1957).
- Information was gathered in anticipation of the decisions that might need to be made. People used informal networks and meetings to share their understanding of the situation prior to making decisions.
- The LC manager and load controller proactively networked to establish cues indicating any changes in conditions. Networks were extensive and at all levels of the organization. (For example, a shop floor operator on one plant often gave the LC team advanced warning of process problems and anticipated downtime for maintenance prior to the event and before his immediate line manager had been informed).
- Although planning rules could be applied to identify the impact of decisions made, the LC manager's 'judgment' was constantly called upon by others.

Table 4.8 Characteristics of the decision making environment at steel manufacturer. Derived from Guinery (2006)

Environment characteristics	Status	Evidence	Reason
State	Unstable	Continuous multiple updates of changing plan status, multiple iterations on planning decisions	Competing and late orders, need to be responsive, poor process reliability, inability to buffer with interim stocks
Clarity of goals	Ambiguous	Shifting trade-offs Referral to BMC to make/agree upon sales decisions	Business areas competing for shared resource
Knowledge domain	Broad	Cross functional decision making required for steel plan	Shared and limited supply resource and fluctuating commodity market
Technical complexity	Low	Steel plan rules have been made explicit and are incorporated into Load Control spreadsheet	Production and operations variables clearly understood
Planning decision complexity	High	Often multiple iterations and multiple player input and negotiation needed to establish agreed plans	There are a large number of options, but with far reaching consequences across all areas of the operation
Timeliness	Short and medium	Decisions need to be made outside the planning cycle to deal with process issues that impact on plan	E.g. reallocation of steel to mills/orders in case of breakdowns

Modes of decision making were also investigated, particularly the impact of decision making mode differing between functions. The modes were established by looking at the characteristics of the decision making environment, and then by verifying whether anticipated modes of decision making were reflected in observed behaviour. The modes were found to differ in different parts of the organization. Differences and their impact are described below.

For the LC manager decision making goals were ambiguous, with many different outcomes being possible and the plan infinitely modifiable. Decision making involved negotiation as well as problem solving. The solutions in themselves were not technically difficult. What was uncertain was whether the decision made was the right one. Here, decision making was 'political'. Likewise, the Business Management Committee (BMC) members' predominant mode of decision making was political, as was that of the business planners in the OMT. Once agreements had been made the steel plan was built up using existing rules embedded in the LC spreadsheet. At this stage, as both the objectives and rules were clear the mode of decision making was rational. Similarly, liquid steel schedulers were observed to predominantly use the rational mode of decision making.

Interestingly, the LC function evolved such that the LC manager and load controller handled different sets of tasks associated respectively with political and rational modes of decision making. The load controller possessed detailed knowledge of the current status of production constraints, and the status of live work. The LC manager, who was senior, was more aware of other peoples' agendas and had a stronger sales awareness. The load controller worked more in the rational mode (Choo 1998). The LC Manager understood and used both political and rational modes of decision making. He understood how to represent and solve problems from both perspectives. As the load controller said: "He does the strategy I do the ...".

Further transcript analysis generated findings on the roles performed by the LC manager and load controller. These are summarised below in Table 4.9.

The load controller was seen to filter the information, he received from his extensive network, to provide the LC manager with only that information essential to decision making. In this way he performed a *hub and filter role*. In the same way the LC manager presented only key facts when he chaired OMT meetings.

The LC Manager clearly acted as an *arbitrator*. He understood different peoples' agendas and therefore acceptable trade-offs and their implications. Where possible he presented issues to individuals then let them negotiate within the limits of acceptable solutions that he provided.

Across the interface with the BMC, the LC manager acted as an *adviser* to the committee in a 'judge-adviser' role (Sniezek and Van Swol 2001). He presented a small number of steel plan alternatives for discussion. This supported pragmatic

Table 4.9 The Load Control knowledge roles at PSC interfaces in the steel manufacturer. Derived from Guinery (2006)

Role	Purpose	Location	Knowledge	Reason
Hub and filter (LC manager & Controller)	Obtaining, filtering and transferring relevant production information	With production and within the LC team	Understanding PSC processes, associated functions and their information needs	Reduction of vast quantities of information to a manageable level
Arbitrator (LC manager)	Resolving resource conflicts involving trust in the arbitrator's objectivity	Within OMT and with all PSC functions	Understanding situation and implications. Aware of other's agendas	Load control the only function with a complete overview of the operation
Adviser (LC manager)	Presenting alternatives and consequences to decision makers	Within OMT and with the BMC	Understanding of situation and implications	High impact of decisions on multiple players
Transformer (LC manager)	Interpreting/ translating requirements and orchestrating decision making across groups	Across all functions	As *all* above	Decision makers with different agendas and modes of decision making

Table 4.10 Translation requirements at steel manufacturer. Derived from Guinery (2006)

Function	Steel Production	Sales (includes Business planners and members of the BMC)
Agenda	Optimise steel production Keep stocks low Sharing of steel supply	Own business area sales performance Own mill efficiency
Decision mode	Rational	Political
Language	Tonnes of steel	Orders achieve due date promised

decision making and the transfer of relatively 'sticky' knowledge between planning levels. Here and in OMT meetings, plan alternatives and consequences (generated on the spreadsheet) were presented in a less than transparent way. This strengthened the influence of the LC team. However, not only did it limit others' influence, it also reduced their ability to contribute to or share knowledge.

His role however was even more sophisticated than the above observations imply. To align the sales requirements of the mill businesses with those of liquid steel production, and help people to understand each others' positions when difficult trade-offs needed to be made, the LC manager met and communicated with individuals (formally and informally) throughout the organization to interpret their requirements and *translate* meanings and implications between functions. In this way he consciously orchestrated problem solving and negotiation across functions. He overcame functional differences in modes of decision making, agendas and languages. (These differences are illustrated in Table 4.10). He transformed requirements into a shared solution applying his knowledge of others' agendas and perspectives, and communicated the issues and requirements in ways appropriate to their different modes of decision making. This role has been defined as that of *transformer*.

The following findings on how planners use and handle knowledge at PSC interfaces were drawn (Guinery 2006); many reaffirm findings from the analysis in previous sections. All were illustrated in the steel manufacturer case:

1. The distribution of knowledge, particularly 'sticky' knowledge associated with any individual decision maker or group must be understood to appreciate those most effective in decision making at interfaces (in the steel manufacturer case, the LC team and OMT had an overview of the impact of decisions across the entire operation)
2. NDM decision making predominates in 'messy' and dynamic decision making environments such as PSC, and this has implications for knowledge use at interfaces. People have to manage with limited information so rely on situation awareness and cues. Planners may put considerable effort into constantly updating their understanding of what is occurring in relation to sales and production
3. 'Sticky' knowledge provides the need to localise experts to where decisions are made, with consequences for organisation and process design. One relevant form of organisation (illustrated in the case study) is a cross functional team. Such a team's knowledge can be extensive. Note: within the team, interactions between people in sales and production were still orchestrated by the planners

4. Ways in which knowledge was shared across PSC interfaces varied. In the case study the load planning spreadsheet was used to present alternatives and describe the impact of decisions on many parts of the business
5. Planners are seen to exhibit a range of roles that support the use of their knowledge and that of others in PSC decisions across organisational interfaces. *Hub and filter*, *arbitrator*, *adviser* and *transformer roles* were identified

4.7 Insights and Implications

From the empirical findings it is clear that we cannot underestimate the contribution of planners and schedulers and the importance of understanding their roles and activities within an organisation. This chapter sought to address many questions associated with these contributions and in doing so has obtained some important insights. The topics examined have included: the nature of PSC tasks and how they are performed, defining and modelling the alignment process between production and sales, ways in which planners and schedulers influence others, and the knowledge they contribute, and how they incorporate knowledge and orchestrate decision making.

The research findings show that planners and schedulers undertake many activities that are not immediately transparent, or acknowledged as part of PSC. In particular, planners and schedulers play a crucial role in information sharing, influencing and orchestrating decision making, and aligning sales and production. This has been illustrated by examples from everyday practice and through the examination of observations made from different perspectives. Clearly, many PSC activities are inter-related, cannot be made explicit, and therefore are infeasible to formalise, consequently inhibiting the extent to which PSC activities can be automated or undertaken through formal procedures.

In particular the research described has focused on the role that planners and schedulers perform in bridging the gap between sales and production. They manage the balance between the business' production capability and customer demands in complex and dynamic environments where functional objectives differ, trade-offs need to be made, negotiation may be required, and there is often a dearth of up to date information.

The findings in all the areas described have practical managerial implications.

A model of the alignment process performed by planners and schedulers has been developed and can be used to support managerial decisions on PSC people, processes and organisational design. Managers can use it as a framework against which they can examine their own business' processes, to help them recognise and visualise elements of the alignment process. The model provides them with an appreciation of the key elements of the process that planners and schedulers are involved in. In particular it can help them to differentiate autonomous and boundary elements, highlighting the latter in which social and influencing skills, and possession of cross domain knowledge may be crucial. From the analysis managers can appreciate the nature of the process and make decisions on planner and scheduler selection, training and induction. Furthermore, they can consider whether necessary competences are in place and the extent to which organisation ethos, design and processes support planners and schedulers. Similarly, analyses of sources of power, the forms of influence that planners and schedulers employ, of knowledge use and sharing, can aid managerial decision making on operational practice and organisation design.

This research clearly shows the need to provide organisational support and recognition to the individual planners and schedulers in many scenarios.

To conclude, this research has established that:

- Planners and schedulers have significant influence across the organisation; typically, although their apparent line authority is minimal, their access to information and their expertise enables them to have significant influence over others
- Planners and schedulers have a key role in orchestrating decision making processes at interfaces where decision makers have different logics and agendas. Furthermore, the field studies reveal that planners and schedulers are aware of their role as mediators between the different logics and agendas of sales and production
- Planners and schedulers often possess a broad knowledge of sales and production practice and issues and therefore may be pivotal as they have the ability, not just to contribute their own knowledge, but to enable others to effectively input, share and transfer knowledge
- Some of the PSC activities support the alignment of sales and production activities. The commonality of observed activities implies that, in combination, they represent a specific process, defined as the alignment process. Planners and schedulers here play a crucial role in developing relationships between functions and improving performance across the organisation
- The discussion has illustrated how a simple PSC alignment process model can be used by researchers to support the analysis of other aspects of the alignment process, and can assist managers in visualising and examining the processes in their specific business operations
- Finally, the contribution of production planners and schedulers must not be underestimated

References

Allison, G. T. (1971). Essence of decision: explaining the Cuban missile crisis. In C. W. Choo, 1998, *The knowing organization: How Organizations Use Information to Construct Meaning, Create Knowledge and Make Decisions*. New York: Oxford University Press.

Augier, M., & Vendelo, M. T. (1999). Networks, cognition and management of tacit knowledge. *Journal of knowledge management, 3*(4), 252–261.

Berglund, M. (2009). *Using tentacles in planning and scheduling work. Activites, roles and contributions*. Doctoral Thesis, TRITA-STH Report 2009:2, Royal Institute of Technology, Stockholm.

Berglund, M., & Guinery, J. (2008). The influence of production planners and schedulers at production and sales interfaces. *Human Factors and Ergonomics in Manufacturing, 18*(5), 548–564.

Berglund, M., & Karltun, J. (2001). *Schedulers'* reality – expectations and dependencies. *Proceedings of the 16th International Conference on Production Research.* Prague, Czech Republic.

Berglund, M., & Karltun, J. (2006). Schedulers' work content – a quantified analysis. *Proceedings of the 16th Triennial Congress of the International Ergonomics Association.* Maastricht, The Netherlands.

Berglund, M., & Karltun, J. (2007). Human technological and organizational aspects influencing the production scheduling process. *International Journal of Production Economics, 110*, 160–174.

Bolisani, E., & Scarso, E. (1999). Information Technology Management: A Knowledge-based Perspective. *Technovation, 19*, 212.

Bolman, L. G., & Deal, T. E. (1991). *Reframing organizations: artistry, choice and leadership.* San Fransisco, CA: Jossey Bass.

Brandon, D. P., & Hollingshead, A. B. (2006). Transactive memory systems in organizational groups: the effects of content, consensus, specialization, and accuracy on group performance. *Journal of Applied Psychology, 88*, 866–878.

Brown, J. S., and Gray, E. S. (1995). The people are the company, *Fast Company* 1 Dec 78-82. In Choo, C. W., 1996, The knowing organization: How organizations use information to construct meaning, create knowledge and make decisions. *International Journal of Information Management, 16*(5), 329–340.

Carballeda, G. (1999). La contribution des ergonomes à l'analyse et à la transformation de l'organisation du travail: l'exemple d'une intervention relative à la maintenance dans une industrie de processus continu, Collection Thèses & Mémoires, Laboratoire d' Ergonomie des Systèmes Complexes, Université Victor Segalen Bordeaux 2 – ISPED.

Carlile, P. R. (2002). A pragmatic view of knowledge and boundaries: Boundary objects in new product development. *Organization Science, 13*(4), 442–455.

Choo, C. W. (1998). *The knowing organization: How organizations use information to construct meaning, create knowledge and make decisions.* London: Oxford University Press.

Clare, D. A., & Sanford, D. G. (1984). Cooperation and Conflict Between Industrial Sales and Production. *Industrial Marketing Management, 13*, 163–169.

Coates, D., Guinery, J., & MacCarthy, B. L. (2005). Coping with Restructuring. *IEE Professional Journal on Manufacturing, 84*(1), 31–35.

Crawford, S., MacCarthy, B. L., Wilson, J. R., & Vernon, C. (1999). Investigating the work of industrial schedulers through field study. *Cognition Technology and Work, 1*(1), 63–77.

Crittenden, V. L., Gardiner, L. R., & Stam, A. (1993). Reducing conflict between marketing and manufacturing. *Industrial Marketing Management, 22*, 299–309.

Cyert, R. M., & March, J. G. (1963). *A behavioral theory of the firm.* Engelwood Cliffs, NJ: Prentice-Hall.

Davenport, T. H. (1998). Putting the Enterprise into the Enterprise System. *Harvard Business Review, 76*(4), 121–133.

Eisenhardt, K. M. (1989). Building theories from case study research. *Academy of Management Review, 14*(4), 532–550.

Etzioni, A. (1966). *Moderna organisationer.* Aldus: Stockholm (In Swedish).

French, W. L., & Bell, C. H., Jr. (1999). *Organization Development* (p. 283). Prentice-Hall: Upper Saddle River.

French, J. R. P., Jr., & Raven, B. H. (1959). The bases of social power. In D. Cartwright (Ed.), *Studies in social power* (pp. 150–167). Ann Arbor, MI: Institute for Social Research.

Gonzalez, M. E., Quesada, G., Mueller, R., & Mora-Monge, C. A. (2004). QFD strategy house: an innovative tool for linking marketing and manufacturing strategies. *Marketing Intelligence & Planning, 22*(3), 335–348.

Grant, R. M. (1996). Prospering in dynamically competitive environments: organizational capability as knowledge integration. *Organizational Science, 7*(4), 379.

Grant, R. M. (1997). The K-based view of the firm: implications for management practice. *Long Range Planning, 33*(3), 450–454.

Griffiths, D., Boisot, M., & Mole, V. (1998). Strategies for managing knowledge assets: a tale of two companies. *Technovation, 19*(8/9), 529–539.

Guérin, F., Laville, A., Daniellou, F., Duraffourg, J., & Kerguelen, A. (1997). *Comprendre le travail pour le transformer*. Lyon: Éditions de l'ANACT.

Guinery, J. (2006). *Knowledge integration in production planning, scheduling and control*. PhD, University of Nottingham, Nottingham.

Guinery, J. and Berglund, M. (2010). Production planning aligning customer requests with production capability, Manuscript, Operations Management Division, Nottingham University Business School.

Guinery, J., & MacCarthy, B. L. (2009). Managing key interfaces in production planning and control. *Production Planning & Control, 20*(1), 40–56.

Gunasekaran, A., Tirtiroglu, E., & Wolstenscraft, W. (2002). Gap between production and marketing functions: a case study. *Management Decision, 40*(5), 428–435.

Hamlin, M., MacCarthy, B., & Guinery, J. (2006). Fast track order fulfilment at Henkel – The PROCHART approach. *Control, 31*(4), 15–19.

Handy, C. B. (1993). *Understanding organisations* (3rd ed.). Harmondsworth: Penguin Business Library.

Hill, T. (1997). Manufacturing strategy - keeping it relevant by addressing the needs of the market. *Integrated Manufacturing Systems, 8*(5), 257–264.

Jackson, S., Wilson, J. R., & MacCarthy, B. L. (2004). A new model of scheduling in manufacturing: tasks, roles and monitoring. *Human Factors, 46*(3), 533–550.

Karltun, J., & Berglund, M. (2002). Production scheduling in practice – a study of managing goal conflicts. *Proceedings of 9th International Conference of European Operations Management Association* (pp. 757–768). Copenhagen.

Katz, D., & Kahn, R. L. (1978). *The social psychology of organisations*. New York: Wiley.

Klein Associates (1989). *Naturalistic Decision Making (NDM) Workshop organized by Klein Associates*, Dayton, OH, September 25–27.

Klein, G. A. (1993). A recognition-primed model of rapid decision making. In G. A. Klein, J. Orasanu, R. Calderwood, & C. E. Zsambok (Eds.), *Decision making in action: models and methods* (pp. 138–147). NJ: Ablex Publishing Corporation.

Konijnendijk, P. A. (1993). Dependence and conflict between production and sales. *Industrial Marketing Management, 22*(3), 161–167.

Lawrence, P. R., & Lorsch, J. W. (1967). Differentiation and integration in complex organisations. *Administrative Science Quarterly, 12*(1), 1–47.

Leplat, J. (1990). Relations between task and activity: elements for elaborating a framework for error analysis. *Ergonomics, 33*(10/11), 1389–1402.

MacCarthy, B. L., & Liu, J. (1993). Addressing the gap in scheduling research: a review of optimization and heuristic methods in production scheduling. *International Journal of Production Research, 31*(1), 59–79.

MacCarthy, B. L., & Wilson, J. R. (Eds.). (2001). *Human performance in planning and scheduling*. London: Taylor and Francis.

McKay, K. N., Buzacott, J. A., Charness, N., & Safayeni, F. R. (1992). The scheduler's predictive expertise: an interdisciplinary perspective. In G. I. Doukidis & R. J. Paul (Eds.), *Artificial intelligence in operational research* (pp. 139–150). Basingstoke: Macmillan.

McKay, K. N., Safayeni, F. R., & Buzacott, J. A. (1995). 'Common sense' realities of planning and scheduling in printed circuit board manufacture. *International Journal of Production Research, 33*(6), 1587–1603.

McKay, K. N., & Wiers, V. C. S. (2001). Decision support for production scheduling tasks in shops with much uncertainty and little autonomous flexibility. In B. L. MacCarthy & J. R. Wilson (Eds.), *Human factors in planning, scheduling and control*. London: Taylor & Francis.

Meredith, J. (1998). Building operations management theory through case and field research. *Journal of Operations Management, 16*(4), 439–452.
Mintzberg, H. (1983). *Power in and around organization* (p. 4). Englewood Cliffs, NJ: Prentice-Hall.
Mukhopadhyay, S. K., & Gupta, A. V. (1998). Interfaces for resolving marketing, manufacturing and design conflicts. *European Journal of Marketing, 32*(2), 101–124.
Nakamura, N., & Salvendy, G. (1994). Human planner and scheduler. In G. Salvendy & W. Karwowski (Eds.), *Design of work and development of personnel in advanced manufacturing* (pp. 331–354). New York: Wiley.
O'Leary-Kelly, S. W., & Flores, B. E. (2002). The integration of manufacturing and marketing/ sales decisions: impact on organizational performance. *Journal of Operations Management, 20*, 221–240.
Orasanu, J., & Connolly, T. (1993). The reinvention of decision making. In G. A. Klein, J. Orasanu, R. Calderwood, & D. E. Zsambok (Eds.), *Decision making in action: Models and methods*. Ablex Publishing Corporation: Norwood, NJ.
Ould, M. A. (1995). *Business processes – modelling and analysis for reengineering and improvement*. New York: Wiley.
Parente, D. H. (1998). Across the manufacturing-marketing interface. Classification of significant research. *International Journal of Operations & Production Management, 18*(12), 1205–1222.
Pfeffer, J. (1981). *Power in organizations*. Cambridge, MA: Ballinger.
Polanyi, M. (1967). *The tacit dimension*. London: Routledge & Kegan Paul.
Raven, B. H. (1965). Social influence and power. In I. D. Steiner & M. Fishbein (Eds.), *Current studies in social psychology*. New York: Holt, Rinehart & Winston.
Salancik, G., & Pfeffer, J. (1977). Who gets power – and how they hold on to it: A strategic contingency model of power. *Organizational Dynamics, 5*, 3.
Shapiro, B. P. (1977). Can marketing and manufacturing coexist? *Harvard Business Review, September–October*, 104–114.
Shin, M., Holden, T., & Schmidt, N. (2001). From knowledge theory to management practice: towards an integrated approach. *Information Processing and Management, 37*, 347.
Simon, H. A. (1957). *Models of man: social and rational*. New York: Wiley.
Skinner, W. (1969). Manufacturing – missing link in corporate strategy. *Harvard Business Review, May-June*, 136–145.
Slack, N., Chambers, S., & Johnston, R. (1998). *Operations management* (4th ed.). London: Pitman Publishing.
Slack, N., & Lewis, M. (2002). *Operations strategy*. London: Financial Times Press.
Sniezek, J. A., & Van Swol, L. (2001). Trust, confidence and expertise in a jugde-adviser system. *Organisational Behaviour and Human Decision Processes, 84*(2), 288–307.
Spencer, M., & Cox, J. F. (1994). Sales and manufacturing coordination in repetitive manufacturing: Characteristics and problems. *International Journal of Production Economics, 37*, 73–81.
Swamidass, P. M., Baines, T., & Darlow, N. (2001). The role of manufacturing and marketing managers in strategy development. *International Journal of Operations & Production Management, 21*(7), 933–948.
Voss, C., Tsikriktsis, N., & Frohlich, M. (2002). Case research in operations management. *International Journal of Operations and Production Management, 22*, 195–219.
Whybark, D. C. (1994). Marketing's influence on manufacturing practices. *International Journal of Production Economics, 37*, 41–50.

Chapter 5
Collaborative Planning in Supply Chains: The Importance of Creating High Quality Relationships

Hannes Günter, Cees De Snoo, Craig Shepherd, Philip Moscoso, and Johann Riedel

Abstract While collaborative planning and relationship quality are considered key contributors to supply chain performance, their mechanisms and linkages remain unclear. In order to help address this issue this book chapter introduces and unpacks the concepts of collaborative planning and relationship quality and investigates their role in supply chains. A multidisciplinary literature review was undertaken to identify conceptual and empirical work on relationship quality and collaborative planning. The chapter reveals a number of shortcomings in the literature and provides suggestions to guide future research on the links between collaborative planning, relationship quality, and supply chain performance. Implications are also provided for practitioners interested in enhancing the quality of interorganizational relationships and collaborative planning in supply chains.

H. Günter (✉)
Department of Organization and Strategy, Maastricht University, Maastricht, The Netherlands
e-mail: h.guenter@maastrichtuniversity.nl

C. De Snoo
Faculty of Economics and Business, University of Groningen, Groningen, The Netherlands
e-mail: c.de.snoo@rug.nl

C. Shepherd
Nottingham University Business School, Nottingham, UK
e-mail: craig.shepherd@nottingham.ac.uk

P. Moscoso
IESE Business School, Universidad de Navarra, Pamplona, Spain
e-mail: pmoscoso@iese.edu

J. Riedel
Nottingham University Business School, Nottingham, UK
e-mail: johann.riedel@nottingham.ac.uk

J.C. Fransoo et al. (eds.), *Behavioral Operations in Planning and Scheduling*,
DOI 10.1007/978-3-642-13382-4_5, © Springer-Verlag Berlin Heidelberg 2011

5.1 Introduction

Supply chain management is the integration and management of supply chain organizations and activities through cooperative organizational relationships, effective business processes, and high levels of information sharing (Handfield and Nichols 1999). This widely used definition of supply chain management, coined by two eminent academics of the field, alludes to two key assumptions guiding supply chain management research and practice. Firstly, a fundamental objective of managing supply chains is to integrate and streamline organizational activities and business processes across organizational boundaries. Secondly, this integration depends on the building and maintaining of cooperative relationships. Supply chain management, according to this view, is at its heart relationship management (cf., Otto and Kotzab 2003; Christopher 1998). Indeed, it has been argued that building and maintaining high quality relationships between suppliers and customers is one of the most important tasks in supply chain management (Chen and Paulraj 2004; Van der Vaart and Van Donk 2008). Despite the importance ascribed to the building of high quality relationships and the streamlining of organizational activities, there is some evidence to suggest that supply chain management in practice has not lived up to its expectations. A more recent international survey, for example, finds that "supply chain integration is more a rhetoric than reality in most industries in Europe" (Bagchi et al. 2005, p. 288).

This chapter builds on the finding that the integration of organizational activities across the supply chain affords organizations to collaboratively plan and execute production, storage, and distribution processes (cf., Barratt 2004; Barratt and Oliveira 2001; Dudek and Stadtler 2005). Collaborative planning, that is in provisional terms, the alignment of planning processes across organizational boundaries is arguably key to supply chain integration (cf., Barratt 2003). While both concepts – collaborative planning and relationship quality – have been suggested to be critcial to supply chain management, their mechanisms and linkages remain unclear. Conceptual work suggests that building and maintaining high quality relationships within and between organizations is crucial for successful collaborative planning. Kilger and Reuter (2002), among others, have observed that collaborative planning requires a collaborative relationship. Seifert (2002), with regard to current collaborative planning and forecasting initiatives, similarly finds, "the demands on the quality of the partnerships are considerably higher than with the classic initiatives in cooperative supply chain management" (p. 28). However, in spite of this early conceptual evidence for the linkages between relationship quality and collaborative planning, further conceptual work is needed to better understand how both concepts interlink and underpin supply chain behaviour.

The purpose of this book chapter is twofold. It firstly introduces and unpacks the concepts of collaborative planning and relationship quality and investigates their role in supply chain management. Secondly, this chapter seeks to understand how relationship quality and collaborative planning interact. In drawing together conceptual and empirical findings from the supply chain management, operations management, marketing, organization studies and applied psychology literatures this chapter provides new perspectives on the difficulties in building high performing supply chains. The chapter is structured as follows. The following section begins with outlining the methodology, that is, a multidisciplinary review on collaborative planning and relationship quality. The results section presents the findings of the reviews of collaborative planning and relationship quality and compares definitions across disciplines. Next, we illustrate relationship quality's decisive role in collaborative planning by referring to one of supply chain management's most stubborn difficulties, that is, the bullwhip effect. On basis of this literature review, the discussion section outlines linkages between collaborative planning and relationship quality. In concluding the chapter, implications for theory and practice are outlined.

5.2 Methodology

A literature review was undertaken by a multidisciplinary team comprising persons with expertise in supply chain management, operations management, marketing, organization studies and applied psychology. Each member of the team began by searching for articles on 'collaborative planning' and 'relationship quality' within a selection of high-impact and key journals from their respective disciplines. This journal searching was complemented by keyword searches on 'collaborative planning' and 'relationship quality' using the online resources ISI Web of Science® and Google Scholar™. These searches were focused on identifying journal articles only from the above disciplines. Moreover, only papers specifically concerned with definitions, antecedents, consequences, and dynamics of collaborative planning and relationship quality were considered for further investigation. Articles investigating these constructs in an interorganizational context were deemed of highest importance. In the following section, we present the findings of an interdisciplinary review of the concept of collaborative planning and outline its impact on organizational performance.

5.3 Collaborative Planning

The term collaborative planning has been referred to in the supply chain and operations management (Barratt 2004; De Snoo et al. 2007a; Dudek and Stadtler 2005; Poundarikapuram and Veeramani 2004), the management (Mitchell and Nault 2007), and the organization studies and applied psychology literatures (Günter et al. 2006; Günter 2007; Marks et al. 2001; Mathieu and Schulze 2006; Weingart 1992; Windischer et al. 2008; Windischer and Grote 2003).

5.3.1 Interdependencies in Planning: A Common Issue Across Disciplines

A common issue across these disciplines is the notion that planning is seen as means to coordinate interdependencies between actors, that is, the extent to which individuals are dependent on each other in fulfilling their tasks and reaching set goals (cf., Thompson 1967; Van De Ven et al. 1976). Bluedorn (2002), for example, argues that the planning literature seems directly linkable to research done on interdependence. Indeed, in the process of collaboratively setting up and carrying out plans, actors must carefully assess and consider their interdependencies in order to reach workable plans. This applies, for example, to sales agents who may assess how their activities influence the planning activities of production managers. Production managers, in turn, may depend on materials processing times provided by engineers, in order to devise production schedules. While these examples denote interdependencies of rather low complexity it is easy to see how complexity can skyrocket if collaborative planning activities not only criss-cross departmental but also organizational boundaries. In order to deal with this complexity in planning, practitioners decompose collaborative planning activities in sub-problems to be solved by multiple people (De Snoo et al. 2007a). In line with this industrial practice, scholars argue that plans can be conceptualised as a number of inter-related planning decisions where each decision eventually constrains the decision space for subsequent decisions (Van Wezel 2001). It is this interdependence of planning decisions, which can make the adaptation of plans a bothersome and often costly endeavour. To help practitioners grapple with the complexity of interdependencies among plans, scholars have devised means for their systematic assessment, as for example, so-called "plan dependence and coordination tables" (cf., De Snoo et al. 2007a).

Figure 5.1 depicts such a plan dependence and coordination table. The table highlights the critical relationships between mutually interdependent planners of different departments, in this example, from the purchasing, production, and sales departments. It illustrates the relationship between plans in a company where the planning of a production order starts with the order acceptance decision. This order acceptance decision is part of the delivery plan (DP). The production planning process results into the plans for three exemplary production departments (PA, PB

		Plan creation logic – Demand chain				
		DP	**PP C**	**PP B**	**PP A**	**PuP**
Plan adaptation logic – Realization chain	**DP**		DP → PC	DP → PB	DP → PA	DP → PP
	PP C	PC → DP		PC → PB	PC → PA	PC → PP
	PP B	PB → DP	PB → PC		PB → PA	PB → PP
	PP A	PA → DP	PA → PC	PA → PB		PA → PP
	PuP	PP → DP	PP → PC	PP → PB	PP → PA	

Fig. 5.1 Plan dependence and coordination table (*DP* Delivery plan, *PP* production plan, *PuP* Purchasing plan

and PC) followed by decisions regarding the purchase of materials that are in the purchase plan (PP).

The demand chain represents the sequence in which plans are developed. To be able to deliver products to the customer, the production department C, for example, has to be ready with its production activities. To be able to produce in department C, materials from department B have to be available, etc. In other words, a plan 'determines' the input (i.e., the demand for material) for another plan. The realization chain represents the sequence in which plans are executed. Basically, plans that could be developed the latest (because of the planning lead time) are executed first. The table also makes visible the possibly detrimental effects of planning changes. Material supply problems, for example, require adaptation of production plans and machine failures eventually may afford to renegotiate delivery dates with customers because products are not ready to be transported. Upward planning changes could cause a snowball-effect throughout the whole demand chain resulting into many plan adaptations due to infeasible commitments and constraints. A change in the customer order for instance could have small consequences for the assembling department, but major consequences for purchasing.

In short, this table makes visible the complexity in collaborative planning arising from interdependencies between different plans and actors carrying out these plans. The AEC-chapter in this book further discusses difficulties related to mutual interdependencies and gives examples on the cooperation between sales agents, production and engineering people. The next sections, firstly, highlight findings on collaborative planning from the supply chain management, operations management, and general management literatures, before then introducing findings from the organization studies and applied psychology domains.

5.3.2 Supply Chain and Operations Management

Reviewing the supply chain, operations, and management literatures, one finds collaborative planning to be of interest to scholars researching, firstly, the supply chain and, secondly, the concurrent design process. The majority of supply chain scholars interested in collaborative planning have investigated the effects of the so-called Collaborative Planning, Forecasting, and Replenishment (CPFR) model conceived in the United States during the 1980s and 1990s (cf., Danese 2007; Petersen et al. 2005). The CPFR model developed by the Voluntary Inter-Industry Commerce Standards (VICS) Association is a standardized business process model that specifies activities to be fulfilled in four consecutive stages: strategy and planning, demand and supply management, execution, and analysis (cf., Chung and Leung 2005; Danese 2006; Terwiesch et al. 2005). The model details activities on the strategic (establishment of the ground rules for the relationship, determination of product mix), the tactical (projection of consumer demand, as well as order and shipment requirements over the planning horizon), and the operational levels (order plans). Research on its effectiveness has shown the CPFR model to reduce inventory costs and to improve forecast accuracy, among others. The very first findings on its effectiveness stem from a pilot project in the retail industry jointly carried out by Wal-Mart and Warner-Lambert. This project resulted in a better use of real-time information in promotion activities, which helped to improve forecast accuracy and reduce inventory. More recently, Terwiesch and colleagues (2005) reported that the introduction of CPFR in the semiconductor industry, led to a 20% reduction in inventory costs and increases in off-the-shelf availability of 12% for retailers. Similarly, Barratt (2004) found that CPFR implementations in the grocery sector in the United Kingdom have significantly improved forecast accuracy.

Researchers exploring the intricacies of the concurrent engineering design process have further contributed to the understanding of collaborative planning from a management perspective. Building on work by Adler (1995) and Loch and Terwiesch (1998), Mitchell and Nault (2007), for example, developed a collaborative planning model in the context of concurrent product engineering. In concurrent design, upstream operations (e.g., business process redesign) and downstream operations (e.g., information technology development) overlap which affords timely communication of design changes and eventually may result in costly rework. Collaborative planning was hypothesized to reduce both the upstream and downstream rework. Indeed, Mitchell and Nault (2007) in their study on 120 concurrent engineering projects found a significant, inverse relationship between collaborative planning and the amount of rework necessary in the concurrent design process. Greater collaborative planning decreases the magnitude of costly rework, both upstream and downstream.

5.3.3 Organization Studies and Applied Psychology

While organization studies and applied psychology disciplines have shown great interest in planning, scholars differ in their philosophical understanding of planning. Whilst some researchers understand plans as structuring action (e.g., Miller et al. 1960) others think of them merely as an orienting device which does not determine behaviour (e.g., Suchman 2007). Advocates of the former view treat plans as 'blueprints for action', that is, mental representations of future realities that precede and determine behaviour (cf., Hacker 1998; Leudar and Costall 1996; Schank and Abelson 1977). Their approach is exemplified by the claim that plans are analogous to computer programs. Just as computer programs control the execution of various processes, plans are thought to control human behaviour. In contrast, advocates of the latter understand plans as a as resource for situated action which help actors to deal with emerging opportunities and constraints in an ad-hoc manner (Suchman 1987, 2007).

More recently, researchers have made efforts in integrating both perspectives (cf., Marks et al. 2001; Mathieu and Schulze 2006; Windischer et al. 2008). Windischer et al. (2008), for example, devised a model of collaborative planning, which has been explored and tested in inter-departmental and supply chain relationships, since its first inception. Their model delineates a set of processes underpinning collaborative planning, that is, the attentive exchange of (preliminary) knowledge, goal agreement, and coordinated, opportunistic planning (cf., Günter 2007). Attentive exchange of preliminary knowledge refers to the situation where individuals take co-actors' working conditions into account when exchanging (preliminary) knowledge. Co-actors are thereby empowered to prepare themselves for disruptions or external variance. Goal agreement increases actors' goal commitment and might strengthen actors' sense of control. Individual opportunistic planning refers to the need for coordinating ad-hoc changes in the planning to ensure synchronized actions. These sub-processes of collaborative planning are not to be thought of as separate entities but as overlapping and recursively interrelated processes. In short, this model specifies collaborative planning as a complex coordination mechanism reflecting deliberate planning (through lateral agreements and exchange of knowledge) and dynamic planning (through opportunistic adaptations). Empirical studies on inter-departmental relationships lend support to the assumption that collaborative planning facilitates logistical performance. More specifically, a study by Windischer et al. (2008) on inter-departmental planning in three manufacturing organizations reveals that the introduction of collaborative planning coincided with higher forecasting accuracy and improved delivery efficiency. Further, empirical findings on 107 supply chain relationships in the forestry and timber industry show positive and significant relationships to exist between collaborative planning and relationship effectiveness (Günter 2007).

5.3.4 Collaborative Planning: A Summary

Despite differences in the conceptualization of collaborative planning, research, as been presented here, has revealed some key insights into collaborative planning. Firstly, research indicates interorganizational collaborative planning to enhance organizational and supply chain performance. Empirical findings show planning to reduce inventory costs, to improve forecast accuracy, reduce rework, enhance relationship effectiveness, and support logistical performance, more generally. Secondly, collaborative planning is not to be seen only as a program to be implemented but also as an ongoing adaptation process. Thus, it is important to not only deliberately define planning goals (e.g., delivery due dates) but also to continuously refine and adapt planning processes as they are being put in practice. Thirdly, by linking collaborative planning to interdependence theory it becomes visible why plans are to be developed and optimized in a joint manner. To view planning processes as interdependent planning decisions carried out by different actors makes clear why actors must strongly cooperate across departmental and organizational boundaries in order to prevent from misfits of plans. In order to minimize costs along the supply chain actors must carefully scrutinize how their planning and changes in planning can affect cooperation partners in the supply chains. In the following section, we present the findings of an interdisciplinary review of the concept of relationship quality. The objective is to compare and contrast definitions of relationship quality across disciplines.

5.4 Relationship Quality

Relationship quality has been researched in the supply chain and operations management (Fynes et al. 2008; Huntley 2006; Ramdas and Spekman 2000), the marketing (Dwyer et al. 1987; Kumar 1996; Naude and Buttle 2000), and the organization studies and applied psychology (Ariño et al. 2001; Garcia-Canal et al. 2003) literatures.

5.4.1 Supply Chain and Operations Management

The relationship that has probably received most attention within the supply chain and operations management literatures is the buyer-supplier relationship (Goffin et al. 2006; Liker and Choi 2004). In theory, the traditional competitive barriers between supply chain members are mitigated by the opportunity to create mutually beneficial relationships. In turn, this should help manufacturers to reduce costs, improve quality and enhance new product development (Goffin et al. 2006). High quality relationships have also been linked to improved information flow, reduced uncertainty, and a more profitable and responsive supply chain (Handfield and Bechtel 2002; Maloni and Benton 1997).

It is therefore surprising that there is a dearth of empirical evidence for the connection between relationship quality and supply chain performance. Arguably, the main reason for this is a lack of conceptual clarity surrounding the concept of relationship quality in the supply chain literature (Huntley 2006). Fynes et al. (2005a) observe that a variety of different constructs have been offered to assess the efficacy of relationships, including partnership success, relationship value, relationship climate, and relationship quality. These concepts are often difficult to distinguish from one another. If used explicitly, the term 'relationship quality' is generally conceptualised as a higher level, multi-dimensional construct. However, there is limited agreement on the dimensions that comprise and constitute it. For example, whilst Parsons (2002) argues it encompasses trust and satisfaction, Fynes et al. (2005a, b, c) claim it includes trust, adaptation, communication, interdependence and cooperation.

Although several scholars have tested how the quality of the buyer-supplier relationship impacts on the performance of supply chains (Fynes et al. 2008; Fynes et al. 2005a; Fynes et al. 2005b, c; Ramdas and Spekman 2000; Tapiero 2001; Toni and Nassimbeni 1999), there is little consistency in the measures used to operationalise supply chain performance (cf., Shepherd and Günter 2006, for discussion). For example, Fynes et al. (2005a) use quality, delivery, cost, and flexibility, whilst

Ramdas and Spekman (2000) consider inventory, time, order fulfillment, quality, customer focus and customer satisfaction. Arguably the most rigorous empirical assessment of the connection between relationship quality and supply chain performance is provided by Fynes and colleagues (Fynes et al. 2005a; Fynes et al. 2005b, c). However, even their research does not demonstrate conclusive evidence for the link between relationship quality and supply chain performance. In their study on customer relationships of 200 manufacturing companies in the electronics sector they found that high-quality relationships resulted in better designed products while reducing costs. Furthermore, adaptation, one of the key ingredients of high quality relationships, is shown to enhance customer satisfaction. However, no links could be established with manufacturing quality, flexibility or delivery performance (Fynes et al. 2005b, c).

5.4.2 *Marketing*

Marketing research most commonly focuses on the quality of the buyer-seller relationship around the concept of relationship marketing, i.e., establishing and building relationships with customers (Dwyer Dwyer et al. 1987; Jap 1999; Naude and Buttle 2000; Storbacka et al. 1994). Relationship marketing describes a marketing strategy where the emphasis is on building long-term relationships with customers for example, rather than one-time transactional-based relations. Studies have also explored the quality of buyer-supplier (Walter et al. 2003), business-to-business (Woo and Ennew 2004), manufacturer-retailer (Kumar 1996) and exporter-importer relationships (Leonidou et al. 2006). Obviously, relationship quality is a central concept also in the marketing literature (Woo and Ennew 2004). While relationship quality has been found to impact on marketing outcomes, either directly or indirectly (Hennig-Thurau et al. 2002; Hewett et al. 2002), this finding is compounded by the fact that, similar to the other literatures reviewed, there is a lack of conceptual clarity in the assessment of relationship quality in the marketing literature. Woo and Ennew (2004), for example, argue relationship quality to comprise cooperation, adaptation, atmosphere, service quality, customer satisfaction and behavioural intention. Lages and colleagues (2005), in contrast, think of relationship quality as a multi-dimensional construct, consisting of four dimensions: amount of information sharing; communication quality; long-term relationship orientation; and satisfaction with the relationship.

Marketing scholars have also invested considerable efforts in assessing antecedents of relationship quality. Findings from different industries suggest following antecedents to be of importance: salespersons characteristics and relational selling behaviour in insurance (Crosby et al. 1990); ethical selling behaviour and expertise in pharmaceutical products (Lagace et al. 1991); perceptions of fairness in automobile dealers (Kumar et al. 1995); relational behaviours (e.g., proactive effort, information exchange, interaction frequency) for industrial purchasing (Leuthesser 1997); same-sex relationships, similar personality and expertise for industrial purchasing executives (Smith 1998); customer skills (i.e., attribution, level, specificity) in retail sales (Hennig-Thurau 2000); and equity in telecommunication business sales (Boles et al. 2000). Interestingly, and in contrast to the supply chain and operations management research, marketing scholars conceptualize relationship quality as a dynamic phenomenon influenced by the development of the relationship over time. While some scholars argue business relationships to develop in a linear and predictable manner, others argue the development to be non-linear and iterative. Inspired by findings on lifecycles in products, brands, and markets, it has been argued that buyer-supplier relationships evolve 'according to a predictable series of events occurring in a fixed order' (Jap and Anderson 2007). For example, the lifecycle, or stage model, developed by Dwyer and colleagues (1987) assumes relationships evolve through five sequential stages; unilateral awareness, exploration, expansion, commitment, and sometimes decline and dissolution. Stage models assume dimensions of relationship quality, such as trust, vary according to the stage of the relationship lifecycle in universal and predictable ways (Jap and Anderson 2007). 'State' theorists, in contrast, propose that business dyads 'move from being new toward being inert' existing in either in a growing, troubled, or static state (Rosson and Ford 1982). Relationships are assumed to move between these three states in an unpredictable and continuous fashion. Hence, in state models, relationship development is not seen as an orderly process (Rosson and Ford 1982) and changes of relationship quality are proposed to be unpredictable and iterative (Batonda and Perry 2003). While both perspectives differ in their conceptual understanding of relationship quality, they complement static models of relationship quality by highlighting its temporally bounded and dynamic character.

5.4.3 Organization Studies and Applied Psychology

Supply chain management is frequently treated in the literature as a synonym of logistics, operations management, and procurement (Lambert et al. 1998). Further, discussions on relationship management are often framed in terms of institutional and inter-firm relations without paying adequate attention to the central role individuals play in shaping supply chain relationships. Therefore, it is not surprising that ideas from other disciplines, such as organization studies and applied psychology, have exerted a limited influence on the concept of relationship quality in supply chains. In this section, we explain how insights from the organization studies and applied psychology literatures may help to augment our understanding of the concept of relationship quality.

Scholars from the fields of organization studies and applied psychology have shown considerable interest in the quality of interorganizational relations, among others in the context of dyadic and multi-partner international joint ventures (cf., Ariño and De la Torre 1998; Garcia-Canal et al. 2003) and inter-organizational relationships among small and medium sized companies (Beugelsdijk et al. 2006). Previous research conducted in these fields acknowledges that relationship quality is an essential for cooperation and performance (Ariño et al. 2001). It is seen to facilitate communication, flexibility, and helping behavior (Venkataramani and Dalal 2007). Furthermore, it buffers interorganizational relations against severe setbacks and increases the chance of alliance survival (e.g., Ariño and De la Torre 1998; Ariño et al. 2001). Conflicts and difficult situations are more likely to be solved satisfactorily if business partners rely on strong relational links (Ariño et al. 2001). Furthermore, an increase in relationship quality fosters partners' willingness to rely on trust; trust, in turn, can act as a substitute for more formal (and often costly) control mechanisms thereby reducing costs. This economic dimension of relationship quality has been highlighted, for example, by Whitten and Leidner (2006). Their study on outsourcing practices explores why companies switch from one IT outsourcing contractor to another. Quantitative analysis of 160 IT managers showed that companies who switch contractors do so mainly because of poor relationship quality. In other words, companies who switch vendors are significantly less satisfied with the quality of their relationship than "non-switchers". Whitten and Leidner (2006), however, found no difference in companies' ratings of the product and service quality of contractors when comparing "switchers" and "nonswitchers". Obviously, relationship quality plays a key role in the decision to maintain or to cease business relationships.

The organization studies and applied psychology literatures are similar to other disciplines in that they use a plethora of different labels to refer to relationship quality. Relational quality (Ariño and De la Torre 1998), interactional richness (Barry and Crant 2000), relationship equity (Scheer et al. 2003), and relationship skills (Beugelsdijk et al. 2006) are among the labels used most often. Differences between these constructs are rarely made explicit. Within the supply chain management and marketing literature relationship quality is generally treated as a higher-order construct comprising dimensions like trust, satisfaction and commitment. In contrast, researchers from organization studies and applied psychology stress different components. They include support, perspective taking, and empathic concern (Settoon and Mossholder 2002) and 'the degree of compatibility of corporate cultures and decision-making styles, a convergence of worldviews, and other organizational characteristics' (Ariño et al. 2001).

These definitions enrich understanding of relationship quality in three important ways. The notions of compatibility and convergence make the inherently relational and two-sided nature of relationship quality more clearly visible. Furthermore, defining relationship quality as a relative – not an absolute – concept clearly shows its value to be dependent on the context and the partners involved. While most of the existing research thinks of relationship quality as an antecedent of performance, Ariño et al. (2001) remind us that relationship quality is "both an input to the success of the venture, and an output of the interactions between the partners" (p. 322). Ariño and De la Torre (1998) emphasise the active role individuals play in shaping relationships. By stressing the importance of human agency and interpretation, they demonstrate that for worldviews to converge, joint learning and sensemaking processes must take place in relationships. Relationships deteriorate if parties do not invest considerable effort in sense making processes and mutual understanding. If insufficient effort is invested in this continuous adaptation and sense making process, incompatibilities will arise. To address these incompatibilities parties reassess and renegotiate equity and efficiency of the relationship. If successful, this can result in a strengthened mutual agreement. If these negotiations are unsuccessful, relationship quality will decline, threatening the proper functioning of the relationship (Garcia-Canal et al. 2003).

5.4.4 Relationship Quality: A Summary

First, this multidisciplinary review of relationship quality revealed a dearth of empirical evidence for the link between relationship quality and supply chain performance. This is mainly due to a lack of conceptual clarity surrounding the notion of relationship quality and its empirical assessment. Second, while research has investigated and identified a wide range of influencing factors of relationship quality, no coherent and integrative framework exists on the antecedents of relationship quality. Third, conceptual research in marketing and organization studies highlights relationship quality's dynamic and temporal aspects. However, with some notable exceptions, this appealing conceptual view has rarely been tested empirically. Fourth, although relationship quality is typically treated as an antecedent of supply chain performance, it can also be seen as an output of the interactions among supply chain partners. This view also accentuates the active role individuals play in creating and constantly shaping supply chain relationships. Indeed, using an interpretive lens reconceptualises relationship quality as a dynamic and emergent phenomenon, existing only through its constant renegotiation by actors. The preceding sections of this chapter have looked at collaborative planning and relationship quality. The following section will be highlighting some of the linkages between relationship quality and planning and their impact on supply chain performance. In order to illustrate some of these linkages this section introduces one of the most persistent difficulties in managing supply chain relationships, that is, the bullwhip effect.

5.5 The Bullwhip Effect in Supply Chains: Illustrating the Interplay Between Collaborative Planning and Relationship Quality

The bullwhip effect is a dynamic phenomenon that essentially substantiates through the amplification of demand order volatility as one moves upstream along the supply chain (Forrester 1961; Lee et al. 1997). The term 'bullwhip effect' was firstly used by the consumer goods company Procter and Gamble (P&G, Lee and Billington 1992). The logistic executives at P&G examined the order patterns of Pampers diapers, one of their best-selling products. They found that the variability of replenishment orders was amplified towards the upstream of the supply chain regardless of the fact that the demand pattern at the customer interface was quite steady. Hence the name: 'bullwhip effect' (Wang et al. 2005). The effect has also been empirically well demonstrated with the so-called “Beer Distribution Game” (cf., Sterman 1989). The purpose of the game is to meet customer demand for beer, through a multi-echelon supply chain with expenditure on back orders and inventory. In the original format of the experimental setup of the game, the beer supply chain involves four players (factory, distributor, wholesaler, and retailer) who make independent inventory decisions without consulting other supply chain members. They rely solely on orders from the neighbouring player as information source. Sterman (1989), by engaging individuals in the Beer Distribution Game, found empirical evidence for the bullwhip effect and shows misperceptions of feedback to cause poor supply chain performance. Other research similarly concludes that the bullwhip effect may lead to an increase in inventory requirements, expediting, customer shortages, and suboptimal capacity utilizations, for instance.

Extensive research on this detrimental effect has identified various causes for its existence (Daganzo 2004; Lee and Billington 1992). Among the most frequently cited causes are: demand signal processing, inventory rationing and shortage gaming, order batching, and price fluctuations (Chen et al. 2000; Lee et al. 1997). Methods to alleviate these operational problems include improved demand forecasting techniques (Chen et al. 1998), better capacity allocation schemes (Cachon and Lariviere 1999), staggered order batching (Cachon 1999) and everyday low pricing strategies (Sogomonian and Tang 1993). However, other scholars have argued and shown empirically that the bullwhip effect might persist even when eliminating operational causes. Croson and Donohue (2006), for example, suggest that as decision makers consistently underweight the supply line when making order decisions, the bullwhip effect is likely to persist even when eliminating all operational causes. They suggest that although inventory information mitigate the bullwhip effect by helping upstream supply chain partners to anticipate and prepare for fluctuations in inventory needs downstream, the effect does not completely disappear. Croson et al. (2005) have provided empirical evidence for this view. Their experimental findings from the Beer Game show the bullwhip effect to persist even after eliminating demand uncertainty. In this experiment, demand was hold constant and made visible to participants. Furthermore, participants were informed upfront about the optimal ordering policy. However, participants still did not take unfilled orders adequately into account, as they perceived a risk of other actors not following the optimal ordering policy. In other words, participants distrusted each other’s abilities and motives, which is why they built up so-called coordination stock thereby triggering instabilities across the supply chain.

These findings on the bullwhip effect aptly reveal, firstly, the paramount importance of collaborative planning in supply chains. If one disables actors from exchanging planning information, the bullwhip effect is inevitable and supply chain performance is likely to suffer. However, even if provided with the opportunity of exchanging planning information, individuals might not make adequate use of it under specific conditions. This is particularly likely to happen if individuals distrust each other’s abilities and motives in forecasting and planning, indicative of a low quality in supply chain relationships. Research on the Beer Game thus suggests that actors must not only engage in collaborative planning activities, but also need to invest in building high quality relationships in order to mitigate the bullwhip effect and to enhance supply chain performance. While it remains debatable to what extent these experimental findings can be directly translated into supply chain management practice they illustrate the potentially frustrating and economically detrimental effects of inadequate planning and low quality relationships. As a final remark, it is interesting to note here that while the horizontal (supply chain) bullwhip effect is the one most studied, it is not the only bullwhip effect reported in literature. Within hierarchical planning systems a vertical bullwhip may also occur. This vertical or planning bullwhip is described in detail in the VBW chapter of this book.

5.6 Discussion

This chapter has attempted to unpack collaborative planning and relationship quality. By juxtaposing conceptual and empirical findings from the supply chain management, operations management, marketing, organization studies and applied psychology literatures, this chapter has revealed, firstly, that there is a lack of common understanding within and between these disciplines over the meaning of collaborative planning and relationship quality. While some scholars explicitly or implicitly think of collaborative planning as a standardized approach (cf., CPFR model), others emphasize the flexible and opportunistic side of collaborative planning. Similarly, while scholars interested in relationship quality commonly agree that relationship quality is a higher-order construct there is little agreement over the dimensions that comprise it. Secondly, while some research shows collaborative planning to reduce inventory costs, to improve forecast accuracy, and to support logistical performance, among others, the majority of findings is cross-sectional, which undermines causal inferences. Similarly, the empirical evidence for the link between relationship quality and supply chain performance is, at best, inconclusive. Thirdly, while relationship quality has often been treated as an antecedent to supply chain performance and collaborative planning, it has also been framed in the literature as an outcome of the interactions of supply chain partners.

In light of these conceptual and methodological ambiguities in the literature, we argue that there is a need for more systematic research on collaborative planning and relationship quality and their linkages. From drawing on the research outlined in this chapter we find three dimensions to be particularly promising candidates for making sense of existing research findings and for building a framework which helps to guide future research. These three dimensions are; firstly, the contextual dimensions of the constructs (*content*); secondly, the level at which the constructs are assessed (*level*); and, thirdly, the importance given to time in researching the constructs (*time*).

The first dimension, *content*, not only refers to the different contextual facets of relationship quality and collaborative planning, such as, trust, satisfaction, knowledge exchange, or mutual goal agreement, but also to the more fundamental assumptions on the nature and empirical assessment of both concepts. Examples on these fundamental assumptions are; planning does not determine behaviour; relationship quality is a dynamic phenomenon shaped continuously through communications; and relationship quality can be adequately understood by assessing its components. The second dimension, *level*, makes clear that collaborative planning and relationship quality concurrently operate at different levels, such as the individual, group, departmental, organizational, and supply chain level. This distinction is particularly warranted and timely, given recent findings that indicate that constructs can have different meanings at different levels (cf., Oosterhuis et al. 2005). The third dimension, *time*, highlights the importance of time in relationships and makes visible that collaborative planning and relationship quality are both dynamic and emergent phenomena, constantly reassessed and renegotiated between supply chain partners. Viewing relationship quality and collaborative planning as multi-facet and multi-level concepts, actively re-negotiated over time through human interaction, provides the basis for a conceptual clarity not often achieved in existing research on relationship quality and its linkages with collaborative planning and for making visible future research opportunities, structured around the three dimensions of content, level, and time.

5.6.1 Avenues for Further Research

With regard to the content-dimension, one finds a lack of common understanding of the content of relationship quality and collaborative planning among scholars. This aggravates comparisons across studies and hampers theoretical progress (cf., Bell et al. 2006). We believe that in order to improve on this unsatisfactory status quo, research is needed that actually assesses properties of relationships, not properties of individuals. Unfortunately, research has often only studied one partner's perspective in a relationship, for example, by asking one informant in an interorganizational relationship, about his or her commitment towards the relationship. While such research reveals important insights on individuals' attitudes, values, and opinions, it does not tap the relational and interactive characteristics of a relationship that arise from a particular combination of two specific individuals (Thompson and Walker 1982). Researchers have thus called for collecting data on relationships from both sides (cf., Gulati and Sytch 2007). In assessing, for example, the difference (or the congruence, respectively) of partners' views and attitudes, researchers can more appropriately tap these emergent properties of any particular dyadic relationship.

With regard to the level-dimension, one finds that while researchers have persuasively argued that relationship properties can have different meanings at different levels, mixed-level approaches are still rare in supply chain research (cf., Mumford et al. 2002; Oosterhuis et al. 2005). Research has not yet paid adequate attention to the fact that relationship quality is a multilevel phenomenon, operating concurrently at the interpersonal, person-to-organization, and inter-organizational level (Palmatier et al. 2007). Mumford and colleagues (2002) have made similar observations on the planning literature, finding that planning research has focused either on the individual, or the group, or the organizational level, but has largely missed out on developing and testing more complex, and probably more lifelike, cross-level hypotheses. Multi-level analyses thus hold the promise of resolving some of the conceptual ambiguity in the assessment of collaborative planning and relationship quality.

Finally, with regard to the time-dimension, there is a need for longitudinal studies to capture the process elements of collaborative planning and relationship quality and compare experiences of actors at different stages of the process. Longitudinal research would allow for example, for investigating how and why partners' views in a relationship converge, and for exploring patterns in interactions of supply chain partners. Indeed, the patterns in the interactions and practices between supply chain partners might be interpreted as what has been referred to above as the emergent property of relationships (cf., Thompson and Walker 1982). Interpretive studies would be particularly welcomed in order to understand how patterns in interactions unfold, change, and eventually cease.

We argue that such longitudinal, relational, and multi-level research, while clearly posing significant conceptual and methodological challenges would not only benefit research on the linkages between collaborative planning and relationship quality and their impact on supply chain dynamics significantly, but would also coincide with managerial concerns. Some potential practical implications are outlined in the coming section.

5.6.2 Practical Implications

Building high quality relationships is an ongoing, resource intensive endeavour. As Ariño and de la Torre (1998) observed, relationships deteriorate if parties do not invest considerable effort in sensemaking processes and mutual understanding. Consequently, the main implication of this paper for practice is that although managers should continue to provide sufficient resources for building strong ties at the intra-organizational level (e.g., between sales and purchasing departments) they also need to invest resources in building mutual understanding with their supply chain partners at the individual level. For example, as well as holding regular meetings, evidence suggests that techniques such as scenario planning can be beneficial in challenging mental models and building mutual understanding (cf., Hodgkinson and Wright 2002). Furthermore, companies might benefit from developing individuals' relationship-building abilities. To create high quality inter-organizational relationships resembles an inter-cultural endeavour individuals are to be prepared for. In the same way that companies send managers overseas to increase intercultural awareness, companies could send purchasing agents to suppliers and vice versa, a management practice used, for example, by Toyota and Honda (cf., Liker and Choi 2004). By spending time "abroad" planning partners can learn about each others' way of doing business, increase their perspective taking abilities, and develop a congruent understanding. These measures can then provide individuals with the necessary means to build high quality relationships.

Finally, the dynamic nature of relationship quality and collaborative planning activities makes it necessary that managers continuously monitor the state of buyer-supplier relationships. Managers could benefit from pre-defining threshold levels for key dimensions of relationship quality, such as trust, commitment and satisfaction. However, these thresholds must be adapted to the stage the relationship is at, as the relative importance of dimensions of relationship quality can change over time. For example, any sign of distrust could be particularly worrying in early stages of a relationship as no institutional scaffolds exist keeping a relationship on track. In order to prevent interorganizational relationship to become overly dependent on single individuals, however, it appears beneficial to invest in institutional and technological scaffolds that tie organizations together. But given the reciprocal relation between collaborative planning and relationship quality, relationship partners must carefully assess the impact that institutional and technological scaffolds have on existing relationships. Technological solutions (e.g., planning software) are to be in line with planning partners' needs and cultural affordances (cf., Loch and Terwiesch 2005).

5.6.3 Limitations

While making some theoretical contributions to the literature on relationship quality and collaborative planning in supply chains, there are some limitations to this chapter. Firstly, while the review clearly delineates the multiple dimensions underpinning and influencing relationship quality and collaborative planning this study has not assessed the relative importance of these different dimensions in a quantitative manner. Secondly, while our threedimensional model (content, level, and time) helps academics and practitioners to make sense of the complex linkages between relationship quality and collaborative planning it still has to be further elaborated to such extent that it allows for the derivation of testable hypotheses. Finally, some fundamental methodological issues remain, mainly with regard to the conceptualization and measurement of relationship quality.

5.7 Conclusion

This chapter has made some important contributions to furthering our understanding of the interplay between relationship quality and collaborative planning and their impact on supply chain dynamics. In order to master the complexity inherent to these linkages, either theoretically or in practice, this chapter has derived three axes along which future research is needed; that is, the content, level, and time axes. We hope that other future research will draw upon this threedimensional framework in order to make theoretical sense and practical use of the linkages between collaborative planning and relationship quality in supply chain relationships.

Acknowledgments Parts of this chapter have been presented during the 14th International EurOMA Conference 2007 in Ankara, Turkey (see De Snoo et al. 2007b), the 11th International HAAMAHA Conference 2007 in Poznan, Poland (see Günter et al. 2007), and the 14th International Symposium on Logistics 2009 in Istanbul, Turkey (see Günter et al. 2009). We thank reviewers and participants for their valuable comments and suggestions.

References

Adler, P. S. (1995). Interdepartmental interdependence and coordination: The case of the design/manufacturing interface. *Organization Science, 6*(2), 147–167.

Ariño, A., & De la Torre, J. (1998). Learning from failure: Towards an evolutionary model of collaborative ventures. *Organization Science, 9*(3), 306–325.

Ariño, A., De la Torre, J., & Ring, P. S. (2001). Relational quality: Managing trust in corporate alliances. *California Management Review, 44*, 109–134.

Bagchi, P. K., Chun, B., Skjoett-Larsen, T., & Soerensen, L. B. (2005). Supply chain integration: European survey. *The International Journal of Logistics Management, 16*(2), 275–294.

Barratt, M. (2003). Positioning the role of collaborative planning in grocery supply chains. *The International Journal of Logistics Management, 14*(2), 53–66.

Barratt, M. (2004). Unveiling enablers and inhibitors of collaborative planning. *International Journal of Logistics Management, 15*(1), 73–90.

Barratt, M., & Oliveira, A. (2001). Exploring the experiences of collaborative planning initiatives. *International Journal of Physical Distribution & Logistics Management, 31*(4), 266–289.

Barry, B., & Crant, J. M. (2000). Dyadic communication relationships in organizations: an attribution/expectancy approach. *Organization Science, 11*, 648–664.

Batonda, G., & Perry, C. (2003). Approaches to relationship development processes in inter-firm networks. *European Journal of Marketing, 37*(10), 1457–1484.

Bell, J., den Ouden, B., & Ziggers, G. W. (2006). Dynamics of cooperation: at the brink of irrelevance. *Journal of Management Studies, 43*, 1607–19.

Beugelsdijk, S., Koen, C. I., & Noorderhaven, N. G. (2006). Organizational culture and relationship skills. *Organization Studies, 27*, 833–854.

Bluedorn, A. C. (2002). *The human organization of time: Temporal realities and experience*. Stanford, CA: Stanford University Press.

Boles, J. S., Johnson, J. T., & Barksdale, H. C. (2000). How salespeople build quality relationships: a replication and extension. *Journal of Business Research, 48*(1), 75–81.

Cachon, G. (1999). Managing supply chain variability with scheduled ordering policies. *Management Science, 45*(6), 843–856.

Cachon, G., & Lariviere, M. (1999). Capacity choice and allocation: Strategic behavior and supply chain performance. *Management Science, 45*(8), 1091–1108.

Chen, F., Drezner, Z., Ryan, J. K., & Simchi-Levi, D. (1998). The bullwhip effect: Managerial insights on the impact of forecasting and information on variability in a supply chain. In S. Tayur, R. Ganeshan, & M. Magazine (Eds.), *Quantitative models for supply chain management*. Norwell, MA: Kluwer Academic. Chapter 14.

Chen, F., Drezner, Z., Ryan, J. K., & Simchi-Levi, D. (2000). Quantifying the bullwhip effect in a simple supply chain: the impact of forecasting, lead times, and information. *Management Science, 46*(3), 436–443.

Chen, I. J., & Paulraj, A. (2004). Towards a theory of supply chain management: the constructs and measurements. *Journal of Operations Management, 22*(2), 119–151.

Christopher, M. (1998). *Logistics and supply chain management*. London: Pitman Publishing.

Chung, W. W. C., & Leung, S. W. F. (2005). Collaborative planning, forecasting and replenishment: a case study in copper clad laminate industry. *Production Planning and Control, 16*(6), 563–574.

Crosby, L. A., Evans, K. R., & Cowles, D. (1990). Relationship quality in services selling – an interpersonal influence perspective. *Journal of Marketing, 54*(3), 68–81.

Croson, R., Donohue, K., Katok, E., & Sterman, J. (2005). "Order stability in supply chains: coordination risk and the role of coordination stock", *MIT Sloan School of Management Working Paper* No. 4513-04.

Croson, R., & Donohue, D. (2006). Behavioral causes of the bullwhip effect and the observed value of inventory information. *Management Science, 52*(3), 323–336.

Danese, P. (2006). Collaboration forms, information and communication technologies, and coordination mechanisms in CPFR. *International Journal of Production Research, 44*(16), 3207–3226.

Danese, P. (2007). Designing CPFR collaborations: insights from seven case studies. *International Journal of Operations & Production Management, 27*(2), 181–204.

Daganzo, C. F. (2004). On the stability of supply chains. *Operations Research, 52*(6), 909–921.

Dwyer, F. R., Schurr, P. H., & Oh, S. (1987). Developing buyer-seller relationships. *Journal of Marketing, 51*(2), 11–27.

De Snoo, C., Wortmann, J. C., Gaalman, G. J. C., & Van Wezel, W. (2007a). Decomposition of the planning function implies coordination of planners. *Proceedings of the 14th International Conference of the European Operations Management Association (EurOMA 2007).* 17–20 June 2007, Ankara, Turkey.

De Snoo, C., Moscoso, P., Günter, H., Shepherd, C., Riedel, J. (2007b). Relationship quality and collaborative planning: an interdisciplinary review. *Proceedings of the 14th International Conference of the European Operations Management Association (EurOMA 2007).* 17–20 June 2007, Ankara, Turkey.

Dudek, G., & Stadtler, H. (2005). Negotiation-based collaborative planning between supply chains partners: Supply chain management and advanced planning. *European Journal of Operational Research, 163*(3), 668–687.

Forrester, J. W. (1961). *Industrial dynamics*. Cambridge: Productivity Press.

Fynes, B., De Burca, S., & Mangan, J. (2008). The effect of relationship characteristics on relationship quality and performance. *International Journal of Production Economics, 111*(1), 56–69.

Fynes, B., de Burca, S., & Voss, C. (2005a). Supply chain relationship quality, the competitive environment and performance. *International Journal of Production Research, 43*(16), 3303–3320.

Fynes, B., Voss, C., & de Burca, S. (2005b). The impact of supply chain relationship dynamics on manufacturing performance. *International Journal of Operations & Production Management, 25*(1), 6–19.

Fynes, B., Voss, C., & de Burca, S. (2005c). The impact of supply chain relationship quality on quality performance. *International Journal of Production Economics, 96*(3), 339–354.

Gulati, R., & Sytch, M. (2007). Dependence asymmetry and joint dependence in interorganizational relationships: Effects of embeddedness on a manufacturer's performance in procurement relationships. *Administrative Science Quarterly, 52*(1), 32–69.

Garcia-Canal, E., Valdes-Llaneza, A., & Ariño, A. (2003). Effectiveness of dyadic and multi-party joint ventures. *Organization Studies, 24*(5), 743–770.

Goffin, K., Lemke, F., & Szwejczewski, M. (2006). An exploratory study of 'close' supplier-manufacturer relationships. *Journal of Operations Management, 24*(2), 189–209.

Günter, H. (2007). *Collaborative planning in heterarchic supply networks,* PhD Thesis, ETH Zurich.

Günter, H., Shepherd, C., Moscoco, P., De Snoo, C., & Riedel, J. (2007). Unpacking relationship quality in supply chains: An interdisciplinary review. In S. Karwowski & S. Trzcielinski, (Eds.) *Value Stream Activities Management – Proceedings of the 11th International Conference on Human Aspects of Advanced Manufacturing Agility and Hybrid Automation (HAAMAHA 2007)* (pp. 229–239). 9–12 July 2007, Poznan, Poland.

Günter, H., Grote, G., & Thees, O. (2006). Information technology in supply networks: Does it lead to better collaborative planning? *Journal of Enterprise Information Management, 19*(5), 540–550.

Günter, H., Shepherd, C., Moscoco, P., De Snoo, C., & Riedel, J. (2009). Understanding relationship quality and its link with supply chain performance – a review. In K. S. Pawar & C. S. Lalwani (Eds.) *Global supply chains and inter-firm networks – Proceedings of the 14th International Symposium on Logistics* (ISL 2009) (pp. 508–516). 5–8 July 2009, Istanbul, Turkey.

Hacker, W. (1998). *Allgemeine Arbeitspsychologie: psychische Regulation von Arbeitstätigkeiten*. Bern: Hans Huber.

Handfield, R. B., & Bechtel, C. (2002). The role of trust and relationship structure in improving supply chain responsiveness. *Industrial Marketing Management, 31*(4), 367–382.

Handfield, R. B., & Nichols, E. L. (1999). *Introduction to supply chain management*. Upper Saddle River, NY: Prentice Hall.

Hennig-Thurau, T. (2000). Relationship quality and customer retention through strategic communication of customer skills. *Journal of Marketing Management, 16*(1), 55–79.

Hennig-Thurau, T., Gwinner, K. P., & Gremler, D. D. (2002). Understanding relationship marketing outcomes: An Integration of relational benefits and relationship quality. *Journal of Service Research, 4*(3), 230–247.

Hewett, K. R., Money, B., & Sharma, S. (2002). An exploration of the moderating role of buyer corporate culture in industrial buyer–seller relationships. *Journal of the Academy of Marketing Science, 30*(3), 229–239.

Hodgkinson, G. P., & Wright, G. (2002). Confronting strategic inertia in a top management team: Learning from failure. *Organization Studies, 23*, 949–977.

Huntley, J. K. (2006). Conceptualization and measurement of relationship quality: Linking relationship quality to actual sales and recommendation intention. *Industrial Marketing Management, 35*(6), 703–714.

Jap, S. D. (1999). Pie-expansion efforts: Collaboration processes in buyer-supplier relationships. *Journal of Marketing Research, 36*(4), 461–475.

Jap, S. D., & Anderson, E. (2007). Testing alife-cycle theory of cooperative inter-organizational relationships: Movement across stages and performance. *Management Science, 53*(2), 260–275.

Kilger, C., & Reuter, B. (2002). Collaborative planning. In H. Stadler & C. Kilger (Eds.), *Supply chain management and advanced planning: concepts, models, software and case studies* (pp. 223–237). Berlin: Springer.

Kumar, N. (1996). The power of trust in manufacturer-retailer relationships. *Harvard Business Review, 74*(6), 92–106.

Kumar, N., Scheer, L. K., & Steenkamp, J. E. (1995). The effects of supplier fairness on vulnerable resellers. *Journal of Marketing Research, 32*(1), 54–66.

Lagace, R. R., Dahlstrom, R., & Gassenheimer, J. B. (1991). The relevance of ethical salesperson behavior on relationship quality: The pharmaceutical industry. *Journal of Personal Selling and Sales Management, 11*(4), 39–47.

Lages, C., Lages, C. R., & Lages, L. F. (2005). The RELQUAL scale: A measure of relationship quality in export market ventures. *Journal of Business Research, 58*(8), 1040–1048.

Lambert, D. M., James, R., & Ellram, L. M. (1998). *Fundamentals of logistics management*. Boston: McGraw-Hill.

Lee, H., Padmanabhan, V., & Whang, S. (1997). Information distortion in a supply chain: the bullwhip effect. *Management Science, 43*(4), 546–558.

Lee, H. L., & Billington, C. (1992). Managing supply chain inventory: Pitfalls and opportunities. *Sloan Management Review, 33*(3), 65–73.

Leonidou, L. C., Barnes, B. R., & Talias, M. A. (2006). Exporter–importer relationship quality: The inhibiting role of uncertainty, distance and conflict. *Industrial Marketing Management, 35*(5), 576–588.

Leudar, I., & Costall, A. (1996). Situating action IV: Planning as situated action. *Ecological Psychology, 8*, 153–170.

Leuthesser, L. (1997). Supplier relational behaviour: An empirical assessment. *Industrial Marketing Management, 26*(3), 245–254.

Liker, J., & Choi, T. H. (2004). Building deep supplier relationships. *Harvard Business Review, 82*(12), 104–113.
Loch, C., & Terwiesch, C. (1998). Communication and uncertainty in concurrent engineering. *Management Science, 44*(8), 1032–1049.
Loch, C. H., & Terwiesch, C. (2005). Rush and be wrong or wait and be late? A model of information in collaborative processes. *Production and Operations Management, 14*(3), 331–343.
Maloni, M. J., & Benton, W. C. (1997). Supply chain partnerships: Opportunities for operations research. *European Journal of Operational Research, 101*(3), 419–429.
Marks, M. A., Mathieu, J. E., & Zaccaro, S. J. (2001). A temporally based framework and taxonomy of team processes. *Academy of Management Review, 26*(3), 356–376.
Mathieu, J. E., & Schulze, W. (2006). The influence of team knowledge and formal plans on episodic team processes-performance relationships. *Academy of Management Journal, 49*(3), 605–619.
Miller, G. A., Galanter, E., & Pribram, K. (1960). *Plans and the structures of behaviour.* New York: Holt, Rinehart, and Winston.
Mitchell, V. L., & Nault, B. R. (2007). Cooperative planning, uncertainty, and managerial control in concurrent design. *Management Science, 53*(3), 375–389.
Mumford, M. D., Schultz, R. A., & Osburn, H. K. (2002). Planning in organizations: performance as a multi-level phenomenon. *Annual Review of Research in Multi-Level Issues, 1*, 3–65.
Naude, P., & Buttle, F. (2000). Assessing relationship quality. *Industrial Marketing Management, 29*(4), 351–361.
Oosterhuis, M. J., Mollemann, E., & Van der Vaart, T. (2005). "Multilevel issues in supply chain management. In H. Kotzab, S. Seuring, M. Müller, & G. Reiner (Eds.), *Research methodologies in supply chain management* (pp. 283–297). Heidelberg: Physica-Verlag.
Otto, A., & Kotzab, H. (2003). Does supply chain management really pay? Six perspectives to measure the performance of managing a supply chain. *European Journal of Operational Research, 144*(2), 306–320.
Palmatier, R. W., Dant, R. P., & Grewal, D. (2007). A comparative longitudinal analysis of theoretical perspectives of interorganizational relationship performance. *Journal of Marketing, 71*(4), 172–194.
Parsons, A. L. (2002). What determines buyer–seller relationship quality? an investigation from the buyer's perspective. *Journal of Supply Chain Management, 38*(2), 4–12.
Petersen, K., Ragatz, G., & Monczka, R. (2005). An examination of collaborative planning effectiveness and supply chain performance. *Journal of Supply Chain Management, 41*(2), 14–25.
Poundarikapuram, S., & Veeramani, D. (2004). Distributed decision-making in supply chains and private e-marketplaces. *Production and Operations Management Journal, 13*(1), 111–121.
Ramdas, K., & Spekman, R. E. (2000). Chain or shackles: Understanding what drives supply chain performance. *Interfaces, 30*(4), 3–21.
Rosson, P. J., & Ford, I. D. (1982). Manufacturer overseas distributor relations and export performance. *Journal of International Business Studies, 13*(2), 57–72.
Schank, R. C., & Abelson, R. P. (1977). *Scripts, plans, goals and understanding: An inquiry into human knowledge structures*. Hillsdale, NJ: Lawrence Erlbaum.
Scheer, L. K., Kumar, N., & Steenkamp, J. (2003). Reactions to perceived inequity in US and Dutch inter-organizational relationships. *Academy of Management Journal, 46*(3), 303–316.
Seifert, D. (2002). *Collaborative planning forecasting and replenishment: How to create a supply chain advantage*. Kevelaer: Galileo Press GmbH.
Settoon, R., & Mossholder, K. (2002). Relationship quality and relationship context as antecedents of person- and task-focused interpersonal citizenship behavior. *Journal of Applied Psychology, 87*(2), 255–267.
Shepherd, C., & Günter, H. (2006). Measuring supply chain performance: current research and future directions. *International Journal of Productivity and Performance, 55*(3/4), 242–258.

Smith, J. B. (1998). Buyer-seller relationships: Similarity, relationship management, and quality. *Psychology and Marketing, 15*(1), 3–21.

Sogomonian, A., & Tang, C. (1993). A modeling framework for coordinating promotion and production decisions within a firm. *Management Science, 39*(2), 191–203.

Sterman, J. D. (1989). Modeling managerial behavior: Misperceptions of feedback in a dynamic decision making experiment. *Management Science, 35*(3), 321–339.

Storbacka, K., Strandvik, T., & Gronroos, C. (1994). Managing customer relationships for profit – the dynamics of relationship quality. *International Journal of Service Industry Management, 5*(5), 21–38.

Suchman, L. A. (1987). *Plans and situated actions: The problem of human-machine communication*. Cambridge: Cambridge University Press.

Suchman, L. A. (2007). *Human–machine reconfigurations*. Cambridge: Cambridge University Press.

Tapiero, C. S. (2001). Yield and control in a supplier-customer relationship. *International Journal of Production Research, 39*(7), 1505–1515.

Terwiesch, C., Ren, Z. J., Ho, T. H., & Cohen, M. A. (2005). An empirical analysis of forecast sharing in the semiconductor equipment supply chain. *Management Science, 51*(2), 208–220.

Thompson, J. D. (1967). *Organizations in action*. New York: McGraw-Hill.

Thompson, L., & Walker, A. (1982). The dyad as a unit of analysis: Conceptual and methodological issues. *Journal of Marriage and the Family, 44*, 889–900.

Toni, A. D., & Nassimbeni, G. (1999). Buyer-supplier operational practices, sourcing policies and plant performances: results of an empirical research. *International Journal of Production Research, 37*(3), 597–619.

Van der Vaart, T., & Van Donk, D. P. (2008). A critical review of survey-based research in supply chain integration. *International Journal of Production Economics, 111*(1), 42–55.

Van De Ven, A. H., Delbecq, L. A., & Koenig, R. J. (1976). Determinants of coordination modes within organizations. *American Sociological Review, 41*, 322–338.

Van Wezel, W. M. C. (2001). *Tasks, hierarchies, and flexibility; planning in food processing industries*. PhD Thesis, University of Groningen, Labyrinth Publications, Capelle a/d IJssel.

Venkataramani, V., & Dalal, R. S. (2007). Who helps and harms whom? Relational antecedents of interpersonal helping and harming in organizations. *Journal of Applied Psychology, 92*(4), 952–966.

Walter, A., Muller, T. A., Helfert, G., & Ritter, T. (2003). Functions of industrial supplier relationships and their impact on relationship quality. *Industrial Marketing Management, 32*(2), 159–169.

Wang, J., Jingdong, J., & Takahashi, K. (2005). A study on the impact of uncertain factors on information distortion in supply chains. *Production Planning & Control, 16*(1), 2–11.

Weingart, L. R. (1992). Impact of group goals, task component complexity, effort, and planning on group performance. *Journal of Applied Psychology, 77*, 682–693.

Whitten, D., & Leidner, D. (2006). Bringing IT back: An analysis of the decision to backsource or switch vendors. *Decision Sciences, 37*(4), 605–621.

Windischer, W., & Grote, G. (2003). Success factors for collaborative planning. In S. Seuring, M. Muller, M. Goldbach, & U. Schneidewind (Eds.), *Strategy and organization in supply chains* (pp. 129–146). Heidelberg: Physica-Verlag.

Windischer, A., Grote, G., Mathier, F., Meunier Martins, S., & Glardon, R. (2008). Characteristics and organizational constraints of collaborative planning. *Cognition Technology and Work, 11*(2), 87–101.

Woo, K., & Ennew, C. T. (2004). Business-to-business relationship quality: An IMP interaction-based conceptualization and measurement. *European Journal of Marketing, 38*(9/10), 1252–1271.

Chapter 6
Measuring Supply Chain Performance: Current Research and Future Directions

Craig Shepherd and Hannes Günter

Abstract This chapter aims to go some way towards addressing the dearth of research into performance measurement systems and metrics of supply chains by critically reviewing the contemporary literature and suggesting possible avenues for future research. The article provides a taxonomy of performance measures followed by a critical evaluation of measurement systems designed to evaluate the performance of supply chains. The chapter argues that despite considerable advances in the literature in recent years, a number of important problems have not yet received adequate attention, including: the factors influencing the successful implementation of performance measurement systems for supply chains; the forces shaping their evolution over time; and, the problem of their ongoing maintenance. The chapter provides both a taxonomy of measures and outlines specific implications for future research.

6.1 Introducing Supply Chain Management

Market globalization, intensifying competition and an increasing emphasis on customer orientation are regularly cited as catalysing the surge in interest in supply chain management (e.g., Gunasekaran et al. 2001; Webster 2002). Against this backdrop, effective supply chain management is treated as key to building a

---

This is a reprint of the paper Measuring supply chain performance: current research and future directions by Craig Shepherd and Hannes Günter published in 2006 in The International Journal of Productivity and Performance Management Vol. 55 No. 3/4, pp. 242–258. The paper was awarded the Emerald Outstanding Paper Award for Excellence 2007.

C. Shepherd (✉)
Nottingham University Business School, Nottingham, United kingdom
e-mail: craig.shepherd@nottingham.ac.uk

H. Günter
Department of Organization and Strategy, Maastricht University, Maastricht, The Netherlands
e-mail: h.guenter@maastrichtuniversity.nl

J.C. Fransoo et al. (eds.), *Behavioral Operations in Planning and Scheduling*,
DOI 10.1007/978-3-642-13382-4_6, © Springer-Verlag Berlin Heidelberg 2011

sustainable competitive edge through improved inter and intra-firm relationships (Ellinger 2000). Supply chains comprise all activities associated with the flow and transformation of goods from the raw material stage through to the end user (Handfield and Nichols 1999). A range of benefits have been attributed to supply chain management, including reduced costs, increased market share and sales, and solid customer relations (Fergueson 2000). However, there is some evidence to suggest this may be hyperbole rather than organizational reality. For example, Deloitte Consulting reported that only 2% of North American manufacturers ranked their supply chains as world class, despite 91% viewing supply chain management as important, or critical, to organizational success (Thomas 1999). Similarly, an international study of modern manufacturing practices reported moderate uptake and perceived effectiveness of supply chain management (Clegg et al. 2002). In view of these modest levels of uptake and effectiveness, one would expect interest in developing measurement systems and metrics for evaluating supply chain performance to be burgeoning. Moreover, it has been argued that measuring supply chain performance can facilitate a greater understanding of the supply chain, positively influence actors' behaviour, and improve its overall performance (Chen and Paulraj 2004, p145). However, as we discuss shortly, until recently this topic has received little attention and significant gaps remain in the literature. In Sect. 6.2, we offer a definition of performance measurement before moving on to outline the methods used in this study.

6.2 Introducing Performance Measurement

Neely et al. (1995) define performance measurement as the process of quantifying the effectiveness and efficiency of action. Effectiveness is the extent to which a customer's requirements are met and efficiency measures how economically a firm's resources are utilised when providing a pre-specified level of customer satisfaction. Performance measurement systems are described as the overall set of metrics used to quantify both the efficiency and effectiveness of action. Neely et al. (1995) identify a number of approaches to performance measurement, including: the balanced scorecard (Kaplan and Norton 1992); the performance measurement matrix (Keegan et al. 1989); performance measurement questionnaires (Dixon et al. 1990); criteria for measurement system design (Globerson 1985); and, computer aided manufacturing approaches. Moreover, they highlight a range of limitations of existing measurement systems for manufacturing, including: they encourage short termism; they lack strategic focus (the measurement system is not aligned correctly with strategic goals, organization culture or reward systems); they encourage local optimisation by forcing managers to minimise the variances from standard, rather than seek to improve continually; and, they fail to provide adequate information on what competitors are doing through benchmarking.

The excellent overview of performance measurement provided by Neely et al. (1995) has been widely cited in recent research into supply chain performance measurement systems and metrics (e.g., Beamon 1999; Beamon and Chen 2001; Gunasekaran et al. 2001, 2004). These, and other studies, have highlighted how the majority of the limitations cited by Neely and his collaborators remain salient in the case of performance measurement systems for supply chains. Moreover, they have stressed the need for new measurement systems and metrics which address these deficiencies. Whilst this represents an important step forward, this chapter argues that there is a need for reflection on contemporary research that has investigated a number of important issues. These include: the factors influencing the successful implementation of performance measurement systems (Bourne et al. 2000, 2002); the forces which shape the evolution of performance measurement systems (Kennerley and Neely 2002; Waggoner et al. 1999); and, how to maintain performance measurement systems over time so they remain aligned with dynamic environments and changing strategies (Bourne et al. 2000; Kennerley and Neely 2003). All of these issues are pertinent to performance measurement in supply chains, yet have received scant attention in the literature.

In the following sections, we outline the methodology used in this study before moving on to present a taxonomy of measures of supply chain performance. Thereafter, the chapter considers in more detail the contributions and limitations of existing research into performance measurement systems for supply chains, before offering possible avenues for future research.

6.3 Methodology

Keyword searches were performed using the online resources ISI Web of Science, Google Scholar™ and PsychINFO to identify articles published between 1990 and 2005 concerned with measurement systems and metrics for evaluating supply chain performance. A total of 362 articles were identified by searching for 'supply chain management' with 'performance' or 'performance measurement'.[1] These results were then sifted through to identify articles specifically concerned with developing performance measurement systems and metrics for supply chains. These articles were obtained and read by the authors. Although the systematic review methodology was initially considered, it was rejected as it argues that researcher bias in traditional narrative reviews can be overcome by adopting more 'explicit and rigorous processes' (Tranfield et al. 2003, p218). The problem with this positivist notion is it assumes it is possible to put aside ones theoretical commitments and step outside of rhetoric, a position robustly contested by post-modern researchers (e.g., Billig 1996).

As Neely et al. (1995) observe, performance measurement systems can be analysed at three levels: the individual metrics; the set of measures, or performance

[1]"Supply chain management" and "performance" were used to search Web of Science whilst "supply chain management" with "performance measurement" were used to interrogate PsychINFO and Google Scholar™.

Table 6.1 Key considerations for analysing a performance measurement system (content abridged from Neely et al. 1995)

Level (1, 2 or 3)	Considerations
Individual performance measures	What performance measures are used? What they are used for? How much they cost? What benefit do they provide?
Performance measurement system	Have all the appropriate elements (internal, external, financial, non-financial) been covered? Have measures which relate to the rate of improvement been introduced? Have measures which relate to the long term and short term objectives of the business been introduced? Have the measures been integrated, both vertically and horizontally? Do any of the measures conflict with one another?
Relationship with internal and external environments	Do the measures reinforce the firm's strategy? Do the measures match the organizational culture? Are they consistent with the recognition and reward structure? Do some measures focus on customer satisfaction? Do some measures focus on what the competition is doing?

measurement system as an entity; and, the relationship between the measurement system and the internal and external environment in which it operates. Some of the principal considerations they offer for analyzing performance measurement systems are illustrated in Table 6.1.

Following Neely et al. (1995) we began by analysing the first level, existing measures of supply chain performance. We did this by compiling a taxonomy of metrics from the articles we had downloaded and recent books concerned with performance measurement in supply chains. The measures were then categorized according to: their applicability to the five supply chain processes defined in the Supply Chain Operations Reference (SCOR) model (plan, source, make, deliver and return or customer satisfaction); whether they measure cost, time, quality, flexibility and innovativeness; and, whether they were quantitative or qualitative. Differentiating measures by business process is useful as it identifies measures which are appropriate at the strategic, operational and tactical levels. Distinguishing between cost and non-cost measures (time, quality, flexibility and innovativeness) is important since relying exclusively on cost indicators can produce a misleading picture of supply chain performance (Chen and Paulraj 2004). Measures of time and quality reflect the ability of a supply chain to deliver a high customer service, whilst flexibility and innovativeness indicate the ability to cope with rapid changes in demand or supply. Within the agility literature, flexibility and innovativeness are considered to be important strategic drivers of supply chain development in the future (Lee 2004; Morgan 2004). Consequently, we argue here it is important to continuously monitor supply chain performance using metrics from all five categories (cost, time, quality, flexibility and innovativeness) and act upon the performance measurement results in order to remain competitive. Finally, the quantitative and qualitative distinction highlights whether measures are objective or rely on the subjective interpretations of individual actors. Having analysed supply chain metrics, levels 2 and 3 were considered by reviewing existing performance measurement systems with special attention to their internal and external environments.

6.4 Findings

The vast majority of articles returned by the database searches could be classified as operational, design or strategic (Huang et al. 2004). Operational studies develop mathematical models for improving the performance of the supply chain (Lin et al. 2005; Smith et al. 2005), whilst design studies aim to optimize performance through redesigning the supply chain. The latter include deterministic analytical models (e.g., Chen et al. 2005), stochastic analytical models (Chiang and Monahan 2005), economic models (e.g., Wu 2005) and simulation models (e.g., Hwarng et al. 2005; Reiner 2005). Finally, strategic studies evaluate how to align the supply chain with a firm's strategic objectives (e.g., Balasubramanian and Tewary 2005). Other researchers focused on how conflict and power affected the performance of supply chain networks (e.g., Bradford et al. 2004; Krajewski et al. 2005). One important contribution of this wider literature is that it emphasizes the need to adopt a systemic approach to performance measurement. For example, modern manufacturing practices such as quality management (e.g., Flynn and Flynn 2005), just-in-time (e.g., Green and Inman 2005) and information technology (e.g., Dyapur and Patnaik 2005) have all been shown to effect overall supply chain performance.

In total, 42 journal articles and books were identified which were directly concerned with performance measurement systems and metrics for supply chains (Artz 1999; Baiman et al. 2001; Beamon 1998, 1999; Bourne et al. 2000, 2002; Cachon and Lariviere 1999; Chan 2003; Chan and Qi 2003; Chen and Paulraj 2004; Dasgupta 2003; De Toni and Tonchia 2001; Fynes et al. 2005; Graham et al. 1994; Gunasekaran et al. 2001, 2004, 2005; Harrison and New 2002; Holmberg 2000; Huang et al. 2004, 2005; Kleijnen and Smits 2003; Lai et al. 2002; Li et al. 2005a, b; Lockamy and McCormack 2004; Lohman et al. 2004; Lummus et al. 2003; Maloni and Benton 1997; Melnyk et al. 2004; Ramdas and Spekman 2000; Schmitz and Platts 2004; Stephens 2001; Talluri and Sarkis 2002; Van der Vorst and Beulens 2001; Van Hoek 2001; Wang et al. 2004, 2005; Webster 2002; Windischer 2003; Windischer and Grote 2003). Five further articles were related to benchmarking (Basnet et al. 2003; Choy and Lee 2003; Cox 2000; Ulusoy 2003; Van Landeghem and Persoons 2001). Unfortunately, we were not surprised by this finding, as it is widely acknowledged that there has been relatively little interest in developing measurement systems and metrics for evaluating supply chain performance (e.g., Beamon 1999; Gunasekaran et al. 2001). In Sect. 6.4.1, we present a taxonomy of measures, before critically evaluating existing performance measurement systems in supply chains.

6.4.1 Performance Measures in Supply Chain Management

There have been relatively few attempts to systematically collate measures for evaluating the performance of supply chains. Moreover, there is dissensus over the most appropriate way to categorise them. For example, they have been grouped according to:

- Whether they are qualitative or quantitative (Beamon 1999; Chan 2003)
- What they measure: cost and non-cost (Gunasekaran 2001; De Toni and Tonchia 2001); quality, cost, delivery and flexibility (Schönsleben 2004); cost, quality, resource utilization, flexibility, visibility, trust and innovativeness (Chan, 2003); resources, outputs and flexibility (Beamon 1999); supply chain collaboration efficiency; coordination efficiency and configuration (Hieber 2002); and, input, output and composite measures (Chan and Qi 2003)
- Their strategic, operational or tactical focus (Gunasekaran et al. 2001)
- The process in the supply chain they relate to (e.g., Chan and Qi 2003; Huang et al. 2004; Li et al. 2005b; Lockamy and McCormack 2004; Stephens 2001)

For example, Chan and Qi (2003) identify six core processes (supplier, inbound logistics, manufacturing, outbound logistics, marketing and sales, end customers) and present input, output and composite measures for each. Similarly, proponents of the supply chain operations reference (SCOR) model, which we will discuss in more detail shortly (e.g., Huang et al. 2004; Li et al. 2005b; Lockamy and McCormack 2004; Stephens 2001) argue that supply chain performance must be measured at multiple levels and assign five categories of metrics to level 1 of this model; reliability, responsiveness, flexibility, cost and efficiency indicators.

The complexity of supply chains makes collating and delineating performance metrics an onerous task. Nevertheless, Table 6.2 presents a taxonomy of measures of supply chain performance, delineated according to: the processes identified in the SCOR model: plan, source, make, deliver or return (customer satisfaction); whether they measure cost, time, quality, flexibility or innovativeness; and, whether they are quantitative or qualitative.

The overall proportion of the measures identified substantiates the argument offered by Beamon (1999) and others, that there remains a disproportionate focus on cost (42%) over non-cost measures such as quality (28%), time (19%), flexibility (10%), and innovativeness (1%). Second, there are relatively few measures concerned with the process of return, or customer satisfaction (5%), in comparison with measures of other aspects of the supply chain process such as plan (30%), source (16%), make (26%) and deliver (20%). Third, the vast majority of metrics are quantitative (82%) rather than qualitative (18%). Finally, as Lambert and Pohlen (2001, p1) observe, one of the main problems with supply chain metrics is that 'they are, in actuality, about internal logistics performance measures' and do not capture how the supply chain as a whole has performed. For example, although measures such as order fill rate are likely to be influenced by activities throughout the entire supply chain, they ultimately measure performance at the intra, rather

Table 6.2 A taxonomy of measures of supply chain performance

Stages in supply chain	Measure	Cost (C), Time (T), Quality (Q), Flexibility (F), Innovativeness (I)	Quantitative or qualitative
Plan	Sales[a]	C	QN
	Profit[b]	C	QN
	Return on investment (ratio of net profits to total assets)[b]	C	QN
	Rate of return on investment[a]	C	QN
	Net profit vs. productivity ratio[a]	C	QN
	Information carrying cost[a]	C	QN
	Variations against budget[a]	C	QN
	Total supply chain management costs[c]	C	QN
	Cost of goods sold[c]	C	QN
	Asset turns[c]	C	QN
	Value added productivity[c]	C	QN
	Overhead cost[d]	C	QN
	Intangible cost[d]	C	QN
	Incentive cost and subsides[d]	C	QN
	Sensitivity to long term costs[d]	C	QN
	Percentage sales of new product compared to whole sales for a period[d]	C	QN
	Expansion capability[d]	C	QN
	Capital tie up costs[e]	C	QN
	Total supply chain response time[c]	T	QN
	Total supply chain cycle time[a]	T	QN
	Order lead time[a,e]	T	QN
	Order fulfilment lead time[c]	T	QN
	Customer response time[b]	T	QN
	Product development cycle time[a]	T	QN
	Total cash flow time[a]	T	QN
	Cash to cash cycle time[c]	T	QN
	Horizon of business relationship[f]	T	QL
	Percentage decrease in time to produce a product[d]	T	QN
	Fill rate (target fill rate achievement & average item fill rate)[b,c,d,g]	Q	QN
	Order entry methods[a]	Q	QN
	Accuracy of forecasting techniques[a]	Q	QN
	Autonomy of planning[f]	Q	QL
	Perceived effectiveness of departmental relations[h]	Q	QL
	Order flexibility[g]	Q	QN
	Perfect order fulfilment	Q	QN
	Mix flexibility[b,d]	F	QN
	New product flexibility[b]	F	QN
	Number of new products launched[d]	I	QN
	Use of new technology[d]	I	QN
Source	Supplier cost saving initiatives[a]	C	QN
		C	QN

(*continued*)

Table 6.2 (continued)

Stages in supply chain	Measure	Cost (C), Time (T), Quality (Q), Flexibility (F), Innovativeness (I)	Quantitative or qualitative
	Percentage of late or wrong supplier delivery		
	Supplier lead time against industry norm[a]	T	QN
	Supplier's booking in procedures[a]	T	QN
	Purchase order cycle time[a]	T	QN
	Efficiency of purchase order cycle time[a]	T	QN
	Buyer-supplier partnership level[a]	Q	QL
	Level of supplier's defect free deliveries[a]	Q	QN
	Supplier rejection rate[a]	Q	QN
	Mutual trust[f]	Q	QL
	Satisfaction with knowledge transfer[i]	Q	QL
	Satisfaction with supplier relationship[j]	Q	QL
	Supplier assistance in solving technical problems[a]	Q	QL
	Extent of mutual planning cooperation leading to improved quality[k]	Q	QL
	Extent of mutual assistance leading in problem solving efforts[l]	Q	QL
	Distribution of decision competences between supplier and customer[m]	Q	QL
	Quality and frequency of exchange of logistics information between supplier and customer[m]	Q	QL
	Quality of perspective taking in supply networks[n]	Q	QL
	Information accuracy[o]	Q	QL
	Information timeliness[o]	Q	QL
	Information availability[o]	Q	QL
	Supplier ability to respond to quality problems[a]	F	QL
Make	Total cost of resources[b]	C	QN
	Manufacturing cost[b,d]	C	QN
	Inventory investment[b]	C	QN
	Inventory obsolescence[b]	C	QN
	Work in process[b]	C	QN
	Cost per operation hour[a]	C	QN
	Capacity utilization as (incoming stock level, work-in-progress, scrap level, finished goods in transit)[a,p]	C	QN
	Inventory cost[d]	C	QN

(*continued*)

Table 6.2 (continued)

Stages in supply chain	Measure	Cost (C), Time (T), Quality (Q), Flexibility (F), Innovativeness (I)	Quantitative or qualitative
	Inventory turnover ratio[p]	C	QN
	Inventory flow rate[g]	C	QN
	Inventory days of supply[c]	C	QN
	Economic order quantity[a]	C	QN
	Effectiveness of master production schedule[a]	C	QN
	Number of items produced[b]	C	QN
	Warehouse costs[g,d]	C	QN
	Stock capacity[g]	C	QN
	Inventory utilization[g]	C	QN
	Stockout probability[b,d]	C	QN
	Number of backorders[b]	C	QN
	Number of stockouts[b]	C	QN
	Average backorder level[b]	C	QN
	Percentage of excess/lack of resource within a period[d]	C	QN
	Storage costs per unit of volume[e]	C	QN
	Disposal costs[e]	C	QN
	Planned process cycle time[a]	T	QN
	Manufacturing lead time[b]	T	QN
	Time required to produce a particular item or set of items[b]	T	QN
	Time required to produce new product mix[d]	T	QN
	Inventory accuracy[g]	Q	QN
	Inventory range[e]	F	QN
	Percentage of wrong products manufactured[d]	Q	QN
	Production flexibility[c]	F	QN
	Capacity flexibility[p]	F	QN
	Volume flexibility[b,d]	F	QN
	Number of tasks worker can perform[d]	F	QN
Deliver	Total logistics costs[e]	C	QN
	Distribution costs[b,d]	C	QN
	Delivery costs[g]	C	QN
	Transport costs[g]	C	QN
	Transport costs per unit of volume[e]	C	QN
	Personnel costs per unit of volume moved[e]	C	QN
	Transport productivity[g]	C	QN
	Shipping errors[b]	C	QN
	Delivery efficiency[e]	C	QN
	Percentage accuracy of delivery[d]	C	QN
	Delivery lead time[a]	T	QN
	Frequency of delivery[a]	T	QN
	Product lateness[b]	T	QN

(*continued*)

Table 6.2 (continued)

Stages in supply chain	Measure	Cost (C), Time (T), Quality (Q), Flexibility (F), Innovativeness (I)	Quantitative or qualitative
	Average lateness of orders[b]	T	QN
	Average earliness of orders[b]	T	QN
	Percent of on time deliveries[b,d]	T	QN
	Delivery performance[a,c]	Q	QN
	Delivery reliability[a,p,c,g]	Q	QN
	Number of on time deliveries[b]	Q	QN
	Effectiveness of distribution planning schedule[a]	Q	QL
	Effectiveness of delivery invoice methods[a]	Q	QN
	Driver reliability for performance[a]	Q	QN
	Quality of delivered goods[a]	Q	QL
	Achievement of defect free deliveries[a]	Q	QN
	Quality of delivery documentation[a]	Q	QL
	Delivery flexibility[b,g]	F	QN
	Responsiveness to urgent deliveries[a,g]	F	QN
	Transport flexibility[g]	F	QN
Return	Warranty/returns processing costs[c]	C	QN
(customer	Customer query time[a]	T	QN
satisfaction)	Customer satisfaction (or dissatisfaction)[b,d]	Q	QL
	Level of customer perceived value of product[a]	Q	QL
	Customer complaints[b]	Q	QN
	Rate of complaint[p]	Q	QN
	Product quality[b,g]	Q	QL
	Flexibility of service systems to meet particular customer needs[a]	F	QL

Note. [a]Gunasekaran et al. (2001)
[b]Beamon (1999)
[c]SCOR level 1 metrics
[d]Chan (2003)
[e]VDI guidelines (association of engineers)
[f]Hieber (2002)
[g]Chan and Qi (2003)
[h]Ellinger (2000)
[i]Sperka (1997)
[j]Artz (1999)
[k]Graham et al. (1994)
[l]Maloni and Benton (1997)
[m]Windischer and Grote (2003)
[n]Parker and Axtell (2001)
[o]Van der Vorst and Beulens (2001)
[p]Schönsleben (2004)

than the inter-organizational level. However, as Chen and Paulraj (2004) point out, it is encouraging that some researchers have developed measures to assess the performance of supply chain relationships or the performance of a supply chain as a whole (e.g., Ellinger, 2000; Fynes et al. 2005; Windischer and Grote, 2003).

6.4.2 Performance Measurement Systems in Supply Chain Management

Perhaps unsurprisingly, criticisms of measurement systems designed to evaluate the performance of supply chains mirror those in the wider performance management literature (e.g., Neely et al. 1995). They include:

- Lack of connection with strategy (Beamon 1999; Chan and Qi 2003; Gunasekaran et al. 2004)
- Focus on cost to the detriment of non-cost indicators (Beamon 1999; De Toni and Tonchia 2001)
- Lack of a balanced approach (Beamon 1999; Chan 2003)
- Insufficient focus on customers and competitors (Beamon 1999)
- Loss of supply chain context, thus encouraging local optimization (Beamon 1999)
- Lack of system thinking (Chan 2003; Chan and Qi 2003)

In recent times, researchers have attempted to respond to these limitations by designing systemic and balanced performance measurements systems. Perhaps the most well known of these is the supply chain operations reference (SCOR) model alluded to earlier. This was developed by the supply chain council in 1997 and has been described as a 'systematic approach for identifying, evaluating and monitoring supply chain performance' (Stephens, 2001). Its guiding principal is that a balanced approach is crucial; single indicators (e.g., cost or time) cannot be adequately taken to measure supply chain performance, which must be measured at multiple levels. Business processes, technology and metrics are all included in the model, which provides five groups of metrics at level 1; reliability, responsiveness, flexibility, cost and efficiency. One of the main limitations of this model is that it does not offer a systematic method for prioritizing measures. However, recently there has been attempts to augment it by combining it with decision making tools such as Analytic Hierarchy Processing, or AHP (Huang et al. 2004; Li et al. 2005b). Nevertheless, there is some disagreement over whether this is the most appropriate technique for selecting measures. For example, whilst Chan (2003) advocates the use of AHP, its efficacy has recently been disputed by Chan and Qi (2003) who favour fuzzy ratios for selecting measures. In summary, there is widespread recognition of the importance of adopting a systemic and balanced approach towards designing performance measurement systems for supply chains. Moreover, in recent times, researchers have attempted to incorporate systematic techniques for selecting measures. Nevertheless, despite these advances, current research has not adequately addressed a number of important issues highlighted by contemporary developments in the wider performance measurement literature.

Firstly, the problem of how to integrate performance measurement systems with human resource management (HRM) and modern manufacturing practices such as total quality management, business process re-engineering, just-in-time, or new information technologies has not been adequately addressed. This is important since as Wood et al. (2004) observe, bundling, or combining, modern manufacturing practices can lead to statistically significant increases in success (also see Flynn and Flynn 2005). Moreover, practices such as just-in-time implicitly privilege certain metrics, which may or may not be aligned with the current strategic objectives. For example, whilst just-in-time encourages low inventory levels, this may conflict with the strategic goal of increased supply chain flexibility. Secondly, existing measurement systems for evaluating the performance of supply chains tend to be static rather than dynamic. So, whilst the need to keep measures aligned with strategy has been well rehearsed within the literature, surprisingly little attention has been paid to the problem of the ongoing management of performance measurement systems, or the forces affecting their evolution over time (Waggoner et al. 1999; Kennerley and Neely 2002, 2003). Therefore, the question of how often measures of supply chain performance should be re-evaluated and when measurement should take place has not yet been given adequate consideration. Thirdly, as Bourne et al. (2002) observe, there have been few empirical studies of the factors influencing the success or failure of attempts to implement performance measurement systems, although some researchers have attempted to address this issue in recent times (e.g., Bititci et al. 2005; Nudurupati and Bititci 2005). This is important, since as they point out implementation failure rates have been estimated at 70%. Unfortunately, this problem is even more pronounced within the supply chain literature where there is a dearth of research into change management issues surrounding their implementation. Fourthly, as highlighted earlier, relatively few researchers have attempted to benchmark the performance of supply chains, despite the repeated calls for a greater focus on competitors (e.g., Beamon 1999). Moreover, where studies have been undertaken, they have invariably been conducted in a single country and within a specific market sector (e.g., Basnet et al. 2003). Therefore, whilst these studies are undoubtedly valuable, there is a pressing need for international benchmarking of supply chain performance, in order that comparisons can be made across countries and both within and across market sectors. Finally, few researchers have explored whether the benefits of supply chain performance measurement systems are outweighed by the cost of implementing and maintaining them in increasingly dynamic business environments. This is likely to be especially pertinent for small enterprises who may lack the resources, time or information to undertake the analyses required to optimise supply chain activities (Morgan 2004). As Morgan observes, one consequence for larger enterprises interested in measuring the performance of their supply chain is they may be forced to consider the developing the capabilities of their suppliers to implement meaningful performance measurement systems.

6.5 Discussion and Implications

In this chapter we have argued that despite the burgeoning supply chain management literature, comparatively few studies have developed performance measurement systems, delineated metrics, or benchmarked supply chain practices. Moreover, we propose there has been limited reflection on important insights from the wider contemporary literature on performance measurement (e.g., Bourne et al. 2000; Bourne et al. 2002; Kennerley and Neely 2002; Kennerley and Neely 2003; Neely et al. 2000; Waggoner et al. 1999). This article has attempted to address these issues by providing a taxonomy of measures, a critical review of metrics and measurement systems used to evaluate supply chain performance, and possible avenues for future research. Nevertheless, despite these contributions, it is important to reflect upon possible limitations of the study. Perhaps the main risk is that the literature review is not exhaustive, since only three online repositories were interrogated (ISI Web of Science; Google Scholar™ and PsychINFO). Whilst they are widely regarded as an excellent data sources, other databases could have been reviewed for completeness. Furthermore, it is important to acknowledge that our introduction to performance measurement systems focuses mainly on the Operations Management literature. There is a significant literature on performance measurement systems in other areas such as strategic management (e.g., Lowe and Jones 2004), human resource management (e.g., Soltani et al. 2005) and management control systems (e.g., Van Veen-Dirks 2005).

As highlighted above, this article carries a number of implications for future research. First, researchers should consider developing measures of supply chain relationships and the supply chain as a whole, rather than measures of intra-organizational performance (Lambert and Pohlen 2001). Moreover, the paucity of qualitative metrics and non financial measures of innovativeness and customer satisfaction should also be addressed. Second, future research needs to explore more thoroughly how to design performance measurement systems which complement HRM and modern manufacturing practices, including JIT, TQM and BPR. Third, the factors influencing the success or failure of attempts to implement measurement systems for supply chains should be investigated. Fourth, it is important to treat measurement systems as dynamic entities that must respond to environmental and strategic changes. Consequently, further work is needed to investigate the factors influencing the evolution of performance measurement systems for supply chains and how to handle their ongoing maintenance. Finally, there is a need to investigate whether implementing measurement systems to evaluate supply chain performance is cost effective, especially for small and medium enterprises.

References

Artz, K. W. (1999). Buyer-supplier performance: the role of asset specificity reciprocal investments and relational exchange. *British Journal of Management, 10*, 113–126.

Baiman, S., Fischer, P. E., & Rajan, M. V. (2001). Performance measurement and design in supply chains. *Management Science, 47*(1), 173–188.

Balasubramanian, P., & Tewary, A. K. (2005). Design of supply chains: Unrealistic expectations on collaboration. *Sadhana-Academy Proceedings in Engineering Sciences, 30*, 463–473.

Basnet, C., Corner, J., Wisner, J., & Tan, K. C. (2003). Benchmarking supply chain management practice in New Zealand. *Supply Chain Management-an International Journal, 8*(1), 57–64.

Beamon, B. M. (1998). Supply chain design and analysis: Models and methods. *International Journal of Production Economics, 55*(3), 281–294.

Beamon, B. M. (1999). Measuring supply chain performance. *International Journal of Operations & Production Management, 19*(3–4), 275–292.

Beamon, B. M., & Chen, V. C. P. (2001). Performance analysis of conjoined supply chains. *International Journal of Production Research, 39*(14), 3195–3218.

Billig, M. (1996). *Arguing and thinking: a rhetorical approach to social psychology*. Cambridge: Cambridge University Press.

Bititci, U., Cavalieri, S., & von Cieminski, G. (2005). Implementation of performance measurement systems: private and public sectors. *Production Planning & Control, 16*(2), 99–100.

Bourne, M., Mills, J., Wilcox, M., Neely, A., & Platts, K. (2000). Designing, implementing and updating performance measurement systems. *International Journal of Operations & Production Management, 20*(7), 754–771.

Bourne, M., Neely, A., Platts, K., & Mills, J. (2002). The success and failure of performance measurement initiatives – Perceptions of participating managers. *International Journal of Operations & Production Management, 22*(11), 1288–1310.

Bradford, K. D., Stringfellow, A., & Weitz, B. A. (2004). Managing conflict to improve the effectiveness of retail networks. *Journal of Retailing, 80*(3), 181–195.

Cachon, G. P., & Lariviere, M. A. (1999). Capacity choice and allocation: Strategic behavior and supply chain performance. *Management Science, 45*(8), 1091–1108.

Chan, F. T. S. (2003). Performance measurement in a supply chain. *International Journal of Advanced Manufacturing Technology, 21*, 534–548.

Chan, F. T. S., & Qi, H. J. (2003). An innovative performance measurement method for supply chain management. *Supply Chain Management: An International Journal, 8*(3–4), 209–223.

Chen, H. X., Amodeo, L., Chu, F., & Labadi, K. (2005). Modelling the performance evaluation of supply chains using batch deterministic and stochastic petri nets. *IEEE Transactions on Automated Science and Engineering, 2*(2), 132–144.

Chen, I. J., & Paulraj, A. (2004). Understanding supply chain management: critical research and a theoretical framework. *International Journal of Production Research, 42*(1), 131–163.

Chiang, W. Y. K., & Monahan, G. E., (2005). Managing inventories in a two-echelon dual-channel supply chain. *European Journal of Operational Research*, 162, 325–341.

Choy, K. L., & Lee, W. B. (2003). An intelligent supplier relationship management system for selecting and benchmarking suppliers. *International Journal of Technology Management, 26*(7), 717–742.

Clegg, C. W., Wall, T. D., Pepper, K., Stride, C., Woods, D., Morrison, D., et al. (2002). An international survey of the use and effectiveness of modern manufacturing practices. *Human Factors & Ergonomics in Manufacturing, 12*, 171–191.

Cox, A. (2000). Benchmarking: A dead end for supply-chain management? *Sloan Management Review, 41*(4), 5.

Dasgupta, T. (2003). Using the six-sigma metric to measure and improve the performance of a supply chain. *Total Quality Management & Business Excellence, 14*(3), 355–366.

De Toni, A., & Tonchia, S. (2001). Performance measurement systems: models, characteristics and measures. *International Journal of Operations & Production Management, 21*(1/2), 46–70.

Dixon, J. R., Nanni, A. J., & Vollmann, T. E. (1990). *The new performance challenge-measuring operations for world class competition*. Homewood, IL: Dow Jones-Irwin.

Dyapur, K. R., & Patnaik, K. K. (2005). Transaction oriented computing (HIVE computing) using GRAM-Soft. *Computational Science, 3516*, 879–882.

Ellinger, A. E. (2000). Improving marketing/logistics cross functional collaboration in the supply chain. *Industrial Marketing Management, 29*, 85–96.

Fergueson, B. R. (2000). Implementing supply chain management. *Production and Inventory Management Journal, March*, 64–67.

Flynn, B. B., & Flynn, E. J. (2005). Synergies between supply chain management and quality management: emerging implications. *International Journal of Production Research, 43*(16), 3421–3436.

Fynes, B., de Burca, S., & Voss, C. (2005). Supply chain relationship quality, the competitive environment and performance. *International Journal of Production Research, 43*(16), 3303–3320.

Globerson, S. (1985). Issues in developing a performance criteria system for an organization. *International Journal of Production Research, 23*(4), 639–646.

Graham, T. S., Dougherty, P. J., & Dudley, W. N. (1994). The long term strategic impact of purchasing partnerships. *International Journal of Purchasing & Materials Management, 30*(4), 13–18.

Green, K. W., & Inman, R. A. (2005). Using a just-in-time selling strategy to strengthen supply chain linkages. *International Journal of Production Research, 43*(16), 3437–3453.

Gunasekaran, A., Patel, C., & Tirtiroglu, E. (2001). Performance measures and metrics in a supply chain environment. *International Journal of Operations & Production Management, 21*(1/2), 71–87.

Gunasekaran, A., Patel, C., & McGaughey, R. E. (2004). A framework for supply chain performance measurement. *International Journal of Production Economics, 87*(3), 333–347.

Gunasekaran, A., Williams, H. J., & McGaughey, R. E. (2005). Performance measurement and costing system in new enterprise. *Technovation, 25*(5), 523–533.

Handfield, R. B., & Nichols, E. L. (1999). *Introduction to supply chain management*. New Jersey: Prentice Hall.

Harrison, A., & New, C. (2002). The role of coherent supply chain strategy and performance management in achieving competitive advantage: an international survey. *Journal of the Operational Research Society, 53*(3), 263–271.

Hieber, R. (2002). *Supply Chain Management: A Collaborative Performance Measurement Approach*. Zurich: VDF.

Holmberg, S. (2000). A system perspective in supply chain measurement. *International Journal of Physical Distribution and Logistics, 30*(10), 847–868.

Huang, S. H., Sheoran, S. K., & Wang, G. (2004). A review and analysis of supply chain operations reference (SCOR) model. *Supply Chain Management: an International Journal, 9*(1), 23–29.

Huang, S. H., Sheoran, S. K., & Keskar, H. (2005). Computer assisted supply chain configuration based on supply chain operations reference (SCOR) model. *Computers & Industrial Engineering, 48*(2), 377–394.

Hwarng, H. B., Chong, C. S. P., Hie, N., & Burgess, T. F. (2005). Modelling a complex supply chain: understanding the effect of simplified assumptions. *International Journal of Production Research, 43*(13), 2829–2872.

Kaplan, R. S., & Norton, D. P. (1992). The balanced scorecard: measures that drive performance. *Harvard Business Review, 70*(1), 71–9.

Keegan, D. P., Eiler, R. G., & Jones, C. R. (1989). Are your performance measures obsolete? *Management Accounting, June*, 134–147.

Kennerley, M., & Neely, A. (2002). A framework of the factors affecting the evolution of performance measurement systems. *International Journal of Operations & Production Management, 22*(11), 1222–1245.

Kennerley, M., & Neely, A. (2003). Measuring performance in a changing business environment. *International Journal of Operations & Production Management, 23*(2), 213–229.

Kleijnen, J. P. C., & Smits, M. T. (2003). Performance metrics in supply chain management. *Journal of the Operational Research Society, 54*(5), 507–514.

Krajewski, L., Wei, J. C., & Tang, L. L. (2005). Responding to schedule changes in build-to-order supply chains. *Journal of Operations Management, 23*(5), 452–469.

Lai, K. H., Ngai, E. W. T., & Cheng, T. C. E. (2002). Measures for evaluating supply chain performance in transport logistics. *Transportation Research Part E-Logistics and Transportation Review, 38*(6), 439–456.

Lambert, D. M., & Pohlen, T. L. (2001). Supply chain metrics. *The International Journal of Logistics Management, 12*(1), 1–19.

Lee, H. L. (2004). The triple-A supply chain. *Harvard Business Review, 82*(10), 102–13.

Li, G., Yan, H., Wang, S. Y., & Xia, Y. S. (2005a). Comparative analysis on value of information sharing in supply chains. *Supply Chain Management-an International Journal, 10*(1), 34–46.

Li, S., Subba Rao, S., Ragu-Nathan, T. S., & Ragu-Nathan, B. (2005b). Development and validation of a measurement instrument for studying supply chain management practices. *Journal of Operations Management, 23*, 618–641.

Lin, F. R., Sung, Y. W., & Lo, Y. P. (2005). Effects of trust mechanisms on supply-chain performance: A multi-agent simulation study. *International Journal of Electronic Commerce, 9*(4), 91–112.

Lockamy, A., & McCormack, K. (2004). Linking SCOR planning practices to supply chain performance: An exploratory study. *International Journal of Operations & Production Management, 24*(11–12), 1192–1218.

Lohman, C., Fortuin, L., & Wouters, M. (2004). Designing a performance measurement system: A case study. *European Journal of Operational Research, 156*(2), 267–286.

Lowe, A., & Jones, A. (2004). Emergent strategy and the measurement of performance: The formulation of performance indicators at the microlevel. *Organization Studies, 25*(8), 1313–1337.

Lummus, R. R., Duclos, L. K., & Vokurka, R. J. (2003). The impact of marketing initiatives on the supply chain. *Supply Chain Management: an International Journal, 8*(3–4), 317–323.

Maloni, M. J., & Benton, W. C. (1997). Supply chain partnerships: opportunities for operations research. *European Journal of Operations Research, 101*, 419–429.

Melnyk, S., Stewart, D. M., & Swink, M. (2004). Metrics and performance measures in operations management:dealing with the metrics maze. *Journal of Operations Management, 22*, 209–217.

Morgan, C. (2004). Structure, speed and salience: performance measurement in the supply chain. *Business Process Management Journal, 10*(5), 522–536.

Neely, A., Gregory, M., & Platts, K. (1995). Performance measurement systems design: a literature review and research agenda. *International Journal of Operations & Production Management, 15*(4), 80–116.

Neely, A., Mills, J., Platts, K., Richards, H., Gregory, M., Bourne, M., et al. (2000). Performance measurement system design: developing and testing a process-based approach. *International Journal of Operations & Production Management, 20*(9–10), 1119–1145.

Nudurupati, S. S., & Bititci, U. S. (2005). Implementation and impact of IT-supported performance measurement systems. *Production Planning & Control, 16*(2), 152–162.

Parker, S., & Axtell, C. M. (2001). Seeing another viewpoint: outcomes and antecedents of employee perspective taking activity. *Academy of Management Journal, 44*(6), 1085–100.

Ramdas, K., & Spekman, R. E. (2000). Chain or shackles: Understanding what drives supply-chain performance. *Interfaces, 30*(4), 3–21.

Reiner, G. (2005). Customer-oriented improvement and evaluation of supply chain processes supported by simulation models. *International Journal of Production Economics, 96*(3), 381–395.

Schmitz, J., & Platts, K. W. (2004). Supplier logistics performance measurement: indications from a study in the automotive industry. *International Journal of Production Economics, 89*(2), 231–243.

Schönsleben, P. (2004). *Integral Logistics Management: Planning and Control of Comprehensive Supply Chains*. Boca Raton, FL: St Lucie Press.

Smith, M. F., Lancioni, R. A., & Oliva, T. A. (2005). The effects of management inertia on the supply chain performance of produce-to-stock firms. *Industrial Marketing Management, 34*(6), 614–628.

Soltani, E., Van der Meer, R., & Williams, T. M. (2005). A contrast of HRM and TQM approaches to performance management: some evidence. *British Journal of Management, 16*, 211–230.

Sperka, M. (1997). Zur Entwicklung eines Fragebogens der Kommunikation in Organisationen. *Zeitschrift für Arbeits- und Organisationspsychologie, 41*(4), 182–90.

Stephens, S. (2001). Supply chain operations reference model version 5.0: a new tool to improve supply chain efficiency and achieve best practice. *Information Systems Frontiers, 3*(4), 471–476.

Talluri, S., & Sarkis, J. (2002). A model for performance monitoring of suppliers. *International Journal of Production Research, 40*(16), 4257–4269.

Thomas, J. (1999). Why your supply chain doesn't work. *Logistics Management and Distribution Report, 38*(6), 42–44.

Tranfield, D., Denyer, D., & Smart, P. (2003). Towards a methodology for developing evidence-informed management knowledge by means of systematic review. *British Academy of Management, 14*, 207–222.

Ulusoy, G. (2003). An assessment of supply chain and innovation management practices in the manufacturing industries in Turkey. *International Journal of Production Economics, 86*(3), 251–270.

Van der Vorst, J., & Beulens, A. (2001). Identifying sources of uncertainty to generate supply chain redesign strategies. *International Journal of Physical Distribution and Logistics, 32*(6), 409–430.

Van Hoek, R. I. (2001). The contribution of performance measurement to the expansion of third party logistics alliances in the supply chain. *International Journal of Operations and Production Management, 21*(1/2), 15–29.

Van Landeghem, R., & Persoons, K. (2001). Benchmarking of logistical operations based on a causal model. *International Journal of Operations & Production Management, 21*(1–2), 254–266.

Van Veen-Dirks, D. (2005). Management control and production environment. *International Journal of Production Economics, 93*(4), 263–272.

Waggoner, D. B., Neely, A. D., & Kennerley, M. P. (1999). The forces that shape organisational performance measurement systems: An interdisciplinary review. *International Journal of Production Economics, 60/61*, 53–60.

Wang, F. K., Du, T. C., & Li, E. Y. (2004). Applying six-sigma to supplier development. *Total Quality Management & Business Excellence, 15*(9–10), 1217–1229.

Wang, G., Huang, S. H., & Dismukes, J. P. (2005). Manufacturing supply chain design and evaluation. *International Journal of Advanced Manufacturing Technology, 25*, 93–100.

Webster, M. (2002). Supply system structure, management and performance: a conceptual model. *International Journal of Management Reviews, 4*(4), 353–369.

Windischer, A. (2003). "*Kooperatives Planen*". Dissertation, University of Zurich, Zurich.

Windischer, A., & Grote, G. (2003). Success factors for collaborative planning. In S. Seuring, M. Muller, & M. Goldbach (Eds.), *Strategy and organization in supply chain* (pp. 131–146). Heidelberg: Physica.

Wood, S. J., Stride, C., Wall, T. D., & Clegg, C. W. (2004). Revisiting the use and effectiveness of modern manufacturing practices. *Human Factors & Ergonomics in Manufacturing, 14*(4), 415–32.

Wu, J. H. (2005). Quantity flexibility contracts under Bayesian updating. *Computers & Operations Research, 32*(5), 1267–1288.

Chapter 7
Planning Information Processing along the Supply-Chain: A Socio-Technical View

Bernard Grabot, Stefan Marsina, Anne Mayère, Ralph Riedel, and Peter Williams

Abstract The increasing focus of manufacturing companies on their core business results in the development of larger and more complex supply chains which are made up of both large companies and small companies. From the perspective of large companies, the quality of coordination in Supply Chains is mainly dependent upon the competence of the human planners involved at each level of the Supply Chain in information processing methods and tools. In this interpretation, coordination problems are commonly considered to be the consequence of poor competence in the SMEs regarding information processing, and especially in operations planning and control. As a consequence, most large companies have launched programs disseminating "best practice", standardised business processes, and software tools aimed at developing efficient planning procedures throughout their supply chains, and so convergence of operational objectives. However, this is picture does not take into account other inherent aspects of the planning problem within a supply chain, such as variety in organisational culture and ways of gathering and interpreting information. The aim of this Chapter is to illustrate the

B. Grabot (✉)
ENIT, Université de Toulouse, Toulouse, France
e-mail: bernard@enit.fr

S. Marsina
University of Economics in Bratislava, Bratislava, Slovakia
e-mail: marsina@euba.sk

A. Mayère
University of Toulouse, Toulouse, France
e-mail: anne.mayere@iut-tlse3.fr

R. Riedel,
Department of Factory Planning and Factory Management, Chemnitz University of Technology, Chemnitz, Germany
e-mail: ralph.riedel@mb.tu-chemnitz.de

P. Williams
University of Limerick, Limerick, Ireland
e-mail: peter.williams@ul.ie

J.C. Fransoo et al. (eds.), *Behavioral Operations in Planning and Scheduling*,
DOI 10.1007/978-3-642-13382-4_7, © Springer-Verlag Berlin Heidelberg 2011

problematic situation outlined above through real industrial examples, and to suggest a framework allowing better representation and understanding of these coordination problems, to inform future system design and improvement activity. It will be emphasised that critical non-technical issues have to be taken into account in the planning process across a Supply Chain. We link these to justified practice and cultural concerns of particular interest to SMEs, a voice not well represented in the literature.

7.1 Introduction

The increasing focus of manufacturing companies on their core business results in the development of larger and more complex supply chains which are made up of both large companies and small companies. The combination of large number and range in size, especially the proliferation of small and medium-sized companies (SMEs), poses challenges for efficient management of these supply chains to ensure that each partner is able to coordinate its activity with those of customers and suppliers. Developments in Information systems and information technologies (IS/IT) are naturally central to enabling this coordination, but they also go hand-in-hand with changes in organisational management which present new opportunities for destructive conflict.

From the perspective of large companies, the quality of the coordination is mainly dependent upon the competence of the human planners involved at each level of the Supply Chain in information processing methods and tools. In this interpretation, coordination problems are commonly considered to be the consequence of poor competence in the SMEs regarding information processing, and especially in operations planning and control. As a consequence, most large companies have launched programs disseminating "best practice", standardised business processes, and software tools aimed at developing efficient planning procedures throughout their supply chains, and so convergence of operational objectives. However, this is an incomplete picture, and does not take into account other inherent aspects of the planning problem within a supply chain, such as variety in organisational culture and ways of gathering and interpreting information.

The aim of this Chapter is to illustrate the problematic situation outlined above through real industrial examples, and to suggest a framework allowing better representation and understanding of these coordination problems, to inform future system design and improvement activity. Through cases in the aeronautical sector in France and in the automotive sector in Germany and Slovakia, it will be emphasised that critical non-technical issues have to be taken into account in the planning process across a Supply Chain. We link these to justified practice and cultural concerns of particular interest to SMEs, a voice not well represented in the literature. Indeed, the dissemination of standard processes by large companies is often considered by SMEs as a means of increased but inappropriate control over them, and fails to consider their cultural and organisational specificities, and can result in less than optimal performance and loss of economic opportunity.

To overcome this, we believe it is necessary to consider the problem from two perspectives: from social and human sciences on the one hand, and from information technology on the other. Both points of view will be considered in this Chapter, with respect to the different, and potentially complementary, approaches of these communities of expertise. It is not expected in this Chapter to achieve a unification of technical and sociological approaches, but to reach a better understanding of the issues underlying the problems considered, assisted by this duality of perspective to which the problems readily resonate. The task in hand is undoubtedly a pre-requisite for addressing very real needs of coordination in supply chains, with respect to the organisational, communication and relational practices of each partner. In particular, we will show that it is important towards that end to understand how small companies, their managers, planners, and the persons who have to interact with the customer, perceive, understand and deal with time management in distributed entities, and how it influences their relationship with business partners.

We feel that there is a need for a construct to capture and act as a flag-of-convenience for development activity in this respect. We suggest, tentatively, alignment with "organisational interoperability" (e.g. of an SME) as offering to captures key aspects such as competence in internal cohesion, external relatability (e.g. with larger companies), and "chameleon-like" adaptation to fit with different emerging customer requirements, but also recognition and respect for a boundary to external control.

In the discussion that follows, we start the thread of our argument with a brief theoretical perspective on how process planning is commonly presented as enabling coordination of partners within a supply chain, i.e. a high-level view on the foundations of accepted wisdom in the domain. We then compare these findings with real coordination practices in Aeronautical and Automotive supply chains based on field research, and highlight some limitations to the applicability of accepted wisdom in practice. From there we proceed to draw out particular socio-technical issues which influence the processes of coordination/collaboration. We reflect on their influence on information sharing and processing capabilities all along the Supply Chain, and especially how this results in conflict between the various parties concerned, culminating in a framework that represents the sometimes contradictory interactions involved. This is aimed at improving the analysis of the coordination practices in Supply Chains, integrating both the technical and human issues in the planning process, especially as regards "organisational interoperability" of small companies.

7.2 Process Planning in Manufacturing Supply Chains

A supply chain consists of a network of organisations, usually legally separated, involved through upstream and downstream contract linkages, with each organisation performing different processes and activities that produce value in the form of products and services provided to the ultimate consumer (after Christopher 1998; Mentzer et al. 2000). A characteristic of successful Supply Chain Management (SCM) is the quality of the coordination and integration of all the activities associated with processing products, from the raw material stage through to the end user (Shore 2001). These activities include systems management, sourcing and procurement, production planning and scheduling, order processing, inventory management, transportation, warehousing, and customer service (Christopher 1998; Romano 2003; Lee and Ng 1997; Stock et al. 1998). The objective of Supply Chain Management is to achieve and ensure competitive supply chain performance by coordinating the organisational units involved, and by integrating material, information and financial flow as well as planning efforts (Freeman and Browne 2004; Pontrandolfo et al. 2002).

Centralisation/de-centralisation and autonomy are key issues. The coordination of a Supply Chain can be performed in a centralised way by the leading partner. This is often the final assembler in the case of complex products like cars and aircraft. To that end, advantage can be taken by using tools like Advanced Planning Systems (APS) (Stadtler and Kilger 2000), which yields a global "optimal" plan covering the actions of all the partners in the chain, often in great detail. Nevertheless, such a centralised approach is in tension with desires for autonomy in the various partners. They may not wish to share confidential information like internal costs or internal cycle times for example. Moreover, companies usually belong to several Supply Chains, which can lead to contradictory constraints when internal manufacturing resources have to be shared between different planning systems. Finally, a key point of the relationship between a company and its sub-contractors is that each sub-contractor is responsible for the whole product he delivers, thus subsuming the activities of his own sub-contractors and suppliers, concerning not only quality but also due dates. This responsibility is clearly not consistent with imposition of a planning process by an external entity as assumed in a centralised framework. Therefore, a point-to-point relationship (each company being responsible of the work of its tier 1 partners) is still the main way to co-ordinate those partners in a Supply Chain who wish to remain independent.

In this general context, coordination is understood in many different ways. At a generic level, an accepted definition of coordination is given by Malone (1987) as a pattern of decision making and communication among a set of actors who perform tasks to achieve goals. Integration aims at bridging boundaries between functions inside a company, and between companies in the case of supply chains. The understanding of the coordination concept can help at two levels of operational activity: at process level, to help managers in their operational decision-making elaborate appropriate solutions, and structurally, to help systems designers define to what extent such action / interventions should pass through organisational boundaries between functions and between companies (Romano 2003). Notions of degree in sharing of mental models of behaviour and goal development in collective work has lead to more refined distinctions between coordination, cooperation and at the highest level collaboration, which are coming of increasing prominence in understanding facility and supply-chain PSC (Nezamirad et al. 2004). For example one might have commissioned a sophisticated level of technical integration, but not achieve in action strong possibilities afforded by technological innovation.

In addition to APS already discussed, tools used to co-ordinate companies in a Supply Chain include Enterprise Resource Planning systems (ERP) (Kelle and Akbulut 2005; Grabot et al. 2008), Web Portals (Carlsson and Hedman 2004) and, at a more global level, the Internet (Garcia-Dastugue and Lambert-Douglas 2003). These technologies are used to manage flows of business and process information and transactions, but do not really specify what information apart from the transactional should be exchanged and how it should be processed all along the chain, especially in the planning process. In practice, the Manufacturing Resource Planning (MRPII) method (Orlicky and Plossl 1994) is usually considered as a suitable frame for providing the operational guidelines for planning-based coordination. Broadly, this works as follows. When the leading partner (usually the final assembler) is at the end of the Supply Chain, he first performs a Sales and Operation Plan (S&OP). This rough plan is built on the basis of forecasts and programmes obtained from final customers, and it is the input to the Master Production Schedule (MPS), which describes in time and place what the company aims at manufacturing according to its capacity and inventory policy. A Material Requirements Planning process (MRP) is then performed according to the bills of materials of the products, resulting in Manufacturing Orders for components internally manufactured, and Purchase Orders for sub-contracted or bought components. Dispatching may follow orders directly (push system) or may be executed by Just-In-Time (JIT) on-demand supply (pull system).

Uncertainty is inherent in the planning process, and in the MRP model it is commonly managed through the consideration of time horizons presenting different degrees of firmness: firm orders in the short term, then flexible orders in the medium term, and looser forecasts further out. This Supply Plan is then transmitted to the suppliers and sub-contractors. For each supplier, this Plan can be considered as a S&OP or a MPS depending on its level of aggregation, and should be processed in order to define the onward Supply Plan for their own suppliers/sub-contractors. An iterative process is so defined, as in Fig. 7.1. This process allows the propagation of the plan upstream in the supply chain, which clearly requires that all partners have the capacity to process the provided forecasts/orders accordingly. In turn, this implies that the following are available:

1. The data required for an MRP calculation, especially an accurate bill of materials including lead times (i.e. time to process internal manufacturing activities and delivery lead time).
2. Production management software able to process orders according to the MRPII logic.
3. Up-to-date follow-up information on both internal activities and expected deliveries.

In practice, gathering and maintaining all these data are critical and difficult for the smallest SMEs (10–50 employees) which frequently have very good technical skills and competences regarding manufacturing activities, but may be less familiar

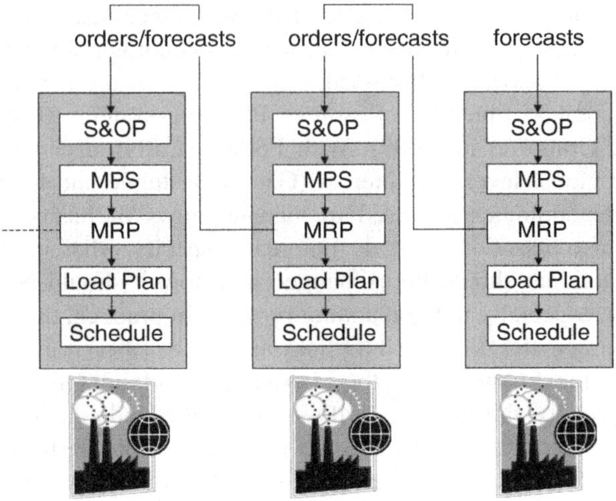

Fig. 7.1 Point-to-point coordination in a Supply-Chain using MRPII

with structured planning techniques and tools. Therefore, large companies, often under the umbrella of quality management systems and certification/accreditation, demand evidence of good usage of production management tools as a criterion for selecting their smallest partners, while for their existing sub-contractors, they mount competence-improvement programmes in production management aspects, and especially planning.

On the basis of studies and interviews of Supply Chain managers both in large companies and SMEs, we also observe in reality that the simple framework outlined above is questionable, especially in the context of coordinating the behaviour of large numbers of companies. In particular, the model is based on the assumption that all companies have the same interests, the same interpretation of information, and the same way to process it. In our opinion, this may explain some of the difficulties of obtaining good global performance in large Supply Chains involving both large and small companies. Some examples demonstrating these hypothesis are described in next section.

7.3 Case Studies from the Aeronautical and Automotive Industries

Automotive Supply Chains are highly emblematic of the contemporary manufacturing arena in general. They combine several characteristics that make their efficient management critical from the perspective of large-scale enterprise coordination. Among the main ones are: complex product with great variety, high number of partners, high imprecision in forecast demand, low financial margins, and the expectation of on-going product-service improvement together with cost reduction. As a consequence, most contemporary production management techniques (including just-in-time and lean manufacturing for instance) have their origin in this manufacturing domain. It has provided the source of inspiration for most other industrial sectors to follow, in domains as diverse as aeronautics, electronics, pharmaceuticals, and healthcare delivery; whether the translation is appropriately executed is naturally another matter, but that point is left for others to elaborate elsewhere. Therefore, when the opportunity of analysing how coordination is performed in an automotive supply chain was presented, we made it a prime focus for this study. In general, automotive supply represents high volume production of similar components (because of a high standardization level), and the system aims to progress parts flow smoothly, even at small-shop level (e.g. by using group technology cells).

Aeronautical supply chains provide an interesting and productive contrasting metaphor, being at the same time close and yet far from the Automotive context. Both sectors deal with the large-scale manufacturing of mechanical parts and their progressive assembly into complex safety-critical products, even if superficially an aircraft is physically quite different from a car. Therefore, fruitful comparisons can be made between the two sectors, yielding fundamental insights that are likely to be of interest to a wider community in other sectors of industrial enterprise.

The Aeronautical industry has its own specific characteristics which enlarge the scope of the study. These are: lower quantities of individual pieces worked in comparison with the automotive sector, higher cycle times, higher standards for quality, higher financial margins, more stable demand through longer planning horizons, but above all, coexistence of very large and very small companies in the same Supply Chains. While there are few SMEs participating in Automotive Supply chains, as a contrast we find that in the Aeronautical sector, companies of 10–50 employees are very common. This size is quite unusual in the Automotive arena, although there is a strategy to widen the supplier base through the device of production networks of small companies in many regions across Europe. Therefore, we considered comparing practices in the two domains to be a good way to make use of consistent but slightly different realities in generating useful fundamental insight.

7.3.1 *Coordination Between Large and Small Companies in the Aeronautical Industry*

The first set of examples comes from companies involved in several aeronautical Supply Chains in the South-West part of France, mainly working on a make-to-order basis. The discussion to follow is drawn from personal interviews with production and supply-chain managers, and from more general observations. Five large companies have been involved in this study, yielding as well twelve subcontractors, with a size varying from 14 to 200 employees.

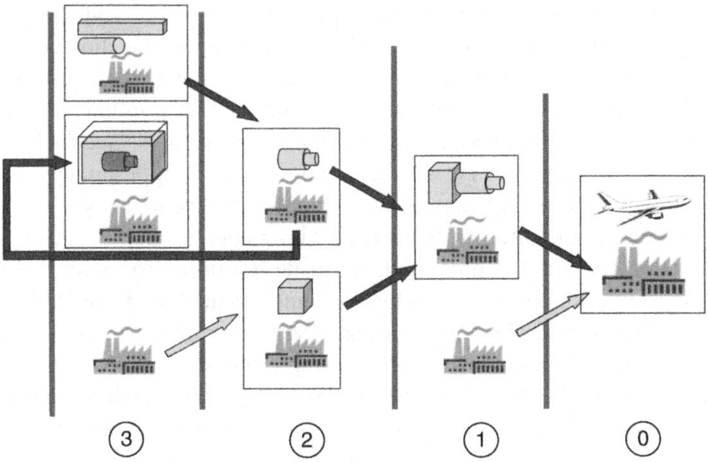

Fig. 7.2 Typical structure of an aeronautical Supply Chain

A typical supply chain in the aeronautical industry can be summarized as in Fig. 7.2. Tier 0 is a large company which designs and assembles the structure of the aircraft. Tier 1 suppliers are mainly large companies which design and manufacture large sub-systems of the aircraft, such as engine, landing gear, air circulation system, large actuators and so forth. Tier 1 companies use smaller companies at tier 2, which are often SMEs, and manufacture precision parts or simple sub-systems. Since these small companies are often specialized in precision machining, they sub-contract thermal or surface treatments to specialized companies at tier 3. The raw material providers, usually of steel plates, are also found at tier 3.

This simple structure is in due course made more complex by the fact that a company can be present at different levels of the supply chain: for instance, companies of tier 0 or 1 also use sub-contractors performing thermal or surface treatments. Therefore, these tier 3 companies are in effect resources shared by partners of the supply chain of different levels, with the inherent result of potential conflicts within the supply chain.

7.3.1.1 The Ideal Planning Process According to the Large Companies and Its Adaptation

Interviews conducted in large companies (tiers 0 and 1 of Fig. 7.2) suggested that, from their perspective, the ideal planning process through the Supply Chain follows the structure described in Fig. 7.1. In general, only tier n-1 suppliers are to be connected with level-n partners, with the only exception that large companies sometimes help their smallest partners in negotiating with their own suppliers of raw materials, which are often also large companies. Even in the event of a severe scarcity of raw materials in the aeronautical industry, some large companies refused to do this, since they considered that this strategy would make it impossible to preserve the responsibility and autonomy of their suppliers to look after themselves.

For large companies, the major problems to address are (1) to increase the reliability of deliveries and (2) to decrease cycle times. As a matter of fact, these two objectives are linked since the safety stocks required by the lack of reliability of the suppliers also results in an increase in cycle times! Similarly, the large companies nurture a suspicion that their suppliers overestimate their lead times when negotiating schedules in order to preserve some internal flexibility.

In order to fix these problems, the large companies have first of all defined rather strict criteria for selecting their partners. Compliance with quality assurance standards and certification is an example of such a criterion. Other examples of criteria include their capability in control protocols to manage their own suppliers, and their capability to manage their internal manufacturing process. If the first criterion can be easily checked, the last two are more difficult to assess. So, in practice, large companies ask their small partners for evidence of a proper use of formal production management tools. To that end, they usually insist on the definition of control loops in the medium term, by a rough checking of the capacity of the supplier and of its own suppliers, then in the short term, by providing clear milestones on each order to allow early problem detection.

Decreasing the cycle times in a continuous improvement process is a different matter, and the large companies usually consider that this can only be obtained if they can exert a coercive influence over their supplier, such as in the case where they represent a large part of the supplier's cash flow. On the other hand, they do not want their suppliers to depend too much on them, so they do not feel too much responsibility when they reduce their orders as when the general business climate becomes unfavourable. Therefore, large companies often consider that representing between 30 and 40% of the cash flow of their suppliers is an optimal value, providing influence without taking too much responsibility.

For one of the large companies, giving more responsibility to the suppliers was considered as a way to obtain better results. In that context, the company would send its present level of inventory together with its orders, so that the supplier could assess the possible consequences of specific late deliveries, and simultaneously plan its internal priorities.

In any event, the use of the ideal planning process described in Fig. 7.1 has clear limits. In particular, customers who do not belong to the same supply chain do not use the same planning representation and parameters, and this has a direct impact on partnership contracts.

A typical example is the definition of firm and flexible periods for the forecasts, which can be very different from one customer to another. A company explained for instance that some customers were sending them firm orders on a two month time horizon, whereas another one imposed on them a contract that orders could be cancelled at any time right up to the date of delivery! Most of the large companies consulted understand that small partners cannot withstand such uncertainty; therefore, they do not propagate these constraints to their suppliers, and instead send to all of them forecasts with typical firm horizons of 1–2 months, and a flexible horizon around 3 months. In that way, they protect their suppliers from demand uncertainty that they could not bear. The degree of flexibility characterising this flexible period can vary, but may reach $\pm 50\%$. In that case, some large companies give more security to their weakest partners: for instance, one of them would take delivery before the end of the year of any production launched on the basis of the flexible period forecasts, even though the corresponding order had been cancelled by the final customer.

This protection has a twofold objective: to ensure that the smallest companies do not wait until the orders are firm for preparing order fulfilment, and to strengthen partnership based on mutual benefit. Anything that increases the delay in ordering of raw materials is of particular concern, as it constitutes the main part of the overall cycle time in the aeronautical industry. Also, many large companies are acutely aware that achieving an alternative source of supply is not easy. Selecting a new supplier from scratch is a long and costly process, and in particular instances has taken several months and resulted in cost of the order of 100k€. So, stability of relationship is of prime interest.

7.3.1.2 The Ideal Planning Process According to the Small Companies and Its Adaptation

The heterogeneity of the SMEs visited during this study regarding their use of IT tools in the planning process is rather high. It is also evident that their reasoning tends to approach that of the large companies as their size increases, and vice-versa. Nevertheless, for many of them, the planning process described of Fig. 7.1 was not considered as a realistic objective, but as an illusion. It can be interpreted as a high-level model with insufficient representational granularity and associated degrees of freedom, especially at lower levels. For most of these small companies, pre-defining stable and reliable internal lead times, which is the very basis of the MRP method, is considered to be impossible since these lead times vary, especially as they respond to an ever-changing load. This reasoning is well known, and is for example the basis of the Theory Of Constraints and OPT/Drum-Buffer-Rope approaches (Goldratt and Cox 1984). Nevertheless, MRP considers that stable average lead times can be defined, and in practice this data has to be continuously updated in order to allow efficient control in the emergent situation.

If internal lead times cannot be pre-defined adequately, the first consequence is of course that finely-tuned internal milestones cannot be pre-set. Therefore, knowing whether an order is late or early usually depends on the feel and expertise of the workshop manager who has "everything in mind". The existence of this kind of "indispensable" person, whose main role is to compensate for an inadequate capability in use of the planning tool, was a common point in all the SMEs visited. The usual consequence is a high level of stress on that person and their immediate collaborators, with quite dramatic situations arising when the person is unavailable during a long period, for example due to vacation or illness, with performance outcomes deteriorating substantially.

Another interesting point is that the distinction between firm, flexible and free periods in the forecasts is not clearly present in the consciousness of all SMEs: some of them take this opportunity to grouping the lots of products which are only needed well out in the planning horizon, in a far and uncertain future. It was interesting to notice that the forecasts coming from one large company were considered as "very good" by one large SME who used a production management tool in a classical way with lots of slack, and as "poorly reliable" by another yielding hazardous order groupings in a tight schedule.

A good summary of the attitude of the SMEs is the statement by one workshop manager: "What we need in fact is to know the orders one year in advance. We can then make a good job out of it". In reality, at several companies we visited, an important objective was to optimise internal production in order to keep prices low. To that end, many small companies extract orders from their formal production management tools, and perform order groupings based on technological considerations using home-grown applications based on Excel™ or Access™. In one of the SMEs for instance, this grouping allows them to process on their lathes all orders obtained from metal bars of similar diameter, which naturally allows a drastic decrease in set-up time and costs. The drawback is that these groupings do not take into account due dates, with the hidden consequence of unnecessary advancement and earliness for some orders, with late processing and tardiness for others.

Inconsiderate pressure is commonly placed on the suppliers by the supply chain managers of the customer companies, and this is a frequent cause of friction. Order cancelling policy emerged as a key consideration. When partnership with their large customers was discussed, the message from the small companies appeared to be rather different from the one of their customers: if partnership agreements were signed, they commonly include a clause to the effect that the agreement may be broken by the large company if they can find a better price. When we mentioned this to the large companies, they answered that this clause was only a way to "motivate" the suppliers for continuous improvement, since validating and adopting a new supplier is very costly and messy in practice. Their strategy is clearly not interpreted in that benign way by small companies.

A second point follows. The SMEs mentioned the case where very urgent orders coming from their customer are finally processed with great effort; however, the orders were either not claimed by the customer or they were subsequently placed in storage by the customer for months. In reflecting with the large companies concerned, one explanation was that their own orders are processed by their ERP which adds slack times depending on many criteria which include criticality of parts, importance of end customer and so on. These slack times are not known by the supply chain managers who instinctively pressurised the suppliers. This clearly discredited the notion of urgency, with indeterminate but serious ramifications for future performance.

Finally, another example was related by one of the SMEs as follows. After ordering rather specific raw materials from a large supplier, they were not delivered on time to the SME. After many fruitless discussions, they informed their customer who replied that he could provide these raw materials. Since this was quite unusual, the supplier made a quick investigation: their larger customer did not trust their ability to source the parts, so he decided to hold a safety stock to secure supply. This problem was compounded by the customer as they obtained the parts from the very same supplier. Being more influential, delivery was prioritised to them instead of their supplier, thus creating the very problem they wanted to solve.

7.3.1.3 Interpretation of Problems

This short description of the two points of view of small supplier and large customer enables important conclusions to be drawn about the position of planning in the coordination of supply chains.

It appears that small companies do not yet have a necessary, let alone complete, understanding of the present time-based competitive context of the aeronautical industry regarding the service to be provided to the customers, after years during which price and quality were nearly the only criteria of selection of a supplier. As suspected by large companies, SMEs demonstrate a lack of understanding of some basic concepts of production management, especially in relation to lead times.

Conflicts often occur between the global objectives of the supply chain and the internal objectives of the suppliers, even if the supply chain can be the indirect origin of the problem by imposing excessive cost reductions. Similarly, large companies suffer from internal conflicts between their strategic objectives such as to decrease the costs by transferring orders to low cost countries, and their medium-term planning objectives such as to develop stable partnerships with present suppliers.

Lack of trust can be both an origin of problems and a result. This lack of trust is often the result of poor communication and mutual understanding (Rousseau et al. 1998). It is also the origin of several vicious circles: e.g., small companies do not give priority to the customers they do not fully trust, while their poor reliability lies at the root of transfers of production business abroad which further feeds mistrust and so on. Conversely, the exercise of trust by the customer can lead to increased self-responsibility and better performance. This is demonstrated in the example of the transmission of information regarding the level of inventories to a supplier.

7.3.2 Coordination Between Large and Small Companies in an Automotive Supply Chain

The following examples draw on research experiences in the Automotive sector in Germany and Slovakia, and aim to show that similar situations arise across apparently different contexts. This lends support to the hypothesis that many of these phenomena are fundamentally inherent in large enterprise networks involving many different companies.

This second set of examples consists of several companies working mainly in the German automotive industry, and includes observations from Slovakian suppliers to German customers.

In this case, the opportunity of examining two manufacturing contexts was presented, and we took it: the first consists of the direct producers of cars and the parts to go into them, the traditional product supply-chain (supply pyramid), and the second consists of suppliers of engineering capital equipment and tooling that are used to make the cars and parts and are based on mechanical and plant engineering as well as production of tools and dies. These are now described.

7.3.2.1 Traditional Product Supply Chain (Supply Pyramid)

In the automotive industry, companies have also attempted to concentrate their businesses on those activities which they know best, dealing with their core competencies. In this strategic model, all other activities are outsourced to specialist producers. This is fundamental for the creation of supply chain. This has resulted in the situation whereby it is expected that the percentage of value added in a car by the Original Equipment Manufacturers (OEMs) will move from nearly 30% today to below 20% by 2010-15 (see Dudenhoeffer 2003, also VDA 2005). This compares with roughly 40% of valued added in the Aeronautical industry. This proportion is set to continue to decrease in the next few years, with implications for supply base challenges and opportunities, as discussed in more detail below.

The traditional supply chain in the automotive industry consists of several levels, as shown in Fig. 7.3.

At the top of the pyramid, tier 0, are the Original Equipment Manufacturers that assemble and sell the cars. Suppliers at tiers 1 and 2 produce the modules and the systems that go to make up a car. The suppliers of single parts and components are at level 3 (see for example International Business Development Corporation 2002). Usually material flow connects neighbouring levels - that means tier 3 suppliers deliver to tier 2 suppliers, tier 2 suppliers deliver to tier 1 suppliers, and the module suppliers supply the OEM. As observed in the case of the Aeronautical industry, the real situation is more confused in practice: there are in addition some suppliers that supply directly to upper levels, skipping one or more levels. Also, what can not be seen in Fig. 7.1 for instance is the class of logistics service providers which are responsible for the activity in collecting, picking, JIT delivery and shipping directly to the line.

While, for instance, each aircraft can be considered as unique, and its progress is managed as a specific project, the diversity of cars is obtained by combination of standard parts, using options and variations. To that end, the structure of a car is made of modules, in turn broken down into sub-modules or systems; the systems consist of parts. Each level of the structure, modules, systems, or parts, is associated with a level of the supply pyramid. This similarity between the Supply Chain structure and a generic bill of materials for the product, which is perhaps less

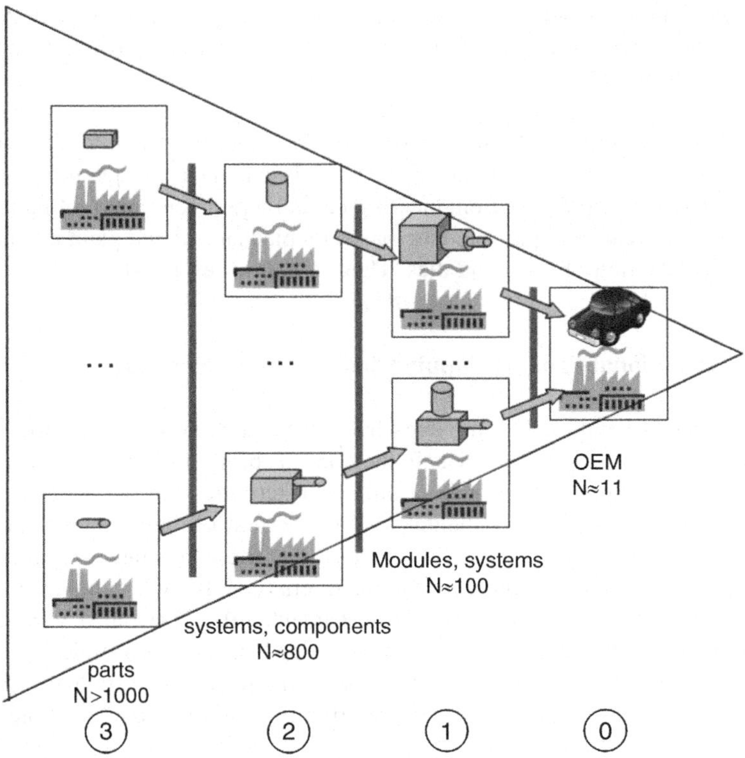

Fig. 7.3 Supply pyramid in the automotive industry

formal in Aeronautical Supply Chains, is illustrated in Fig. 7.4 (see for instance Marktl 2006).

When looking at some trends in the automotive industry globally, not only in Germany and Slovakia, the situation in relation to cooperation between partners can be summarised as follows. As stated above, OEMs are increasingly off-loading the proportion of value added on to suppliers, especially at tiers 1 and 2. As a consequence, coordination with partners becomes a vital competitive and survival issue, and of course a source of opportunity.

OEMs are also shifting design and development activities to their suppliers, so that in the future, tier 1 and tier 2 suppliers will be more completely responsible for the development and design of whole modules. In this model, the OEM provides design guidelines and sets the framework, for instance by specifying the external product shape, the look and feel factors, and technical performance. This could result in having less and less "small" SMEs involved in Automotive Supply Chains at tiers 1 and 2, since these types of companies have seldom high competences on their own for designing sub-systems.

It is expected that the module concept will be extended so that suppliers, especially at level 1 and 2, are responsible for the full set of product management

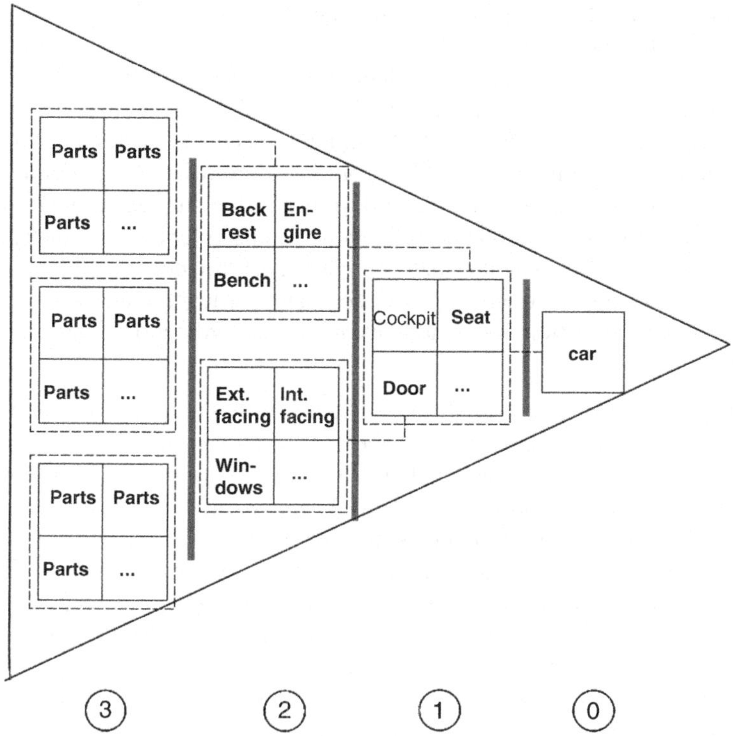

Fig. 7.4 Module concept

activity life cycle management, JIT delivery, project management, supplier management, complexity management, and even for pre-financing of new product and process innovation. This points to a Supply Chain managing flows of components more as continuous flow in the automotive sector, rather than the larger discrete items we find in the Aeronautical case. Therefore, it is expected than JIT dispatching and flow control will be remain in strong complementarity with MRP planning, while MRP is still the basic model for Aeronautical Supply Chains. In that context, it can be expected that the various partners of the chain will seek more autonomy on product design and product flow management.

When we look at the present planning process within the traditional Automotive supply chain, different roles and actors can be identified.

(a) At the customer level: product planning, who define the product, also the module; technology planning, who define how the product is made and also how the modules are assembled to become the whole car; a purchasing department, who selects the supplier and in some cases also the sub-suppliers, i.e. the suppliers for their suppliers; a production planner, who defines the amount of products that are to be produced in and their time-line; a logistics department, who is responsible for serving the production plan in relation to movement, i.e. that the production plan is communicated to the suppliers and that they deliver to the due times.

(b) At the supplier level: the product planner, who transforms the requirements of the customer into the design of the module or part; the technology planner, who defines the technology by which the module or part is made, the purchasing department, which may select the suppliers according to the guidelines of the customer, and negotiates contracts and so forth; a logistics department, which is responsible for the Just-in-Time delivery to the customer and is also responsible for product quality and due date delivery performance of the suppliers; a production department, with a production manager/team-leader who is responsible for the production of the modules or parts according to the demands of the customer.

There are many degrees and levels of relations between the different actors, within the organisation and between the organisations, and we are especially concerned with the relations across boundaries. Usually, the planning process is quite hierarchical, as illustrated in Fig. 7.3, which means that similarly to the Aeronautical industry, the OEM gives their planning data to the 1st tier supplier, and the 1st tier supplier receives and transforms this information, giving in turn their own planning data to their supplier which is located at level 2. The suppliers usually get a rough forecast for a longer period of time which is firmed up in one or even several steps as the impending production date is approached.

The technology for operationalising the planning process and for communicating the planning data is common to both sectors, and includes complex information technology architectures, including the latest in web portals and so forth.

Over the past few years, the experiences of two particular companies have been studied closely. Each is part of a traditional supply pyramid in that they are module and system suppliers established in different locations across Germany, and they also have assembly factories around the world, depending on location of their customers. For instance they have plants in Slovakia, Mexico, the United Kingdom, Canada, South Korea, and the United States.

They have experienced a higher demand for additional coordination tasks, which are over and above that required for maintaining direct tier-to-tier relations. The problem is that module suppliers used to have lean assembly plants with only a few "overhead" employees. Managing sub-suppliers creates a demand for extra management activity and the acquisition and development of competencies that were not formerly available to them. This is quite similar to experience in the Aeronautical industry, in that supplier coordination now becomes part of the work expected from tier n-1 partners, together with the traditional production of parts. However, this new requirement on supply-chain system design is more apparent in the Automotive sector.

The operational freedom that suppliers have for coordination is very tight. Suppliers at lower levels are still selected and evaluated by the OEM. As a result, the sub-supplier is chosen according to optimisation criteria of the OEM, such as a focus on quality and price, and not those of the module supplier. For the latter, other additional criteria would be applicable, for instance prior experience, coordination expense, compatibility and so forth. It is interesting to notice that such aspects are already taken into account in the Aerospace industry, where production management is paradoxically made more complex by the lower quantities of product, in particular by discrete management of small quantities of parts which are subject to chaotic logistical behaviours that are not found in high-volume manufacturing of relatively low-cost items.

Coercive influence is common: there are those suppliers who will even go so far as to scare their sub-contractors into submission with threats to substitute them with another supplier if they do not meet tougher performance criteria, especially in the cost area. It has to be recognised, however, that the customer is the partner operating in the highly competitive business environment, who is in direct contact with the "heat" of the market and has the immediate imperative to anticipate and respond to market changes. There are many examples where production of components, and even the production of whole cars, was removed to low-cost areas of Central Europe, Turkey, Tunisia, or South East Asia. Again, this phenomenon is only at its beginning in the Aeronautical industry, with the consequence that the relationships between customers and suppliers have been considered to be less tense for the time being. However, as will be seen below that is changing.

The flow of data on demand may occasionally cross from two or more tiers. So, it could occur that suppliers at a lower level get two different planning information inputs: one from their direct customer, the module supplier at level 1, and another from the OEM that distributes the forecast directly for a certain period of time to all suppliers at lower levels such as tier 2. Those planning requirements could differ, because module suppliers try to optimise their own production programme. This problem is still an exception in the Aeronautical industry where assemblers are still reluctant to send plans to suppliers beyond the next tier.

As in Aeronautical Supply Chains, the infrastructure and competencies for information exchange, collaborative production planning, optimisation, and so forth of suppliers at lower levels are often considered to be weak by the assemblers. Thus, the module supplier becomes a translator or mediator that gets planning information from the OEM, and must transcribe this into the language of their own suppliers. The general lack of planning competency at those levels is considered as a big limitation for flexibility and optimisation. Even if this role of mediator is less clear in Aeronautic chains, the assumption that problems in processing the plans come from a poor competence of the supplier is the same.

We shall see in the following that reality is perhaps not as clear as the perception. Because many module suppliers have more than one customer, e.g. at OEM level, it is difficult for them to standardise processes and organisation. Despite the fact that there are standards like ISO TS 16949 (2002) and guidelines from the VDA (Association of German Automotive Industry) for information exchange, every customer has its own preferences and self-grown procedures for supply chain planning, operating and optimisation. The module suppliers are usually in a weak position to get their standards accepted by their customers. This is also the case in the Aeronautical industry, and prevents suppliers from developing their own capability and culture, as development of procedures has been led by the final assemblers and the manufacturers of large systems.

If the module or system supplier applies a diversification strategy, such as by operating plants that serve different customers, then planning and scheduling become even more difficult. Although there are good information flows from customers in terms of obliging forecasts and requests regarding shipping time and quantity, it is common that those requests are changed right up to the day before delivery. For the module supplier, it is then difficult to react and to optimise their own production because there are additional restrictions from the other customers. In that case it is nearly impossible to pass those changes on to the suppliers at lower levels because their ability to react flexibly is even less. It is to be noticed that such diversification is common in the Aeronautical sector, since the quantities manufactured do not justify dedicated plants, or even dedicated workshops, for each customer.

Tier 3 suppliers are continuously placed under stress by customers at tiers 1 and 2. Discussions with tier 3 suppliers have revealed that their customers test them continually, not just in the normal operational factors of delivery time, quantity and quality, but also for their loyalty and reliability, even in the face of incessant on-going cost reductions. We have seen that the fear engendered by these tests, and the threat of substitution, is also coming into the Aeronautical sector, with the consequence of breeding mistrust.

7.3.2.2 Extended Networks: The Case of Including Equipment Suppliers

The large quantities of parts and products produced in the Automotive sector drives demand for dedicated tools, machines and whole production lines as the production supply-chain is subjected to continual change and improvement. As a consequence, there are extended networks of companies supplying various capital elements of the production system. In this sector these companies have a much closer link with the core supply chain than their counterparts in the Aeronautical sector. They mainly involve standard machines which are configured for specific use and are engineered to order. This offers another perspective on the broad supply-chain coordination arena, so the case of equipment manufacturers in the automotive supply chain is now examined.

In this section, we report on observations from another two companies who are equipment manufacturers. They are based in the South-East and in the South of Germany and are small and medium sized companies having approximately 150 and 350 employees respectively. These companies produce machinery, tools and dies have customers at all levels. They provide equipment for assembly, for the production of modules and for the production of parts. The relationships of those enterprises to each other are shown in Fig. 7.5.

As depicted, one problem they have to struggle with is the complex and filtered information flow, especially when getting design specifications for equipment designated for parts production. It is common that suppliers at lower levels of the pyramid

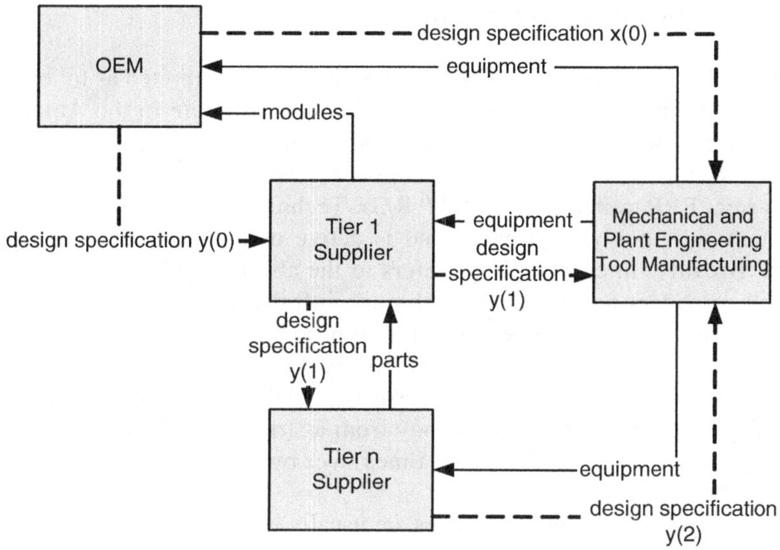

Fig. 7.5 Relations between automotive industry equipment suppliers and their customers

constitute their direct customers, but these companies receive their design specifications from others at higher levels. To complicate matters, those specifications are based on the specifications of the OEM. The problem is that there is no direct information flow between the OEM and a tier n supplier, and also no information flow between the OEM and the equipment manufacturer, at least in terms of detailed component specifications. So, the information the equipment manufacturer gets is, in the worst case, filtered and transformed two or more times. This can lead to substantially sub-optimal solutions that have to be improved during the ramp-up phase, at considerable economic cost. This problem is made worse by the "black box" manner in which detailed activity costs are not revealed between tiers, so only a weak sense of the situation is ever available formally.

Furthermore, equipment manufacturers have their own supply chains consisting not only of material suppliers but also of enterprises that work in design, manufacturing, heat treatment, metal working and so forth, which require coordination.

There is an emerging trend in which OEM and tier 1 suppliers increasingly use their equipment manufacturers also for project management. Whereas in the past the module supplier or OEM themselves selected and coordinated several equipment manufacturers such as for the whole interior of a car, it is now becoming common that there is an equipment manufacturer who is responsible for the whole interior package, quite a change. Because those companies are not able to produce the whole equipment by themselves, they are forced to involve and to collaborate with other partners, who may also be their competitors. Thereby, those companies are forced to build up additional competencies for instance in project and programme management. The coordination activity of such complex networks is increased by the modification loops which are often necessary when a new car model is launched.

7.3.2.3 Interpretation of Problems

The observations described above are summarized and interpreted as follows.

In former times, there was a heterogeneous IT structure in the Automotive supply chains, and this is still present to some extent in the Aeronautical sector. This has changed in the Automotive sector, and many enterprises use the same system (e.g. ERP systems, often SAP R/3). Technical interoperability between IT systems is quickly improving, and is active on the international research agenda. Broadly 'interoperability' refers to the ability of IT applications from different providers to inter-connect through infrastructure from different vendors. Despite this fact, the sector is far from achieving an integrated general solution in practice. Different releases, different customization of the software, and scarce financial and technical resources make it difficult to utilise the potential of the information technology from a strictly technical point of view. However, this is compounded many times over by organisational, cultural and sociological issues.

Even though the smallest companies are usually larger than in the Aeronautical sector, there is also a great disparity between tier 1 and tier 2 suppliers, and so on for suppliers at lower levels. Companies at lower levels of the supply pyramid are usually very small and control their production manually or with the help of spreadsheets. So, the enterprises at upper levels can be seen as an interface between the sophisticated IT world of the OEM and the pragmatic way of doing business of the small company. This role covers not only technical capability but also social capability. The culture of information sharing, dealing with failures, decision making and so forth is usually quite different. This characteristic phenomenon is exactly the same in the Aeronautical case.

If coordination problems have to be solved in such supply chains, then different integration strategies collide. On the one hand, there are formal rules, standardized processes, and formalized information flow, and on the other hand there are personal relations, informal processes and informal information exchange. The enforcement of formalism and standardization is perceived in the large companies as a necessary and normal means to work efficiently, whereas in the SMEs it is often seen as indicating a lack of trust. This is in our opinion also true in the Aeronautical sector, and suggests that SMEs of different industrial sectors have perhaps more in common that they have with large partners of their own manufacturing arena.

7.3.3 Summary of Findings from the Cases

In spite of differences in their specific form, we have seen in previous sections that there are many similar underlying phenomena across both the Automotive and Aeronautical Industries. We have summarised major points of similarity and

Table 7.1 Commonality and differences in the coordination problem found in Automotive and Aerospace Supply Chains

	Sector	
Dimension	Aerospace	Automotive
Share of value-added in supply chain by a tier0 company	Now 40% In future less	Now about 25% In future less
Type of supply chain structure	Large companies at beginning and end of chain Many small or very small companies	Large companies all along the chain Some small companies possible especially for components or production machines
Expected planning methods	Mainly MRP-based In recent years: efforts for implementing lean techniques	Mainly JIT / lean / synchronised production Equipment: programmes
Access to planning requirements beyond the next tier	Insufficient	Sufficient
Additional factors of complexity	Small quantities managed heterogeneously: chaotic demand within the chain Presence of very small companies (10–30 employees)	Design activities allocated to suppliers Integration of equipment suppliers in the chain
Need for cooperation	Increasing with complexity of supply chains	Increasing but compensated by simpler flow management through lean manufacturing
Limits for efficient cooperation – point of view of large companies	Competences of small companies regarding planning issues	Competences of small companies regarding planning issues
Limits for efficient cooperation – point of view of small companies	Pressure Lack of trust	Pressure Lack of trust
Protection of smallest partners by larger ones	Yes, by not transmitting them all the constraints received from the customers	No
Stable partnership	Still preferred, but set into question by externalisation of production to low cost countries	Lower prices preferred to stable partnership Threat of the supplier substitution by a company from the low cost country

dissimilarity between the two industrial sectors, and these are presented in Table 7.1.

In both cases, it is interesting to notice that:

(1) The strategic model of supply chains within both the industries indicates a tendency for the Original Equipment Manufacturers to concentrate more and more effort on developing their core competencies and outsourcing all other

activities to specialist producers. The proportion of value added by tier-0 companies currently lies in the region of 20-40%, being considerably lower in the automotive than the aeronautical sector.

(2) The need for coordination within the supply chains is increasing with concomitant supply chain complexity, and strength of competitive pressure.

(3) From the larger company perspective, small companies seem to be considered as the "weak links", to the small company's detriment as selected supplier. The diagnosis of the large companies is mainly that SMEs suffer from a particular lack of competence with regard to planning capability, which does not match their technical capability.

(4) From the perspective of the small company, based on interviews with managers of small companies, the origin of this real poor use of planning methods and tools seems to be much more complex. The poor performance that is attributed by the large company customers to the small companies demonstrates in their opinion a lack of interest in these methods. However, there is a much deeper cause in that there is generally a poor congruence of these methods with the range of practical problems faced by the SMEs. This is compounded by the decreasing trust shown by many large company customers, who small companies suspect to only seek lower prices from them, and this reaction prevents satisfactory engagement with the type of drastic changes which are required on all sides to move on.

From this, we can readily see that we have in an important sense, strong resonances between the sectors. Differences in size (though exaggerated in the aeronautical cases), presence of hierarchical planning structures, complicating factors like demand patterns, strong perceived need for cooperation, recognition of major problems to be resolved even if the views on means are not agreed, the constraints placed on progress by commercial pressure and trust (or the lack of it), the desire for stable relationships.

This serves as a point of departure to examine more deeply some of these characteristics with respect to interaction and company-level issues.

7.4 Coordination in Planning as a Socio-Technical Process

We now reflect on the analysis of coordination and planning with reference to both social science and technological dimensions.

7.4.1 Problematic Situation and Possible Approaches

The extended enterprise raises three inter-related issues and these are related to structural change and design, concern with flexibility and lean production requirements, and supplier selection.

With respect to structural change and design, it appears that planning methods, rules and tools have been designed with reference to the need for coordinating workshops internally. When applied to a supply chain with external partners, these methods and tools have been assumed to be sufficiently congruent with such a new context. This could be so in certain cases, such as when the orders refer to the logic of routine batch production. However, the framework is very different when the economic relationships are not stable anymore, and with different decision-making processes. The objectives and constraints tend to be more specific to each partner, with possible conflicts of priorities and constraints both at a global level, and also in important local details. Decision-making applied in larger companies tend to be based on more generic know-how, while in small companies it will tend to be based on more practical and local know-how.

With respect to the concern with flexibility and lean production requirements, demand has become highly hierarchic, and at the highly disaggregated level especially of the smaller SME, product presentation in terms of, quantity and timing may be highly variable and even chaotic. This is related to the phenomenon associated with slowing any flow to very small levels, such as one can demonstrate by closing a faucet until it drips. Units commonly have to "nurse" each unit through production. The combination of all these specific demands raises additional contradictory constraints and priorities at a more local level. This means that there is no possible "one best way", and that there is a need for negotiation and communication at the different levels and scales, and across enterprise boundaries.

With respect to supplier selection, it is desirable that when customers in the supply chain want to be able to change suppliers, this change is easy and smooth, and the rules transparent to facilitate opportunities for business. It would therefore be desirable that this be carried out in a "plug in" way. From the point of view of the small company navigating their way in the market to supply the larger companies, this requires capability in interoperability with respect to organisational processes, and not just from the technical information processing point of view (e.g. as in CIMOSA, ARIS and so on).

These characteristics imply that a satisfactory cooperation in Supply Chain cannot be limited to a simple exchange of information and a common technical way to process it, even if that is a necessary pre-condition.

There are at least two implicit hypotheses which sustain the classical solutions proposed by the main companies and these are problematic:

The first one is that there could be a "one best way" which is related to the project of optimization, and that this best way is to design according to the main company characteristics.

The second one is that small firms should possess the same capability to behave similarly and have the same competences as large ones, just with less capacity (e.g. headcount). Regarding production planning, this would mean for instance using MRPII tools for processing a smaller quantity of forecasts and orders than a bigger firm, but according to the same conceptual framework.

These findings are consistent with at least some of the conceptual framing suggested in the literature, a small sample of which follows.

For instance, Bensaou and Venkatraman (1995) develop a conceptual model of inter-organisational relationships in which the exercise of these relationships depends on the fit between information processing needs and information processing capabilities. In their model, the information processing needs arise from three types of uncertainty (environmental, partnership, task) and the information processing capabilities are provided by three types of mechanisms (structure, process, information technology). They use distinct but complementary theoretical perspectives to support this model and derive five different relationships:

1. Remote relationships, that emerges in a low uncertainty context and gives rise to low levels of information processing requirements; correspondingly, there are low or minimal levels of information processing capabilities
2. Electronic control, which is characterized by high task analysability and low task variety; the information processing capabilities of such relationships reveal a particular emphasis on control combined with a low frequency of information exchange
3. Electronic interdependence, which is clearly a configuration for the high uncertainty contingencies that require important and rich information processing capabilities; the environment consists usually of a highly dynamic, complex and growing market for high tech products
4. Structural relationships, that face an environment characterized by low capacity (i.e., limited growth) and low dynamism (stable technology with few changes in products), but high complexity (i.e., market for products with a high level of customization)
5. Mutual adjustment, which is restricted to high tech, new and complex products quickly changing in their design and performance

More recently, Palaneeswaran et al. (2003) categorize the integrative forces in a supply chain into hard and soft factors. Their hard factors cover binding needs to achieve contractual agreements, short-term benefits, and contractual commitments for liability. Their soft factors are the relational binding forces, which can be developed from sources such as trust, involvement, mutual respect, sharing of risks and rewards, effective communication, transparency, ethics and discipline and other non-contractual measures. This can also be observed in supply chains in automotive or aeronautical industries. Beside the hard facts, soft factors are necessary to make the supply chain work. This analysis represents a wider interpretation beyond an information processing model. The richness of personal interaction dimensions are central.

In these two frameworks, it is clear that coordination is much more than a process based on information exchange, but on the organisational interoperability of two organisations. For inspiration, we refer to Vernadat (1996). He refers to interoperability and interworkability. We feel the first is more appropriate to the prospective need – this is a holistic company-level property.

7.4.2 Towards an "Organisational Interoperability"

Let us develop a little further this notion of "Organisational interoperability" and its implications.

Technical interoperability concerning software, data standards and networks is of course a pre-requisite to facilitate exchange of data such as process plans, CAD files and so forth.

However, the main issue we wish to address deals not with the transaction level sharing of drawings, process plans and orders etc, but with sharing a common understanding of the situation and context required for coordinating different activities and priorities all along the supply chain. It deals with the capability and capacity for cooperating in an efficient way for both partners over time, including the way to face unexpected events.

The two case studies have demonstrated that large companies try to define rather strict criteria for selecting their partners according to such concerns. ISO 9000/14000 certification, risk management procedures, project management, process analysis, problem solving and the linked organisational methods and tools can be framed as setting the basis of an organisational interoperability based at least partly on the combination of such organisational rules, values and methods.

This organisational framework helps to define, at least partly, the same organisational 'vocabulary' and use the same 'grammar'. However, sharing such basis in common does not mean that the priorities and objectives are the same. They have to be discussed and regularly re-evaluated according to the on-going emergent process and its situated context. The main point of language is so the actors can "see" a situation in a common frame. Seeing is dependent on having appropriate schema to attach ideas to, and also to serve as domains which can be developed further by developing proximal knowledge for example by training and experience. An appreciation of representations for time and space is a pre-requisite for more advanced rationales.

Looking back to the large companies and their selection criteria, this is a way to evaluate whether the possible partners share a common organisational language and common values. However, it seems that the large firms would like that the integration or 'plugging-in' of new suppliers could work like technical networks: just look to the commonly accepted formalised standards and all will be fine.

Yet, the partners are social organisations, they involve individuals with specific ways of thinking, objectives and values, and communities of their own for sustenance. The need for negotiation is central. In some way, the planners and schedulers sustain the burden of the non-discussed decisions acting as translators, buffers and mediators between the organisations.

In this environment, change and transformation are necessary for survival and growth, so there needs to be an available and exercised capacity for development, especially in the smaller companies and of a form that reflects their specificities. It would appear that they frequently lack capacity for self-development of capabilities other than technical.

A fuller characterisation of Organisational Interoperability must encompass these aspects.

However, we must now bring the argument forward by discussing some other points related to companies and coordination.

7.4.3 SME Characteristics and Related Issues

Research programmes in the social sciences concerning SMEs have demonstrated that small and medium firms should not be considered as big firms at a smaller scale (Julien 2005). In particular, it has been shown that a majority of SMEs do not grow beyond a certain size, and this is because their managers do not want them to develop and lose their SME characteristics. In addition, this is a definite strategic decision, even if it is implicit. These managers place great emphasis on being able to decide the way their firm will develop, and on keeping a relative freedom regarding their environment. In such a context, the control of the planning process appears to go much further than questions regarding either technical issues or human capabilities related directly to supply-chain performance.

Organisational independence is highly valued. In many respects, the issue of keeping companies relatively independent meets the goals of the clients very well: we have seen that the large companies, in their pursuit of flexibility, ask their suppliers to be economically sustainable under widely varying demand, even to such a degree that the large company does not order anything from them on occasion. This places pressure on the SME to diversify by being involved in different supply chains as a risk reduction strategy. This is coherent with the desire of the SME to keep independent of the influence of particular clients. To complicate matters, we have shown the contradictory constraint that on occasion the main company looks to optimise its supply chain without taking into account the supplier's involvement in different supply chains, thereby creating a planning disconnect.

Another characteristic of SME deals with what has been called the 'magnifying effect' of reduced size on specialist power and influence within a company (Torres 2003): the less people you have in a firm, the more specific and important their standing in decision-making and getting things done. This importance will be even higher if so-and-so is the only person in charge of a core function, as was seen many times in the case studies. Thus, SMEs are highly dependent on individuals. While this can result in a positive empowerment for such employees, but it can also result in strong pressure on the 'indispensable' person, as mentioned in the aeronautical case study.

Research programs have also drawn attention to a so-called 'proximity logic' that characterizes SME ways of dealing with exchange relationships (Torres, ibid). Proximity logic in this sense is related to the propensity of each person or organisation to see herself or himself in the centre of her or his world. In such a centric view of the world, the nearer are the persons, the things or the events, the more important they appear. Proximity according to geographical and/or social criteria is often mobilised to deal with exchanges. Therefore, the way production planning is designed and carried out has to be congruent with such a way of dealing with relationships. The proximity logic deals also with the choice regarding priorities (e.g. the nearer the more important) and related decision processes. Proximity logic needs to be explained further.

Another important notion is the ‘wall’ that tends to separate the inner view and its related priorities and constraints, from the outside. This operates both between firms and between different functions in a firm. In the SME, one single person is often in charge of different functions, and therefore the proximity between the functions and the persons in charge of them is often high. Management of functions is thus naturally quite informal. This situation can be compared with that of the large firms which are often characterized by strong division between functions, each group of persons being dedicated to their specific domain, with often fairly strong ‘walls’ between each function. Of course, this cohesion can be changed by design, with more focus on product over function.

Such an organisational difference between SMEs and large firms can have a strong impact on cooperation capabilities, especially the relationships between their respective planners and schedulers across the boundaries at the front line. The SME planner will possibly search for combining the constraints and priorities of the commercial and the production components of their own enterprise, because of the functional proximity that helps them being concerned with such issues. Meanwhile, the large firm planner will tend to focus on optimizing planning methods as a measure of “professionalism”, because the walls are fairly high with the other functions in their firm. They may consider that the SME planner is fairly inconsistent with regard to planning methods and criteria as such, and therefore “unprofessional”.

This is coherent with what we observed in the automotive case study, with the main firm sending forecasts to their subcontractors that were substantially different from the forecasts sent by their customers.

The automotive case study also brings a very interesting finding regarding this issue, whereby the module supplier performs a role as translator and mediator between the main firm and the subcontractors. Developing this role more formally could be part of the solution. However the case study showed that such coordination activity is generally not valued on the contractual side, and that this issue is a major obstacle to development.

It would be an interesting research question to evaluate the hidden costs of bad coordination, with a view to encouraging large firms to pay more attention to coordination capability factors in defining their partner selection criteria.

It would also be an interesting research task to design a diagnostic tool describing these proximity domains, linked priorities and concerns, to help defining the scale of the negotiation teams to be set up.

Martina Berglund and Jane Guinery explore in greater detail associated aspects in Chapter 4, where they look at the relationships across the organisational interface between commercial, planning and production functions, and issues of influence and so forth. They discuss how the proximity between these functions can be very different between the firms according to the combination or specialisation of functions. Therefore, the issues that have to be clarified, negotiated and jointly solved can have different scope, and require mobilising different professional competencies not just according to the situation but also its organisational context.

7.4.4 SME Organisational Slack and Concerns of the Main Company for Control Thereof

James March has showed that organisational slack was commonly mobilised in organisations so as to cope with uncertainty and ambiguity (March 1991; Mayere and Vacher 2005). Just in Time and lean Supply Chain Management tend to remove the slack within and between firms. The way SMEs tend to escape from the constrictions of production planning prescription can be seen as a way for including a new slack, so as to deal with their reality of contradictory constraints. These contradictory constraints result notably from being involved in different supply chains, as we have seen, and from the growing needs for system level flexibility and adaptability, that do not combine easily with tight coupling of forecasting and planning.

Underlying this evolution in the removal of organisational slack, is a growing concern for the main companies regarding their organisational control, and regarding the efficiency and effectiveness of the services and information delivered at all points along the supply chain. These companies try to deal with this issue by imposing tools and methods. However, as we have seen, key parts of these tools and methods appear to be either inefficient, or worse still irrelevant, to SME characteristics. Some of the particular tools imposed appear to create more problems than they solve in terms of reducing requisite decision freedom at lower levels. The question moves therefore to a different level. Could it be more relevant in supply contracting to define guidelines at the organisational level, targeting these services and information delivering processes at a more aggregated level, empowering the company to decide key operational details locally?

7.4.5 Coordination and the Requirement for a Common Interpretation

According to the above mentioned issues, important questions are raised regarding information processing:

First of all, coordination through exchange of information requires a common interpretation "language" which is far from being always the case. This cannot be obtained by imposing a unified vocabulary, as this would be a negation of the SME's specificities, but by the definition of a common framework dedicated to communication which would act to mediate between the company and its customers and sub-contractors.

Secondly, it is clear that even in the case of common understanding, trust in the exchanged information is a key factor for a common sharing of information. "If I do not think that you really need the parts at the required date, I will use my capacity for working for other customers... ". Creating trust may require a transparency which can have positive but also negative effects. In any case, creating trust requires removing some of the barriers which protect the company from its environment. This is also related to the 'proximity rule' mentioned above. More generally, trust in the information flow requires first trust in the partner. Again, the usual means used by a company to protect itself in its contracts with its sub-contractors may have an opposite effect to the one expected in cases where this protection decreases trust.

A company is not a unique free-standing entity, and decisions made locally by its various departments within may have perverse opposing effects through lack of shared understanding and vision. A better identification of the global consequences of local decisions on the Supply Chain is a pre-condition for better coordination and/or collaboration processes.

There is an opportunity and need for capability-building in this regard. At one level, a participative process of involving the different actors of the company in the definition of unique coordination/collaboration processes would enable the clearer identification of possible overlapping or conflicts between decisions. At a higher level, improvement in consciousness capability is necessary in order that a critical mass of relevant people "see" the picture better from a logistical point of view, through competence-building involving staff education / training and at times selection.

7.4.6 Consequences Regarding Planning as a Collaborative Process

These findings have important consequences for the way information should be processed in a Supply Chain:

The exchange not only of information, but of knowledge between partners of a Supply Chain contributes strongly to reduction of misinterpretation. In this case knowledge is in the sense of information to which context and interpretation are added (Tsuchiya 1993), but also the competencies to "see" more in the data. In one example above, sending their inventory level to their supplier is an attempt to send them a richer description of the situation than the one expressed only by their orders or forecasts. In the same sense, knowledge is not present in the data of ordinary order transactions such as in SOP, and so it is not catered for directly by technological platforms of integration; the active human presence is required.

Sources of power and influence are to be found at both customer and supplier sides. The relationships between two companies in a chain are more complex than the common assumption "the customer has the power". While power can be on the customer's side, it can also be on the supplier's side. For an instance, the reader is referred to Marcotte et al. (2009) for an attempt to adapt planning techniques to reflect the reciprocal influences between customer and supplier and to their mutual interest. In that case, seriously pretending to impose tight due dates is inappropriate (especially when they are not strictly required), and it is more realistic to negotiate delays. To address this, it has to be accepted in the information system design specification that constraints may pass downstream to the larger company as well as upstream, and that the necessary functionality is provided to afford this flow.

The process of sending orders to the sub-contractors may be improved by taking into account considerations other than purely technical, such as their vulnerability, which could enable positive moves by the customer to protect the supplier from unexpected disturbances. Some support should be provided for managing such benign protection activity.

Planning is a basis for negotiation and on-going adjustments between actors. Plans processed according to the available data and existing methods can be considered as a way to present the questions to be solved and the contradictory choices to be addressed in a further negotiative and collaborative step, with shared consideration of goals not just one-way compliance, and involving a wider range of people both within and between firms than those formally charged with the role. This implies that planning knowledge and activity has become a cognitively and socially shared and distributed organisational process, involved both in the minds of individual people and in the organisational framework, rules and methods as well as in the information systems.

7.5 A Frame of Reactive Conflicts in Imposed Coordination Improvement Effort

We now have a more complete picture of the range of tensions, and the contextual characteristics of the parties concerned. We try to summarise some of these key tensions as they arise at the interface between large and small companies, an analysis which points to key "pressure points" for systematic intervention to improve the situation.

7.5.1 Conflict Influence Diagram

The case studies have shown the importance of shared objectives, shared interpretations, mutual understanding, and trust, in productive co-ordination between partners within a supply chain. They show these concepts colliding with an unrealistic wish to make the co-ordination integration process as simple as the technical connection of two information systems.

The divergences between the perspectives of small companies and their larger customers appear to be enduringly characteristic and inherent. We are of the view that these are not well understood and have been poorly articulated in the literature.

To articulate the findings we particularly focus on the dyadic link between the large company and the small company in the figure below. On the left of the diagram is pictured the large company (LC) domain, and on the right, the small company (SC) domain. The common space of interaction lies in the intersection of their boundaries, running vertically down the middle of the diagram. The inter-connected effects are described in the text below.

To demonstrate the inter-connectedness and conflicting feedback paths of cause and effect, let us start with an example of a service quality issue identified by the star just to left of bottom-centre (point 1 in Fig. 7.6):

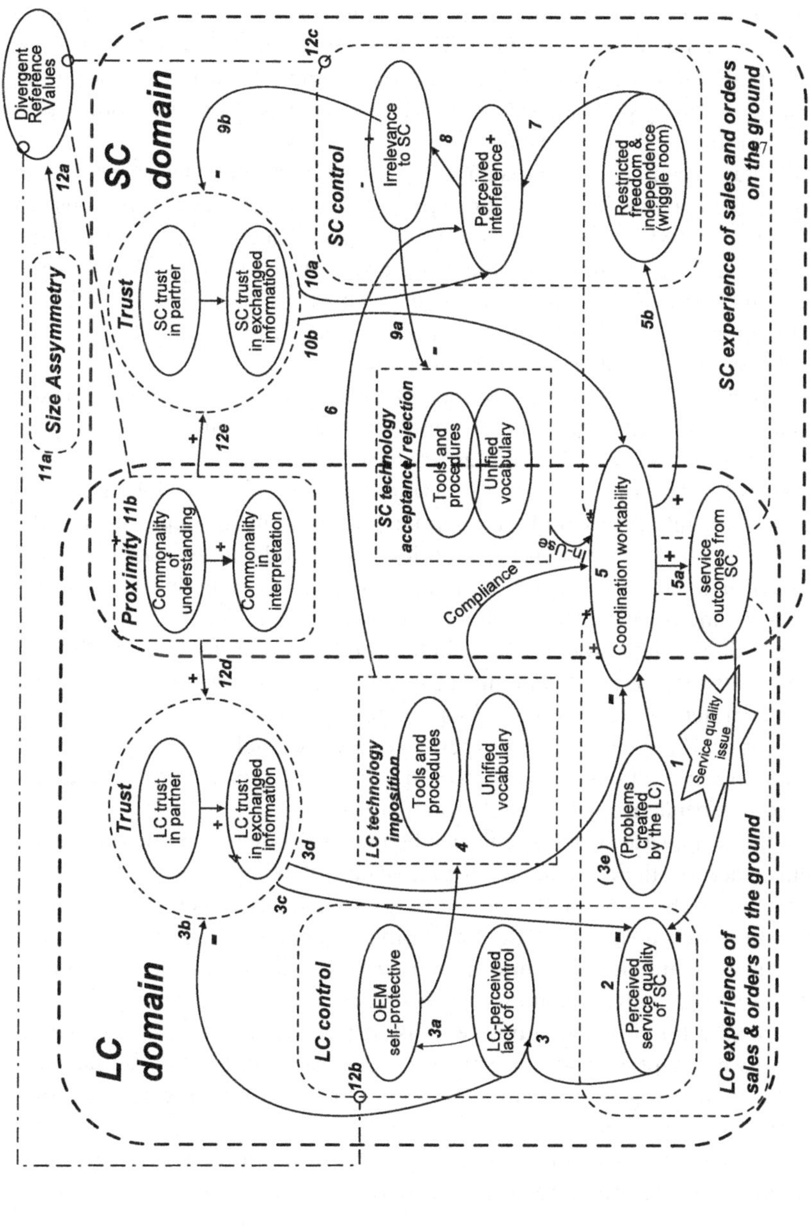

Fig. 7.6 Some aspects of large company-small company dyadic tensions

The small company fails to achieve compliance with requested due dates, a key performance indicator of service outcomes (1).

This is perceived by the large company in negative terms (2). This negative perception of experience by the large company raises a perception of systematic lack of control by the large company management (2) in the small company's capacity to plan and control their affairs. The perception is reinforced by a lack of trust in the small company (3b), both in the small company's capacity and in the information provided by it. This further contributes to deterioration in coordination workability (5) in a reinforcing cycle.

Note. We have seen earlier that such a delivery performance deterioration can easily arise in the reaction by a supplier to produce larger batches as a response to large company demands to reduce costs (for example projecting further out in the planning horizon for parts requiring the same raw material size). This upsets the production schedule, and leads to not having the right part at the right time. Without a systematic improvement (e.g. to competence), reducing cost can only be achieved with a reduction in some other area of performance.

In addition, we have already seen examples of where some of the large company activities can inadvertently contribute to small company problems (3e).

In response to the perceived poor performance, management at the large company implements self-protective activity (3a), and makes moves to tighten control mechanisms at the interface with their supplier (4) – in terms of imposing specific structural changes from a range of possibilities including IT tools, policies/procedures, "agreed language"/"unified terminology" and so on, in the interest (as they see it) of improving the workability of activity coordination (5).

Doing this, they achieve some success in terms of better service quality (5a), but this may be short-term or local, and thus of limited impact compared to the effects of conflict generated by other reactive loops as follows.

On the ground, these changes are experienced by the small company to restrict the freedom to carry out their work within what they consider their requisite freedom and independence, or "wriggle room" (5b).

Note. In the worst case scenario, this can be perceived by the small company as an outright attack on their fundamental philosophy of being! It should be noted again that commonly, the small size of these companies underpins their core business competence in terms of the flexibility, adaptability and responsiveness that complements their narrow technical focus, and that this lies at the core of their very usefulness in the supply chain in the first place.

Again, we saw this in both supply-chains. An example is to impose stricter adherence to weekly plans and corresponding penalties, without consultation to establish the underlying reasons for non-compliance. Underlying reasons may be temporary, for example machine capacity constraints due to several high-priority orders coming together at the same time, or unavailability of particular raw materials due to competing orders placed with the same suppliers by the customer.

This constriction (5b) combined with the unilateral and impositional nature of changes (6) mandated by the large company contributes to a perception of inappropriate interference in their internal affairs, which contributes to an attribution of irrelevance of the changes to the small company (7).

The behaviour of the small company operates at two levels: compliance (4→5) and usage (9a→5). The compliant component will give the explicit appearance of successful usage, but the willingness to use the solutions in everyday practice is implicitly dependent on goodwill, and perception of relevance. Hence the tools suffer from the perception of irrelevance (9a).

Furthermore, the small company control loop (5b, 7, 8) contributes to a lack of trust in the large company partner, and in turn the exchanged information for PSC activities.

This contributes a negatively reinforcing feedback loop that amplifies the perception of interference (10a) and deteriorates coordination workability (10b), and so reinforces the performance problems which the moves were intended to resolve in the first place!

Note. Commonly, this is compounded by a bad past history of inter-company relations, and a small company can be particularly sensitised by their negative experiences of being let down by their customer in the past.

An example symptomatic of such mistrust and its effect on workability was when a small company failed to cooperate with a request for a favour by the large company at a critical time for them.

From the case studies, two important influencers appear: asymmetry in relative company size (11a), especially for smaller companies, and the related notion of organisational proximity (11b). The desired closeness is especially at the level of two-way communication and goal-sharing collaboration, not just cooperation in mechanical coordinative plan/schedule execution. Proximity is related to mutual understanding of divergent values (12a) used as reference for interpretations by both parties 12b, 12c). Proximity can attenuate (12d, 12e) the mistrust and thus damaging outlined above.

Such paths of conflicting reaction phenomenon appear to be inherent in the large company-small company dyad in the enterprise network context. To a greater or lesser degree this elucidation seems relevant for interactions between individual companies and groups or networks of companies in a more general sense.

7.5.2 Reflection Towards Design Improvement

The phenomenon mapped above appear to be inherent in the large company-small company dyad in the enterprise network context, and highlights the importance of mutual understanding and the need for constructive respect for differences.

However, as noted earlier it is the larger companies who more directly face the "heat" of the broader market-place, and it is necessary for the network to adapt their structure, and grow into the new possibilities in time or else fall by the wayside and disappear. Of course, companies at all levels do not have the luxury to avoid this imperative. The operational and economic ecology is changing, and so production systems within must adapt to survive or else contract and risk disappearance.

For example, new possibilities derive from the trend in strategy towards outsourcing, and off-shoring of non-core competencies. As we have seen above, there is a growing trend in which each tier is being given more knowledge/service work, such as in design and development, material sourcing, life-cycle management, and so on, but under the watchful eye of the OEM, and under a hierarchy of watchful eyes down as far as the small company. As we saw also, not all the watchful eyes agree on what should be done and how especially by the lower levels. This was well demonstrated in the case of the equipment suppliers.

What are the implications? This analysis suggests that improving respectful communication between the partners would be beneficial. Earlier discussion focused on the possibilities of proximity – operational and at the knowledge level - of the partners by structural support to small networks of companies and company level competency development along certain lines (yielding a common understanding between them of each others ways of working e.g. processes, constraints, required freedoms and their justification operative language and mental models or ways of seeing) which in turn would improve common interpretation, and thus improve both directly performance and indirectly trust, and that this would prove a counter to improve the overall position with respect to service quality, yielding both cost and service gain.

In short, improving capability in small companies has the potential to give more flexibility or "wriggle room", to enable them to be more organisationally interoperable, and so to be more responsive and resilient, enabling more routine flexibility in their commitments, and the evolution of responsive and resilient supply-chains as a whole.

7.6 Conclusion

In a commonly – held view, coordination in Supply Chains is considered as a purely technical and economic process linking entities implicitly considered as similar. To that end, only the definition of a clear contract, and the transactional processing of the information flow along and across the chain seem to be necessary for addressing the planning issue. For large companies, the accepted way to generate this succession of plans is to follow iteratively the logic of the MRPII method (possibly including JIT dispatching), i.e. to allow for generating a supply plan for upstream partners, then organizing internal work through finite capacity planning.

Through some examples taken from real cases, both in aeronautical and automotive Supply Chains, we have attempted to show that coordination is indeed a complex socio-technical process based on characteristics of the relationships between two partners that include power, trust and mutual understanding. These parameters are greatly influenced by the size and culture of the companies, and we have emphasized the idea that small companies are not simply "reductions of large companies". They participate in tightly-coupled intra-enterprise activities, closely guard their internal independence, and have a justified need for internal freedom from administrative imposition, yet must be able to plan for and to be held accountable for their overall actions.

Stereotypical ways of performing the planning process can be grossly inadequate regarding their socio-technical specificities. Moreover, small companies usually belong to several supply chains, which makes their interests only partially consistent with those of each customers. Therefore, trying to impose a unique way of processing planning information all along supply chain leads to critical problems in these small companies, in that the techniques, and the tools which support them, may be inadequate. The consequence is that there are very substantial avoidable conflicts between on the other side customers, thinking that their smaller partners do not want to progress, and on the other, SMEs thinking that the demands of their customers are irrelevant. By neglect of socio-technical aspects of the type identified above, these avoidable conflicts such as represented in Fig. 7.6 are inevitable. Such modelling can lead to better understanding of the secondary impacts and their dynamics.

Coping with these problems requires an increase in the quality of the exchanges between companies: indeed, forecasts or supply planning are only data, and they require to be related to a given context for becoming exploitable information. At this small scale, this task requires human interpretation to cope with the organically-rich forms of situations that can arise regularly. Furthermore, the introduction of knowledge-enrichment into the exchanged flow of information will allow each partner of the Supply Chain to much better understand the consequences of its own decision making on the global supply chain performance.

For future work, in the context of enterprise networks, an interesting research theme arises for both engineers and social scientists acting in a joint program: to analyze the planning process across a supply chain in terms of requirements, with respect to local needs and specificities, rather than in terms of solutions to be implemented in a forced manner, especially with small companies in mind.

References

Bensaou, M., & Venkatraman, N. (1995). Configurations of inter-organisational relationships: A comparison between US and Japanese automakers. *Management Science, 41*(9), 1471–1492.

Carlsson, S., Hedman, J. (2004) From ERP systems to enterprise portals. In Adam (Ed.), *The future of ERP systems: Technological assessment* (pp. 254–277). Hersey, PA: Idea Group Publishing.

Christopher, M. (1998). *Logistics and supply chain management – strategies for reducing cost and improving service*. London: Prentice Hall.

Dudenhoeffer, F. (2003). *Trends und Entwicklungen in der Automobil- und Zulieferindustrie*. Emden, Germany: CAR Center Automotive Research, SESAMES Kooperations-Messe.

Freeman, S., & Browne, E. (2004). The influence of national culture on dissolution communication strategies in Western versus Asian business relationships: A theoretical model. *Supply Chain Management, 9*(2), 169–182.

Garcia-Dastugue, S. J., & Lambert-Douglas, M. (2003). Internet-enabled coordination in the supply chain. *Industrial Marketing Management, 32*(3), 251–263.

Goldratt, E. M., & Cox, J. (1984). *The Goal*. Great Barrington, MA: North River Press.

Grabot, B., Mayère, A., & Bazet, I. (2008). *ERP systems and organisational change – A socio-technical insight* (Springer Series in Advanced Manufacturing). London: Springer.

International Business Development Corporation (2002). SAE Media Briefing February 22. Bloomfield Hills, MI

ISO/TS 16949 (2002) *Quality management systems - Particular requirements for the application of ISO 9001:2000 for automotive production and relevant service part organizations*. International Organisation for Standardisation.

Julien, P. A. (2005). *Les PME: bilan et perspective* (3rd ed.). Québec: Presses InterUniversitaires.

Kelle, P., & Akbulut, A. (2005). The role of ERP tools in the supply chain information sharing, cooperation, and cost optimization. *International Journal of Production Economics, 93–94*, 41–52.

Lee, H. L., & Ng, S. M. (1997). Introduction to the special issue on global supply chain management. *Production and Operations Management, 6*(3), 191–192.

Malone, T. W. (1987). Modeling coordination in organizations and markets. *Management Science, 33*(10), 1317–1332.

March, J. G. (1991). Décision et organisation. Dunod.

Marcotte, F., Grabot, B., & Affonso, R. (2009). Cooperation models for Supply Chain Management. *International Journal of Logistics Systems and Management, 5*(1), 123–153.

Marktl, A (2006). *Die Koordination und Konfiguration der globalen Wertschöpfungskette in der Automobilindustrie*. Bachelor Thesis, University of Graz.

Mayere, A, Vacher, B (2005), Le slack, la litote et le sacré. Revue Française de Gestion, hors série Dialogues avec James March: 63-86.

Mentzer, J. T., Min, S., & Zacharia, Z. G. (2000). The nature of interfirm partnering in supply chain management. *Journal of Retail, 76*(4), 549–568.

Nezamirad K, Higgins PG, Dunstall S (2004) Collective work in dynamic inter-organizational scheduling. *Proceeding of the Asia Pacific Industrial Engineering and Management Systems Conference 2004 and The Seventh Asia-Pacific Division Meeting of the International Foundation of Production Research*, Gold Coast, Australia, December 12-15, 2004.

Orlicky, J., & Plossl, G. (1994). *Orlicky's material requirement planning* (2nd ed.). New York: McGraw Hill.

Palaneeswaran, E., Kumaraswamy, M., & Ng, S. T. (2003). Formulating a framework for relationally integrated construction supply chains. *Journal of Construction Research, 4*(2), 189–205.

Pontrandolfo, P., Gosavi, A., Okogbaa, O. G., & Das, T. K. (2002). Global supply chain management: a reinforcement learning approach. *International Journal of Production Research, 40*(6), 1299–1317.

Romano, P. (2003). Coordination and integration mechanisms to manage logistics processes across supply networks. *Journal of Purchasing and Supply Management, 9*(3), 119–134.

Rousseau, D. M., Sitkin, S. B., Burt, R. S., & Camerer, C. (1998). Not so different after all: A cross-discipline view of trust. *Academy of Management Review, 23*, 393–404.

Shore, B. (2001). Information sharing in global supply chain systems. *Journal of Global Information Technology Management, 4*(3), 27–50.

Stadtler, H., & Kilger, C. (2000). *Supply chain management and advanced planning*. Berlin: Springer.

Stock, G. N., Greis, N. P., & Kasarda, J. D. (1998). Logistics, strategy and structure. A conceptual framework. *International Journal of Operations and Production Management, 18*(1), 37–52.

Torres, O. (2003). *Petitesse des enterprises et grossissement des effets de proximité*, Revue Française de Gestion, n° 144, pp. 119-132.

Tsuchiya, S. (1993). *Improving knowledge creation ability through organizationallearning*. International Symposium on the Management of Industrial and Corporate Knowledge, UTC Compiègne, France, October 27–28.

VDA Verband der Deutschen Automobilindustrie. (2005). *Annual Report*. Frankfurt: VDA.

Vernadat, F. B. (1996). *Enterprise modelling and integration: Principles and applications*. London: Chapman and Hall.

Chapter 8
The Planning Bullwhip: A Complex Dynamic Phenomenon in Hierarchical Systems

Philip Moscoso, Jan Fransoo, Dieter Fischer, and Toni Wäfler

Abstract Instabilities in production planning and control have received considerable attention due to their negative impact on planning performance. However, extant research has been limited to theoretical (e.g. simulation) settings and has focused on specific methodologies (e.g. mathematical) to overcome instabilities. The objective of this chapter is to make two contributions to the theory development on production planning instabilities. First, it aims to make an empirical contribution through an in-depth case study, and second, it introduces a holistic framework that supports analysis of hierarchical planning systems and their potential instabilities.

The in-depth case study is carried out on an industrial company that has difficulty to meet its customer deadlines and faces a significant order backlog. Planners of the company at different hierarchical levels and order chasers on the shop floor end up rescheduling open orders and updating lead times continuously

---

This book chapter is a revised and extended version of a previously published article: Moscoso, P. G., Fransoo, J. C., & Fischer, D. (2010). An empirical study on reducing planning instability in hierarchical planning systems. *Production Planning & Control*, 21(4), 413–426.

P. Moscoso (✉)
IESE Business School, Universidad de Navarra, Pamplona, Spain
e-mail: pmoscoso@iese.edu

J. Fransoo
School of Industrial Engineering, Eindhoven University of Technology, Eindhoven, The Netherlands
e-mail: J.C.Fransoo@tue.nl

D. Fischer
Institute for Business Engineering, University of Applied Sciences Northwestern Switzerland, Brugg, Switzerland
e-mail: dieter.fischer@fhnw.ch

T. Wäfler
School of Applied Psychology, University of Applied Sciences Northwestern Switzerland, Olten, Switzerland
e-mail: toni.waefler@fhnw.ch

J.C. Fransoo et al. (eds.), *Behavioral Operations in Planning and Scheduling*, DOI 10.1007/978-3-642-13382-4_8, © Springer-Verlag Berlin Heidelberg 2011

when trying to meet deadlines, but eventually are not able to improve order fulfillment. Only after the introduction of an Advanced Planning System and centralization of planning decisions in a single department, on-time delivery was significantly improved and order back log drastically reduced. This case study allows studying of the underlying mechanism of such planning instabilities, with a particular focus on the impact on stability of human and organizational factors. On the basis of our findings and additional conceptual research we have then developed a framework constituted by six key planning systems attributes. By taking into consideration these factors, a firm can address the root causes of planning instabilities, rather than merely focus on its symptoms.

8.1 Introduction

In industrial practice, production planning is typically done following a hierarchical approach (Meal 1984; Qui et al. 2001). There are good reasons for organizing planning processes in this way, such as decreasing the complexity of the planning problem by decomposition, leaving control opportunities at a place as low as possible in the decision hierarchy to encourage fast response, and allowing the planning systems to make as much use as possible of local information (e.g., Anthony 1963; De Kok and Fransoo 2003).

However, also anecdotal situations have been reported in practice, where in such hierarchical structures, decisions by planning entities at the various levels may lead to significant planning instability and order backlog. This instability phenomenon may be called the planning bullwhip, and may occur because the process of generating detailed plans on the basis of given aggregated plans, may not only introduce sub-optimality, it may also lead to inconsistencies and infeasibilities (Gelders and Van Wassenhove 1982), or even become instable (Selcuk et al. 2006). Inconsistencies may arise because of conflicting objectives at different decision levels. Infeasibilities usually occur because of aggregation. But there may be also other reasons for such deficiencies. Overreactions of a higher planning level on deviations on the shop-floor level may initiate a vicious circle causing even more deviation on the shop-floor level (Fig. 8.1).

This may happen because a modification in a key planning parameter such as, for example, the lead time may immediately cause substantial changes in the planning and release policy, likely to have substantial effects on the actual lead time on the shop floor. Furthermore, as all planning systems in practice are socio-technical systems that comprise both human and technical subsystems, any analysis of its behaviour needs to analyze not only both parts, but also their interaction (Wäfler 2001; Moscoso 2007). Typically, IT-systems may potentiate overreactions of planning entities, for example, due to the fact of not having fully up-to-date or complete information. At the same time human operators, which still are indispensable in industrial practice, can also influence significantly the planning process through their particular behaviour.

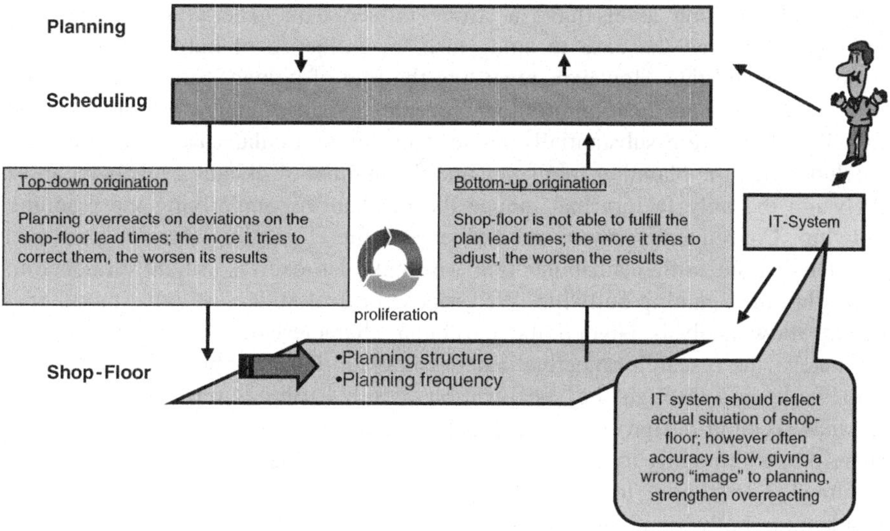

Fig. 8.1 Mechanisms underlying origination of planning instabilities

Another complication is that it is not obvious, from a theoretical point of view, what should be the proper value of the lead time, and if and how it should be updated best (Fransoo 2004; Selcuk et al. 2006). In general, most practitioners, as well as academics of the field, have been concerned about the symptoms derived from such planning instabilities, but have not yet studied empirically the underlying mechanisms of origination and propagation of the phenomenon.

In this chapter, we present a number of insights into the reality of planning instabilities and introduce the term "planning bullwhip", under which we subsume any kind of planning "instabilities" generated primarily by planning policies and internal actions. Those insights are based primarily on a single but extensive case study at a Swiss discrete manufacturing company, supplying very sophisticated parts for the air defence industry. The company had been having substantial problems with the very same symptoms that can be observed when a planning bullwhip occurs. We found empirical support on how poor estimates of the lead time and poor actual performance can reinforce each other, and eventually result in significant planning instability and order backlog. After substantial changes in the planning structure and organization were introduced the observed problems could be discontinued. The case study provides a clear context to empirically investigate some aspects of the planning bullwhip that have not yet been reported in the literature. To have this case study as playfield for research was of crucial importance, as it provided an ideal basis for studying and exploring the relationships and patterns behind the planning bullwhip. However, the fact that it was only a single case study imposes obvious limitations to the generalization of our conclusions.

Our findings from the case study indicate support for the existence of planning bullwhip, as well as for the hypothesis that both the planning frequency, and the number of planning levels have a strong influence in generating the planning bullwhip. We used the case to study also the influence of human planners on planning instabilities. However, regarding the role of humans, our study was not fully conclusive, as the opportunities for the human planners to mitigate any bullwhip effect were substantially reduced in this particular case. Therefore, we concluded that the planning frequency and the number of planning levels are most likely not the only factors that induce the mechanisms underlying the planning bullwhip. Consequently, we wanted to extend these insights from the case and postulated some further attributes that we think also exert a critical influence in generating the planning bullwhip. We proposed therefore a conceptual model that should allow analysis beyond the particular characteristics of our case study. Ultimately, the research objective was to integrate those attributes into a holistic framework, and through the specification of the attributes provide support for the analysis (and design) of hierarchical planning systems in industrial practice. Therefore, additionally to the case study, and based primarily on conceptual work, we aimed at developing in a second step a conceptual framework for the analysis of the planning bullwhip phenomenon, which is also presented in this chapter.

This chapter is organized as follows. In Sect. 8.2, we review the relevant literature about planning instabilities and related phenomena. In Sect. 8.3, we will present and discuss our research questions and methodology. Section 8.4 provides a brief description of our case company, and in Sect. 8.5 we describe the analysis of the case and present our results. In Sect. 8.6, we discuss our findings and summarize them as an integrated framework, which is composed of six key attributes that we think determine fundamentally the occurrence of a planning bullwhip. Finally, we summarize our findings and conclusions in Sect. 8.7.

8.2 Literature Review[1]

A hierarchical planning approach divides a large, complex planning problem into simpler sub-problems. One then solves the set of problems in a sequential fashion, increasing the level of detail and decreasing the planning horizon, whereby the higher planning level imposes constraints on the lower planning level. The Hierarchical Production Planning (HPP) approach and its benefits are well documented in the literature (see, e.g., Anthony 1963; Bitran and Hax 1977; Bowers and Jarvis 1992; De Kok and Fransoo 2003; Leong et al. 1989; Neureuther et al. 2004). Extant research on hierarchical planning has been primarily concerned with technically solving, logically or mathematically, clearly delimited planning problems and in theoretical (e.g. simulation) settings. However, empirical research on hierarchical

[1]We do not include in this section a literature review of the field of human and organizational factors in planning and scheduling, as it has been covered extensively in other chapters of this book.

planning and its practical utilization has been limited. Some exceptions are an early study by Gelders and Van Wassenhove (1982), and a more recent study by Berglund and Karltun (2007), as well as articles by Jonsson and Mattsson (2003) and Olhager and Rudberg (2002).

The process of computing detailed plans on the basis of given aggregate plans, supported or not by human experts, is challenging. For example, it may not be always possible to ensure mathematical optimality. But the process can also lead to inconsistencies and infeasibilities (Gelders and Van Wassenhove 1982), or even a loss of stability (Selcuk et al. 2006). Inconsistencies can arise because of conflicting objectives at different decision levels. Infeasibilities usually arise because of aggregation. But the planning system can even lose stability. In previous research the terminology of "lead time syndrome" has been introduced for specific instances of production planning instabilities. However, the best known and most studied "nervousness" in operations management probably is the supply chain bullwhip. We briefly review the two concepts, and eventually suggest introduction of a generic term of "planning bullwhip", which we define later on.

8.2.1 The Lead Time Syndrome

Within a production planning system, the lead time[2] is a key control parameter in the planning hierarchy. While some authors argue that the lead time is an exogenous parameter to planning (De Kok and Fransoo 2003), other state that the lead time should be endogenous to the planning system and reflect the current state of the shop floor (e.g. Zijm and Buitenhek 1996). Most of the planning systems deployed in industry are based on some kind of MRP-logic, and then the lead time is an exogenous parameter to the planning system. That is, the user inputs a value for the lead time into the ERP-system for each of the levels in the bill-of-materials. However, as mentioned before, it is theoretically not fully clear, what the proper value and updating frequency of the lead time should be (Fransoo 2004; Selcuk et al. 2006).

These decisions, however, are critical for the planning performance, as in dependence of the updating policy of the lead time the planning system may lose its stability. This effect has been coined the "lead time syndrome" (Mather and Plossl 1978) and only much later studied in a more formal and quantitative manner by Selcuk et al. (2009). The authors demonstrate the existence of the lead time syndrome subject to certain assumptions regarding the updating behaviour of the planning parameters by the planner operating the ERP system. Following a number of more conceptual papers on the lead time syndrome, Selcuk et al. (2006, 2009) completed a series of formal model studies investigating the effect of the lead time syndrome.

[2]For a better understanding of the chapter we will differentiate here conceptually between the planning lead time (MRP information) and the production lead time (performance measure).

8.2.2 Supply Chain Bullwhip and the Notion of a Planning Bullwhip

In operations and supply chain management, however, probably the best established and most studied instability phenomenon is the supply chain "bullwhip". This effect is also very well known among practitioners. The bullwhip effect is a dynamic phenomenon that essentially substantiates through the amplification of demand order volatility as one moves upstream along the supply chain (Forrester 1961; Lee et al. 1997). The term 'bullwhip effect' was firstly used by the consumer goods company Procter and Gamble (P&G) (Lee and Billington 1992). The logistic executives at P&G examined the order patterns of Pampers diapers, one of their best-selling products. They found that the variability of replenishment orders was amplified towards the upstream of the supply chain regardless of the fact that the demand pattern at the customer interface was quite steady. Hence the name: 'bullwhip effect'. The effect has also been demonstrated experimentally with the so-called "Beer Distribution Game" (Sterman 1989).

Later, many scholars have studied this phenomenon because it can be a big practical concern for supply chains as it may lead to increased costs and reduced efficiency. Such issues may result from increased inventory requirements, expediting, customer shortages, or suboptimal capacity utilizations, for instance. In general, research has focused primarily on the operational causes of the bullwhip effect. Several contributing factors to this instability have been identified already but it is not clear whether this list is exhaustive (Daganzo 2004). The causes that are most frequently discussed include: demand signal processing, inventory rationing, order batching, and price variations (Lee et al. 1997; Chen et al. 2000). In other words, the lead times of information and materials are the primary reasons for the bullwhip effect. The reaction of a supply chain on a change in end-customer demand is delayed firstly because it takes time to pass on information to suppliers, and secondly, because the suppliers have to adjust their outputs.

Ways to alleviate these operational problems include improved demand forecasting techniques (Chen et al. 1998), capacity allocation schemes (Cachon and Lariviere 1999), staggered order batching (Cachon 1999), and everyday low pricing (Sogomonian and Tang 1993). However, recent research into the behavioural effects of supply chain bullwhips by Croson and Donohue (2006) have concluded that given decision makers consistently underweight the supply line when making order decisions, the bullwhip effect still may exist even when normal operational causes (e.g., batching, price fluctuations, demand estimation, etc.) are removed (cf. also Chap. 5). This is in line with previous research by Sterman (1989), who was the first to test the existence of the bullwhip effect in an experimental context playing the well known beer distribution game, where he relied on a simple, non-stationary retail demand function unknown to supply chain members.

The supply chain bullwhip has some interesting similarities with the "lead time syndrome". The most important similarity we consider is that all these dynamic phenomena depend much more on the specific planning policies and planning structures than on the real demand processes. For the supply chain bullwhip this has been proved extensively (Daganzo 2004). Ultimately, all these phenomena are originated by some kind of "overreaction" or partial optimizations of planning entities. Additionally, information systems and human planners can potentiate or mitigate the dynamics of all these effects. Finally, the three phenomena have also in common that their resulting effects can be considered to impact negatively the company's performance, and at the same time the attempt to improve performance contributes to the effects even more. A key difference, however, is that while the supply chain bullwhip propagates horizontally along the supply chain, the "lead time syndrome" propagates rather more vertically along planning hierarchies, both top-down as well as bottom-up. Given the similarities, we therefore introduce the concept of a "planning bullwhip", under which we subsume any kind of planning "nervousness" generated primarily by planning policies and actions "internally". We introduce the following definition:

> *Definition.* A "planning bullwhip" is a complex dynamic phenomenon where a planning system ends up with an erratic ordering and updating behaviour in response to changing workload levels, resulting in uncontrolled order release patterns, generating eventually a larger variability in the workload levels and lead (flow) times.

Based on this definition planning bullwhip should be understood as a category of planning instabilities characterized by certain characteristics. A planning bullwhip situation is a situation where lead time variances have been amplified as a result of planning actions. Those planning actions appear to be erratic both in magnitude and time (frequency), but are the response to changes in the workload levels. Typically also backlog will increase. The "lead time syndrome" could be considered a specific instantiations of a "planning bullwhip", but also any other situation of planning instability fulfilling this definition could be considered a planning bullwhip. For our purpose of empirically studying the impact of human and organizational factors on planning stability, we can operate with such an ample delimitation of the terminology. However, the aim of our empirical study is to contribute to a better understanding of the underlying mechanism, something which eventually will help narrow down the definition more. Nevertheless, we are fully aware that this understanding, same as it occurred with behavioural factors in the supply chain bullwhip, will be a long process requiring substantial efforts by researchers.

It is important to note that we do not claim to have discovered a fundamentally new type of dynamic phenomena in hierarchical planning systems. We rather think it could be of interest to introduce an aggregated category of planning instabilities, utilizing for its denomination the bullwhip term, given similarities with the traditional bullwhip. Furthermore, we hope that relaying on this well known term may allow traditional production planning stability issues to benefit from the attention the bullwhip phenomena has received both by researchers as well as practitioners.

8.3 Research Contribution: Questions and Methodology

The aim of our research is to empirically study the underlying mechanism of a planning bullwhip and the effect of particular measure to overcome it, with particular focus on human and organizational factors. There are several reasons why we think such an empirical study is useful.

First, previous papers on the lead time syndrome assume a specific planning structure and also a specific way in which the main planning parameter (lead time) is updated. Due to the large-scale implementation of ERP systems, it may be the case that the updating procedures in industrial practice are different. As previous research on MRP has pointed out, it is not unreasonable to assume that due to the complexity of parameterizing large scale systems, planning parameters are not updated very frequently in industrial practice in general, and at our case site.

Second, the information infrastructure is expected to be crucial for the size and effect of the planning bullwhip. Modern information technology, such as Manufacturing Execution Systems (McClellan 1997) enable higher level echelons to have fairly detailed status information while the absence of such or similar systems may decrease the visibility of higher level planners on the actual process on the shop floor and thus enhance the planning bullwhip effect.

Third, as already mentioned, in industrial practice human planners play a key role in planning systems, and thus it is reasonable to expect that a planning bullwhip may be influenced by the actions of humans in the system. The research challenge when studying the influence of human planners on the planning bullwhip is to adequately understand how human planning takes place, and is affected by the planning context. Extant research suggests that human planning is affected, for example, by the environment or the organizational structure (McKay 1992). Theoretical studies have postulated, for example, that decentralization of planning can have negative impact already in very simple production systems, and that specific coordination arrangements are required to leverage decentralization (Bernstein and Federgruen 2005). But also the weighting and priority given by planners to different sources and types of information in determining a course of action, including the effect of visual inspection and observation can vary depending on the expressions of those factors. Previous empirical work at a chemical plant found that human planners systematically and largely neglect the information system's planning recommendations and that the extent of neglect is larger if the planning problem is more complex (Fransoo and Wiers 2006).

The aim of the research reported here was to develop an understanding of the planning bullwhip that would lead to the development of an integrated framework that supports the analysis and design of hierarchical planning systems in industrial practice, and eventually improves quality of design output. In this chapter, we are interested in developing some insights on these topics. Our research objectives can be synthesized in the following two research questions:

1. Can we observe the planning bullwhip in industrial practice and do theoretical explanations for the existence hold?

2. What role do human and organizational factors play in planning bullwhip situations?

The first question focuses on finding empirical evidence that establishes a clear link between the lead times, the planning frequencies, the resulting backlog and deterioration of on-time delivery. Earlier studies on the lead time syndrome (Mather and Plossl 1978; Selcuk et al. 2006; Selcuk et al. 2009) provide us the theoretical basis to formulate the following hypotheses that will form the basis for analysis:

Hypothesis 8.1. *Under a hierarchical planning structure, very frequent rescheduling and lead time updating of open orders to cope with work load changes can eventually lead to erratic and very long lead times, order back log and not delivering on time.*

In simulation models it has been possible to demonstrate that under certain conditions frequent updates of lead times can cause increases in lead times and, as a result, unstable, inefficient decision making (Selcuk et al. 2006). Contrarily, decreasing the planning frequency decreases the variability in lead times and the average lead time (Selcuk et al. 2009). This is the basic premise of the lead time syndrome. But the studies were only theoretical and presuppose an active updating procedure by higher-level planners based on the operational lead time performance on the shop floor. When the planner "observes" lead time changes he adjusts the planning parameters. In industrial practice this may be different. Furthermore, for settings with low degree of variability lead time updates even provides better cost performance (Selcuk et al. 2006).

With regard to the second research question, the human and organizational factors that have an impact on the planning bullwhip, we are particularly interested in the effects of particular organizational planning structures (in particular the number of planning levels), of having complete and detailed information at the higher level (provided by modern IT systems for manufacturing execution), and the effect of humans interfering with system decisions. For all these aspects an empirical analysis is very valuable. For this question, the analysis was based on testing the following hypothesis:

Hypothesis 8.2a. *Humans may have a different kind of information than computer systems, and may have more local control options than modelled in a system, and hence can attempt to overrule any planning decision taken by computer systems. As a consequence, human planners' behaviour can have a substantial effect on the planning bullwhip.*

In reality, it is reasonable to assume that there is more planning flexibility than any system can model. Operators can make modifications in the machine allocation, planners and foremen can prioritize in a smart manner. Also, the planners and operators physically observe the backlog and will attempt to reduce this by taking additional measures. On the one hand, this may lead to an increase in the planning bullwhip effect due to further overreaction; alternatively, on the other hand, it can be argued that because of additional local flexibility, this will reduce the planning bullwhip effect. For all this reasons an empirical test is very valuable.

Hypothesis 8.2b. *Decreasing the number of planning levels decreases the planning bullwhip effect since the system will have less feedback loops to cope with, and delays in the system will decrease.*

One of the major factors affecting a planning bullwhip is delay in information and delay in response. Decreasing the number of planning levels will generally decrease these delays, as for example, the number of feedback loops will be reduced. Feedback loops play an important role in the supply chain bullwhip, as they may potentiate amplification of variability given their control characteristics (eg. these loops feedback control signals and can generate delays in response). Any MRP technique itself tends to implement feedback on stock levels, but often is only partly reactive, because it combines closed-loop measurements with open-loop data on forecasted demands and requirements. Chung and Krajewski (1987) were the first ones that modelled a rolling horizon feedback procedure to coordinate aggregate planning with master scheduling. However, they ignored the stochastic nature of the manufacturing systems, which is, in our concern, a significant characteristic of industrial practice. Moreover, there will be an adverse effect in terms of planning complexity. The fact that human planners may interfere with planning decisions can lead to a planning structure and feedback mechanism that de facto are different from what was formally specified in an organization or model.

We investigate these hypotheses using a single in-depth case study (Eisenhardt 1989; Voss et al. 2002; Yin 1994). The case company selected was Oerlikon Contraves AG, a discrete parts manufacturer in Switzerland that supplies complex parts to the air defence industry. A description of the case site and observations is included in the next section.

We have collected data on the case by several methods. First, we have conducted extensive semi-structured interviews (Lindlof and Taylor 2002) with the production manager, the manager of the production control centre, the IT manager responsible for the ERP system, and with the two project managers responsible for the APS/MES (planning) system. A transcript was made of each interview and was reviewed by the interviewee for accuracy. Furthermore, the work of three planners has been observed extensively and our observations have been discussed with them. Also, we have observed extensively the work of the foreman and discussed the observations with them. Observations were done along works shifts, with the researcher making protocols of the observed persons' work complemented again by semi-structured interviews (Wäfler 2002). The protocols of these interviews and the observations were then analyzed and validated by the observed person in order to identify possible evidences for the hypothesis, for example, in terms of information flows (e.g. bilateral information flow with a defined source (e.g. the scheduler) and a defined addressee (e.g. a foreman).

Second, we collected documents describing the functionality and performance of the planning and production system. The main work was conducted during three full days and follow-up of the initial findings has been done by phone, follow-up meetings and email. The field work was conducted by one of the authors. The combination of observations, interviews, discussions, and documentary study enabled us to triangulate many of the findings from the case, and have several sources for each of the findings. Nevertheless, we are aware that limitations still exist, given it is a single case study we conducted, and therefore more general conclusions should be drawn carefully.

8.4 Case Description: Oerlikon Contraves AG

We now describe the production planning characteristics of the case under consideration that are relevant with regard to the research questions and hypotheses formulated in Sect. 8.3. First, general production and planning characteristics of the case are described, and then the key problems of the initial planning system are reviewed. Finally, a new planning approach that eventually improved planning performance is summarized.

8.4.1 Production and Planning Characteristics of OCAG

Oerlikon Contraves AG (OCAG) is a Swiss manufacturer of discrete parts for the air defence business, and belongs to a European multi-billion industrial group. The company has 1,760 employees and generates a turn-over of 330 Mio; of which 6% are produced in Switzerland, 94% are produced abroad. The core capability of OCAG is the development and manufacture of advanced air defence systems as well as simulators and training systems.

60% of the products offered are standardized. However, they are quite complex in terms of the number of components they are built of. In fact, OCAG manages around 600 client orders a year that on the shop-floor revert to around 15,000 production orders. Manufacturing of an order typically requires going through 5–7 different work units for completion, with an average lead time of around 2.7 months. Stocks are primary kept at the level of subcomponents and raw materials, and orders are only released when a client demand exists. Fulfilment of due dates is very critical as typically the OCAG products are subassemblies of even larger defence systems.

8.4.2 Initial Planning System and Perceived Problems

Until the year 2002, production planning at OCAG was done end-to-end with an integrated ERP (SAP), but organizationally following a decentralized approach. The planning tasks were distributed across four organizational levels:

(a) At the beginning dispatchers did a rough production plan, determining key dates for start and end of a client order.
(b) These plans were then detailed at a next level by the work preparation unit, adjusted for the different shop floor working units.
(c) Those operational production units were allowed to do some final adjustments in the production plans if required. They were able, for example, to choose a particular machine among a set of equals.
(d) In order to certainly meet the agreed due dates of key client orders, the role of order chasers was introduced. 24 of these operators physically pushed high priority orders through the shop-floor. For this they negotiated which each of the different working unit the best possible schedule for their orders.

Thanks to a significant effort of the order chasers OCAG was able to fulfil 87% of its due dates, but generating a significant backlog (Fig. 8.2) of orders in the range of several thousand working hours. The order chasers had in fact created two types of order flows at OCAG.

High priority orders were pushed physically through the shop-floor but at the expense of regular order been delayed repeatedly.

Additionally, higher level plans were flawed as the ERP system in place was permanently not up-to-date with respect to the shop-floor. That happened because the tremendous effort by the planners to systematically and regularly update the

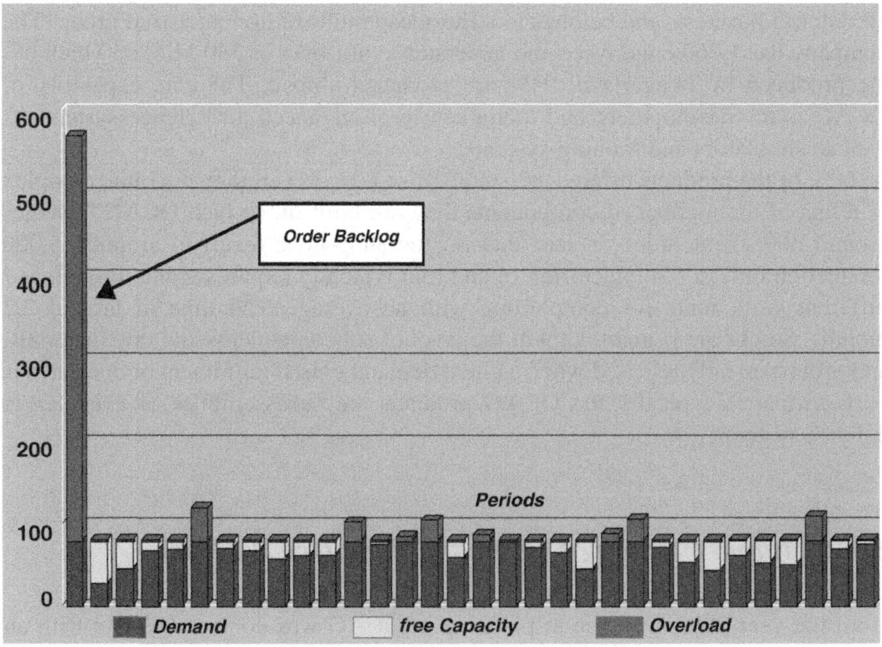

Fig. 8.2 Order backlog

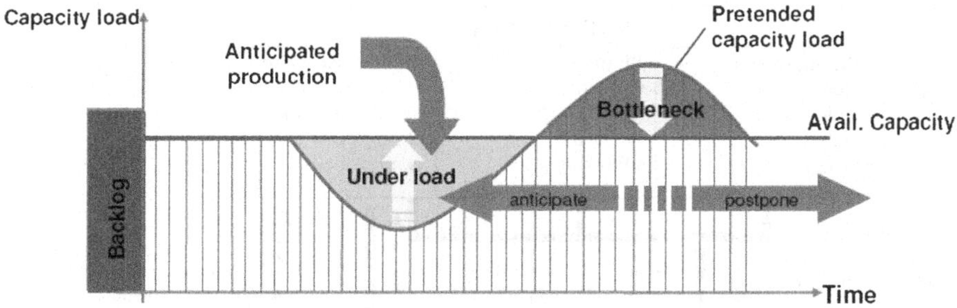

Fig. 8.3 Capacity load: effects of backlog (Wassermann 2004)

comprehensive ERP was not considered to be worth the effort. Consequently, for example, despite the tremendous backlog, the lead times remained essentially unchanged in the ERP, and consequently in the production plans released (Fig. 8.3). Moreover, once an order chaser took charge of an order, planners did lose control of it until the chaser reported its final completion.

8.4.3 A New Approach to Solve Planning Problems

OCAG decided to revise the described production planning approach, and completed implementation of a new one in 2003. In this new model, the planning decisions were completely centralized in a production control centre with ten qualified planners that had gained experience in previous OCAG assignments. Order chasers were discontinued, and all remaining operation units had to stick strictly to the plans provided by the control centre.

Additionally, OCAG installed an Advanced Planning System (APS). In this new planning system, once a customer order enters, material requirements and base data are elaborated in the ERP (Fig. 8.4). Then the orders are transferred to the APS for the shop-floor planning (backward termination), including a daily 2-step capacity harmonization. Permanently work progress is fed directly into the APS by the shop-floor operators. Further, every 14 days, the control centre informs the local operating units about the medium-term capacity requirements. On this base, these units generate their resource action plans (including shift work planning).

With this new planning system planning results at OCAG were significantly improved. Backlog, for example, was drastically reduced in around 3 months (from mid November 2003 to mid February 2004), as depicted in Fig. 8.5, and a year after implementation of the new APS there was practically no more backlog in the system. But it also could improve accomplishment of due dates to 97%, and planners had at every moment a very accurate view of work in progress.

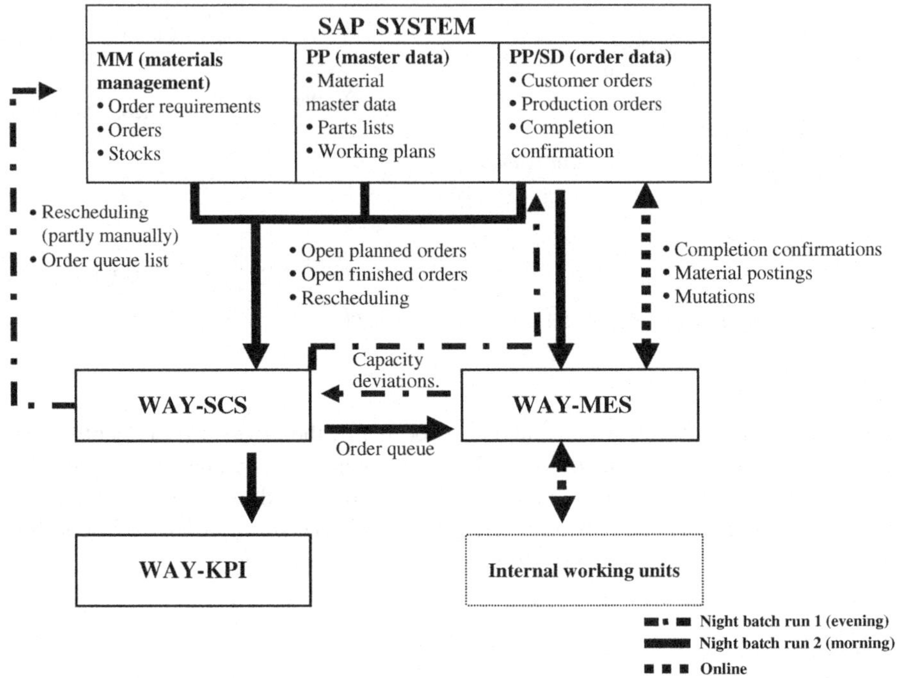

Fig. 8.4 IT systems for planning and control

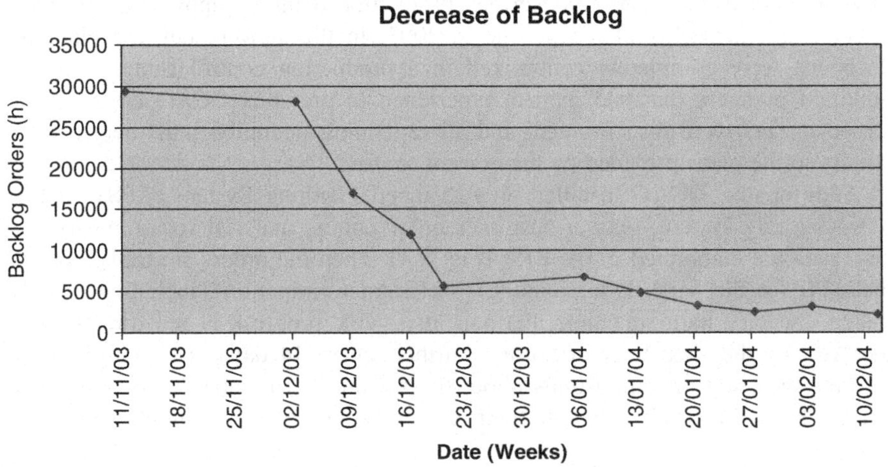

Fig. 8.5 Backlog at OCAG

8.5 Case Analysis and Results

In this section, we will now analyze the case following the structure provided by the hypothesis introduced in Sect. 8.3.

Hypothesis 8.1. *Under a hierarchical planning structure, very frequent rescheduling and lead time updating of open orders to cope with work load changes can eventually lead to erratic and very long lead times, order back log and not delivering on time.*

First of all, it is interesting to note that, formally, planning lead times were not updated in the initial situation. This was not an explicit policy decision, but changing the lead times of thousands of items in a comprehensive ERP system like SAP was considered too cumbersome for the planner to conduct in a systematic and frequent way. The system lead times thus remained unchanged. Effectively, however, open orders were rescheduled on a continuous base as well as planning lead times informally updated. Consequently, effective production lead times were impacted by the actions of the order chasers. The order chasers changed the sequence of the production orders on the shop floor, effectively reducing the production lead time for a small number of high-priority customers and increasing the production lead time for a large number of other customers. As a result the average production lead time increased by about 15%, and with very high variability. In addition, a considerable backlog was built up in the system, as has been discussed and illustrated in the previous section. The case is therefore a clear example for a practical situation where the different lead-times in the system have to be distinguished carefully (i.e., theoretical and real planning lead time, and production lead time). On the one hand there were differences between the formal planning lead times code in the IT system, and the planning lead times assumed by the order expediters, and on the other hand those planning lead times differ from effective lead times (as a performance measure). To this extent, feedback loops in terms of information management procedures were not working properly.

The detrimental effect of rescheduling and updating that is found in the theoretical studies thus was also found at our case site. However, the mechanics of the effect look to have been different. While in the updating procedures studied in the literature, the main driving force for the updating is the overall queuing behaviour on the shop floor, at the case site the updating was done informally and differentiated: some customers received shorter and controlled lead times, while the majority of customers received longer lead times. Simulation studies in the literature have demonstrated that it is possible, even under high utilization and backlog, to give preference to a small number of orders with high priority (Bertrand and van de Wakker 2002). Apparently, human decision makers tend to act more differentiated than the rules suggested in the literature.

This differentiation and creativity of human planner has also been pointed out in a study by Fransoo and Wiers (2006), who conducted a quantitative field study and stated that planners deploy a larger action variety in their decision making once the complexity of the planning problem increases. We thus find partial support for hypothesis 8.1: we observe a planning bullwhip situation characterized by very frequent rescheduling and updating, but the mechanics are more subtle and differentiated than suggested in the modelling literature. Given this conclusion, one could ask if simply discontinuing expeditors with no other change would have already improved the situation. Our conclusion, and the one of the people at OCAG, is that then probably backlog would have been reduced, but delivery on-time would also have been affected. This kind of trade-off is well known in scheduling, where there exist heuristics aiming at minimizing specific performance goals, such as, for example, minimum orders late, or minimize maximum lateness, or minimize average lateness, but generally can not achieve all goals at the same time. What probably would have improved performance while keeping expediters is if they would have acted more in a more coordinated manner. This, in fact, appears to be the better way to benefit from the high visibility of shop floor conditions that order chaser have. A more centralized planning, would have allowed for a better trade-off between the benefits and penalties derived from the expediters actions when rescheduling open orders.

It is interesting to note that both under the old and the new planning structure, the formal planning frequency was identical: plans are revised daily. However, the work of the order chasers and the authority they had to continually adjust the planning by giving preference to certain orders and changing the announced sequence of production meant that under the old system the effective rescheduling was much higher. On an almost continuous basis, plans were being adjusted. Under the revised planning structure, this effective planning frequency was decreased. Capacity planning, however, was very rudimentary in the original approach, and then improved with the new system, where it was done centrally and at different levels. This effect is also likely to have helped to reduce "nervousness". The number of new messages SAP issued was also reduced significantly, for example.

Hypothesis 8.2a. *Humans may have a different kind of information than computer systems, and may have more local control options than modelled in a system, and hence can attempt to overrule any planning decision taken by computer systems. As a consequence, human planners' behaviour can have a substantial effect on the planning bullwhip.*

It is very clear from the case analysis that the main role of the new software was to have complete, updated, and accurate status information. Furthermore, the planners now have far more accurate status information. Note that this information is not used in more frequent planning decisions, as decisions are made daily and not revised in between.

In the old situation, an extensive number of local control options existed for the planners. These local control options were executed by the order chasers, who continually updated the sequence of orders based on instructions from the planners at the higher level. This caused a substantial planning bullwhip. It can therefore be concluded that in this particular case, the planning bullwhip was reinforced rather than mitigated by the action of the human planners.

Hypothesis 8.2b. *Decreasing the number of planning levels decreases the planning bullwhip effect since the system will have less feedback loops to cope with, and delays in the system will decrease.*

This hypothesis had strong support for validation in our case study. The reduction of planning levels from the initial situation at OCAG until 2002 (cf. Sect. 8.2) to the new one afterwards (cf. Sect. 8.3) was significant. The new planning approach also reduced delays drastically. For example, in the new situation the APS made planners immediately aware of planning conflicts, something which was not the case before, or through its link with the work units malfunction of machinery was instantly detected when reported by the units.

In fact, the decrease in the number of planning levels has probably been the main cause, alongside with the reduction in the planning frequency and improving capacity planning, in reducing the planning bullwhip. This is in line with the insights from theoretical studies. Moreover, as Gelders and Van Wassenhove (1982) stated in their theory review, planning performance depends not only on having good decision models at the different planning level of the decision hierarchy, but that the different models are carefully integrated. The planning integration at OCAG was also significantly improved with the implementation of the new APS and the organizational centralization.

8.6 An Integrated Planning Bullwhip Framework

As mentioned in the introduction of this chapter, the goal of our research effort was not only to study existence of the planning bullwhip in industrial practice, but furthermore to understand better the factors that may exert a key influence on the likeliness of the planning bullwhip to occur. Our findings from the described case study indicate strong support for the assumption that both (a) the planning frequency, and (b) the number of planning levels have a strong influence in generating the planning bullwhip. Furthermore, as mentioned before, we also studied the influence of the human planners in relation to the planning bullwhip in the OCAG case study. However, as already mentioned, our study was not fully conclusive regarding the role of human planners, as the opportunities for the human planners to mitigate any bullwhip effect were substantially reduced by the constraints in place.

After conducting this extensive case analysis we concluded that the planning frequency and the number of planning levels are most likely not the only factors determining the mechanisms underlying the planning bullwhip. Consequently, we wanted to extend these insights and propose some further attributes that we think also exert a critical influence in generating the planning bullwhip. We envisioned a conceptual model that furthermore should not be limited by the constraints of a particular case study. Ultimately, the research objective was to integrated those attributes into a holistic framework, and through the specification of the attributes provide support for the analysis (and design) of hierarchical planning systems in industrial practice.

The planning bullwhip is a very complex dynamic phenomenon that can best studied through observation of the effective operating behaviour of the system. We will have to consider therefore, on the one hand, the particular contingencies

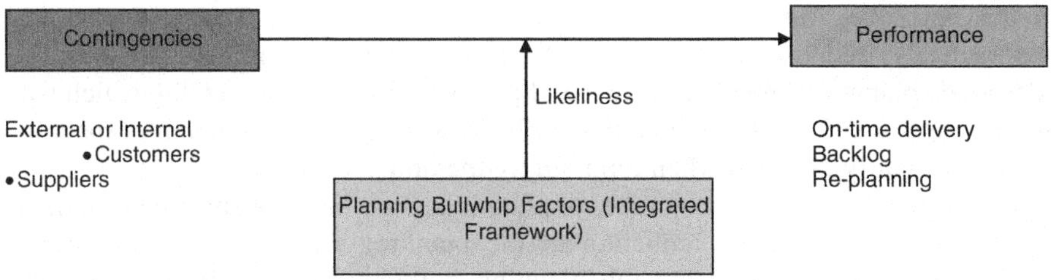

Fig. 8.6 Schematic model of planning bullwhip analysis

a planning system may be confronted with, and on the other hand, we have to measure the planning outcomes (cf. Fig. 8.6).

The first step in developing the integrated framework is to specify those key attributes of a hierarchical production planning system, which may influence the emergence of the planning bullwhip. The difficulty of developing a terminologically precise planning bullwhip framework is that such planning systems comprise a large number of very different and interrelated constituents. Table 8.1 summarizes the six key attributes postulated for our planning bullwhip framework, which will be described more in detail later on.

8.6.1 Planning Frequency

Planning decisions are typically taken periodically, i.e., decisions are grouped together. This grouping usually occurs around regular calendar time scales. Typically planning frequencies in industry are daily, weekly, monthly or quarterly. There is surprisingly little theory available on the optimal planning frequencies. As discussed above, we hypothesize that the planning frequency has an effect on the planning bullwhip, but there is yet only partial and inconclusive evidence for it in empirical studies. In highly stylized modelling environments, research has been conducted which proves that the variance of inventories and lead times increases monotonically with the planning frequency (Selcuk et al. 2006). The empirical study reported in this chapter also found this relationship, albeit that the theoretical explanation is different, in the sense that the parameters are not updated in the planning logic per se, but were de facto part of the planning decision as a consequence of the follow-up by the order chasers.

The work of Selcuk et al. (2006) assumes a stationary demand distribution. With regard to the effects when the demand distribution is non-stationary, we can only speculate and rely on some preliminary simulation studies, which suggest that updating does make sense, as long as the response is moderated (dampened). This bears similarities to principles in control theory. In comparison to situations that are typically modelled in process control, using control theoretic models, production systems and supply chains have three substantial differences. The first is that the

Table 8.1 Planning system attributes influencing the emergence of a planning bullwhip

Planning Frequency	Time interval between formal revisions of plans. When different planning levels operate with different frequencies, in principle the planning scope under study will define the frequency relevant for the bullwhip study. Otherwise, it should be the lowest planning level, which typically will have the highest frequency
Number and structure (organization) of planning levels	Formal number of levels at which planning entities take planning decisions; be it IT-systems and/or human planners. They may be organized more or less centralized, but in general higher planning levels determine the planning options of the levels below
Inertia of the system and system delays	The following kinds of delays in planning systems are of interest:
• Time to notice action is required	• Time span between an event which requires a planning action (update) and its recognition by the planning entity
• Time to take a decision	• Time that is needed for the planning entity to update the plans. In fact, these first two time categories are related to the planning frequency. It is possible, that both categories depend on exogenous factors
• Time for implementation	• After the plans have been updated, they need to be implemented on the shop floor. First, the lower planning levels will have to adjust for the specific planning level details. Secondly also on the execution level actions may be required
• Reaction time or inertia	• Time span that is required by the operating system to implement an updated plan. For example, if the raw material supplier has a lead time of 3 weeks, planning updates that require changes in raw materials (e.g. more quantity or different material) would have to include a horizon of at least these 3 weeks
Degree of interrelations	
• Coupling	• Degree of "hard-wired" interrelations in the planning system; i.e. technically programmed or organizationally determined interrelations (e.g. effects of changes in parameters/variables on other parameters/variables). It can be differentiated between a loose and a close coupling. In the latter case, for example, a certain decision makes immediately other decisions necessary
• Interaction	• Degree of unintended interactions between the effects of planning decisions taken independently of each other (e.g. incompatibility of decisions, unintended mutual amplification leading to unexpected escalation)
Ease of representation	Easiness of representing the planning situation in all its aspects. This includes, for example, characteristics of customer orders, product structures resources or capacity profiles. It may be differentiated between mental models coded in the brains of human operators and the models coded in the IT-systems in place:
• IT-Systems	• Effort to keep accurate process data (orders, master, ...) and rules of optimization (constraints, priorities, ...)
• Mental models	• Effort to generate and manage the mental models the people involved in planning have with regard to production management
Planning scope and action variety	The planning scope includes elements like the planning horizon or the elements covered (e.g. resources). Action variety refers to the degree of freedom a planning entity has in a specific situation. This is again influenced by the planning structure

exact status of the system is often not known. Planners do not have a full view of the status of the production floor, and schedulers do not have a full view of the status at each of the work stations. The second is that the response times in production systems and supply chains tend to be relatively long. One of the main reasons for this is that people are involved in these systems as decision makers at various levels in the planning hierarchy. These people, as indicated above, organize their work in periods, which leads to delays in the decision making process that have been studied extensively in the system dynamics literature. Finally, the response of the operational system to certain instructions (control decisions) is not known, and may even be strategic. Since human decision makers are involved, it is next to impossible to model this behaviour completely. Moreover, the response may depend upon the specific instruction and decision makers may show strategic (gaming) behaviour). Extensive further research, both theoretically and empirically, is needed to increase our understanding of these phenomena.

8.6.2 Planning System Structure and Levels

Closely related to the issue of planning frequency is the (hierarchical) organization of the various planning levels. Planning frequencies are usually decided based upon the typical response time needed or possible at a specific hierarchical level. For instance, capacity expansion decisions may, dependent on the industry, take several weeks, months, or years to be realized. It does not make sense to review a capacity expansion decision daily, if it takes two years to plan and build a new plant. Across different levels in the organization, different planning frequencies are used and implemented. The alignment of planning frequencies is an important design decision when designing the planning structure.

8.6.3 Inertia of Planning Systems and Delays

For both academics and practitioners alike it is very clear that understanding and planning complex production systems requires mastery of concepts such as nonlinearity, feedbacks, and particularly time delays. Research has shown that these concepts are highly counterintuitive and poorly understood (Sterman 2002). The research area of system dynamics has been particularly focused on studying this kind of phenomenon, in general, but also specifically in business systems. System dynamics is fundamentally interdisciplinary and is grounded in the theories of nonlinear dynamics and feedback control developed in mathematics, physics and engineering. But as systems are studied where human and organizational factors are of crucial relevance, like production systems, system dynamics draws also knowledge from cognitive and social psychology, as well as organization theory, and other social sciences.

Systems' thinking considers that the world is a complex system where everything is connected in some way (cf. Forrester 1961, and also Sect. 8.6.4). Additionally to structural complexity, i.e. the complexity in terms of the number of components or decisional possibilities, system dynamics is also concerned about the dynamic complexity of systems which arises from the interaction of agents over time. In fact, the previously mentioned "Beer Game" (Sect. 8.2) was developed to illustrate the concept of dynamic complexity with the specific example of the supply chain bullwhip.

Among the different characteristics of dynamic complexity there are some we are particularly interested in this section about inertia of planning systems. The first is that in production systems cause and effect are often very distant in time and space. In general time delays between taking a decision and its effect on the state of a system are very common in the production context but al so quite troublesome. For example, if the time delays exist in feedback channels, higher leverage policies may sometime cause "worse-before-better" behaviour, while low leverage approaches may result in transitory improvements. Additionally, such delays in feedback loops of planning systems create instability and increase the tendency of the production system to oscillate. As a result, planning entities often continue to intervene to correct apparent discrepancies between desired and actual state of the production long after sufficient corrections have been taken to restore the envisioned working level. Next, the planning bullwhip has also the typical "home-made", self-organizing or emergent character of dynamic complexity: dynamics that arise spontaneously out of the internal structure of the system. Often small, typically random perturbations are enough to initiate important alterations that are amplified over time.

Reviewing these general characteristics helps us to better understand the challenges behind managing a planning system's inertia in relation to the planning bullwhip. Let's complete this analysis looking at the different components of the delays in the (re)planning process:

- First of all there is the time span between something happens in the production that requires action by the planning entity and this entity becoming aware of the necessity for action. Here we could have the temptation to opt for real-time control, so that we continuously monitor production in order to reduce dramatically this time span. However, this would still leave us the problem addressed under the planning frequency dimension of finding the appropriate planning frequency, i.e. even if we recognize a perturbation, we should not always act immediately otherwise such a high frequency may be counterproductive. Therefore we may set-up a system where control is continuous, but not all "deviations" are necessarily leading to re-planning immediately.
- Next, we should analyse the time required to define a new plan, or modifications of an exiting plan. The reasons for delays here are generally routed in the gathering of all the inputs required to update the plan. For example, if third parties such as suppliers are also affected by changes or many decision levels have to be consulted, to include all required inputs into the decision making may take significant time. Such feedback delays may cause undesired effects at the production site as mentioned before.

- Once an update of the plan has been decided, it has still to be implemented on the shop floor. This often may require a stepwise concretion of the plan along the different planning levels, down to the shop-floor, something which again may take some time in practice.

8.6.4 Degree of Interrelations

Since it is never one single planning decision that causes the planning bullwhip, but rather a qualified combination of many different planning decisions, the bullwhip can be considered an emerging phenomenon of the planning system. In safety research similar phenomena have been observed. Perrow (1999) investigated several major accidents in high risk industries. He found that accidents may occur because of the interrelatedness of an organization's sub-systems. Thereby he differentiates between the concept of coupling and the concept of interaction:

- Coupling describes the degree of elasticity or the extent of the buffers between two interrelated parts. Coupling can be loose or close. Following are examples for close coupling: Production processes do not allow for temporal flexibility, production methods do not provide alternative ways for achieving objectives, production delays in one organizational unit has an immediate impact on the performance of subsequent organizational units. As these examples show, close coupling hampers flexible reacting on and local regulation of variances and disturbances and hence promotes the uncontrolled spread of problems throughout an organization.
- Interactions are occurrences that originate from the combination of two or more events, which themselves occur independently of each other. These events coincidentally occur in a temporal relation that allows for an interaction and as a consequence causes an effect which none of the events could cause individually. Perrow (1999) differs between linear and complex interactions. Whereas the former are known and perceptible, the latter occur unexpectedly and may even be unrecognizable. Complex interaction therefore may remain undiscovered even after a cause analysis. Furthermore, undesirable interactions may even occur, when each of the interacting events for itself is completely sensible. Therefore major accidents can occur although no individual failure happened. This is the case for example, when two independent decisions, each of which is fully justified for itself, lead to problems because they follow irreconcilable objectives. In such situations it may not be foreseeable from the single decision-makers perspective that his decision may contribute to the occurrence of the undesirable interaction.

Perrow (1999) argues that the probability of interactions increase in closely couples systems. Therefore close coupling is to be avoided especially when complex interactions may occur.

Against the background of these concepts, the planning bullwhip can be considered an unwanted interaction of planning decisions. When different decision-makers on different levels of the planning hierarchy (planners, schedulers, dispatchers, shop floor supervisors, operators) are taking planning decisions in an uncoordinated way, the probability of such unwanted interactions increases. There are different means to reduce the probability of uncoordinated planning decisions.

First of all, to reduce coupling needs as much as possible, in order to provide the single decision-maker with decision latitude that is as much as possible disconnected from the decisions of other decision-makers. However, contemporary IT-solutions rather provide closer coupling. This is for example due to the fact that algorithms and models aim at taking more information into account, or due to the mere enhancement of computer performance allowing for real time computing, which - in comparison with batch processing - increases coupling.

However, although it is the aim of many organizational concepts to reduce dependences at the organizational interfaces as much as possible, there will always be some remaining dependency. Generally, regarding these remaining dependences, there are two possibilities to reduce unwanted interactions: centralization or coordinated decentralization. On the first glance centralization may seem to be the more obvious solution. Centralization makes sure, that decisions are taken by one single entity only. This decreases the probability of unwanted interaction among the planning decisions. However, centralization has a number of disadvantages: it reduces a system's flexibility for local regulation of variances and disturbances, and it cannot take detailed information into account. Decentralization on the other hand allows for local flexibility but runs the danger to produce unwanted interactions due to uncoordinated local decisions. Therefore decentralization required measures that ensure for the compatibility of locally taken decisions (Wäfler 2002).

8.6.5 Ease of Representations

Production planning in industrial practice requires accurate process data (orders, master data, etc.) as well as the definition of clear rules for optimization (constriants, priorities, etc.). This kind of information has to be coded in the IT systems used for planning purposes. But also in the brains of the human planners a planning model has to be feeded with data in order they may take planning decisions. The effort to generate and manage the mental models in the brains of the human planners is strongly related to the complexity of the system to be planned. One way to define the complexity of a system can be, for example, by the number of possible states that a system can take during a specific time period. These different states as well as the number of elements involved in the operation are therefore also determining the easiness of representing the planning situation in all its aspects (Fig. 8.7).

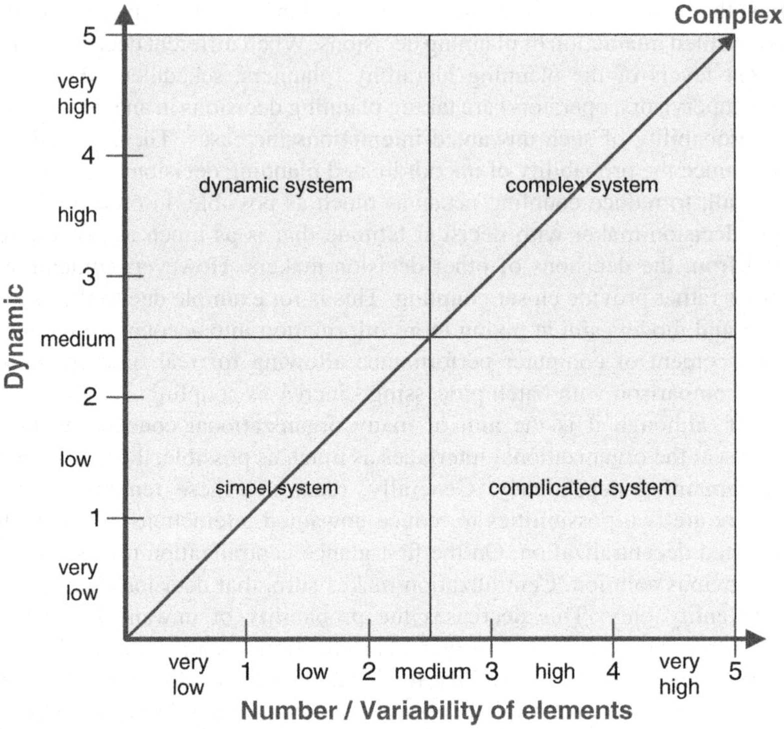

Fig. 8.7 Portfolio of Complexity, according to Bucher (2002)

In the context of planning, scheduling and control such items and system elements as 'customer order', 'product structure', 'resources' and 'production order' are highly determining the effort to keep accurate process data, the rules for optimization as well as keeping the mental models for decision making. Table 8.2 shows a possible operationalization of these aspects.

8.6.6 *Planning Scope and Action Variety*

A final design decision is the scope of each of the planning decisions. The most important design characteristic is the length of the planning horizon. Again, this is related to the lead-time characteristics of the system. The planning horizon does not need to be longer than the time it takes for the current decisions to be implemented and effectuated. On the other hand, the planning horizon needs to be sufficiently long to allow the planner to resolve the planning problem he is faced with: he needs to have sufficient degrees of freedom to make the plan or the schedule (Fransoo and Wiers 2006).

Table 8.2 Operationalization of the easiness of representation

	Effort of representation		
	Less	Medium	High
Customer orders			
Number of customer	<100	101–1,000	>1,000
Number of positions	1–2	3–10	>10
Number of modifications (incl. date	0–10%	11–30%	>30%
Demand predictable	Constant	Predictable	Not
Repeat rate	>40%	11–40%	0–10%
Planning scope (time horizon)	4 months	1 month	Week
Deviation in lead times	High significant	Medium significant	Low significant
Lead times	Cant	Cant	Cant
Product			
Product life cycle (durability)	>10 years	1–10 years	<1 year
Number of sales articles	<1,000	1,001–5,000	>5,000
Bill of material (number of levels)	1–2	3–5	>5
Resources			
Number of different technical methods (procedures)	1–5	6–9	>9
Number of different suppliers (for the order fulfilment)	<10	10–100	>100
Human skills	Polyvalent	Oligovalent	Monovalent
Production orders			
Number of operations/order	1–2	3–5	>5
Variety of capacity load (unpredictable)	Constant	Medium constant	Not constant
Batch size (quantity of same pats)	>10	6–10	1–5
Relation between process time and lead time	<10%	10–25%	>25%

Ultimately, the aim of the developed framework is to provide support when profiling a particular case by means of its analysis along the dimensions described in it (Table 8.1). However, several challenges arise when modelling practical examples. Firstly, a high degree of interrelation exists between some of the framework dimensions. For example, reaction time of planning is influenced by the formal planning frequencies of different planning levels. Also coupling within the planning systems would be highly influenced by the planning structure. Therefore, analyzing a particular case requires an integrated and holistic view of the planning system in order to provide insight allowing for the reduction of the likeliness of a planning bullwhip. Isolated countermeasures approaching only individual aspects of the planning system may be even counterproductive. Secondly, some of the dimensions can be quantified exactly, such as for example the planning frequency or the reaction time. But other dimensions, such as for example the ease of representations or the coupling can be only assessed qualitatively. This will result in an overall semi-quantitative evaluation. But we think that this is still of significant industrial interest, as the results may be rough but most likely good enough to screen out possible structural and operational actions to reduce planning bullwhip likeliness.

8.7 Conclusions and Insights

The planning bullwhip is a complex phenomenon of planning instability. Similarly as the supply chain bullwhip it can affect negatively efficiency and service levels of production. Typically it will lead to such undesirable effects as increased lead times, order backlogs and a general unpredictability. Consequently, the planning bullwhip can be observed best by measuring the planning performance and particularly these symptomatic outcomes, as for example, order backlog. In this chapter, our objective was to study the existence of the planning bullwhip in industrial practice and to investigate the underlying factors influencing the occurrence of this phenomenon. Our findings indicated support for the main mechanisms underlying the planning bullwhip. However, we did not find strong support in our case study for the mitigating abilities of human planners. Apparently, the opportunities for human planners to mitigate any effect are substantially limited by the planning structure, and the availability of up-to-date information.

On the basis of our findings and more conceptual research we have developed a conceptual framework for analysis and design constituted by six key planning systems attributes. By paying careful attention to these six factors, a firm can address the root causes of planning bullwhip, rather than merely focus on its symptoms. For example, it is important to keep in mind that from a more theoretical point of view, it is not fully clear how often planning lead times should be updated, since neither too frequent nor too sporadic updates are always appropriate. Similarly, the appropriate number of hierarchical planning levels is something any company should analyze carefully, as more of them can have its pros and cons.

However, the developed framework should not be understood as leading to a single straight-forward solution. Rather, it is intended to support analysis and design of planning systems. The six attributes are not easy to model and are still somewhat conceptual. Further research will therefore aim at further operationalizing the dimension in more detail, and at testing the framework through its application in further case studies and empirical analysis. Such practical validations will also help to identify opportunities for further improvement of the postulated planning bullwhip framework. Success in mastering better the planning bullwhip may result in significant economic benefits for the companies, as they will improve their on-time delivery, their cost position, and ultimately their customer service.

References

Anthony, R. N. (1963). *Planning and control systems; a framework for analysis*. Boston, MA: Graduate School of Business Administration, Harvard University.

Berglund, M., & Karltun, J. (2007). Human, technological and organizational aspects influencing the production scheduling process. *International Journal of Production Economics, 110*(1–2), 160–174.

Bernstein, F., & Federgruen, A. (2005). Decentralized supply chains with competing retailers under demand uncertainty. *Management Science, 51*(1), 18–29.

Bertrand, J. W. M., & van de Wakker, A. M. (2002). An investigation of order release and flow time allowance policies for assembly job shops. *Production Planning & Control, 13*, 639–648.

Bitran, G., & Hax, A. C. (1977). On the design of hierarchical production planning systems. *Decision Sciences, 8*, 28–55.

Bowers, M., & Jarvis, J. (1992). A hierarchical production planning and scheduling model. *Decision Sciences, 23*, 144–157.

Bucher, H. (2002). *Planung im turbulenten Umfeld, Konzeption idealtypischer Planungssysteme für Unternehmenskonfiguration*. München: Verlag Franz Vahlen.

Cachon, G. (1999). Managing supply chain variability with scheduled ordering policies. *Management Science, 45*, 843–856.

Cachon, G., & Lariviere, M. (1999). Capacity choice and allocation: Strategic behavior and supply chain performance. *Management Science, 45*, 1091–1108.

Chen, F., Drezner, Z., Ryan, J. K., & Simchi-Levi, D. (1998). The bullwhip effect: Managerial insights on the impact of forecasting and information on variability in a supply chain. In S. Tayur, R. Ganeshan, & M. Magazine (Eds.), *Quantitative models for supply chain management*. Norwell, MA: Kluwer.

Chen, F., Drezner, Z., Ryan, J. K., & Simchi-Levi, D. (2000). Quantifying the bullwhip effect in a simple supply chain: the impact of forecasting, lead times, and information. *Management Science, 46*(3), 436–443.

Chung, C. H., & Krajewski, L. J. (1987). Interfacing aggregate plans and master schedules via a rolling horizon feedback procedure. *OMEGA International Journal of Management Science, 15*(5), 401–409.

Croson, R., & Donohue, D. (2006). Behavioral causes of the bullwhip effect and the observed value of inventory information. *Management Science, 52*(3), 323–336.

Daganzo, C. F. (2004). On the stability of supply chains. *Operations Research, 52*(6), 909–921.

De Kok, T. G., & Fransoo, J. C. (2003). Planning supply chain operations: Definition and comparison of planning concepts. In A. G. De Kok & S. C. Graves (Eds.), *Supply chain management: Design, coordination and operation. Handbooks in operations research and management science* (pp. 597–675). Amsterdam: North-Holland.

Eisenhardt, K. M. (1989). Building theories from case study research. *Academy of Management Journal, 14*, 532–550.

Forrester, J. W. (1961). *Industrial dynamics*. Cambridge: Productivity Press.

Fransoo, J. C. (2004). *Lead time technology*. Inaugural Lecture. Eindhoven: Technische Universiteit Eindhoven

Fransoo, J. C., & Wiers, V. C. S. (2006). Action variety of planners: cognitive load and requisite variety. *Journal of Operations Management, 24*(6), 813–821.

Gelders, L. F., & Van Wassenhove, L. N. (1982). Hierarchical integration in production planning: Theory and practice. *Journal of Operations Management, 3*(1), 27–35.

Jonsson, P., & Mattsson, S.-A. (2003). The implications of fit between planning environments and manufacturing planning and control methods. *International Journal of Operations & Production Management, 23*(8), 872–900.

Lee, H. L., & Billington, C. (1992). Managing supply chain inventory: pitfalls and opportunities. *Sloan Management Review, 33*(3), 65–73.

Lee, H., Padmanabhan, V., & Whang, S. (1997). Information distortion in a supply chain: the bullwhip effect. *Management Science, 43*, 546–558.

Leong, K., Oliff, M., & Markland, R. (1989). Improved hierarchical production planning. *Journal of Operations Management, 8*(2), 90–114.

Lindlof, T. R., & Taylor, B. C. (2002). *Qualitative communication research methods*. Thousand Oaks: Sage.

Mather, H., & Plossl, G. W. (1978). Priority fixation versus throughput planning. *Production and Inventory Management Journal, 3*, 27–51.

McClellan, M. (1997). *Applying manufacturing execution systems*. Boca Raton, FL: The St. Lucie Press.

McKay, K. N. (1992). *Production planning and scheduling: A model for manufacturing decisions requiring judgement*. Ph.D. Thesis, University of Waterloo, Waterloo, ON.

Meal, H. C. (1984). Putting production decisions where they belong. *Harvard Business Review, 62* (2), 102–111.

Moscoso, P. G. (2007). A design-oriented framework for modelling production management systems. *Journal of Engineering Design, 18*(6), 599–616.

Neureuther, B., Polak, G., & Sanders, N. (2004). A hierarchical production plan for a make-to-order steel fabrication plant. *Production Planning & Control, 15*(3), 324–335.

Olhager, J., & Rudberg, M. (2002). Linking manufacturing strategy decisions on process choice with manufacturing planning and control systems. *International Journal of Production Research, 40*(10), 2335–2351.

Perrow, C. (1999). *Normal accidents*. Princeton NJ: Princeton University Press.

Qui, M., Fredendal, L., & Zhu, Z. (2001). Application of hierarchical production planning in a multiproduct, multimachine environment. *International Journal of Production Research, 39* (13), 2803–2816.

Selcuk, B., Fransoo, J. C., & de Kok, A. G. (2006). The effect of updating lead times on the performance of hierarchical planning systems. *International Journal of Production Economics, 104*(2), 427–440.

Selcuk, B., Adan, I. J. B. F., de Kok, A. G., & Fransoo, J. C. (2009). An explicit analysis of the lead time syndrome: Stability condition and performance evaluation. *International Journal of Production Research, 47*(9), 2507–2529.

Sogomonian, A., & Tang, C. (1993). A modeling framework for coordinating promotion and production decisions within a firm. *Management Science, 39*, 191–203.

Sterman, J. D. (1989). Modeling managerial behavior: Misperceptions of feedback in a dynamic decision making experiment. *Management Science, 35*, 321–339.

Sterman, J. D. (2002). All models are wrong: reflections on becoming a systems scientist. *System Dynamics Review, 18*(4), 501–531.

Voss, C. A., Tsikriktsis, N., & Frohlich, M. (2002). Case research in operations management. *International Journal of Operations & Production Management, 22*, 195–219.

Wäfler, T. (2001). Planning and scheduling in secondary work systems. In B. L. MacCarthy & J. R. Wilson (Eds.), *Human performance in planning and scheduling*. London: Taylor & Francis.

Wäfler, T. (2002). *Verteilt koordinierte Autonomie und Kontrolle*. Dissertation, Zentralstelle der Studentenschaft der Universität Zürich, Zürich.

Wassermann, O. (2004). *Das Intelligente Unternehmen Prozesse beschleunigen, Menschen begeistern*. Berlin: Springer.

Yin, R. K. (1994). *Case study research: Design and methods* (2nd ed.). Sage: Thousand Oaks.

Zijm, W. H. M., & Buitenhek, R. (1996). Capacity planning and lead time management. *International Journal of Production Economics, 46*(47), 165–179.

Chapter 9
Product Centric Organization of After-Sales Supply Chain Planning and Control

Jan Holmström, Naoufel Cheikhrouhou, Gael Farine, and Kary Främling

Abstract Research on after-sales supply chain management has commonly emphasized strategic business aspects and the application of generic operations control principles. The adoption of Internet and identification technology in after-sales operations opens up new and interesting perspectives that may enable a radical reorganization of after-sales supply chain planning and control. This chapter introduces the product centric service concept, which is the particular organizational perspective that is the basis for the reorganization of the after-sales supply chain. A two-layer approach for organizing after-sales supply chain planning and control is described and the potential benefits explained.

9.1 Introduction

A large number of manufacturing companies have reorganized their business models according to a 'customer-focused perspective', representing a shift from a focus on the products they manufacture to a concentration on their customers and the value that their customers derive from ownership and use of these products (Cohen et al. 2006b). These business models are not only more lucrative when dealing with complex assembled products along the supply chains but also provide competitive differentiation by capturing profits at the customer's end of the value chain (Dennis and Kambil 2003; Wise and Baumgartner 1999).

Today, after-sales services – i.e. the services supporting products – contribute to about 25% of all revenue and 40–50% of all profits for manufacturing companies.

J. Holmström (✉) and K. Främling
Aalto University, Helsinki and Espoo, Finland
e-mail: jan.holmstrom@tkk.fi, Kary.Framling@tkk.fi

N. Cheikhrouhou and G. Farine
Ecole Polytechnique Fédérale de Lausanne, Lausanne, Switzerland
e-mail: naoufel.cheikhrouhou@epfl.ch

J.C. Fransoo et al. (eds.), *Behavioral Operations in Planning and Scheduling*,
DOI 10.1007/978-3-642-13382-4_9,

Moving downstream towards after-sales supply chain activities results in revenue and profit enhancement, but it also increases customer acquisition and retention, and competitive advantage (Dennis and Kambil 2003). The complex structure and organization of an after-sales supply chain cannot be easily replicated and hence provides a sustainable source of competitive advantage.

Even the best preventive maintenance policies cannot eliminate breakdowns; the focus of development actions is then put on exploring the conjoint improvement of after-sales supply chain planning and servicing processes of installed bases – which represent products on customer site, target of the industrial after-sale services (Auramo and Ala-Risku 2005). Going downstream rises different questions related to the way information need to be controlled, updated and shared between the supply chain actors, the coordination of different planning sources and identifying the enabling technologies for handling after-sales supply chain planning and control. Some work has focused on the strategic importance of aftermarket supply chain organization and benchmarking without attention to the technical enablers to achieve the expected reorganization (Lewis and Naim 1995; Harland 1995). In particular, Farris et al. (2005) considers the importance of addressing the distributed information management issue as a key factor for organizing aftermarket support processes. This is the organizational planning and control issue discussed in this chapter.

A critical technical problem – which is related to organizational design – is updating information related to a single product (unique product identity) as well as composite products (bill-of-material made products) along the after-sales supply chain. Product instance specific information updating is one of the basic features required to improve planning and control in the after-sales supply chain (Kärkkäinen et al. 2003). For a traceability purpose with a unique product identity, Jansen-Vullers et al. (2004) have developed the concept of the traceability decoupling point in the particular field of food industry, fostering the downstream information push as the most important criteria in building traceability projects within supply chains. For composite complex products, Främling et al. (2006) propose a multi-agent system to manage distributed information system along in supply chains. From this promising starting point, the Product Centric Service is here directed towards after-sales supply chain planning and control, particularly for product data management.

The objective of this chapter is twofold: the identification of what type of after-sales supply chain planning and control is needed for high logistic performance and proposing an organizational design for the needed planning and control. The proposed solution is a two-layer organization where the managerial as well as the technical aspects are treated. The purpose of the proposed organization design, based on product centric services, is to improve productivity, efficiency and quality of after-sales service operations. In the first part of this chapter the basic research questions and major challenges are presented. An approach for organizing the planning and controlling processes of after-sales supply chain is described in the next part, followed by an illustrative example and the discussion of benefits and constraints of introducing product centric service organization in the after-sales supply chain. The chapter ends with a summary and concluding remark.

9.2 Organizing After-Sales Supply Chain Planning and Control

9.2.1 *Characteristics and Problematic of After-Sales Supply Chains*

After-sales supply chains are defined as the network of resources that includes the appropriate material (service parts, service utensils), people (central after-sale managers, servicing technicians, call centre staff, report depot staff, warehouse and transportation staff) and infrastructure (for material movement and storage, repair, transportation, information system, and communication) required to service installed bases (Cohen et al. 2006a; Auramo and Ala-Risku 2005). Such supply chains are significantly more complex than traditional manufacturing supply chains. This is principally due to more extensive information and product flows and to the complicated hands-on logistics of spare parts. In addition, other factors can complicate after-sales compared to manufacturing supply chains as indicated by Dennis and Kambil (2003). First, inventories complexity is increased by an inconsistent and uncertain demand for spare parts. This is directly tied to equipment breakdowns and therefore maintenance needs, which tend to be unpredictable and differs dramatically from manufacturing schedule plans found in usual manufacturing supply chains. Spare-parts inventories require up to 20 times more stock keeping units than what is needed for current-product manufacturing. Second, spare-parts logistics leads to a challenge through stocking locations more distributed than for manufacturing operations. This increase in stocking locations combined with varying customer service requirements – i.e. huge differences between the worth of the many different spare parts – make resource requirements and warranty costs and other parameters extremely difficult to predict. Finally, non-routine product failures often appear, making the supply chain service slow and inefficient as long as information about location, product and service providers is missing or difficult to retrieve. Sharing knowledge and information on resources is therefore necessary to keep a high quality of service.

9.2.2 *Organizational Design Challenge in After-Sales Supply Chains*

The dominant view on how to manage inter-organizational processes is to focus on the sequential chain of actors and actions. For example, the objective in the manufacturing supply chain management is to manage material and information flows between actors linked to locations swiftly and fluently (Schmenner and Swink 1998; Lummus and Vokurka 1999; Mentzer et al. 2001). On the most fundamental level planning and control rests on inventory management solutions that use material specific accounts to keep track of on-hand stock in specific inventory locations (see e.g. Vollmann et al. 2004, pp. 224–235).

Organizing the information management for tracking product state, product location and planning service deployment is a challenging task if the information management is set up to mirror the supply chain sequence. The organization then has to cope with the following challenges:

- A company belonging to the supply chain may not need the information for its own purpose but it is required to receive and send information to other companies for their own purposes
- If a company belonging to a supply chain is unable to communicate with the other companies, the information flow is interrupted on the downstream as well as on the upstream direction
- All the information about product maintenance, control and reporting is difficult to store and to keep up-to-date for each company of the supply chain
- Transmitting information downstream may cause information overflow, especially in the manufacturer end of the supply chain where information on the final product, the subassemblies and the components each have to be managed

However, in an after-sales service operations all products can be considered to be individuals based on their use and maintenance history (Callon et al 2002). The consequence of this is that the sequential flow configuration of the sequential supply chain needs to be challenged for planning and controlling after-sales operations (Sampson and Froehle 2005). The reason is that for customized service delivery the individual product or its resource inputs may need to be located and redirected according to customer specification simultaneously at many different points of the supply chain. Currently this type of responsive service is feasible within the organizational boundary of a specific service provider, say a contract manufacturer or a logistics service provider like UPS. But accomplishing this between independent service providers – if the organization is sequential – requires a cumbersome sequential use of several different planning and control solutions.

From a theoretical perspective of organization design (Galbraith 1972), alternatives can be found to the chain configuration in an inter-organizational context. There are three basic configurations for inter-organizational systems (Kumar and van Dissel 1996), of which a sequential chain configuration is only one. The other alternatives are the pooled and reciprocal configurations. The benefit of moving away from a sequential organization towards a pooled configuration is that information loss can be reduced, i.e. in a pooled organization uncertainty absorption (March and Simon 1958) can be reduced through sharing the same information rather than sequentially passing on information. Uncertainty absorption only occurs between the collection and use of product individual information, not between each actor in the supply chain. In this chapter, we discuss an easy to deploy pooled configuration – the product centric service organization that is centred on the product individual – for inter-organizational planning and control.

9.3 Proposed Organization Design: Product Centric Services for After-Sales Supply Chain

Considering supply chains as distributed networked companies, the idea is to use the Internet as the communication vector to fulfil the after-sales supply chain needs and to distribute product-centric information. Product centric services is based on a multi-agent system where the software agents are associated to their physical counterparts (Främling et al. 2006).

With a purpose of improving productivity, efficiency and quality of service operations in after-sales supply chain service delivery, the proposed organization is based on building two interconnected layers, the instance layer and the provision layer (see Fig. 9.1 below). These two layers use Internet as a vector for information access, sharing and updating through unique identification of every considered item (product, sub-assemblies and components). Standardization initiatives for item reference standardisation, such as the Global Trade Item Numbers and Electronic Product Code for RFID-based identification (Gs1 2007) are building the organizational foundations for the instance layer of product centric services. The instance layer organizes information related to an individual product, its components, use location, and use and maintenance history. The instance information can be distributed between the physical product and application components. The organizing principle is the unique identity of the instance, based on which the information on the instance is collected and accessed.

The development of the provision layer is the means to improve the performance of the after-sales supply chain. The challenge is to identify those specific after-sales supply chain planning and service delivery activities (provision layer) that can be enhanced with product centric information (instance layer). The product centric data can be conceptually regarded as an instance layer that contains the information

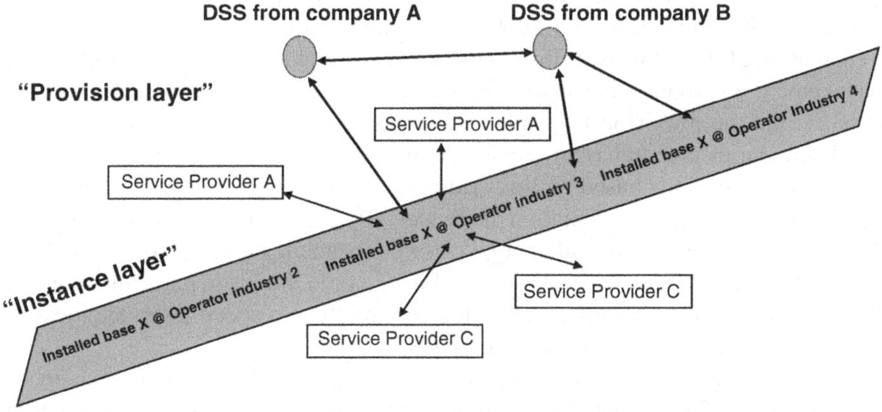

Fig. 9.1 Two-layered organizational design combing product centric data management and after-sales service supply chain management

on individual products, their service history and requirements (Parsons and Wand 2000). For improving supply chain planning and service delivery this information needs to be transferred and gathered in different decision support systems, which conceptually make up the provision layer.

Provision layer Decision Support Systems (DSS) act as systems for helping decision makers to carry out after-sales supply chain planning by giving more insight of resources and installed bases status according to the used service scheme, or service composite. The decision support systems also allow dynamic creation of the knowledge resources needed for effective planning execution. The mechanism behind improved planning performance is reduced uncertainty absorption (March and Simon 1958) and pooling of resources (Galbraith 1972). The tracking based data collection on the instance layer retains accurate and reliable information on the individual products, which can then be used to organize and allocate resources according to actual requirements rather than aggregated estimates. For example, enabling more accurate positioning of spare parts in distributed service locations when the installed base is highly concentrated to a limited number of places, and in centralized locations when the installed base is spread out between many places.

The integration of product centric data and service provision to planning activities is a major organizational challenge. It is difficult to incrementally develop provision level applications if all potential uses of instance information on the provision layer need to be in place when the instance level is created and deployed. To create a full-grown provision level solution without trial and error is also very difficult. A way out is the proposed two-level architecture, where applications on the instance and provision layer can be linked both in initially well-defined ways and more loosely in ways that have not been defined from the beginning.

The proposal builds on the approach that has been outlined by Främling et al. (2006). A generic instance layer where information can be both directly and indirectly gathered and accessed are separated from decision support systems on the provision layer. The approach is similar to the asset life cycle management approach which is frequently deployed to improve the reliability of large industrial systems (Liptrot and Palarchio 2000) and complex equipment, such as airplanes (Pátkai et al. 2007). However, the scope of interest is wider than in asset life cycle management. In product centric services the instance layer cover multiple plants and systems, and the provision layer planning tasks cross geographical and organizational boundaries in search of improved planning and resource allocation efficiency.

The direct access is based on links from the instance layer to application on the provision layer where information is needed. Product data can be sent as text, HTML, XML or by some other application-dependent message, or be updated through a direct database to database link. For example, for spare parts inventorying the individual product instance can be linked directly to a specific service location responsible for the availability of spare parts.

The indirect approach is based on the identity of the product being known on the provision layer. The unique identity of an instance is the key to the database where instance information is stored. Data is fetched by visiting the instance data storage locations in the same way as search engines visit Internet pages. Gathering the data for provision layer decision support systems applications can be done manually by visiting the instance level information stores, but such operations are slow and prone-to-error. The use of specific instance crawlers represents the best solution for fetching indirectly instance data. This way, provision level applications can browse the installed base information in the same way as search engines browse Internet pages in a methodical and automated manner, periodically updating the requirements for different spare parts in the inventory locations based on changes in the installed base.

9.4 Example Implementation

In the proposed organizational design, the notation ID@URI is used to uniquely identify an item (and so to the corresponding agent) and link it to the party responsible for information management. The URI side identifies the company responsible for managing information related to the product, and this identifier also refers to the server in which the management information is stored. The ID part is the identifier of the product that the URI location is responsible for.

For both the direct and indirect linking between the instance and provision layers, an organizational challenge is how to ensure that both software components indeed have the same interface so that they use compatible information identification when the components are programmed by different companies or when they use different versions (Anke and Främling 2005) Messages content and interface are therefore best standardized to succeed in relevant data fetching and this represents one of the principal constraints of such a system. Also, the fetching and update frequency of information between layers leads to big challenges. Some instance level information may indeed need to be automatically updated on the provision level immediately when changes occur, e.g. when contractual obligations exist for maintaining the availability of critical spare parts or service technicians. Whereas others, need only be updated periodically, such as for inventory management in centralized inventory locations. The update in frequency in indirect links must therefore be examined with care and depend on information and products types and on the system capacity (Främling et al. 2007).

An example from practice is the manufacturer of home appliances that HUT is working with just now. An installed prototype implementation demonstrating this organizational design exists with functionalities on both organization layers, focusing on predictive maintenance and spare parts provision.

A set of home appliances (such as refrigerators) is the first layer, i.e. the collection of instance-objects. Parts of the instance level information can be located in the appliance itself, some parts at the manufacturer's information service and some parts may also be in the computer of the local maintenance provider or local shop that sold the appliance. All these links can be handled directly (as described in Främling et al. 2007) or indirectly by some lookup mechanism. This is still first-layer functionality, though it is distributed in numerous locations and systems. Since the product agent, or at least some of the product information, is distributed, we need a mechanism to retrieve it and update it if needed.

What is then the second layer? Or said in a different way, what dynamic views can there be on the first layer information? The important point to note here is that a first layer object is unique and that this unique identity is used to manage all information about the instance, or object. On the second layer there may be any number of different views of the information related to a number of objects, but these views all (except maybe one "total information view") only use parts of the available information. The same view is repeatedly applied to many different objects of the same type when planning and control tasks are to be performed, for example when updating inventory requirements for a local repair shop based on the appliances present in the local area. This is the major purpose of the organizational design proposal: if object properties and references are done according to identical patterns, then you can also design different views or algorithms that handle different objects, for example appliances by different manufacturers, in the same way.

To sum up, interesting second layer views in the setting of appliance repair example are (1) local repairs and service shop view that collects maintenance-related information for appliance in a specific area; (2) the end-user view for the appliance owner that presents information about appliances based on ownership in an uniform way; (3) recycler/dismantler view providing the relevant information for appliances that have reached the end of the life-cycle.

The instance layer in Fig. 9.2 consists of all the product information on the appliances accessible over digital communication lines. The distributed instance layer is indicated with bold font, and the second layer is in normal font. The instance layer information can be accessed by using the ID@URI identifier in any of the (Internet-)connected provision layer nodes. The appliance itself might not be

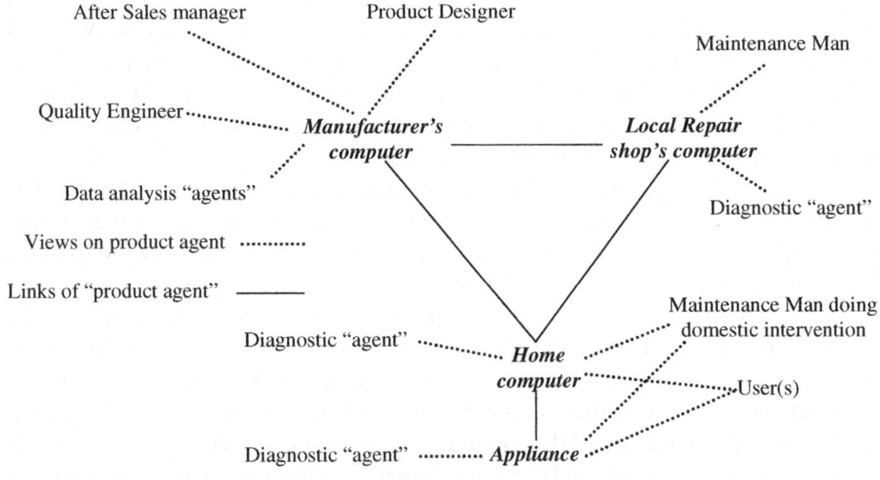

Fig. 9.2 Illustration of the product centric service organizational design

Internet-connected, it is only essential that a instance level node exists that tracks the information related to the appliance. User interfaces corresponds to views, which are in this case the software components of the service shop, the user, or recycler that query the needed information about the product item and presents it in a useful form to a human user or an automated agent.

That the decision support views on the provision layer can use instance data that has been collected indirectly, means that there can be ad-hoc providers and new service providers introduced over time. Different routing, problem locating and defect rectification systems can integrate product data as an input and drive service management steps in ways that were not envisioned when the instance layer was created. This way, based on the proposed organization design the specific services that are created and in which order does not need to be pre-planned, but may be provided based on market need and opportunity. Depending on the supply chain position of a company and the time horizon for improvement the information accessed from the instance layer can be used to improve product design, enhance the service product, reposition service assets, update the location-spare parts distribution map, make better spare parts replenishment decisions, improve demand fulfillment actions, and customer contract management. The two-layered organization can support each of these after-sales service supply chain initiatives.

9.5 Feasibility Evaluation

Adopting new modes of operation, such as product centric service in the after-sales supply chain require efforts work, and for an operation distributed in many different locations and many organizations the required efforts are greater than in a centralized and tightly controlled organization. Technology adoption research and the analysis of technological frames of potential adopters (Orlikowski and Gash 1994) indicate that in this situation it is critical that the perceived effort to adopt a new way of working is low and the potential benefit is easy to understand. This means that the design of the instance layer must be kept simple and open, in order to keep the perceived effort of implementing the instance level low. Furthermore, it is imperative that there are clearly communicated provision layer benefits, such as spare parts inventorying based on the instance layer status.

Product centric systems are new and it will take time for the appropriate and most useful applications to emerge. In this situation it is important that future uses are not restricted by first uses. Agent interaction analysis (Stuit and Meyer 2009) has been proposed as a formal approach to determine the openness of a product centric system. If the two layered approach is used the product centric service system can be organized in loosely coupled and open way, and can potentially be used changed over time with less work than current conventional and centralized proprietary systems. The proposed two-layered organization makes explicit the role of the product instance as the coordinating entity in the delivery of customized products and services on both the local and supply chain level. The key is that changes in one part of the provision layer do not affect other parts. However, it is important to note the constraints imposed when different provision level planning activities require coordination.

Finally, security is a challenge, which will remain one of the biggest stumbling-blocks for open and distributed supply chain systems (Kärkkäinen et al. 2003). Implementing an instance layer that relies on the Internet exposes it to server and network down-time, as well as to intrusions and virus attacks.

9.6 Summary and Discussion

In this chapter, the two-level product centric organization to support after-sales supply chain planning and execution has been presented and discussed. It consists of an instance layer representing the product instances through their evolutions through supply chains, managed and controlled by a provision layer of the different supply chain actors. Its implementation via the ID@URI concept has the benefit of better responsiveness, efficiency and productivity along supply chain actors without specifying a rigid implementation sequence or roadmap. In addition, thanks to the ability of managing composite product data, rather "market opportunity" and innovation can be supported.

A reliable instance level information system is not a solution for problems in service resources planning and dispatching, but a tool for dealing more effectively in situations where human resources may be unavailable for a service visit or not skilled enough, and in situations when parts need to be located on the fly when unexpected problems surface during the corrective phase. The product centric organization fulfils important requirements of the after-sales supply chain planning and control in:

- Accessing the product data everywhere internet is available without a need of physical availability on the installed bases
- Updating product information such as recording new maintenance operations and making it available to the concerned actors of the supply chain
- Remote maintenance controlling and reporting
- Locating quickly the product along the supply chain and identifying the nature of the service needed and the adequate resources

Different new research directions can be identified, depending on the focus. On the technical side, generalization of the ID@URI concept by adapted standards of communication protocols is a big challenge. From an organizational point of view, and thanks to the product centric enabling technologies, different questions rise concerning information sharing processes and mechanisms in supply networks for equitable aftermarket business profit sharing. Finally, trust and security along after-sales supply chain s is a matter of further research.

References

Anke, J., & Främling, K. (2005). Distributed decision support in a PLM scenario. *Proceedings of Product Data Technology Europe 14th Symposium* (pp. 129–137). 26–28 September 2005, Amsterdam, The Netherlands.

Auramo, J., & Ala-Risku, T. (2005). Challenges for going downstream. *International Journal of Logistics: Research and Applications, 8*(4), 333–345.

Callon, M., Méadel, C., & Rabeharisoa, V. (2002). The economy of qualities. *Economy and Society, 31*(2), 194–217.

Cohen, M., Agrawal, N., & Agrawal, V. (2006a). Achieving breakthrough service delivery through dynamic asset deployment strategies. *Interfaces, 36*(3), 259–271.

Cohen, M., Agrawal, N., & Agrawal, V. (2006b). Winning the aftermarket. *Harvard Business review, May*, 2–13.

Dennis, M. J., & Kambil, A. (2003). Service management: Building profits after the sale. *Supply Chain Management Review, 7*(1), 42–49.

Farris, T., Wittmann, M., & Hasty, R. (2005). Aftermarket support and the supply chain. *International Journal of Physical Distribution & Logistics Management, 35*(1), 6–19.

Främling, K., Ala-Risku, T., Kärkkäinen, M., & Holmström, J. (2006). Agent-based model for managing composite product information. *Computers in Industry, 57*(1), 72–81.

Främling, K., Ala-Risku, T., Kärkkäinen, M., & Holmström, J. (2007). Design patterns for managing product life cycle information. *Communications of the ACM, 50*(6), 75–79.

Galbraith, J. (1972). Organization Design: An information processing view. In J. W. Lorsch & P. R. Lawrence (Eds.), *Organization planning: Cases and concepts* (pp. 49–72). Homewood, IL: Irwin.

Gs1 (2007). www.gs1.org, accessed June 11, 2007.

Harland, C. (1995). The dynamics of customer dissatisfaction in supply chains. *Production Planning & Control, 6*(3), 209–217.

Jansen-Vullers, M. H., Wortmann, J. C., & Beulens, A. J. M. (2004). Application of labels to trace material flows in multi-echelon supply chains. *Production Planning & Control, 15*(3), 303–312.

Kärkkäinen, M., Ala-Risku, T., & Främling, K. (2003). The product centric approach: a solution to supply network information management problems? *Computers in Industry, 52*(2), 147–159.

Kumar, K., & van Dissel, H. (1996). Sustainable collaboration: Managing conflict and cooperation in inter organizational systems. *MIS Quarterly, 20*(3), 279–300.

Lewis, J. C., & Naim, M. M. (1995). Benchmarking of aftermarket supply chains. *Production Planning & Control, 6*(3), 258–269.

Liptrot, D., & Palarchio, G. (2000). Utilizing advanced maintenance practices and information technology to achieve maximum equipment reliability. *International Journal of Quality & Reliability Management, 17*(8), 919–928.

Lummus, R., & Vokurka, R. (1999). Defining supply chain management: a historical perspective and practical guidelines. *Industrial Management & Data Systems, 99*(1), 11–18.

March, J. G., & Simon, H. A. (1958). *Organizations* (2nd ed.). New York: Wiley-Blackwell.

Mentzer, J., DeWitt, W., Keebler, J., Soonhoong, M., Nix, N., Smith, C., et al. (2001). Defining supply chain management. *Journal of Business Logistics, 22*(2), 1–25.

Orlikowski, W., & Gash, D. (1994). Technological frames: Making sense of information technology in organizations. *ACM Transactions on Information Systems, 12*(2), 174–207.

Parsons, J., & Wand, Y. (2000). Emancipating Instances from the tyranny of classes in information modeling. *ACM Transactions on Database Systems, 25*(2), 228–268.

Pátkai, B. et al. (2007). *Requirements for RFID-based sensor integration in landing gear monitoring – a case study, Auto-ID Labs.* Available at: http://www.aero-id.org/research_reports/AEROID-CAM-016-MessierDowty.pdf [Accessed July 20, 2009].

Sampson, S., & Froehle, C. (2005). Foundations and Implications of a Proposed Unified Services Theory. *Production and Operations Management, 15*(2), 329–343.

Schmenner, R. W., & Swink, M. L. (1998). On theory in operations management. *Journal of Operations Management, 17*(1), 97–113.

Stuit, M., & Meyer, G. (2009). Agent Interaction Modeling Based on Product-Centric Data: A Formal Method to Improve Enterprise Interoperability. In K. Fischer et al. (Eds.), *ATOP 2005 and ATOP 2008, LNBIP 25* (pp. 197–219). Berlin: Springer.

Vollmann, E., Berry, W., Whybark, D., & Jacobs, F. (2004). *Manufacturing planning and control for supply chain management*. New York: McGraw-Hill.

Wise, R., & Baumgartner, P. (1999). Go Downstream: The New Profit Imperative in Manufacturing, Harvard Business Review, September, pp. 133–141.

Chapter 10
Human Control Capabilities

Toni Wäfler, Rüdiger von der Weth, Johan Karltun, Ulrike Starker, Kathrin Gärtner, Roland Gasser, and Jessica Bruch

Abstract This chapter has been triggered by the experience that the implementation of new information technology (IT) supporting planning, scheduling, and control – although being more sophisticated than earlier systems – does not necessarily result in better control. Also, the experience was made that the implementation of the same IT leads to different results in similar organisations. Against this background, we introduce a process model of control (Sect. 10.2). The model proposes a set of interrelated factors determining control. At its core it assumes that control results as a fit of control requirements and control behaviour. The former is determined by operational uncertainties the latter by control opportunities, control skills and control motivation. Since the implementation of a new IT can have an impact on all these factors it can lead to a misfit of control behaviour and control requirements and hence to low control – even if the new IT itself is more powerful than the old IT. Furthermore, we also discuss motivational influences these changes

T. Wäfler (✉) and K. Gärtner
School of Applied Psychology, University of Applied Sciences Northwestern Switzerland, Olten, Switzerland
e-mail: toni.waefler@fhnw.ch, kathrin.gaertner@fhnw.ch

R. von der Weth
HTW, University of Applied Sciences, Dresden, Germany
e-mail: weth@wiwi.htw-dresden.de

J. Karltun and J. Bruch
Department of Industrial Engineering and Management, Jönköping University, Jönköping, Sweden
e-mail: johan.karltun@jth.hj.se, jessica.bruch@jth.hj.se

U. Starker
University of Bamberg, Bamberg, Germany
e-mail: ulrike.starker@uni-bamberg.de

R. Gasser,
Department of Mechanical and Industrial Engineering, University of Toronto, Toronto, ON, Canada
e-mail: roland.gasser@utoronto.ca

J.C. Fransoo et al. (eds.), *Behavioral Operations in Planning and Scheduling*,
DOI 10.1007/978-3-642-13382-4_10,

may have on human behaviour (Sect. 10.3). Finally we derive some practical dos and don'ts when implementing new IT (Sect. 10.4).

10.1 Introduction

Manufacturing operations are continually exposed to a large number of disturbances and fluctuations both from outside and inside the organisation that decrease their potential output. Since the ideal conditions for maintaining a stable, high-quality and low cost output in manufacturing are those associated with stability, manufacturing organisations strive for establishing such conditions by various techniques and ways of organizing.

One of the methods used in manufacturing history was standardising the product, a technique that became a real success in the beginning of the twentieth century. A good example of such thinking is the expression that is ascribed to Henry Ford concerning the Ford Model T: "Any customer can have a car painted any color that he wants as long as it is black." Other examples of methods for creating stability in manufacturing are standardization of procedures, production levelling and Just-In-Time, all being essential parts of "lean production". However, manufacturing systems are characterized by a variety of disturbances and fluctuations that must be managed. Under certain conditions these can be decreased by methods and techniques mentioned above. For a large majority of cases though, manufacturing firms use more or less advanced administrative systems to regulate and control their operations in order to handle the disturbances and fluctuations in an acceptable or favourable way.

Thus, control of operations is central in manufacturing and huge efforts are put on improving the efficiency and effectiveness of the control system. The complexity in the control task is high and the amount of data needed and used is extremely large. Together with human decision-making, Information Technology (IT) has been *the* tool for improving control for decades now. Software developments still lead to more and more sophisticated information systems, often termed Enterprise Resource Planning (ERP) systems. Newer ERP systems are modularized and incorporate a range of modules from finance and accounting to manufacturing and logistics, and some are even providing supply chain management solutions (Dery et al. 2006). These systems can cover the control needs of the whole organisation, from sales to manufacturing, delivery, and invoicing. The investments in ERP systems are high. The worldwide market for ERP systems has reached 79 billion US$ annually (Gefen and Ragowsky 2005).

These systems function very well in many applications and are believed to lead to significant cost savings and increased profitability by delivering operational advantages like reduced procurement costs, smaller inventories, more effective sales strategies, lower administration costs and reduced direct and indirect costs (Dery et al. 2006). Furthermore, they are expected to improve decision making because of their ability to provide specifically designed "real time" information to assist different management functions and procedures. However, they are also considered to bring unforeseen costs of large magnitudes as well as leading to unintended outcomes. Implementation costs often exceed cost savings and revenue gains. Between 60 and 90% of implementations failed to achieve the projected return on investment (Stedman 1999; Trunick 1999). These shortcomings during the 1990s have resulted in a growing literature on implementation success factors and better implementation knowledge. Wang et al. (2008) found that the consistency among the success factors had a significant positive impact. Ngai et al. (2008) identified 18 critical success factors and more than 80 sub factors for successful implementation of ERP systems in their literature review of 48 publications. They also concluded that it is not only implementation that determines the success. But the measurement of success remains very difficult for such projects. Some authors argue that the implementation should be seen as the first step towards a more or less continuous process of developing the fit between the operations and the ERP system. This post implementation process needs to be carefully designed to make the ERP system profitable (Nicolaou and Bhattacharya 2008). Yeh et al. (2007) concluded that the service quality of semiconductor-related industries in Taiwan was improved by ERP implementation. But they also remarked that the failure rate of the implementation of ERP systems in Taiwan is very high. In general, ERP implementations are still considered to be a high risk venture with uncertain outcome.

The design and functionality of ERP systems are such that they transform the nature, structure, and management of work throughout the entire organisation (Dery et al. 2006). End-users of ERP systems, being those who are dependent on their interaction with the software to fulfil their task, differ in needs and conditions for using the system. Calisir and Calisir conclude that there is a need to define user categories broadly in order to understand the end-user satisfaction as defined by the learnability and the perceived usefulness of the system (Calisir and Calisir 2004).

Dery et al. (2006) found that ERP publications are focussing mainly on the functional/technical aspects of the software and their likely impact on business performance in terms of efficiency gains, improved information flows, data processing, and profitability. Another main focus is laid on the implementation of ERP systems. Recently, some studies have been addressing the usage and maintenance of such systems. However, the literature is mainly managerially oriented, and there is a lack of studies that examine aspects related to work organisation and workplace performance as post-implementation responses, as well as effects on workplace control, power, and resistance (Dery et al. 2006). Moreover, these systems incorporate models of operations that were shaped not by the users, but by the designers of the systems – some of them reach back in history, as they are building up on older models during the development process – and these models may well be different from the work system where the ERP system is supposed to be used (Benders et al. 2006; Light and Wagner 2006; Locke and Lowe 2007).

Thus, the problem of getting value out of ERP implementations is not well understood due to shortcomings of current ERP research. From a professional practitioner's point of view questions like "Why does the same IT system work differently in highly similar production systems?" and "Why does not a more sophisticated IT system automatically lead to more control?" are expressions of these shortcomings that need to be explained and further investigated.

In this chapter, we will suggest an alternative approach for elucidating the problems of the interaction between IT system and workplace performance. We present a view of the control system in manufacturing that is taking into account humans as actors for control. It is a view that considers control of a manufacturing system to be performed by a large number of actors in the system. It is furthermore a view that takes into consideration different means for control that are available. This implies not only using the technology for control, the IT system, but also using the organisational opportunities as well as the human capabilities of each employee.

The view is based on the socio-technical system approach and hence on the idea of considering a manufacturing system as a work system composed of a technical system and a social system, which interact and are mutually dependent. When changing the system (e.g. due to the implementation of an ERP system) the social as well as the technical system need be analysed and changed in accordance with each other and the overall goals of the change. Our position is thereby not to regard ERP systems as being unable to satisfactorily control operations, but to present how their potentials can be exploited by taking into account the power of human control. This means that the work system is considered to be controlled by the interplay of humans and technology. Control is executed in a distributed way in the system as a whole. Many different actors in the work system are – to some extent – involved in decision making with respect to control. A more elaborate discussion on how the chapter is related to socio-technical theory can be found in Sect. 10.2.6.

One characteristic of a work system that clearly demonstrates the difficulty of centralised planning is the fact that the degree of detailing is at its highest at short term shop floor level and decreases with distance in time from present (cf. Fig. 10.1). Since centralized planning is most likely not able to take into consideration all detailed information available at the short term shop floor level, control needs to be distributed throughout the organisation.

Control is thus constituted by human control behaviour of all humans involved in planning and scheduling activities - in one way or another, formally or informally. However, their control behaviour is conditioned by the availability of different kinds of resources, with the ERP system being one of the most important. In general, these resources are determined by the way work systems are (socio-technically) designed and organized. Any change (e.g. implementation of an ERP system) will consequently alter the available resources and hence the human control behaviour. If the change leads to a poor fit of the conditions and the demands for control, the control within the system will decrease instead of improve. This usually happens unexpectedly and control loops that were unknown up to date are revealed by changes.

An example of this is the machine operator whose schedule was updated daily instead of weekly after a change. His ability to locally optimise his machine concerning setups and loading was thus heavily impaired, because of frequent schedule changes. At next period's degree of machine utilization assessment, he

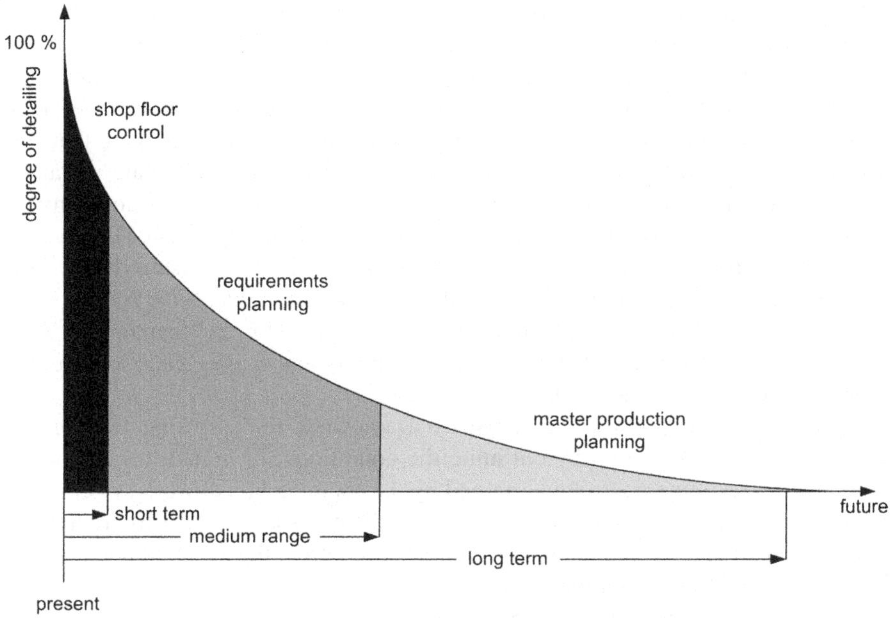

Fig. 10.1 The temporal range of planning activities in production (modified from Scherer 1998)

was blamed for decreased performance, although he had no longer any influence on the schedule and thus could not affect the performance. It was not until after the change that it was discovered that the implementation of the new IT hampers local control. This is because the implementation project focused on the IT's ability to control rather then on the impact the implementation of the IT has on the humans' ability to control – especially those humans who's main task consists not of planning and scheduling are normally not considered at all.

Furthermore, research on control behaviour suggests that human control requires the notion or feeling of being in control. Consequently, the objective fit between the demands for control and the organisational resources is not sufficient to improve control. A subjective perception of being in control is also necessary to improve human control behaviour and hence system control performance.

The need for a personal feeling of being in control is recognized at the top management levels. Consequently most ERP systems provide sophisticated functions to instil the perception of being in control to management. A tempting strategy sometimes applied for this purpose is to (mis-)use integrated ERP systems for a very detailed data monitoring. The expectation is that in doing so, an in-depth description and analysis of production and logistic processes is generated and hence an improvement of operational control as well as perceived control is achieved. However, we assume that in many cases this strategy increases the management's perceived control only, whereas its actual implication on the control capability of the work system remains unclear. There are several reasons for this assumption.

One is that managers suffer from the same human limitations as everybody else concerning the ability to use large quantities of data for decision purposes. Another is that the data in the system may not fully correspond to reality, which means that the detailed monitoring is done on a more or less invalid description of the real situation. Moreover, other effects of ERP system implementation may be that the flexibility of the work system decreases, that work design principles are violated, and that the basic ideas behind such an IT system and the way a company is organized may be in conflict (Benders et al. 2006; Koch and Buhl 2001).

A further problem is the increasing complexity of the ERP system itself. The objective of a complete IT-based model of the organisation makes the ERP system very complex and thus difficult to understand, to interact with, and to maintain. This development increases the demands on the humans and makes their work more difficult instead of facilitating it.

Against this background, we suggest a framework for assessing the control capability of a work system by outlining the conditions for human control behaviour and the resulting control determined by the socio-technical work system. We furthermore suggest how control capability can be analysed in order to predict the influence of different means taken to improve control, the most important one probably being implementation of new IT.

The chapter is arranged in the following way:

- The second section outlines a system design model and its underlying perspectives.
- In the third section, the behavioural perspective and its influence on the control capability of a work system is developed and underpinned.

10.2 Control Model

Control can be seen as a part of all coordination activities within a work system, with the goal to smoothen operations to enable order fulfilment as good as possible. From a technological perspective, the concept of control in work systems can be modelled using control theory for depicting the entire system (cf. Fig. 10.2).

Figure 10.2 shows that the interplay between the information system and the physical system is highly integrated. The control input to the system consists of the objectives to be reached concerning output. Moreover, Fig. 10.2 gives information on sources of disturbances and unforeseen events. As can be seen these disturbances emanate both from within the system and from outside (Grote 2004), and they can be of informational or physical nature. Furthermore, the control system must also be able to handle the natural variances in operations as well as the variations imposed by the complexity of the product and production system as such (McKay and Wiers 2004). Disturbances, unforeseen events, variations, and variances together form the operational uncertainty of a work system. They all contribute to the control requirements, which can be defined as the sum of proactive and reactive actions

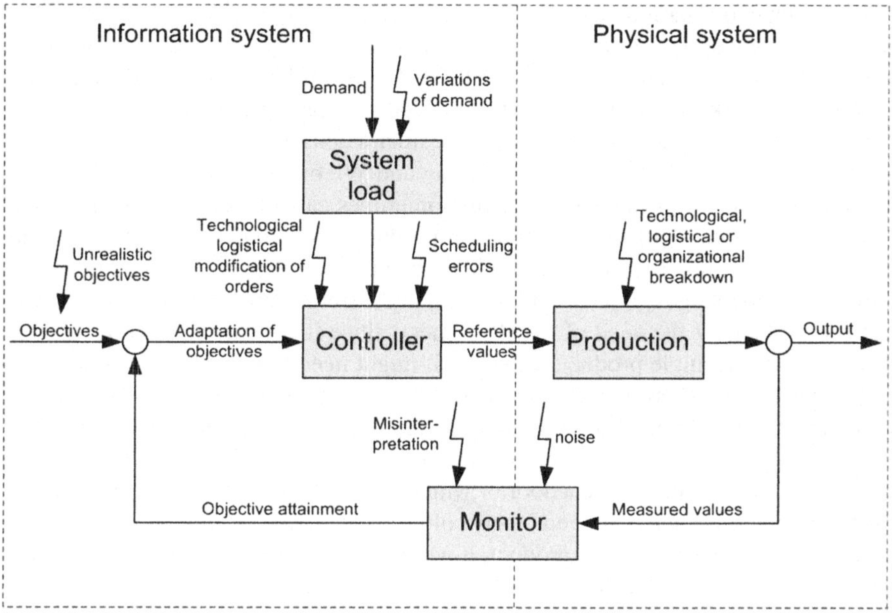

Fig. 10.2 The manufacturing work system depicted as a technical control system (modified from Scherer 1998)

required to reach the objectives. A more elaborate discussion of operational uncertainty and control requirements can be found in the next section of this chapter.

However, the model in Fig. 10.2 may be interpreted as a strictly technical model. This might lead to the wrong conclusion that control of manufacturing is a strictly technical issue, ignoring the fact that control of a work system necessitates people acting for control both within the information system and within the physical system. Consequently, control can be described as being composed by people exerting control in a network within the work system. In such a view, the control system is distributed and interlinked with the work system. However, the control system and the work system do not necessarily divide similarly into subsystems, i.e. one control subsystem may be interlinked with several work subsystems and vice versa.

Moreover, the control system as well as the work system is composed by mutually dependent technical subsystems and social subsystems. Control can thus be exerted by actors using the technical subsystem (mostly the ERP system), the social subsystem through direct interaction with other employees or – more likely – by a combination of both.

In contemporary manufacturing, the controlling actors are coupled by the ERP system, which often is supposed to be the overall coordination system. A problem related to this is that some couplings are tight and some are loose or very loose and in some cases the coupling does not exist. Furthermore, control actions are performed by human actors and each human actor has his/her own relation to the overarching coordination system (ERP system) and thus chooses to what extent his/her actions should be coordinated through the system. The coordination through the system may even violate the ability of the actor to fulfil his/her task. Research on scheduler's work has been demonstrating that this problem is frequent, because apparently it is often a vital part of the scheduler's activities to make sure that shop floor operators follow the schedule (McKay and Wiers 2004; Scherer 1998).

Another problem is that ERP systems sometimes cannot be configured to fit the reality. Interlinked bills of material or products that must be described with different units are often very difficult to handle. For example, a manufacturer using the same standard sheet metal for many different products may have difficulties in planning the need of the sheet metal since it is dependent on the mix of sales and not on single products. A sawmill might need to describe the same piece of lumber in the ERP system as a length (in meters), as a volume (in m^3), as a part of a pile for drying which in turn is different from the same piece of lumber described as a part of a wrapped delivery pile. Another example is a low-value customized product in large volumes like a door or window that can be equipped with different customized fittings and delivered in any colour. If the ERP system requires a unique structure for each customized product, it will be very costly to control the customization through the ERP system. However, more recent ERP systems might offer a configuration module that can handle these problems.

In cases where the ERP system is not sufficient, schedulers often use self-made spreadsheets as an extension. Control exerted through the use of these spreadsheets is another example of control that is not or at least loosely coupled to the overall coordination system.

Control actions are thus performed more or less coordinated by the ERP system. How to describe the total control of the system is an unanswered question.

If the system is considered to be composed of a structure for control and of actors using this structure, the structure and the actors can be analysed separately concerning their contribution to the system control. Against this background, we suggest a process model for control basically consisting of six variables (cf. Fig. 10.3). In the following subsections these variables are described.

10.2.1 Control Requirements

Control requirements are determined by the complexity of the subject matter of control. Ashby's Law of requisite variability relates to actors trying to reduce the variety in output with the aim of keeping the system stable (Ashby 1957). According to Ashby's law the actor's variety is the only possibility to destroy variety in the system to be controlled. The law of requisite variety has two important consequences:

1. The amount of information available determines the amount of appropriate actions that can be performed.

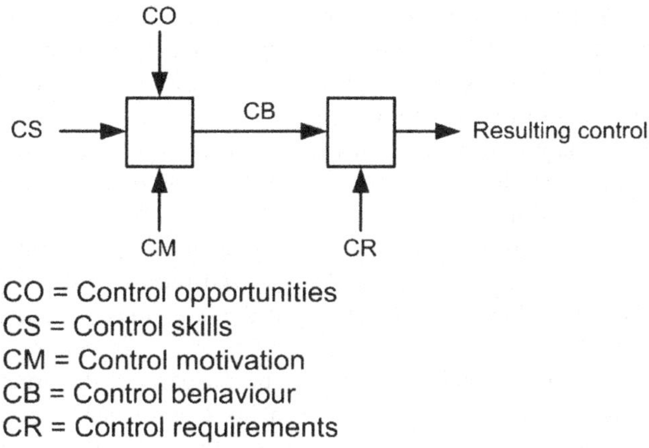

Fig. 10.3 Process diagram describing the process of executing control

2. To be able to control a situation the variety of the controller must be equal or greater than the variety in the system to be controlled.

The law of requisite variety thus describes that the variety of actions possible in the control system must as a minimum be as large as the variety of changes that need to be compensated in order to achieve control. The larger the variety of the controller's options is, the larger the ability to cope with changes through regulation. A control system that is tightly optimised for a limited variety might be more efficient as long as the variety does not change. However, it cannot survive should the variety increase above the controller's variety.

The law of requisite variety thus has important implications on flexibility issues because of its focus on possible outcomes within a set of defined situations. Consider for example the task of manufacturing various products with different assembly times at one assembly line. In this case, there is no possibility to adopt the manufacturing to the different demands of the products. Would there be at least two different assembly lines, the planner would have the possibility to control the manufacturing in a better way by splitting the products with long assembly times from products with short assembly times.

10.2.2 Control Opportunities

To be able to cope with operational uncertainties and to reach desired objectives, the actors need to be provided with control opportunities. Control opportunities are the sum of all available opportunities any actor may apply to perform any action of control. That is, control opportunities define the possible control actions available to an actor to change the state of the system. Control opportunities should permit the actor to reach the desired goals as well as to define means and standards to reach them. Control opportunities are determined by the structural properties of the system and can thus be deliberately designed to a certain extent. Information availability, organisational solutions, and technology are introduced below as being important parts of the structural properties.

Information availability concerns the type of information needed for the actor to perform the control task. In order to manage task related uncertainty and to reduce equivocality, organisations have to process information (Daft and Lengel 1986). The actor needs access to relevant, understandable, accurate, and timely information to be able to control the situation. Actors need relevant information such as information about the objectives, the status of the situation, the actions required, and the planned schedule.

Further, it is also necessary to look at different aspects of the organisation that may limit the ability of the actor to perform his control task successfully. The organisational solution will either enable or hinder the actor to act according to the demands of the situation. This has been a central tenet in the design of socio-technical solutions for planning and scheduling (Slomp and Ruël 2001). Examples would be the negotiation of additional manpower, a different shift pattern, or a resource changeover. As a result, it could be necessary to alter the authority of the actor and the needed support from the organisation. Within high uncertainty contexts it is more likely to observe control benefits when individuals or teams are able to make use of autonomy and collective problem solving than in low uncertainty environments (Parker and Wall 1998).

Finally, control opportunities are also depending on technology in different ways. There is the controll*ed* system as well as the controll*ing* system and its different control characteristics that must be considered. The complexity of these technologies and the extent to which they are understandable for the actors will affect many control opportunities.

To summarize, the control opportunities of a work system are the sum of all designed possibilities that the actors controlling the system can use in order to reach a system state that corresponds to the overall objectives of the system.

10.2.3 Control Skills

In order to make use of the control opportunities as described above, the individual actor must have adequate control skills. Control skills are thus defined as the ability of a human to produce control by acting within a certain situation using the given control opportunities of that particular situation. Professional knowledge of control methods as well as competencies to make use of control opportunities – such as available instruments and tools – are at the core of the actor's possession of control skills.

However, detailed knowledge about the system to be controlled is also required. Hence, it is a fundamental precondition for skilful control that an operator has the necessary (tacit or explicit) knowledge of the cause-effect relationships that can be used for control. Based on the writings of Von Wright (1971), Petersen (2004) distinguished between what an actor is doing and what change she or he is bringing about. The actor performs control actions in order to bring about system changes. The desired change may however not be a direct consequence of the action but something it is brought about by a causal relation inherent in the controlled system (cf. Fig. 10.4).

The causal relationship between doing and bringing about might be easy to define and realise in physically proximal and simple control situations. However, the causal relation might also be mediated by the ERP system or other technical or organisational means, and there might be time delays involved that make the causal relation difficult to foresee or understand. Regardless of the relation, the control skill needed is depending on the content of the relation and the control agent's knowledge about the relation and his/her ability to perform the actions needed.

Another fundamental aspect of skills in control tasks is what can be called situation awareness. This is a concept that was originally developed for aircraft pilots but that has spread to other domains. One of the most comprehensive definitions was provided by Endsley: "Situation awareness is the perception of the elements in the environment within a volume of time and space, the comprehension of their meaning, and the projection of their status in the near future" (Endsley 1988). For operators maintaining a system or operating complex equipment, situation awareness generally means the sensory, perceptual, and cognitive activity that prepares the user to make a decision, but it does not include the action as such (Pew and Mavor 2007).

The process of exerting control skills could thus be defined as – in the first step – assessing the situation to upgrade the situation awareness which would provide the control agent with the necessary cognitive background for making the right decision, and then – as a second step – continuing with doing what is necessary in order to bring about the desired change of the system. It may be noted that seldom all

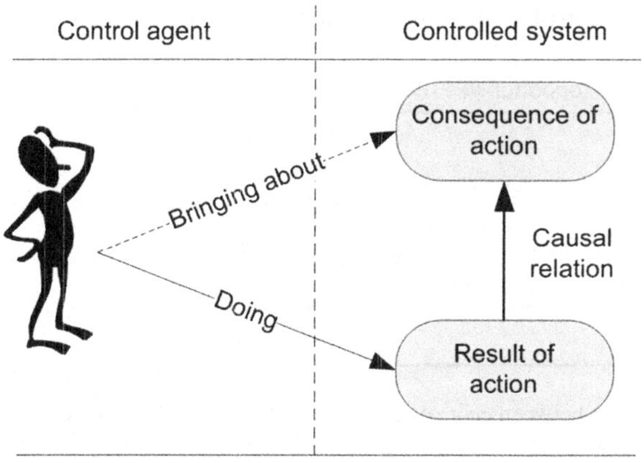

Fig. 10.4 The doing and bringing about aspects of control actions (Petersen 2004)

available control skills in a work system can be used. There might be control skills that do not correspond to any control opportunity.

10.2.4 Control Capability

As control capability we define the intersection of control opportunities and control skills, i.e. control capability is given when both, control opportunities as well as adequate control skills exist at the same time and the same place. This means that the total control capability of a work system can be regarded as the fit between the control opportunities and the control skills. Hence, control capability is the maximum control that can potentially be executed by the system (cf. Fig. 10.5). Whether or not this control capability results in control depends on two aspects. On the one hand needs to fit the control requirements. On the other hand it manifests in control behaviour only, when the humans are motivated. The following section elaborates on the relation between motivation and behaviour in control.

10.2.5 Control Motivation and Control Behaviour

Control behaviour incorporates all control actions the human controllers actually perform. These actions are determined by the control opportunities and the control skills (i.e. by the control capability). Whether or not the human controllers make use of the whole scope of possible actions as determined by the control capability is dependent on their control motivation. Consequently we take into account that human planning and scheduling activities are determined by the objective means available for being in control (i.e. the control capability) on the one hand and the subjective control motivation on the other hand. As for example described by Scherer (1998), control decisions in practice are influenced by different rules.

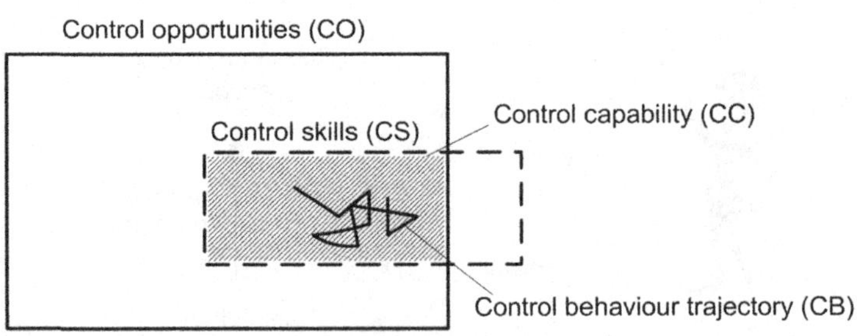

Fig. 10.5 The relation between control opportunities (CO) given by all the structural possibilities for control of the work system (solid line), control skills (CS) given by the summarized abilities to exert control (dashed line), control capability (CC) which is the sum of all usable possibilities for control (shaded area), and the resulting control behaviour trajectory (CB), i.e. the utilized control

Some of them are defined by the control opportunities described above, but some are also determined by informal rules inherent in the social situation of the actor. These informal rules are depending on the cultural characteristics of the organisation.

The control motivation depends on emotional and motivational processes connected to the individual behaviour of each human actor. The decisions and actions taken can be seen as a behaviour trajectory, which describes the sequence of actions according to the actor's actual motivation (cf. Fig. 10.5). The psychological processes and mechanisms that are underlying control motivation are further developed in Sect 10.3.

So far, the core of our process model of control has been outlined (cf. Fig. 10.3). It suggests on the one hand that control behaviour results from an interaction of control opportunities, control skills and control motivation. On the other hand it postulates that control behaviour needs to meet control requirements in order to maintain control. In the following, this model is elaborated further, mainly by differentiating the control requirements as well as the control behaviour, and by integrating feedback loops.

10.2.6 Operational Uncertainty

As described above, the law of requisite variety (Ashby 1957) assumes that the amount of variability in control behaviour should always exceed the amount of variety in the processes to be controlled in order to enable the system to cope with occurrences. These dynamic occurrences might be expected, unexpected or inexperienced and are causing uncertainties that have to be regulated. The quality and effectiveness of all activities concerning the regulation of uncertainties are critical for the outcomes and profits of a work system. In fact, already more than forty years ago uncertainties were seen as a fundamental problem for complex organisations (Thompson 1967). The ability to manage uncertainties successfully is one of the key factors to sustain and even expand a business. Uncertainty management therefore has become a research interest. Consequently, the amount of uncertainty embodied in the control requirements defines the variability in control behaviour that is necessary to control the situation.

In planning and scheduling, a huge variety of uncertainties needs to be regulated. These uncertainties can be due to many different contributing factors. Generally it can be differentiated between external and internal factors. External factors are situated outside an organisation. Typically, market dynamics fall into this category. Examples are the limited predictability of customer demand as well as unexpected changes in existing orders. However, there are many more uncertainties caused by external factors: unreliability of suppliers, supplies with bad quality, new laws and other legal prescriptions, strikes and other occurrences suitable to disturb logistic processes, just to name a few.

Uncertainties can also arise internally. Examples of such uncertainties are technology and production process related uncertainties, distributed information, information loss at organisational interfaces, missing or ambiguous specifications and instructions, unavailable key persons, lack of rules or standards for collaboration, machine breakdowns, and bad product quality which causes additional effort.

Furthermore, as planning decisions are highly interrelated, even planning decisions themselves can potentially lead to uncertain situations. Today's solutions may easily implicate tomorrow's problems. Sometimes these problems do not affect the original decision-maker, but somebody subsequent in the stream of actions. This does not necessarily require improvidence or even bad intentions of the original decision-maker. It might happen just because of ignorance towards system complexity. Maybe the original decision-maker did not at all have a chance to appraise the impact of his decisions on other people's work. Hence, the mere design of planning and scheduling structures and processes including the allocation of planning and scheduling tasks can be a source of uncertainties (Grote 2004).

In particular, there is a dilemma regarding allocation of planning and scheduling tasks to centralized planning departments and the shop floor respectively (Wäfler 2001). Allocating autonomy to the shop floor causes uncertainties for the planning department, which only with immense effort can keep an up-dated picture of the shop floor's dynamics and therefore runs into the danger of planning on an outdated information basis. On the other hand, allocating as much decision-making authority as possible to the centralized planning department is not well suited for taking into account detailed local information when making planning decisions. Consequently a well-balanced allocation of decisions-making competencies among planners and schedulers as well as between the shop floor and the planning department is required. In order to allow for flexible (re-)acting on uncertainties, it might also be required to design the allocation of decision-making authority in an adaptable way (cf. below 'structural control').

Figure 10.6 summarises the consequences of the statements made so far for the control model: in planning and scheduling, control requirements (CR) incorporate operational uncertainties (OU), which arise from environmental dynamics (ED) as well as from the internal structures (IS) of planning and transformation processes.

Since the control model depicted in Fig. 10.6 suggests that the scope of required control behaviour should fit the control requirements given in a concrete situation, a measure for control requirements or for operational uncertainties would be very helpful when designing planning and scheduling structures. However, well defined and operationalized propositions for conceptualisations of operational uncertainties are rather rare in literature.

Control of operational uncertainties is a core issue in the socio-technical system design approach (for a comprehensive description of this approach cf. Van Eijnatten 1993). This approach takes into account that business organisations consist of a technical as well as a social sub-system. The former incorporates all machines, technical resources, prescriptions, regulations, condition set by the factory layout, and the like. In contrast, the latter incorporates the humans as individuals as well as groups of humans with their needs and behavioural patterns. Since the two sub-systems

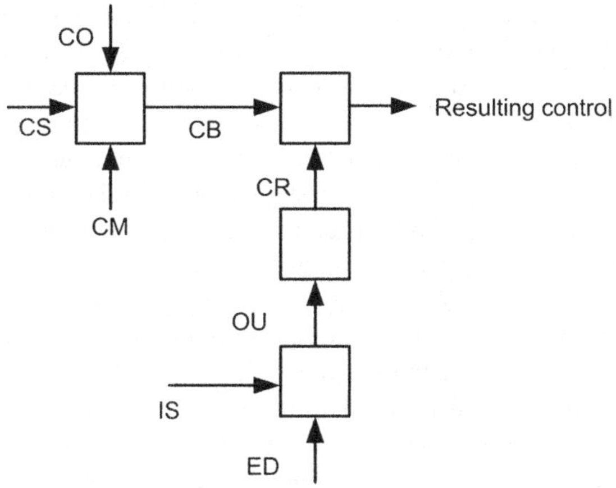

Fig. 10.6 Internal structures (IS) and environmental dynamics (ED) are defining operational uncertainties (OU) which lead to control requirements (CR)

are following different logics, the socio-technical system design approach postulates that only a combination of the two allowing for a joint optimization is suitable for the successful functioning of the system as a whole. Proponents of the approach criticize that in practice mostly a one-sided optimization of the technical sub-system can be observed. This is especially true with regard to the design of planning and scheduling processes, which mostly are driven by IT-implementation. However, such a one-sided optimization is considered to be sub-optimal not only since it runs the risk to hamper human flexibility and creativity required for a successful mastering of operational uncertainty. Moreover, it might even destroy respective human competencies because it deprives employees of working conditions allowing for the development of required know-how, experiences, and motivation.

However, the socio-technical system design approach considers the successful functioning of a system as a result of competent coping with continually changing conditions and hence with variances or disturbances with respect to the plan.

In general, the socio-technical system design approach favours a system design that allows for a local regulation of variances and disturbances. Therefore, it promotes a decentralization of decision-making into autonomous organisational units, requiring as little cooperation as possible at the organisational interfaces between these units. However, applying this concept to the design of planning and scheduling structures proofs to be rather difficult (Wäfler 2001), since it presupposes that the task can be modularized and allocated to organisational units in a way making the units independent of each other. In contrast, planning and scheduling incorporates coordination of the order flow through the organisation and hence requires regulation at the interfaces between organisational units. Therefore, with regard to planning and scheduling, organisational units remain dependent of each other. Often it is aimed at mitigating this dependency by introducing buffers at the interfaces. However, such buffers often increase lead times and therefore hamper efficiency.

In fact, the socio-technical system design approach gives hints on how to deal with uncertainty (by allowing for local regulations on the basis of autonomy and independency). But it does not really define what uncertainty is or how it could be measured. It considers every variance or disturbance of the normal or planned workflow that could critically influence the performance of an organisation as an operational uncertainty. However, this definition of uncertainties is very general and does not provide detailed information about the features or the measurement of occurring problems in the planning and scheduling context.

Wall et al. (2002) propose a definition with implications for measurement. They suggest that operational uncertainty is defined in terms of the number and difficulty of problems, key variances or exceptions that have to be accommodated. This definition refers to problems as countable entities of a specific source with a certain "difficulty" for their regulation.

In addition, some definitions highlight further psychological aspects of uncertainties and their complexity. Jackson (1989) for example defines operational uncertainty as a lack of knowledge about production requirements like the occurrence of problems and how best to deal with them. Wall et al. (2002) illustrate that this concept represents a lack of understanding about cause and effect between knowledge and uncertainty. Where such uncertainty is high, knowledge is incomplete and therefore problem-solving requirements are high.

Milliken (1987) distinguishes between three dimensions of uncertainty based on lacking information: state uncertainty (concerning the future development of the environment), effect uncertainty (regarding possible effects of changes on the organisation) and response uncertainty (regarding alternative responses and the prediction of their consequences).

With respect to the work task, Clegg et al. (1989) describe task uncertainty as a lack of predictability over matters as the timing of, and demand for, a particular task, the meaning of inputs that trigger the need for a response, and the nature of the required actions or responses.

Summing up these views, operational uncertainties can be considered as situational changes caused by occurring variances and disturbances. Uncertainties are characterized by a lack of knowledge about the new situation. The lacking information can refer to the "when" and "why" of an uncertainty (constrained predictability) and/or to the further development of the situation and the effects of possible actions (constrained transparency). Uncertain situations are triggered by external or internal variances and disturbances and need to be regulated adequately in order to avoid or decrease negative outcomes. For that purpose control opportunities and control skills are needed which allow for control behaviour. The following section describes structural and operational control as different forms of control behaviour.

10.2.7 Structural and Operational Control

A crucial premise for successful coping with an operational uncertainty is the fit between control requirements (as consequences of operational uncertainties) and control behaviour. This control behaviour is a result of control opportunities,

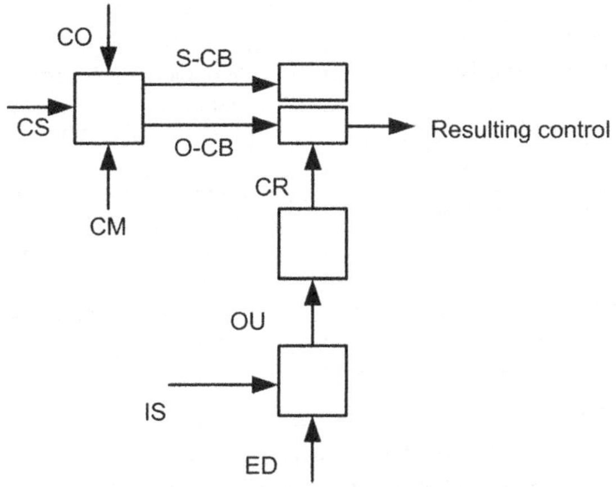

Fig. 10.7 Control behaviour (CB in Fig. 10.6) is differentiated into two types: structural control behaviour (S-CB) and operational control behaviour (O-CB). The latter refers to all actions aiming at coping with control requirements and hence is directly influencing 'resulting control'. However, O-CB is determined by an organization's structural conditions. The structure limits the scope of possible O-CB. Changing such structural limitations requires a change of an organization's structure. S-CB refers to such behaviour, resulting in a change of structures and hence modifying the frame of O-CB

control skills, and control motivation (cf. Figs. 10.3 and 10.6). However, two different kinds of control behaviour can be identified: structural and operational control (cf. Fig. 10.7). Whereas the latter refers to control behaviour within given structures, the former refers to changing the structures themselves. Hence, the structures of an organisation as defined for example by organisational and job design, by provided instruments and tools, and by the human planners' qualification and skills provides the conditions for the operational control behaviour that can be deployed to cope with uncertainties. Structural control surmounts such limitations. It refers to changing given structures and therefore for changing limitations of operational control.[1] In the following, these two types of control behaviour are described in more detail.

Control opportunities are possibilities to influence a situation in order to reach certain goals, to set these goals as well as to define the means and regulations to reach them. Control opportunities can be designed to a certain extent. Grote (2004) proposes to design the control opportunities in a way balancing dependency and autonomy (cf. Fig. 10.8). Dependency results from a centralized planning system that reduces local degrees of freedom. Such design is aiming at avoiding

[1]This definition of structural control differs from other definitions that can be found especially in industrial engineering (cf. e.g. Reveliotis et al. 1998). There, structural control is rather understood as control *through* the structures, e.g. by designing a manufacturing system in a way avoiding deadlocks or by designing regulations and procedures prescribing how a system should react on different disruptions. In contrast, structural control is understood here as control *of* structures.

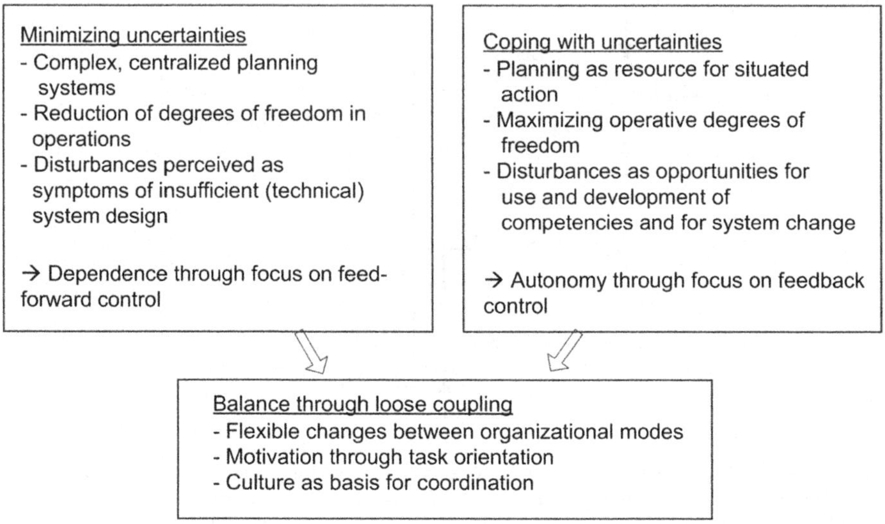

Fig. 10.8 Basic principles of uncertainty management underlying organisation design (adapted from Grote 2004)

uncertainties by feed-forward control. The overall aim of plans is to minimize uncertainty by preventing from or (at least) by preparing for future uncertainties. This is achieved for example by an exact timing of production processes or by reservation of production resources. In such systems, plans are prescriptions to be followed. Deviation from the plans would cause additional uncertainty for the centralized planner. The prescriptive character of the plans creates dependency. The problem of such a strategy is its inflexibility to react to unforeseen events. However, the more complex production becomes, and the more dynamic markets are, the more likely is the occurrence of unforeseen events.

In order to be able to cope with such unforeseen events, feedback control is necessary. In line with the assumptions of the socio-technical system design approach (see above) effective feedback control requires opportunities for local regulation of uncertainties. Consequently, local independency and autonomy is needed in order to allow for local coping with unforeseen events and hence for preventing uncontrolled spread of problems throughout an organisation. In such systems, plans are not prescriptions but rather resources for local acting by providing transparency and orientation.

As mentioned above, Grote (2004) suggests an organisational design – and hence a design of control opportunities – that balances autonomy and dependency by a loose coupling of organisational units. However, she also pleads for flexibility regarding the balance of autonomy and dependency. Such flexibility should allow an organisation to flexibly adjust its organisational mode (i.e. its design of autonomy and dependency) in order to allow for an optimal coping with actual variances and disturbances.

Control of the organisational mode is a core aspect of structural control since it involves the (re-)assignment of decision-making responsibility. Dependent on the organisational mode chosen, more or less decision-making responsibility is assigned to the shop floor or to the centralized planning department. However, structural control involves more than controlling the organisational mode. It involves all behaviour that influences an organisation's structure and hence also its control opportunities and control skills (cf. Fig. 10.9). This includes activities like designing products and production, designing the control system and especially the control tools and instruments, but also qualifying the humans in order to develop their skills.

Figure 10.9 also shows another important property of structural control behaviour. It does not only influence an organisation's overall control ability by determining control opportunities and control skills. It also influences the control requirements. This is due to the fact that operational uncertainties are partly depending on an organisation's structure, which in turn is influenced by structural control behaviour. De Sitter et al. (1997) argue that a suitable structuring of production process can reduce control requirements up to 80%. In their concept, "suitable" mainly means parallelization and segmentation of production in order to avoid dependencies at the organisational interfaces. Thereby organisational interfaces are to be designed in a way reducing the need for boundary spanning cooperation as much as possible. Although total independency is not possible (see above, Grote 2004), structural design creates more or less dependency and

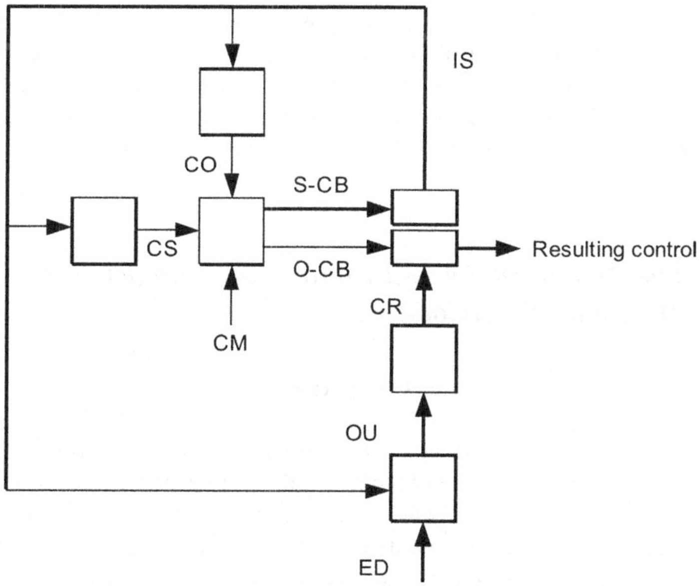

Fig. 10.9 Structural control changes internal structures (IS) in order to optimize control opportunities (CO) and control skills (CS). However, the internal structure also influences operational uncertainties (OU)

hence more or less operational uncertainty and control requirements. With reference to planning and scheduling, the following are examples for such homemade control requirements (a) two planners need to access the same production resource, (b) continuation of production is delayed due to a missing approval from external persons (e.g. quality managers or constructing engineers), (c) centralized planners lack detailed situational knowledge and decentralized dispatchers lack overview, (d) local optimization is no more possible since the global optimization works in real-time and therefore creates uncontrollable local dynamics.

So far, a process model for work system control has been outlined (cf. Fig. 10.9). It defines a works system's control capability as an overlap of control opportunities and control skills (cf. Fig. 10.5). However, whether or not the actors within the system are carrying out the potential control behaviours determined by the control capability is depending on their motivation. We understand actual control behaviour of humans as trajectories within the control capability field of possible actions. The following section describes psychological processes that influence these trajectories.

10.3 Emotional and Motivational Influences on Control Behaviour

In this section, we will explain important emotional and motivational processes, which are underlying individual behaviour. We will show how these mechanisms are connected to human control behaviour and hence to resulting control (cf. Fig. 10.9). The last part consists of proposals for improvement of technical innovation projects and for measurements of work design, which are concerned with the ability of the members of a work system to cope with critical situations.

10.3.1 The Role of Individual Control and Competence in Human Behaviour

10.3.1.1 Individual Control and Competence

When humans learn how to meditate, at first they have to learn to stop thinking. It is very difficult to stop the inner monologue. It is also very difficult or even impossible to stop emotions. Humans are always in a certain emotional state: angry, happy, hateful, calm... This emotional condition influences our whole stream of activities. Being angry means to be concentrated on the object of anger, it means also to be full of tension and ready to act. Possible actions taken into account have aggressive character. Happy people don't have this readiness to act, they think about positive things and they are rather absentminded than concentrated. Happiness and anger are embedding our thinking and acting in a different way, leading it in a certain direction and influencing the quality of the results. By the way, the probability that extreme anger and extreme happiness are leading to good performance is low. Accordingly, the control behaviour trajectory described in our model (cf. Fig. 10.5) is strongly influenced by the emotional state of the humans.

Along to most psychological theories, the most important variable triggering our emotional state is individual control (Lazarus and Launier 1978). Individual control from a psychological point of view is the actual probability of a human to influence a situation according to his or her own goals[2]. If individual control has the value "1", one has the perfect ability to reach all intended goals. If individual control is "0", there is no chance to reach the goals at a certain moment. Different values of individual control are related to certain moods and behaviour patterns. Individual control close to zero for instance leads to a behaviour pattern which is called "learned helplessness" (Seligman 1975). If this is the case for a longer period, human beings and highly developed animals loose the ability to act autonomously and begin to stay in a state of complete inactivity.

The "classical" reaction to an actual loss of individual control is an increase of arousal, which is a biologically useful reaction. The organism becomes more activated and thus ready for fighting or fleeing, depending on what is most useful in the actual situation. Humans, especially in the context of modern work environments, typically do not react in that way. The normal reaction in such situations is unhealthy stress, because the arousal cannot be transformed into fighting or fleeing.

In situations with loss of individual control people can react in different ways, also at work, where individual control is to a large amount determined by the design of the human's work task. However, the actual emotional state is not only influenced by the amount of individual control determined by the design of the work task. It is also influenced by the importance the work task has in relation to the individual's goals, and by other aspects not necessarily related to work. There is a different effect on the emotional state when a very important part of work has gone bad in a situation with additional private problems than in a situation of perfect happiness. Therefore it is necessary to distinguish between 'individual control' as related to the actual work task, and a much more general belief in one's own 'actual competence' (Dörner 1999), related to the estimated success to cope with all actual and future demands. According to most theorists, humans have a general motive to acquire such competence. This motive causes curiosity and empowers for learning.

[2]Please note: The psychological concept of individual control is different from the control concept in our model (cf. Fig. 10.9). Whereas Control in our model is the ability of a work system to reach business objectives, individual control refers to the ability to influence one's own situation in order to reach personal objectives. These personal objectives may or may not correspond with the employer's business goals. However, the ability to influence one's own situation – or even the belief to have the ability to influence one's own situation – has a major impact on motivation and hence on human behaviour. Since human behaviour is an important variable in our control model (cf. control behaviour in Fig. 10.9) the psychological concept of individual control (belief) has an impact on a work system's ability to reach its objectives.

Actually, subjective beliefs regarding 'individual control' and 'actual competence' are rather influencing a human's behaviour and his/her emotional reactions than an objective evaluation of the two. Furthermore, individual control and actual competence are connected, especially if work is an important part of one's life. However, the same disturbance in a work system can provoke very different individual emotional reactions, because the actual competence of its members can be different. Each disturbance that leads to a decrease of individual control also leads – differently for different individuals – to a loss of actual competence.

10.3.1.2 Control and Competence in Complex Work Situations

Psychological research describes typical patterns of behaviour which are related to impaired performance in very complex situations (Dörner 1996; Jansson 1994). Most of them are dysfunctions in the use of information during the process of planning and acting, for example:

- 'Encapsulation' is the tendency to avoid problematic fields of action and to concentrate on aspects where great skills and competencies exist, disregarding the importance and necessity of these activities.
- There is also a tendency to use mental models that are too simple for the adequate description of actual problems (e.g. only one reason for all problems). Hence, one's cognitive limitations might be underestimated.
- A typical mistake in prognosis is the tendency to believe that major actual constraints of a complex situation will not change in future.

Such critical behaviours happen in very different professional fields (engineering design, management, spatial planning), and they also happen to people knowing about the risks to act in that way. Dörner (1996) calls these and other similar behaviour patterns '*intellectual emergency reactions*'. They are typical for complex work tasks where the relation between human acting and its outcome is not clear. In such situations, the outcome of acting can be interpreted in different ways (e.g.: management strategy and success; product design and reaction of the market; the real quality of some typical regional planning activities can be assessed twenty years later).

In situations with non-existing, unclear, or long term cause-effect relations, it is possible to select specific information which allows the human to perceive a high amount of individual control although real individual control is low (control illusion). Encapsulation allows to avoid negative information by concentrating on own strengths, and hence leads to overestimating one's individual control. Too simple mental models strengthen the belief in the own ability to understand the complexity of a problem, and so on. Research results and case studies show that a low level of perceived actual competence increases the probability of these intellectual emergency reactions because they satisfy the motive to feel competent for the actual demands. Under certain conditions the intellectual emergency reactions

Fig. 10.10 Positive feed back loop with negative consequences (please note: 'actual competence' is always the subjective perception of one's own competence)

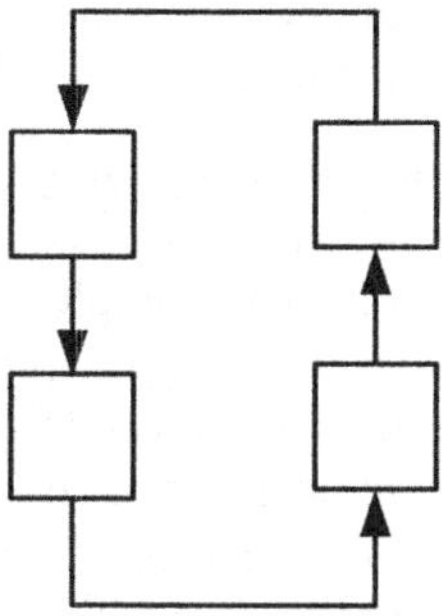

lead to a feedback loop which is amplifying itself concerning its negative consequences (cf. Fig. 10.10).

According to this model, a low level of perceived actual competence leads to intellectual emergency reactions. The consequence of this behaviour is an inadequate mental model leaving out important information to cope adequately with the complex work task. This lowers performance, and coming from that decreases actual competence again. This positive feedback loop can have very important negative consequences.

In a work system, disturbances may decrease individual control (see above). However, also changes and innovations may decrease individual control. Hence changes and innovations may increase the risk of inadequate problem solving behaviour. This is especially true for people with a belief of low actual competence, because they are more likely to show intellectual emergency reactions.

As a consequence, the vicious circle depicted in Fig. 10.10 can decrease the resulting control of a work system (cf. Fig. 10.9), since inadequate problem solving behaviour is most likely to lead to inadequate control behaviour.

10.3.2 Competence Regulation in the Context of Teams and Organisations

The model of competence regulation was originally related to individual acting. In industrial contexts normally people have to work within groups and organisations and this leads to division of labour. Work is related to and depending on others to achieve a final collective result. Moreover, its consequences are not under individual control of the acting person. Therefore individual control depends on the control capability of the whole work system. Competence and control are generated also by additional factors, e.g.

- Confidence in the regularity of work processes
- Confidence in the organisation's ability to cope with extraordinary demands

- Trust in the personal qualities of the management
- Trust in cooperation partners (colleagues, customers, suppliers)

Confidence and trust normally are generated by former experiences, i.e. typical behaviour in similar situations and resulting knowledge about resulting processes. But if people have to cope with complex and new demands, there is – by definition – not much knowledge about former processes and organisational resources. So people have to estimate the control capability of their teams and organisations in another way. This estimation in many cases is based upon trust in the general abilities and integrity of cooperation partners and leaders. This is the reason for the common experience that coping with major changes and complex disturbances in work systems depends on transparency and an atmosphere of trust. This is described broadly in management, leadership and organisational development literature (e.g. for ERP system implementation by Kwasi 2007; Holland and Light 1999; Krause and Gebert 2005; and on a more general level in this book by Günter et al. in Chap. 5). If trust and confidence is low, the positive feedback loop which has been described for individuals in Fig. 10.10 can also lead to a crisis of the whole work system (cf. Fig. 10.11).

Loss of confidence lowers the level of perceived individual control and actual competence. This increases the probability of intellectual emergency reactions. In turn, that leads to more mistakes, lower performance, and less trust and confidence. The consequences are positive feedback loops leading to very negative developments and results.

These considerations may show the impact of emotional regulation processes on the control capability and performance of the whole work system. To understand fully how work systems cope with complex demands, emotional and motivational

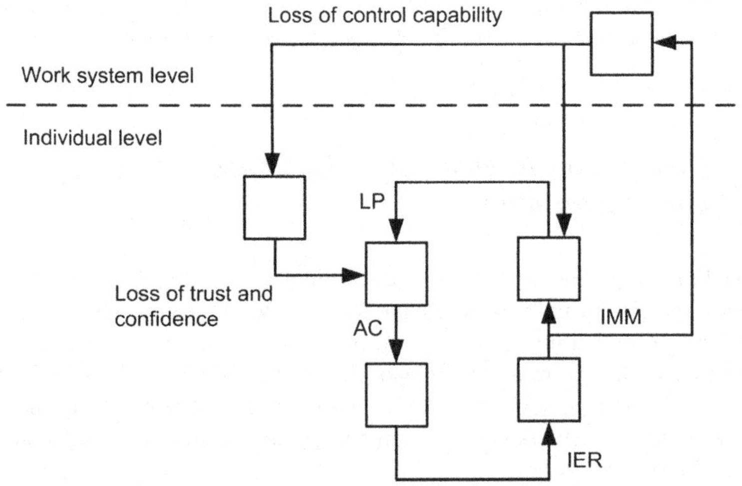

Fig. 10.11 Individual emotional processes and control capability. *LP* Loss of performance, *AC* actual competence, *IER* emotional emergency reactions, *IMM* inadequate mental models

processes have to be integrated into the analysis. We will show this by the following examples.

10.3.2.1 Examples

Example 1: The Role of Fear in 'Technical' Innovation Processes

Implementing new software often leads to major disturbances and new structures. Work and business processes, work tasks, and the relations in teams are changing. Sometimes technical change can lead to loss of job. All this causes increasing uncertainty, and coming from that a loss of individual control. In many cases, implementation strategies do not consider these emotional aspects (von der Weth and Spengler 2007). The resulting behaviour patterns are, amongst others:

- A negative attitude to the change process, because existing qualification is partly not needed anymore
- No discussion about problems and mistakes, because people fear to be regarded as incompetent
- Risk avoidance in learning to handle the new software, because experimenting with the new software possibly generates loss of face by mistakes
- Strategies for individual improvement of the own work process, not regarding the demands of cooperation within the work system
- Intellectual emergency reactions because of decreasing control

Behaviour of this type decreases the actual performance within the implementation process and influences the individual control as well as the emotional state of the implementation project team. In turn this leads to a higher probability of wrong planning and decision making of these people and diminishes the probability of success.

Example 2: Conflict Escalation

If control capability of the work system and individual competence is perceived as low, the quality of conflict resolution in the work system decreases. This can be exemplified with a common conflict between a new superior and members of his/her team. A new superior often feels the challenge to establish new processes and forms of communication because he/she wants to leave his/her marks on the team. The result of these activities is highly correlated to his/her own perceived competence in the new job. Any objections in this situation often lead to stress and a decrease of perceived control. This can cause aggressive or fearful reactions.

On the other hand we also have a critical situation for the team members. Their perceived competence is low because the innovative ideas of the new superior cause uncertainty vice versa. This leads to a positive feedback loop with decreasing control on both sides and can cause aggressive behaviour. Typically, both parties increase their efforts and involve allies in the organisation if they want to enforce their goals. In situations of that type, conflicts can start from minimal disturbances in the work process and spread like wildfire, causing serious damage in an organisational culture.

Escalations like this can happen in every work system, but their probability increases when the basic level of the work system's control capability is low. A low control capability does not contribute substantially to the work system members' perceived competence. On the other hand, a high control capability can be a source of trust and confidence and can support the individual competence of the parties involved in a conflict. With a higher perceived competence on both sides in the beginning of the conflict, the chance increases to solve the team conflict before an aggressive and irreversible escalation starts.

Example 3: Trust, Communication, and Knowledge Development

Long term developments in work systems are also related to emotional processes. The control capability of an organisation depends strongly on its knowledge. Motivation and strategies for individual knowledge acquisition are connected to the emotional state of the work system members. Like knowledge management on the level of organisations, individual knowledge handling strategies can be described in the terms of the knowledge management strategy of Probst et al. (2003): knowledge identification, knowledge acquisition, knowledge development, knowledge sharing, use of knowledge, and knowledge preserving. The quality of these activities is also connected to the control capability of the whole organisation. If the perceived individual competence is high and connected to an actually high control capability of the whole organisation, then the readiness to share knowledge in many cases is also high: The members of the organisation have seen many times before that sharing knowledge and collaborative work are useful. In addition, the members of the organisation have the chance to learn appropriate strategies for cooperative knowledge management for the benefit of the whole work system. This increases the control capability, because more integrated knowledge structures exist. Work systems of this kind have actual and intensively used information infrastructure.

Low perceived control capability of the work system is connected with many forms of fear, e.g. loss of job and payment. A more aggressive organisational climate is probable. In that case we have two phenomena which are counterproductive for knowledge development in the work system:

- Fear prevents an open discussion of problems and mistakes, which is necessary for a good quality process and effective innovation. If people don't share knowledge about mistakes and problem solving processes, they cannot learn from each other and the quality of knowledge and hence control capability decreases.
- Fear also causes a behaviour which is called "knowledge retention" (Hacker 2005): If people are in a very competitive situation with their colleagues sharing

knowledge is not useful individually, because it improves the chances of the competitors. Individual knowledge retention also has an influence on knowledge infrastructure. Tools are not maintained and hence loose actuality. Because of that, they are not used anymore and become more and more irrelevant. This also leads to decreasing control capability of the whole work system.

10.4 Conclusions

This chapter reflects on the complexity of control capability. It addresses the subject matter from two points of view:

- A process oriented design model of control (cf. Sect. 10.2)
- Emotional and motivational influences on control behaviour (cf. Sect. 10.3). The former refers mainly to objective aspects of control capability. At its core it defines a work system's control capability as an overlap of control opportunities and control skills. This overlap limits the scope of possible control behaviour, and hence of resulting control. Resulting control is high, when it fits with control requirements as determined by operational uncertainty. However, control behaviour is differentiated into two types: operational control behaviour and structural control behaviour. Operational control behaviour on the one hand aims at immediate coping with operational uncertainties and hence has a direct impact on resulting control. On the other hand structural control behaviour influences an organization's structure and hence the frame, which is determining the limits of operational control behaviour. Consequently structural control behaviour does not have an immediate impact on resulting control, but changes the frame of operational control behaviour and therefore has an impact on resulting control mediated by operational control behaviour. Such structural control behaviour allows for feedback loops in our process oriented design model of control.

However, as all human behaviour also control behaviour is influenced not only by objective aspects but also by subjective motivational and emotional processes. Control capability solely determines the objective frame for operational and structural control behaviour. No behaviour out of this frame is possible. However, not all the behaviour possible within the frame must necessarily become manifest. Subjective motivational and emotional processes influence the manifestation of concrete control behaviour.

Both, the objective given frame of control behaviour as well as the subjective processes influencing the manifestation of actual control behaviour need to be considered for understanding control. Following the implications of both for practice are discussed.

The main implications of the design model of control (cf. Fig. 10.9) are the following:

- Control results from an interaction of control opportunities, control skills, and control motivation on the one hand and control requirements on the other hand.

- Control behaviour can affect operations or structures. The former results in process control, the latter results in structural changes and hence influences control opportunities as well as control skills. Control therefore is self referential.
- Furthermore, since control requirements are not independent from structures, structural control even impacts operational uncertainties.
- Any change (as e.g. the implementation of a new ERP-system) is a structural change and therefore impacts control opportunities, control skills as well as operational uncertainties directly. If these changes result in an inadequate fit of control behaviour and control requirements, the resulting control is low although the new IT is better than the old one. An inadequate fit can be caused for example by the following consequences:
  - New regulation requirements might occur (e.g. due to change in information or control structure like required fixing of dates)
  - (Structural) control opportunities might be lost (e.g. due to proprietary software)
  - (Structural) control skills might be missing (e.g. due to lack of experience of in-house IT staff)
  - (Operational) control opportunities might be lost (due to centralisation, loss of transparency like daily or intra-daily changing job lists)
  - (Operational) control skills might be lost (e.g. because an experienced planner's know-how is not applicable any more)

The main implications of our reflections on emotional and motivational influences on control behaviour are the following:

- Individual thinking and behaviour are influenced by emotional processes.
- The emotional state is influenced by the belief to be competent (to have actual competence). This belief includes individual control of the actual situation and more general competence for future tasks and demands. High belief in ones own competence is connected to positive emotions, low actual competence to negative emotions and stress.
- A low level of competence is closely related to certain behaviour tendencies. The classical biological reactions are aggression or flight. Modern reactions especially in complex and new situations are so called intellectual emergency reactions (see above). In many cases, the resulting behaviour has the consequence that only positive information is selected and actual competence can be perceived as high, although it is low for the objective demands. Fear, aggression, and intellectual emergency reactions on the one hand and positive emotions on the other hand influence the behaviour trajectory substantially.
- This has also an influence on the control capacity of a whole work system, its work processes and results. Negative emotions generate inadequate behaviour on the individual level and decreasing control capability on the work system level. There is also an influence of the control capability on emotional and motivational processes. In normal work situations, individual performance depends to a certain degree on the performance of the whole work system.

Therefore, the control capability of a work system is an important source for the individual competence to cope with work related problems (e.g. innovation, disturbances).
- A low control capability increases the probability of negative emotions and intellectual emergency reactions. This interdependency between individual competence and control capability can be the starting point for dangerous developments in an organisation, described in several examples of positive feedback loops in this chapter. Especially in situations with great uncertainties on the individual as well as on the work system level, the danger of such aberrations is high because uncertain demands – like technical and organisational innovation or disturbances – lower the actual competence of the members of a work system. In situations like that, it is necessary to act psychologically in a careful way. Project managers and superiors should avoid behaviour, which causes additional loss of control.

Our reflections have the following implications for the design of planning, scheduling, and control systems:

From a process oriented design model perspective as well as from a psychological perspective, planning, scheduling and control is complex, incorporating multiple feedback loops. Therefore the complex interplay of humans, organisational structures, and technology needs to be carefully considered when designing or changing the control system. It is especially required that the humans – who are the actual 'producers' of control – are empowered to really perform control. In general, this presupposes the following:

- Objective conditions empowering the humans need to be provided, especially a control capability (as a result of control skills and control opportunities) that fit the control requirements. This allows for required structural and operational control behaviour.
- To assure such objective conditions, comprehensive instruments for work system analysis and design need to be applied. These instruments are still to be developed.

However, as discussed in the section on motivational and emotional influences on human behaviour, the provision of the objective conditions is far from being sufficient. What is required in a complementary manner to the objective conditions is a belief of the humans to actually be in control. The following attributes of work systems support individual development of such beliefs:

- Clear processes and tasks which are adapted to the individual's performance level
- Tasks which stimulate learning processes and interest in interdependencies within the work system
- Information about processes in which the individual's work is embedded
- Chance for participation in activities improving the control capability of the whole work system (Kaizen, participative knowledge management)

- An organisational climate reducing fear by productive procedures and methods for learning from errors and mistakes

Moreover, when changes occur (e.g. due to the implementation of new IT), the beliefs of being in control are endangered. For change management this means for instance:

- Clear defined goals, quality criteria, and milestones for change projects
- Participation in work design as early as possible in the process
- The members of the work system should be informed clearly about future tasks and demands
- The innovation process should be connected with qualification programmes which are adaptable to the specific individual qualification level
- Emotional development should be observed carefully on individual and work system level. There should be "sensors" in the work system for fear, aggression, and intellectual emergency reactions

References

Ashby, W. R. (1957). *An introduction to cybernetics*. London: Chapman & Hall.

Benders, J., Hoeken, P., Batenburg, R., & Schouteten, R. (2006). First organise, then automate: A modern socio-technical view on ERP systems and teamworking. *New Technology, Work and Employment, 21*, 242–251.

Calisir, F., & Calisir, F. (2004). The relation of interface usability characteristics, perceived usefulness, and perceived ease of use to end-user satisfaction with enterprise resource planning (ERP) systems. *Computers in Human Behavior, 20*, 505–515.

Clegg, C. W., Ravden, S. J., Corbett, J. M., & Johnson, G. I. (1989). Allocation functions in computer-aided manufacturing: A review and a new method. *Behaviour and Information Technology, 8*, 175–190.

Daft, R. L., & Lengel, R. H. (1986). Organizational information requirements, Media richness and structural design. *Management Science, 32*, 554–571.

De Sitter, L. U., Den Hertog, J. F., & Dankbaar, B. (1997). From complex organizations with simple jobs to simple organizations with complex jobs. *Human Relations, 50*, 497–534.

Dery, K., Grant, D., Harley, B., & Wright, C. (2006). Work, organisation and Enterprise Resource Planning systems: An alternative research agenda. *New Technology, Work and Employment, 21*, 199–214.

Dörner, D. (1996). *The logic of failure*. New York: Metropolitan Books.

Dörner, D. (1999). *Bauplan für eine Seele*. Reinbek b.H.: Rowohlt.

Endsley, M. R. (1988). Design and evaluation for situation awareness enhancement. *Human Factors Society 32nd Annual Meeting*. Santa Monica, CA: Human Factors Society.

Gefen, D., & Ragowsky, A. (2005). A multi-level approach to measuring the benefits of an ERP system in manufacturing firms. *Information Systems Management, 22*, 18–25.

Grote, G. (2004). Uncertainty management at the core of system design. *Annual Reviews in Control, 28*, 267–274.

Hacker, W. (2005). *Allgemeine Arbeitspsychologie*. Bern: Huber.

Holland, C. P., & Light, B. (1999). A critical success factors model for ERP implementation. *IEEE Software, 16*, 30–36.

Jackson, S. E. (1989). Does job control control stress? In S. L. Sauter, J. J. Hurrell Jr., & C. L. Cooper (Eds.), *Job control and worker health* (pp. 25–53). Chichester: Wiley.

Jansson, A. (1994). Pathologies in dynamic decision making: Consequences or precursors of failure. *Sprache und Kognition, 13*, 160–173.

Koch, C., & Buhl, H. (2001). ERP-supported teamworking in danish manufacturing? *New Technology, Work and Employment, 16*, 164–177.

Krause, D. E., & Gebert, D. (2005). Effekte von interpersonalen Konflikten und Widerstand von Führungskräften auf die Implementierung von Verfahrensinnovationen. *Zeitschrift für Personalforschung, 20*, 102–120.

Kwasi, A.-G. (2007). Perceived usefulness, user involvement and behavioral intention: an empirical study of ERP implementation. *Computers in Human Behavior, 23*, 1232–1248.

Lazarus, R. S., & Launier, R. (1978). Stress related transactions between person and environment. In L. A. Pervin & M. Lewis (Eds.), *Perspectives in interactional psychology*. New York: Plenum.

Light, B., & Wagner, E. (2006). Integration in ERP environments: Rhetoric, realities and organisational possibilities. *New Technology, Work and Employment, 21*, 215–228.

Locke, J., & Lowe, A. (2007). A biography: Fabrications in the life of an ERP package. *Organization Science, 14*, 793–814.

McKay, K., & Wiers, V. C. S. (2004). *Practical production control: A survival guide for planners and schedulers*. Boca Raton, FL: J. Ross Publishing.

Milliken, F. J. (1987). Three types of uncertainty about the environment: state, effect and response uncertainty. *Academy of Management Review, 12*, 133–143.

Ngai, E. W. T., Law, C. C. C., & Wat, F. K. T. (2008). Examining the critical success factors in the adoption of enterprise resource planning. *Computers in Industry, 59*(6), 548–564.

Nicolaou, A., & Bhattacharya, S. (2008). Sustainability of ERPS performance outcomes: The role of post-implementation review quality. *International Journal of Accounting Information Systems, 9*(1), 43–60.

Parker, S., & Wall, T. D. (1998). *Job and work design: organizing work to promote well-being and effectiveness*. London: Sage Publications.

Petersen, J. (2004). Control situations in supervisory control. *Cognition, Technology & Work, 6*, 266–274.

Pew, R. W., & Mavor, A. S. (2007). *Human-System integration in the system development process: A New Look*. Washington, DC: National Academies Press.

Probst, G., Raub, S., & Romhardt, K. (2003). *Wissen managen*. Gabler: Wiesbaden.

Reveliotis, S. A., Lawley, M. A., & Ferreira, P. M. (1998). Structural control of large-scale flexibly automated manufacturing systems. In C. T. Leondes (Ed.), *Computer aided and integrated manufacturing systems: Techniques and applications*. London: Gordon & Breach.

Scherer, E. (1998). *Shop floor control – a systems perspective*. Berlin: Springer.

Seligman, M. E. P. (1975). *Helplessness – on depression, development and death*. San Francisco, CA: W.H. Freeman.

Slomp, J., & Ruël, G. C. (2001). A socio-technical approach to the design of a production control system: towards controllable production units. In B. MacCarthy & J. Wilsson (Eds.), *Human performance in Planning and Scheduling*. London: Taylor & Francis.

Stedman, C. (1999). PeopleSoft speeds ERP installations, upgrades. *Computerworld, 33*, 78.

Thompson, J. D. (1967). *Organizations in action*. New York: McGraw-Hill.

Trunick, P. (1999). ERP: promise or pipedream. *Transport and Distribution, 40*, 1–39.

van Eijnatten, F. M. (1993). *The paradigm that changed the workplace*. Stockholm: Arbetslivscentrum/van Gorcum.

Von der Weth, R., & Spengler, R. (2007). Human factor resources in ERP-system implementation. In W. Karwowoski & S. Trzcielinski (Eds.), *Value stream activities management*. Madison: IEA.

von Wright, G. H. (1971). *Explanation and understanding*. Ithaca, NY: Cornell University Press.

Wäfler, T. (2001). Planning and scheduling in secondary work systems. In B. L. MacCarthy & J. R. Wilson (Eds.), *Human performance in planning and scheduling*. London: Taylor & Francis.

Wall, T. D., Cordery, J. L., & Clegg, C. W. (2002). Empowerment, performance, and operational uncertainty: a theoretical integration. *Applied Psychology, 51*, 146–169.

Wang, E. T. G., Shih, S.-P., Jiang, J. J., & Klein, G. (2008). The consistency among facilitating factors and ERP implementation success: A holistic view of fit. *Journal of Systems and Software, 81*(9), 1609–1621.

Yeh, T.-M., Yang, C.-C., & Lin, W.-T. (2007). Service quality and ERP implementation: A conceptual and empirical study of semiconductor-related industries in Taiwan. *Computers in Industry, 58*, 844–854.

Chapter 11
Building Decision Support Systems for Acceptance

Ralph Riedel, Jan Fransoo, Vincent Wiers, Katrin Fischer, Julien Cegarra, and David Jentsch

11.1 Introduction

Production planning and control fulfill a crucial role in enterprises. Planning and scheduling activities are very complex, and take place within the enterprise and across the entire supply chain in order to achieve high quality products at lower cost, lower inventory and higher levels of customer service. Since the information that has to be processed in planning and scheduling functions is very complex information technology is used extensively to support these functions. In the field of manufacturing planning and control Decision Support Systems (DSS) are used. Those are also known as Advanced Planning Systems (APS).

Planning determines what and how much to manufacture and to purchase in order to satisfy future demand for end products. Scheduling takes place at the execution level of the plans and covers a step by step work or activity list, specifications of time at which every activity should start and end as well as the sequencing and re-sequencing of job orders; it is internally focused. Advanced planning and scheduling covers simultaneous coordination of material and capacity

R. Riedel (✉) and D. Jentsch
Department of Factory Planning and Factory Management, Chemnitz University of Technology, Chemnitz, Germany
e-mail: ralph.riedel@mb.tu-chemnitz.de, david.jentsch@mb.tu-chemnitz.de

J. Fransoo and V. Wiers
School of Industrial Engineering, Eindhoven University of Technology, Eindhoven, The Netherlands
e-mail: J.C.Fransoo@tue.nl, v.c.s.wiers@tue.nl

K. Fischer
School of Applied Psychology, University of Applied Sciences Northwestern Switzerland, Olten, Switzerland
e-mail: katrin.fischer@fhnw.ch

J. Cegarra
Université de Toulouse, Toulouse, France
e-mail: julien.cegarra@univ-jfc.fr

J.C. Fransoo et al. (eds.), *Behavioral Operations in Planning and Scheduling*,
DOI 10.1007/978-3-642-13382-4_11, © Springer-Verlag Berlin Heidelberg 2011

constraints at an operational level to best meet market demand. Advanced Planning is the synchronization of constrained material and resources to independent demand. Its purpose is to create a plan that is feasible with respect to all resources with sufficient operational slack to permit re-sequencing of work orders. Advanced scheduling covers the detailed sequencing of operations and material. Its purpose is to provide properly sequenced work orders under a lot of restrictions (sequence dependent set up times, maintenance schedules, machining constraints, operator restrictions, etc.). (Musselmann and Uzgoy 2001).

In a lot of cases DSS are not used as originally intended. This results in increased planning efforts and therefore in inefficient planning processes and – mostly at the same time – in poor planning solutions. The investment caused by the acquisition of a DSS will lead to no return or to a return much below expectations. The question is therefore what characteristics DSS for planning should have and how should the modeling and implementation process be organized in order to foster the acceptance and usage of the system and to gain the maximum success. In this chapter, our focus is on this acceptance process of DSS. We will extensively review the literature on DSS with a focus on user acceptance. Based on the Technology Acceptance Model, we will specifically address the dynamic development, with user involvement, of Decision Support Systems. Later in this chapter, we will specifically focus on the role of trust in accepting Decision Support Systems.

11.2 Decision Support Systems and Advanced Planning Systems: Terminology

Decision Support Systems are a class of computer-based information systems or knowledge based systems that, in very different manner, support decision making activities (Druzdzel and Flynn 1999). A Decision Support System is an interactive, flexible, and adaptable computer- based information system, especially developed for supporting the solution of non-structured management problems for improved decision making. It utilizes data, provides an easy-to-use interface, and allows for the decision makers own insights. (Turban 1995)

A Decision Support System is a system that supports technological and managerial decisions making by assisting in the organization of knowledge about ill structured, semi-structured or unstructured issues. A DSS provide support to decision makers to increase the effectiveness of the decision making effort. DSS support humans in formal steps of problem solving: formulation of alternatives, analysis of their impacts, interpretation and selection of appropriate options. The primary purpose of a DSS is to support cognitive activities that involve human information processing and associated judgment and choice (Sage 2001). Support is given for instance for (Sage 2001) strategic planning decisions, management control decisions, operational control decisions, and operational performance decisions.

Decision Support Systems usually consist of three components: a database management system, a model based management system, and a dialogue generating and management system. The development of DSS is a multidisciplinary activity with contributions from disciplines such as computer science, management science, operations research, organizational behavior, behavioral and cognitive science, and systems engineering.

A common classification of DSS is:

- Model-Oriented DSS: contain models like regression and optimization routines that are developed to aid in decision making a specific problem domain
- Data-Oriented DSS: are like database management systems; they focus on providing information and facilities for data storage, retrieval, and update
- Decision-Oriented DSS: focus on supporting a specific decision process
- General DSS: support multiple decision areas or domains

Within the field of production planning and control, decision support systems are now mostly described as so-called Advanced Planning Systems (Stadler and Kilger 2000). APS's have developed into multi-module systems that each addresses specific planning and scheduling problems within a company, and in some cases even along the supply chain beyond the company borders. Many of the APS's are based on advanced optimization logic in order to provide solutions to the user of the system. Examples of modules are:

– Strategic network design
– Demand planning
– Master planning (of an aggregated production and distribution plan for all supply chain entities)
– Demand fulfillment & available to promise
– Production planning and scheduling
– Material requirements planning
– Distribution and transport planning

Originally ERP systems (Enterprise Resource Planning; usually based on the Materials Requirements Planning logic) were meant to provide decision support in planning and scheduling. Due to the substantially better capabilities provided by dedicated planning and scheduling systems such as APS's, ERP systems have now more and more developed as large data capture and storage warehouses that keep track of any imaginable operational process. ERP systems hence provide the data needed to plan, while APS systems provide a planning and scheduling solution to the user.

11.3 General Framework of Decision Support Systems

The expression Decision Support System (DSS) might have been used with declining frequency during the last years. Other terms like business intelligence and on-line analytical processing (OLAP) took the stage (Carlson and Turban 2002).

However, the general concepts behind these phrases strongly refer to the ideas of DSS, which will be explored throughout the subsequent paragraphs.

DSS are computer-based solutions intended to assist complex decision-making and problem solving. A rather simplistic example gives an impression how easily the necessity for computer assistance may arise in the field of production scheduling: Image the sequencing of four different jobs whereas each job needs five operations. The number of possible schedules for this task is (4!)5 = 7,962,624. Apparently, the classical MABA/MABA (men are better at / machines are better at) paradigm holds for this small example aiming at the necessity to process and store a large amount of data where machines i.e. computers are certainly better at (Karwowski 2006). Apart from mathematical and data issues, two major fields of science and research contributed to the growth of DSS in the very beginning. First, the investigation of organizational decision-making realized by Simon, Cyert and March during the 1950s and 1960s. Second, the methodological and technical advancement by Gerrity and Ness that addressed the requirements of DSS development and was carried out in the 1960s. (Gerrity 1971; Hogue and Watson 1983; Keen and Morton 1978; Shim et al. 2002).

Referring to organizational decision-making, Simon (1960) distinguished in The New Science of Management Decision two different kinds of people's problem solving called "programmed" and "unprogrammed". These categories gave the first major input to the fundamental framework of DSS by Gorry and Scott Morton published in 1971 (Gorry and Morton 1971). Gorry and Scott Morton renamed Simon's categories to "structured" and "unstructured" in order to emphasize a more general approach to decision-making and to decrease the one-sided connection to computers.

The second main influence for the first DSS framework were Anthony's three categories of managerial activity (Anthony 1965). Therefore, the core of the suggested framework consists of a two-dimensional matrix. The first dimension addresses structured, semi-structured, and unstructured decisions. The second dimension distinguishes operational control, management control, and strategic planning in accordance to Anthony. Within the framework, the authors argued that DSS "ought to be centered around the important decisions of the organization, many of which are relatively unstructured." (Gorry and Morton 1971, p. 63) This sheds light on the first of three main concerns attributed to the usage of DSS from a managerial perspective (Carlson and Turban 2002):

1. Improved efficiency when dealing with semi- and unstructured problems
2. Sound decisions even without employing sophisticated optimization tools and complex modeling
3. Interactive problem solving and systematic application of knowledge

This vision is realized with advanced technology consisting, in abstract terms, of three interrelated building stones: a database management system, a model base management system, and last but not least a dialog generation and management system (McNurlin and Sprague 2002). A possible database for a general management level within an organization could contain corporate objectives, given constraints, long-term plans, and environmental data (Ramesh and Sekar 1988). Concerning the production environment in specific, engineering, and production control data might be stored in the database (Tsubone et al. 1995). The database is often relational in nature. A model base could provide forecasting models, simulation models, and mathematical programming models (Ramesh and Sekar 1988; Tsubone et al. 1995). Dialogs are to enable the user working interactively and are furthermore subject to considerations of usability, which will be discussed in Sect. 11.7. Especially in the historical context, where computers lacked multitasking capabilities to provide real-time interaction and powerful graphical user interfaces were far from what we know today, it is not surprising that this element was said to be the most important (Sprague and Carlson 1982). In addition, the user interface tends to represent the entire DSS for a user who might be less familiar with the other system components working in the hidden background (Saxena 1991).

The actual realization of these three building blocks differs considerably in real world environments. Pearson and Shim (1995) identified five distinct DSS structures that were significantly influenced by specific combinations of environmental factors . These variables are (i) task structure (structured, semi-structured or unstructured), (ii) supported management level (strategic, managerial or operational), (iii) usage pattern (subscription, terminal, clerk or intermediary), (iv) the number of supported users (single user or multiple users), (v) the user's computer skills (high or low), and (vi) the interaction with other information systems (no interaction, internal systems or external systems). Observations of how these factors apply to a given situation can be integrated into to the design process of a DSS. Thus, the design process needs to take at least these six variables into account.

Furthermore, in the context of DSS development, Moore and Chang (Bennet 1983) suggested a differentiation between weak and strong design. Weak design considers the needs and personal attributes of the DSS user, while strong design refers "to manipulate or refine the user's approach to problem solving." (Bennet 1983, p. 188) Henceforth, the user is seen as reluctant to changes within the strong design approach. The authors concluded that the strong design is rather usual in real world settings and, obviously, technical concerns play a major role in this attempt to build DSS. Unsurprisingly, numerous types of technological applications are to be found in field of DSS: data mining, data warehousing, spreadsheets, neuronal networks, visual programming, intelligent agents, various applications for mathematical programming, and a great deal of software functions more (Alter 2004; Beyon et al. 2002).

The emphasis of mathematical models applied in DSS reveals according to Courtney (2001) a common process of decision-making using DSS, which is displayed with Fig. 11.1. After recognizing a specific problem, it is defined in a manner that enables the development of quantitative i.e. mathematical models, which are further evaluated based on the creation of alternatives. Supported by the results of mathematical model analysis, a selection of the appropriate alternative is made and the solution is implemented. On a more general level, the described process has strong roots in Simon's work (Simon 1960) and the, possibly surprising,

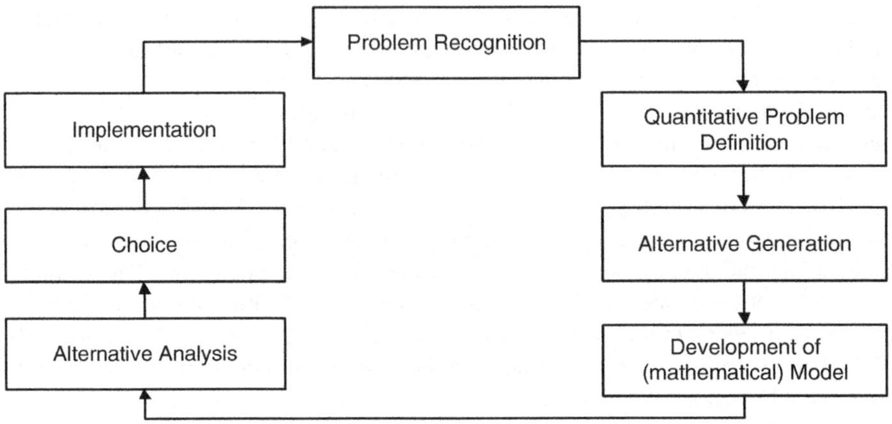

Fig. 11.1 Common DSS decision-making process. Borrowed from Courtney (2001)

inclusion of DSS in the final implementation phase of the decision process can be found in the earlier work of Sprague and Carlson (1982).

After outlining a decision process that utilizes a DSS, it is necessary to take a closer look on how to develop the desired computer system. As stated earlier, the environment within which a DSS is integrated might play a role to decide whether to follow the "strong" or "weak" intentions to introduce a support system. However, a comprehensive approach to realize DSS is given with the Decision Support Engineering (DSE) method, which was proposed by Saxena in the early 1990s (Saxena 1991). Indeed, various other proposals describing how to develop a DSS are available; just to pick two of them: the Decision-Oriented DSS Development Process by Stabell (Bennet 1983) or A Nine-Step Prototyping Design Blueprint by Andriole (1989), which was proposed for military applications.

DSE has a core of six stages, which address a software life cycle model and focus on tangible support needs of future software users. The life cycle model serves to highlight the evolutionary nature of the DSS, which will lead in turn to several prototypes, which serve for testing purposes. The DSS development is regarded as a negotiation process between the DSS builder and the user, which underlines the crucial role of user's participation during all the developmental stages. For a further disquisition of participation and user involvement, refer to Chap. 4, which will provide theoretical considerations as well as practical insights derived from ample implementation projects.

The six steps of DSE according to Saxena (1991) are (1) problem definition and feasibility assessment, (2) decision task analysis, (3) requirements engineering, (4) system design, (5) prototyping, and (6) user evaluation and adaptation.

1. Problem Definition and Feasibility Assessment
   A DSS provides decision makers with the ability to support the solution of complex and ill-structured problems. If DSS development requires use of a

special methodology, it has to be known in advance that the decision task is large, complex and/or ill-structured enough to warrant a DSS solution. The problem domain can be classified in terms of size, complexity, and degree of structure. Size of the problem domain refers to the number of elements that must be considered (within a certain time limit); complexity relates to the number of inter-relationships between elements; and structuredness means the degree of uncertainty about the precise nature of the relationship between elements. Once the need for a DSS-solution is established, the study assesses its technical feasibility (hardware/software tools, and the DSS builders' expertise) taking into account the specific characteristics of the decision task, economic feasibility, and the feasibility of integrating the system into the operating environment.

2. Decision Task Analysis
   The objective of this stage is to provide a realistic understanding of the decision task. This phase, therefore, includes identification of decision subtasks and acquisition of decision task knowledge. Complex decision tasks of any respectable size are too difficult to comprehend as a whole, so the tasks should be decomposed into understandable subtasks. Small subtasks make for small prototypes that are easier to build than larger ones; and could be integrated later. In order to model the support process in a DSS, it is essential to know the problem domain, how the decision maker thinks about the decision situation and what mental data manipulation procedures are used. The type of knowledge required to solve a problem is influenced by the degree to which the task has been formalized. As a problem domain becomes better understood, formal theories or normative models can be constructed. In the absence of this formalization, problem solving and understanding are more likely to depend on informal, intuitive and possibly heuristic models. We refer to Chap. 13 for an extensive discussion on task analysis.
3. Requirements Engineering
   Requirements specify capabilities that a DSS must provide in order to support a decision task. Requirements engineering is, therefore, concerned with identifying the functional and performance characteristics of a DSS. These characteristics in turn depend on the characteristics of the decision task, the decision maker(s), and their perception of decision support. This stage therefore includes user analysis, support analysis, decision model analysis, knowledge base analysis, data base analysis, user interface analysis, hardware/ software environment determination, and usability analysis as its sub-stages.
4. DSS Design
   Though the purpose of this stage is to provide detailed DSS specifications which could enable the DSS builder to build a DSS prototype, it is different from the design phase in conventional system life cycle in that it is intensively directed towards usability goals. This stage therefore includes decision modeling, knowledge base modeling, user interface modeling, data base modeling, and DSS architectural design as its sub-stages.

5. System Prototyping
   Given the prototypes of decision model, knowledge base, and user interface, the final task for the DSS builder is to "package" them together and load the DSS data base in order to develop a prototype DSS ready for user evaluation.
6. User Evaluation and Adaptation
   During this stage the prototype DSS is evaluated by the user and its performance compared with the usability goals using the test data for all the decision scenarios. In case of significant variation in performance the design stage is reopened for modification. In case the user wants to reset some of the usability goals, the requirements and design stages are reopened and the whole life cycle repeated.

Decision task analysis, requirements engineering, and DSS design are creative activities characterized by a large number of complex and inter-related tasks, flexible temporal sequence, trial and error, learning and discovery, interruptions and resumptions. Thus, the DSS life cycle is not a typical waterfall model but an evolutionary life cycle where different stages only define their relative sequence but do not impose any constraint in going back to any of these stages in case of a goal failure at a later stage. This preserves the evolutionary nature of DSS development and avoids the maintenance stage as DSS development always takes place with the users, never without them!

As a conclusion of this methodology, the evolution of the DSS takes place together with the user. Always. Nonetheless, the widespread application of successful methodologies seems rather sparse when switching attention to present challenges of DSS.

Carlson and Turban (2002) proposed that the majority of past DSS-claims are still unfulfilled, i.e. complex problems are still solved manually. They see the reasons for this ongoing trend related to humans, who are still part of the problem solving process. These "people problems" refer to cognitive constraints, lack of understanding the computer support, favor of past experiences, preference of other people instead of inanimate computers, as well as frustration due to incomprehensible and complex algorithms, which are implemented in the DSSs. LaForge and Craighead (2000) found the latter among other aspects and suggested more appropriate training for the people involved. Therefore, the importance of appropriate training will be theme of Sect. 11.7.

The presence of a vast number of interrelated variables causes sometimes counterintuitive results e.g. in automatically generated schedules. These, from the perspective of a human scheduler, suspicious results might suffer from being accepted, since the scheduler simply does not trust those (McKay and Buzacott 2000). Henceforth, McKay and Black (2007) conclude after a 10-year case study the importance of system's robustness because "a system that is not robust will not be trusted or utilized." (p. 769) The roles of trust for actual usage as well as acceptance of technology and DSS in particular will be examined into greater detail in Sect. 11.7. For the moment, it can be stated that the construct of trust has caused tremendous interest in research and it could be accentuated as a major key for the future success of DSS.

Considering a constant amount of work e.g. a scheduling task, a shift towards an ever-increasing portion of automated or computerized task fulfillment by employing some of the previously outlined technological applications will lead to a declining role of humans carrying out the job. To follow this line of reasoning might lead to possible dilemmas of human-machine cooperation, which will be dealt with in Sect. 11.5.

There is ample evidence indicating that successful decision support is not exclusively bounded to technology. Steven Alter, a pioneer in the area of DSS since the 1970s, recently suggested a revised view on DSS employing a work system oriented approach in order to overcome the merely limited view on the computerized information system, which averts, according to his experience, flourishing decision support (Alter 2004). A fundamental definition of a work system can be found in (REFA 1984) and further discussion of work systems concerning production planning and control in particular is available in (Strohm 1996).

Accordingly, work systems are sociotechnical systems influenced by the environment, in which people and technology collaborate in an organizational framework to accomplish a specific task. The system's inputs are tools, information, energy, and people. These inputs are transformed within the system in order to generate desired output according to the given task as products and/or services. It is noteworthy that DSS could be considered as technological system element, which is embedded in the work system. Therefore, it has relations to the other system elements, influences other elements and is reciprocally affected by other elements within the system.

On an organizational level, this systems theory can be traced back to Katz and Kahn (1966). Moreover, the detected influence of (new) technology on people in a work-context was already studied at the Travistock Institute of Human Relation in the 1940s and became known as the sociotechnical approach (Landy and Conte 2007).

Necessarily, the role of DSS in the decision-making process shall be reconsidered to face the outlined obstacles e.g. system rejection. Based on "wicked" problems identified by Rittel and Webber (1973) and the Multiple Perspective Concept of Unbounded Systems Thinking (Mitroff and Linstone 1993), Courtney (2001) suggested an advanced decision making process for DSS, which is, with slight adaptations from the original, depicted by Fig. 11.2.

The striking features of Fig. 11.2 in comparison to Fig. 11.1 is the definite consideration of human attributes. **T** as technical perspective is found just as a part next to the organizational or societal **O**, and the personal or individual **P** perspective (Mitroff and Linstone 1993). Similarities between these three perspectives and the characteristics of the work system should be easily recognizable. Nevertheless, Mitroff and Linstone specifically highlight the influence of mental models influencing each process step. Mental models are said to determine "what data and what perspectives we examine in a world of overabundant data sources and plethora of ways of viewing data." (Courtney 2001, p. 31) Additional elements to be mentioned, which contribute to the development of rich perspectives, are ethics and aesthetics.

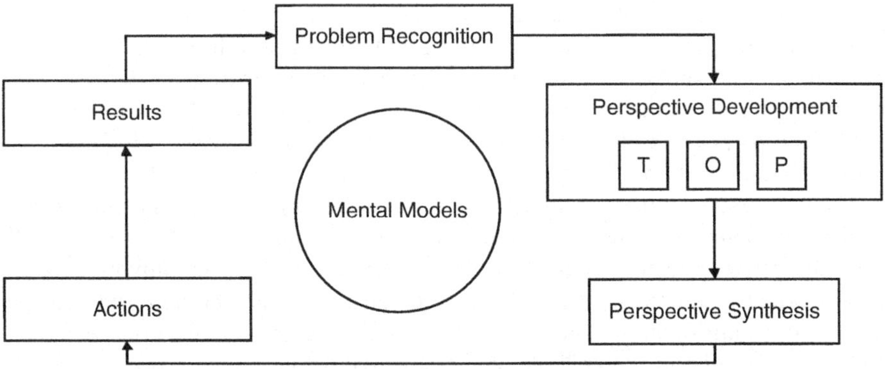

Fig. 11.2 New DSS decision-making process. Borrowed from Courtney (2001)

11.4 Planning Hierarchies

The production scheduling function is typically part of a planning hierarchy. This implies that the orders that a production scheduler receives are orders that have been generated at another level in the planning hierarchy. If an ERP system is used, orders are usually generated by the ERP system using Material Requirements Planning Logic (MRP). MRP logic uses the structure of the bill-of-materials and its associated lead times to determine when and how much of a certain component needs to be ready, and – based on standard lead times – when production orders need to be released. The task of a production scheduler is then to complete that order by the date allocated. MRP decisions are typically taken weekly, but sometimes are also take on a daily basis. In a limited number of companies, also advanced planning logic is deployed to make the planning decision at the higher level, either at the plant level or across multiple plants and inventory locations. We denote this as Supply Chain Operations Planning (SCOP). In the SCOP function, the release decision is essentially identical to the output of an MRP decision, but different logic is applied, which explicitly coordinates the release decisions, sometimes even taking (aggregate) capacity constraints into account. Note again that MRP logic does not coordinate the decisions, but just extrapolates the needs from the customer upstream without explicitly coordination material and resource availability. For a further discussion on this item, we refer to de Kok and Fransoo (2003).

What is important to realize in the realm of this chapter is that ordering decision are hence not autonomous random processes, but are typically the result of deliberate decisions, often supported by other software. Thus, if we study the user involvement in the construction of decision support systems for production scheduling, we need to be aware of the context in which the schedulers operate. If schedulers operate in an MRP environment, their task may be more complicated, due to the fact that they face challenges in material or resource availability and hence may not be able to construct a schedule. If schedulers operate in a SCOP environment, schedules have a higher likelihood of being feasible, but simultaneously the requirements on the schedule may be more extensive, for instance in the utilization of resources assumed.

All this leads to a need to spend time in properly defining and agreeing on the exact planning problem, the types of present constraints, and the exact objectives under which to operate. This puts requirements both onto the production scheduler, and on the entire development process of the Decision Support System for production scheduling.

Furthermore, this also should specify the interface between the (supply chain operations or materials) planning level and the scheduling level. In many DSS projects, this interface is purely defined as a technical (computer science) interface, setting up definitions of which data to exchange between the various planning modules or functions. However, it is even more critical to properly define the interface between the planning and scheduling level from a process perspective. This would include issues such as the aggregation of resource information (e.g., capacity) from the detailed scheduling level to the aggregate planning level, and decisions on frequency, such as the planning frequency. For a further discussion on this issue we refer to Chap. 8.

11.5 Human Factors in Software Design

It is often assumed that human factors are only concerned with ergonomic issues such as furniture design, human-machine interfaces and workplace environmental conditions. These are all important considerations but only cover a relatively small proportion of the influences on human behavior. Other more important influences include:

- The availability of "best practices" that define practical and efficient methods of performing tasks
- The quality of training and the way in which individuals maintain their competence
- Allocating functions appropriately between people and automatic systems such that people are neither over or under occupied at any time
- Availability of information concerning system operation that provides cues for action and feedback on task success
- Risk perception and its influence on how activities are carried out
- Communication of priorities and other organizational factors.

The engineering approach to human factors problems often involves technical solutions. This may involve "engineering-out" human intervention through increased automation and protection devices. In many situations this is a sensible approach but it rarely removes people from the system, it simply changes the function they perform. This can lead to a situation where people are expected to diagnose and respond to unusual situations, but are not given the opportunity to maintain the skills necessary to achieve this.

People are involved at all stages of a systems life cycle and they have to be considered as an essential part of the overall systems design. Their errors can cause failures whilst their ability to diagnose problems and act appropriately can lead to the successful recovery of technical failures.

Systems engineering should encompass human factors in order to account for the positive and negative impact people have on systems reliability. People perform quite diverse functions. It is this diversity that makes them particularly useful components in systems but it introduces many opportunities for failure. It must be accepted that it will be almost impossible to get the human factors aspects of a system correct at the first attempt. Also, the nature of systems is that they continually evolve and a static solution is unlikely to result in a system achieving its full potential. It is important, therefore that the incorporation of human factors is considered as part of the systems evolution over its full life cycle.

There is a wide field of concepts that are concerned with the incorporation of human factors into system design. The following subsections provide an overview.

11.5.1 Interaction Design (IxD or IaD)

Interaction design is the discipline of defining and creating the behavior of technical, biological, environmental and organizational systems, see for instance (Sharp et al. 2002; Cooper et al. 2007; Saffer 2006). Examples of these systems are software, products, mobile devices, environments, services, wearables, and even organizations themselves. Interaction design defines the behavior (the "interaction") of an artifact or system in response to its users over time. Interaction designers are typically informed by user research, design with an emphasis on behavior as well as form, and evaluate design in terms of usability and emotional factors.

There is a general process that most interaction designers follow. A key element in this process is the idea of iteration, where the aim is to build quick prototypes and test them with the users to make sure the proposed solution is satisfactory.

1. Design Research
   Using design research techniques (observations, interviews, and activities) designers investigate users and their environment in order to learn more about them and thus be better able to design for them.
2. Concept Generation
   Drawing on a combination of user research, technological possibilities, and business opportunities, designers create concepts for new software, products, services, or systems. This process may involve multiple rounds of brainstorming, discussion, and refinement.

3. Creation of Scenarios/Personas/Profiles
   From the patterns of behavior observed in the research, designers create scenarios (or user stories) or storyboards, which imagine a future state of the product or service. Often the designer will first create personas or user profiles from which the scenarios are built.
4. Wireframing and Flow Diagrams
   The features and functionality of a product or service are often outlined in a document known as a wireframe ("schematics" is an alternate term). Wireframes are a page-by-page or screen-by-screen detail of the system, which include notes ("annotations") as to how the system will operate. Flow Diagrams outline the logic and steps of the system or an individual feature.
5. Prototyping and User Testing
   Interaction designers use a variety of prototyping techniques to test aspects of design ideas. These can be roughly divided into three classes: those that test the role of an artifact, those that test its look and feel and those that test its implementation. Sometimes, these are called experience prototypes to emphasize their interactive nature. Prototypes can be physical or digital, high- or low-fidelity.
6. Implementation
   Interaction designers need to be involved during the development of the product or service to ensure that what was designed is implemented correctly. Often, changes need to be made during the building process, and interaction designers should be involved with any of the on-the-fly modifications to the design.
7. System Testing
   Once the system is built, often another round of testing, for both usability and errors ("bug catching") is performed. Ideally, the designer will be involved here as well, to make any modifications to the system that are required.

11.5.2 Human–Computer Interaction, Man–Machine Interaction (MMI), Computer–Human Interaction (CHI)

The focus of human–computer interaction (HCI) is the study of interaction between people (users) and computers, see for instance (Jacko and Sears 2003; Baecker et al. 1995). It is an interdisciplinary subject, relating computer science with many other fields of study and research. Interaction between users and computers occurs at the user interface (or simply interface), which includes both software and hardware, for example, general purpose computer peripherals and large-scale mechanical systems, such as aircraft and power plants.

A basic goal of HCI is to improve the interaction between users and computers by making computers more usable and receptive to the user's needs. Specifically, HCI is concerned with:

- Methodologies and processes for designing interfaces (i.e., given a task and a class of users, design the best possible interface within given constraints, optimizing for a desired property such as learnability or efficiency of use)
- Methods for implementing interfaces (e.g. software toolkits and libraries; efficient algorithms)
- Techniques for evaluating and comparing interfaces
- Developing new interfaces and interaction techniques
- Developing descriptive and predictive models and theories of interaction

A long term goal of HCI is to design systems that minimize the barrier between the human's cognitive model of what they want to accomplish and the computer's understanding of the user's task.

Researchers in HCI are interested in developing new design methodologies, experimenting with new hardware devices, prototyping new software systems, exploring new paradigms for interaction, and developing models and theories of interaction.

Human–computer interaction is an interdisciplinary field, combining aspects of several major fields including

- Computer science – for ideas concerning how to design algorithms
- Psychology and related fields – for knowledge concerning the capabilities of human memory, motor skills, and perception; how people communicate with each other and work in groups; and social dynamics
- Artificial intelligence and related fields – for ideas concerning how to automate more work, or make computers that behave more like intelligent assistants
- Computer graphics – for ideas concerning how to output visual information
- Design – for example, graphic design of visual output, industrial design of mice and keyboards, etc.

A number of diverse methodologies outlining techniques for human–computer interaction design have emerged since the rise of the field in the 1980s. Most design methodologies stem from a model for how users, designers, and technical systems interact. Early methodologies, for example, treated users' cognitive processes as predictable and quantifiable and encouraged design practitioners to look to cognitive science results in areas such as memory and attention when designing user interfaces. Modern models tend to focus on a constant feedback and conversation between users, designers, and engineers and push for technical systems to be wrapped around the types of experiences users want to have, rather than wrapping user experience around a completed system.

User-centered design (UCD) is a modern, widely practiced design philosophy rooted in the idea that users must take center-stage in the design of any computer system. Users, designers and technical practitioners work together to articulate the wants, needs and limitations of the user and create a system that addresses these elements. Often, user-centered design projects are informed by ethnographic studies of the environments in which users will be interacting with the system.

11.5.3 Participatory Design

Participatory design (PD) is an approach to design that attempts to actively involve the end users in the design process to help ensure that the product designed meets their needs and is usable, see for instance (Kensing 2003; Muller 2007; Schuler and Namioka 1993). It is also used in urban design, architecture, landscape architecture and planning as a way of creating environments that are more responsive and appropriate to their inhabitants and users cultural, emotional, spiritual and practical needs. It is one approach to place making. It has been used in many settings and at various scales, and in the United Kingdom is known as community architecture. It is important to understand that this approach is focused on process and is not a design style. For some, this approach has a political dimension of user empowerment and democratization. For others, it is seen as a way of abrogating design responsibility and innovation by designers. In several Scandinavian countries of the 1960s and 1970s, it was rooted in work with trade unions; its ancestry also includes Action research and socio-technical Design.

In participatory design, end-users are invited to cooperate with researchers and developers during an innovation process. Potentially, they participate during several stages of an innovation process: they participate during the initial exploration and problem definition both to help define the problem and to focus ideas for solution, and during development, they help evaluate proposed solutions.

Participatory design can be seen as a move of end-users into the world of researchers and developers, whereas empathic design can be seen as a move of researchers and developers into the world of end-users.

There has been some participatory design in the United States in the Scandinavian style and widespread use of design techniques that are based on participatory design.

Greenbaum and Kyng (1991, p. 4) identify four issues for design:

1. The need for designers to take work practice seriously – to see the current ways that work is done as an evolved solution to a complex work situation that the designer only partially understands
2. The fact that we are dealing with human actors, rather than cut-and-dried human factors – systems need to deal with users' concerns, treating them as people, rather than as performers of functions in a defined work role.
3. The idea that work tasks must be seen within their context and are therefore situated actions, whose meaning and effectiveness cannot be evaluated in isolation from the context
4. The recognition that work is fundamentally social, involving extensive cooperation and communication

These principles apply in all workplaces, regardless of the specific interactions between workers and management.

11.5.4 Usability Engineering

Usability Engineering (UE) is a subset of human factors that is specific to computer science and is concerned with the question of how to design software that is easy to use. It is closely related to the field of human–computer interaction and industrial design.

The term Usability Engineering describes a pragmatic approach to user interface design which emphasizes empirical methods and operational definitions of user requirements for tools. Extending as far as International Standards Organization-approved definitions (see e.g., IS0 9241 part 11) usability is considered as a context-dependent agreement of the effectiveness, efficiency and satisfaction with which specific users should be able to perform tasks. Advocates of this approach engage in task analysis, then prototype interface designs and conduct usability tests. On the basis of such tests, the technology is (ideally) redesigned or (occasionally) the operational targets for user performance are revised. (Dillon 2000)

11.6 Human Factors when Designing and Implementing DSS: Modeling Dynamics

11.6.1 Introduction

In this section, we investigate the modeling process. Our main hypothesis is that the modeling process takes place as a dynamic interactive process between the modeler (consultant) and user (scheduler). We are interested in better understanding this process, and investigate this in four different cases. Our objective is to develop a number of generic designer guidelines. While substantial research has been conducted on technology acceptance, this chapter is specific in two ways. First, we address a technology that is developed and customized for a specific professional user. Second, we study the process of building the model as a dynamic process that is conducted in interaction with the user.

11.6.2 Literature review

11.6.2.1 User Participation and User Involvement

In terms of user participation and user involvement a lot of research regarding input and outcome variables as well as moderators has be done so far, (Barki and Hartwick 1994; Blili et al. 1998; Davis and Kottemann 1995; Doll and Torkzadeh 1991; Foster and Franz 1999; Hartwick and Barki 1994; Hawk and Dos Santos

1991; Hunton and Beeler 1997; Ives and Olson 1984; Kim et al. 1998; Lawrence et al. 2002; Lin and Shao 2000; Palanisamy 2001; Tait and Vessey 1988; Xie 2003), especially for the field of DSS see (Guimaraes et al. 1992; Igbaria and Guimaraes 1994). User participation is considered to be critical for system quality and system use, (Ives and Olson 1984). Hwang and Thorn (1999) provide a good overview and also a meta-analysis of the effects of user participation on system success.

First, we have to define the terms of user participation and user involvement. Palamirany states that in the MIS (and DSS) context user involvement is a subjective psychological state of the individual user in terms of importance the user attaches to a given system. Participation, on the opposite is an observable behavior respectively a behavioral engagement, (Kappelman 1995.) In the literature, there is often no clear distinction between user influence, engagement, participation in the process and user involvement. This influences also the measurability of the mentioned constructs and makes the verification of the effects user participation or involvement have more difficult, (Barki and Hartwick 1994).

Considering the existing research results it can be subsumed that most of the studies done so far make a clear distinction between involvement and participation and refer to such outcome variables like user satisfaction and (perceived) benefits. More or less several moderating variables are taken into account, see Table 11.1.

According to Barki and Hartwick (1994), referring to a lot of other authors, participation can be described to be direct or indirect, formal or informal, alone or shared, actual or perceived and according to its scope. Participation in information systems development, according to its importance, needs to be direct and indirect, formal and informal, alone and shared. Barki and Hartwick (1994) clearly distinguish between participation and involvement and present a questionnaire for analyzing and describing participation consisting of three subscales (user-IT relationship, responsibility, hands-on activities) in different stages (systems definition, physical development, implementation and overall). Other (mostly similar) measures and measurement instruments (on the basis of questionnaires) are provided by Blili et al. (1998), Foster and Franz (1999) – based on Franz and Robey (1986) – Guimaraes et al. (1992), Hunton and Beeler (1997), Ives and Olson (1984), Lin and Shao (2000), McKeen et al. (1994), Lawrence et al. (2002).

Integrating these findings it can be stated that for our purpose:

- We need to look at participation, defined as activities or behaviors that users show, as well as to user involvement, defined as psychological state (acc. to Hartwick and Barki (1994)), because both seem to have significant impacts
- We need to distinguish between different stages of the development process, that means we have to consider the dynamics
- We assume, in congruence with major findings, that user participation and user involvement have a direct significant impact on the attitudes toward the system, which in turn leads to system acceptance, and on user satisfaction (Hartwick and Barki 1994; Ives and Olson 1984; McKeen et al. 1994; Lin and Shao 2000) and that this relation depends on several variables like characteristics of the

Table 11.1 Former research in the field of user participation and user involvement

Author(s)	Input variables	Output variables	Moderating, other influencing variables
Baroudi et al. (1986)	User involvement	System usage, information satisfaction	
Blili et al. (1998)	User involvement	Success (satisfaction, impacts)	Task uncertainty, competence
Doll and Torkzadeh (1991)	Congruence between desired and perceived involvement	User satisfaction	
Guimaraes et al. (1992)	User involvement	System success (satisfaction, perceived benefits)	Characteristics of decision maker, task, system, user training
Hunton and Beeler (1997)	User involvement, user attitude, user self efficacy, user participation	User performance	
Lawrence et al. (2002)	User participation	DSS use, decision accuracy, satisfaction	
Lin and Shao (2000)	Participation, attitudes, involvement	System usage, acceptance, satisfaction	System impact, system complexity, development methodology
McKeen et al. (1994)	User participation	Success	Task complexity, system complexity, user influence, user – developer communication
Tait and Vessey (1988)	User involvement	System success	User system (impact, attitudes), technical system (complexity), development process (resource constraints)

system (complexity, of the user (competence, training) and the process (stage), (Blili et al. 1998; Guimaraes et al. 1992; McKeen et al. 1994) (Fig. 11.3)

11.6.2.2 Technology Acceptance

The performance and satisfaction resulting from the usage of an information system depend largely on the acceptance of the particular system respectively the technology. For the explanation of technology acceptance serves the Technology Acceptance Model (TAM) proposed by Davis (1986). The TAM is based on intention models from social psychology, especially the Theory of Reasoned Action (TRA). The TRA assumes that behavioral intentions (BI) are prior to every behavior and

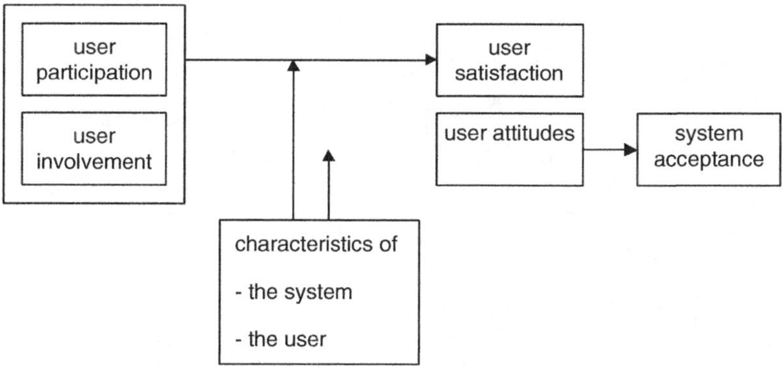

Fig. 11.3 User participation and user involvement technology acceptance

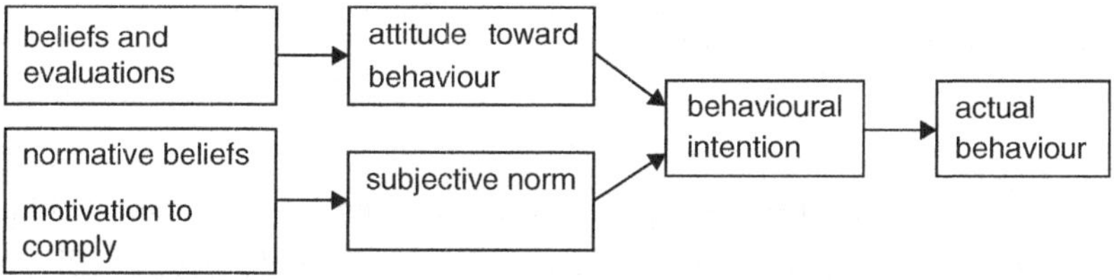

Fig. 11.4 Theory of reasoned action

that attitudes and subjective norms are decisive for the intentional behavior, (Davis et al. 1989) (Fig. 11.4).

The TAM is an adoption of the TRA, specifically tailored for modeling user acceptance of information systems, (Davis et al. 1989). The main concept underlying this model is, that two constructs, the perceived usefulness and the perceived ease of use determine the attitudes toward using a certain technology and the behavioral intention to use, (Davis et al 1989).

Perceived Usefulness (PU) is defined as the degree to which a person believes that using a particular system could enhance his or her job performance: it is the extent to which an individual believes that using the system enhances his/her performance. Perceived Ease of Use (PEU) can be described as the degree to which a person believes that using a particular system is free of effort, (Saade and Bahli 2005).

Davis et al. (1989) theorized perceived usefulness to be a more important determinant of intention when compared to perceived ease of use. They justified the argument on the basis that in a workplace setting, which typically emphasizes productivity, a rational factor that is an individual's assessment of the performance outcomes associated with technology use (i.e., perceived usefulness) will be the single most important determinant of usage intentions and usage behavior. Empirical studies, spanning a range of different systems and user populations, have found perceived usefulness to be a stronger determinant of intention/usage than perceived ease of use (Davis and Venkatesh 2004) (Fig. 11.5).

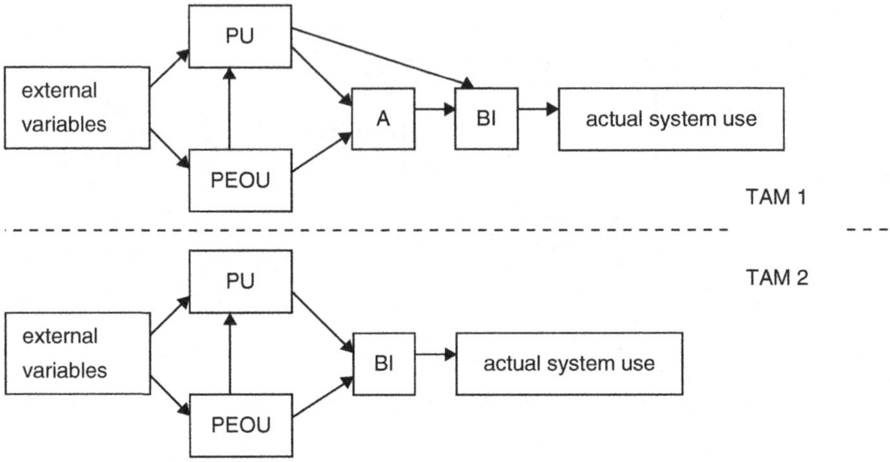

Fig. 11.5 Technology acceptance model

PU, PEOU and BI are usually captured by a questionnaire (Davis and Venkatesh 2004; Doll et al. 1998; McHaney and Cronan 2001; Mathieson et al. 2001; Szajna 1994).

A lot of empirical studies confirmed the stated relationships, (Davis 1993; Doll et al. 1998; Legris et al. 2003; Ma and Liu 2004; Szajna 1994; Szajna 1996). Others extended the model in different ways, see Table 11.2.

Another relation was introduced by Igbaria and Tan (1997), which state that IT acceptance leads to both user satisfaction and system usage. System usage and user satisfaction are interrelated as well (see Fig. 11.6).

Interesting for our approach is also the recent consideration of some dynamic developments, (Davis and Venkatesh). The objective of this is the ability to predict technology or system acceptance after the implementation by presenting design specifications to potential users prior to implementation, (so called "pre prototype user acceptance testing"). The related model is shown in Fig. 11.7.

The importance of this consideration is given by higher development or adjustment costs and a smaller modifiability when the development process goes on (Davis and Venkatesh 2004).

The result of their study was that only the user perceptions at the first state (early user reactions) and the usage behavior at the previous state had significant relations to the usage behaviors at later states.

This approach is also important for DSS. If the system and the inherent model is not accepted by the user or leads to poor results a modification process is induced, which causes high expenses and influences the overall profitability of the systems development and implementation project.

Table 11.2 Overview of former research in the field of technology acceptance

Author(s)	External variables (according to the TAM)	Moderating and other influencing variables
Amoako-Gyampah and Salam (2004)	Communication, training, shared belief in benefits	
Chau (2001)	General attitude toward computers, self efficacy	
Davis (1993)		System design features
Davis et al. (1989)		Self efficacy
Dishaw and Strong (1999)	Task-technology fit	Task-technology fit
Elbeltagi et al. (2005)	Task, culture, environment, system characteristics, internal and external support, organization characteristics, decision maker characteristics	
Gefen and Keil (1998)	Perceived developer responsiveness	
Hardgrave and Johnson (2003)	Subjective norm, perceived behavioral control	Subjective norm, perceived behavioral control
Hodgson and Aiken (1998)	Implementation gap, transitional support, organizational change, general attitudes	
Hubona and Kennick (1996)		Education, employment category
Igbaria and Iivari (1995)	Computer experience, organizational support, self efficacy, computer anxiety	
Jackson et al. (1997)	Arguments for change, prior use	Situational involvement, intrinsic involvement
Karahanna and Straub (1999)	Social presence, social influence, physical accessibility, support	
Kleintop et al. (1994)	Experience	
Klopping and McKinney (2004)	Task-technology fit	Task-technology fit
Mathieson et al. (2001)		Perceived personal resources
Ndubisi et al. (2005)		Innovation, perseverance, flexibility, risk taking propensity
Rawstorne et al. (2000)		Behavioral attitude, subjective norm, perceived behavioral control
Saade and Bahli (2005)	Cognitive absorption	
Straub et al. (1997)	Cultural dimensions (individualism/collectivism, uncertainty avoidance, short-term-/long-term orientation)	Cultural dimensions (power distance)
Taylor and Todd (1995)		Experience, subjective norm, perceived behavioral control
Venkatesh (2000)	Computer self efficacy, perceived external control, computer anxiety, perceived playfulness, perceived enjoyment, objective usability	
Venkatesh et al. (2002)	Intrinsic motivation, extrinsic motivation, training, control	
Venkatesh and Davis (1996)	Computer self efficacy, objective usability	
Venkatesh and Davis (2000)	Subjective norm, image, job relevance, output quality, result demonstrability	Experience, voluntarism

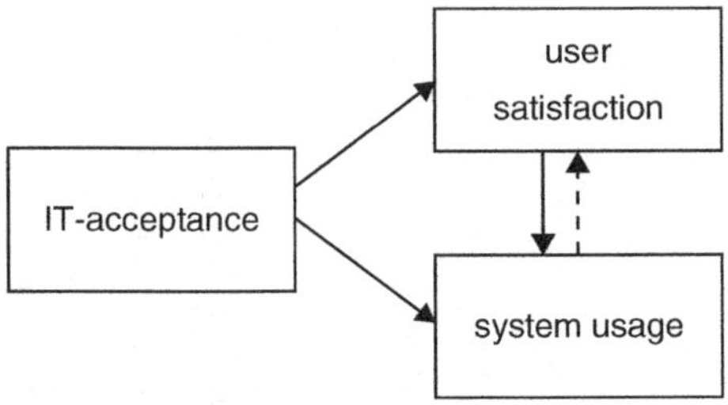

Fig. 11.6 Acceptance, satisfaction and usage

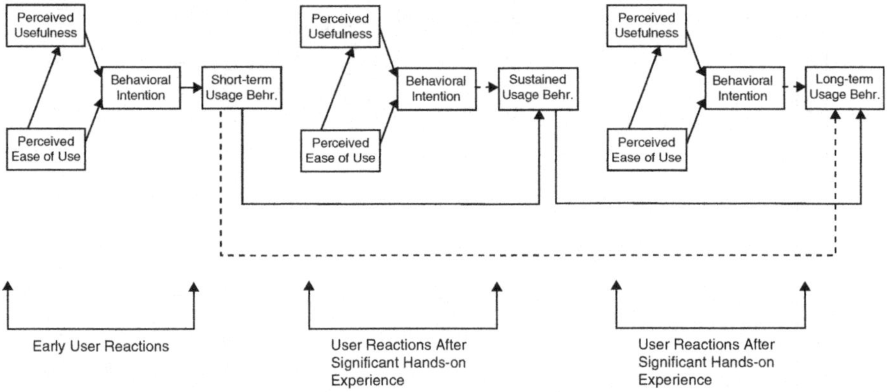

Fig. 11.7 Dynamics in the TAM

Subsuming the theory of technology acceptance we can conclude that:

- Perceived usefulness and perceived ease of use are predictors for technology acceptance (BI within the TAM)
- Acceptance leads to system usage and user satisfaction which in turn will positively influence the obtained performance
- There are dynamic developments within system usage that are mainly controlled by initial built attitudes and former experiences
- There are a lot of influencing variables that can be influenced by organizational and technical design (Fig. 11.8)

11.6.2.3 Control Theories

Control theory (Fig. 11.9) has a close relationship to the research conducted in the field of participation. Participation in organizations implies, by definition, that workers enter into decision making and that workers thus exercise legitimate control, (Bartoelke et al. 1982).

Control itself is defined as the need to demonstrate ones competence, superiority and mastery over the environment, (Faranda 2001). Mostly control has been

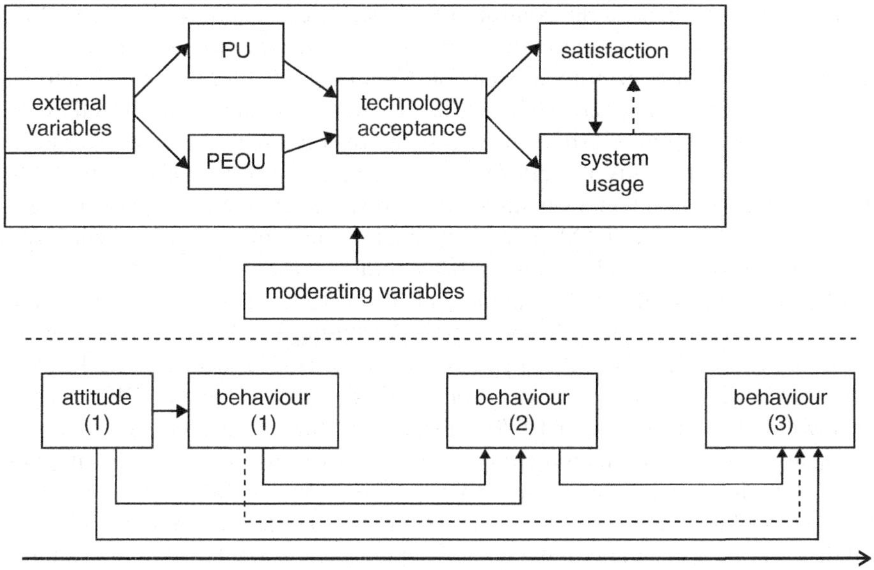

Fig. 11.8 Dynamics in technology acceptance

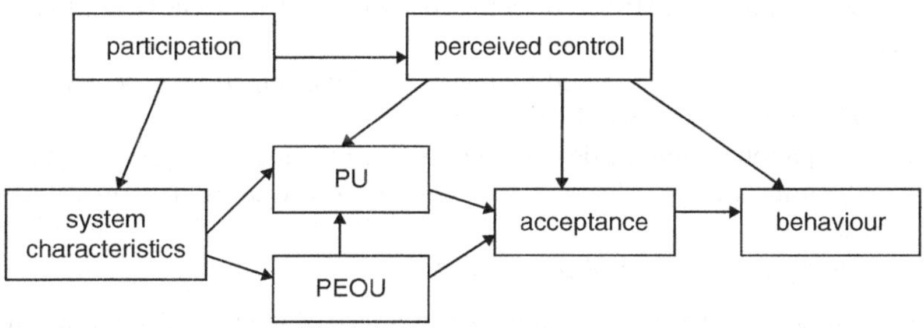

Fig. 11.9 Control theory

operationalized in three different ways, (Faranda 2001; see also Morris and Marshall 2004):

- Behavioral control, i.e., the availability of response which may directly influence or modify the objective characteristics of an event
- Decisional control, i.e., the opportunity to choose among various courses of action
- Cognitive control, i.e., the way an event is interpreted, appraised or incorporated into a cognitive plan

Control plays a decisive role within the theory of planned behavior (TPB) which is an extension of TRA). Within TPB intention is determined by cognitive evaluations of the behavior (i.e., attitudes), perceptions of social pressure (i.e., subjective norm), and perception of behavioral control (PBC), (Kidwell and Jewell 2003). PBC specifies the likelihood of successful performance and will vary as a function of controllability toward performing behavior. Controllability can be divided into an internal and external perspective, (Kidwell and Jewell 2003). Internal controllability describes the perception that someone has control over personal resources like skills, confidence, and the ability to perform a special behavior. External controllability refers to the perception that a situation is relatively free from extrinsic influences.

Coming back to the TAM, perceived control, i.e. internal resp. behavioral control, is considered as influencing the behavioral intention as well as the behavior itself (Rawstorne et al. 2000; Taylor and Todd 1995 and also the perceived usefulness (Hardgrave and Johnson 2003). According to Davis and Kottemann (1994) and Kleingeld et al. (2004) perceived control is also related to the level of participation within a design or decision making process. Davis and Kottemann (1994, with reference to Langer 1975) argue that in situations the more one actively participates in the event, the more control one has over the outcome. That implies that by allowing decision makers to manipulate decision variables and assumptions within a model and to observe the effects on predicted outcomes this should create confidence in the model and in the related outcome and should lead in the end to a higher level of perceived control, (Davis and Kottemann 1994).

So it can be subsumed that:

- Perceived control is a decisive factor for perceived usefulness, attitudes and behavior in the former described model
- Perceived control is determined by the level of participation, especially within the design process of a technical system
- The participation within the design process also influences the system characteristics, which in turn are a factor influencing perceived usefulness and perceived ease of use

  Participation from a practical perspective

Specifications for Advanced Planning and Scheduling systems come from two main sources: the primary process that has to be planned or scheduled, and the planners that are carrying out the task. It could be argued that information about the primary process, including demand and supply patterns, should be enough to design planning and scheduling decision support. However, many primary processes contain a massive amount of details. Although scheduling is the most detailed production control level, even in designing scheduling systems decisions must be made what to include and what to exclude in the scheduling model. So, to guide in these decisions, it is practically inevitable to study the scheduling task. Furthermore, the scheduling task is important in determining the sequence of subactivities – schedule machine 1 and then machine 2 or the other way round, first assign material and then sequence or do it the other way round? Of course, what is discovered in the scheduling task should not be blindly copied, but starting with a task model makes designing decision support much more a guided activity than to start from scratch.

Hence, in designing planning and scheduling decision support, it is not a question whether users should participate. Instead, an implementation approach rather should indicate *how* the users should be involved in designing the system. There are five dimensions on setting the user participation in designing production control decision support systems:

Representation. In most implementations, not all intended users of the system are equally involved in the project. Instead, it is more efficient to make a selection where some users are heavily involved in setting the specifications – the so-called key-users – and other users are involved from a distance. The other users can be informed about the project status by the key users, and when needed, the key users can involve other users.

Facilitation. Intended users of a decision support system are usually not very well able to evaluate the consequences of design decisions. This is because they typically lack the experience in designing such systems, and the ability to conceptualize everyday problems. Therefore, all input from users should pass through an experienced system designer before it is realized. If this is not facilitated, user participation might lead to overly detailed and complex systems, as the user cannot see through the forest that grew out of the trees of all functional requirements.

Involvement level. There are various levels of involvement that can be offered to users. On the one extreme, their tasks are studied in the beginning and they are allowed to comment on the end result. On the other hand, the users are allowed to give extensive input, comment on design decisions, test prototypes, supply and clean up data.

Timing. An implementation project usually consists of several phases, where subsequently a design is made, modeled, tested and then implemented. For every phase it can be decided to involve the users to a certain extent. The involvement of key users is especially important in the design and in the testing of the system. In the design phase, the main design decisions are made, such as: what to include/exclude, what level of support will the system offer. And in the test phase it becomes clear what details have been overlooked (typically a lot) and how the system can be made more usable, by fine-tuning the GUI.

Relationship. Apart from the abovementioned 'rational' factors in systems design, it can be useful to let users participate to give them a feeling of involvement and to establish a relationship between the system designers and the user. For that reason, it may even pay off to have the users with the strongest objections involved intensively, so that they feel responsible for the end result.

11.6.2.4 Apprenticeship Model (Beyer and Holtzblatt 1995)

Looking outside the computer field, the relationship between master and apprentice stands out as a useful model. Just as an apprentice learns a skill from a master, designers want to learn about their customers' work from the customer. The authors of this concept see the designer taking on the role as an apprentice, learning from the user.

The critical aspects of the relationship are as follows:

- Teaching ability is not needed.
  Teaching in the context of doing the work obviates formal presentations and course materials. It eliminates any need for craftspeople to think in advance about the structure of the work they do. As they work, the structure implicit in the work becomes apparent because both master and apprentice are paying attention to it.
- Seeing the work reveals what matters.
  People are not aware of everything they do. Each step of doing a task reminds them of the next step; each action taken reminds them of the last time they had to take such an action and what happened then. Some actions are the result of years of experience and have subtle reasons; other actions are habit and no longer have a good justification. Nobody can talk better about what they do and why they do it than they can while in the middle of doing it.
- Seeing the work reveals details.
  Talking about work while doing it protects the master from the human propensity to talk in generalizations. Customers do this, too. Rather than remember all the details of a task or event, they mix up many similar events and then talk about them in the abstract as though they were all one. This abstraction is divorced from any of the reasons for choosing one action over another and may not match any real instance at all. A design built on such generalizations may not meet anyone's needs.
- Seeing the work reveals structure.
  In the same way, designers observing multiple events and multiple customers learn to see the common strategies underlying the work. Once they understand the basic strategies, they can start to imagine a system that would support those strategies.
- The apprentice can learn from the master's experience.
  Designers typically have less time to spend with their customers than the years needed for an apprenticeship. But in the same way an apprentice can learn from the master's experience, designers can learn about events that occurred in the past. Events that occur while the designer is present remind customers to talk about events that happened previously.

While apprenticeship defines an attitude for designers to adopt, their motive in observing work is not that of an apprentice. Where apprentices want to know how to do the work, designers must determine what a system might do to support the work. This leads to changes in the customer-designer relationship:

- The designer must be responsible for seeing work structure.
  A designer must understand structure and implication: the strategy to get work done, constraints that get in the way, the structure of the physical environment as it supports work, the way customers divide work into roles, and the recurring patterns in their work – and what all this implies for any potential system. The customer is not an expert in seeing work structure, and does not naturally articulate it.

- Designers must articulate their understanding.
  Any system is based on a chain of reasoning starting from a work observation and leading to an aspect of the design. Designers will build different systems depending on how they understand their customers' work.
  The only way to ensure an interpretation is correct is by sharing it with the customer. We fail in the entire purpose of working with customers if we do not share and validate these interpretations.
- The designer's job is to improve work.
  The designer expands his or her understanding of the work in this way – if the idea does not work out, there is some aspect of the work the designer is not accounting for.
  And the customer enters into the design conversation – learning what technology can do and how it might be applied. This gives the customer the power to shape the initial perspectives that will eventually result in a full design.
  The designers' focus determines the scope of the information they need. The apprenticeship model allows designer and customer to guide the conversation into areas of work that are relevant to the design. The customer, by revealing all the aspects of the work, broadens the view of designer beyond the initial assumptions. The designer, by focusing on particular aspects, draws the customer's attention to the parts of the work that can be affected in the design.
- The designer has a specific focus.
  Designers know about technology, have skill in applying it to real-life problems, and naturally respond to the customer's activities by designing improvements to them. The apprenticeship model allows both designer and customer to introduce design ideas as the work suggests them. The customer can respond to the idea while doing the work the idea supports.

Using the apprenticeship model as a base, designer and customer can develop an interaction that allows them to explore and understand the customer's work together. The customer is the expert in the work and how to do it; the designer is the expert in seeing the structure of work and the technology available to support it. Although the designer is not an apprentice, the model is an effective form for interacting with customers.

In practice, the customer might work for a while, with the designer looking on. The customer is immersed in the work, thinking about content. The designer, as apprentice, looks for patterns, for structure, and for things he or she does not understand. At some point the designer makes an observation or asks a question. This interrupts the flow of work. In order to respond, the customer has to stop working, step back, and think about the question.

The customer now responds at two levels: First, he or she addresses the question and a conversation about the work ensues. Second, the question is an example of seeing strategy where before the customer saw only actions. Customers soon learn to see strategy, and start interrupting themselves to make observations about the work they do.

11.6.3 Conceptual Model

Integrating these findings into our model, we consider participation, defined as activities or behaviors that users show, and user involvement, defined as psychological state (Hartwick and Barki 1994), since both appear to have significant impacts. Furthermore, we need to distinguish between different stages of the development process. And we propose, in line with previous findings, that user participation and user involvement have a direct impact on the attitudes toward the system – which in turn leads to system acceptance – and on user satisfaction (Hartwick and Barki 1994; Ives and Olson 1984; McKeen et al. 1994; Lin and Shao 2000). Furthermore, we propose that this relation depends on several variables like characteristics of the system complexity, of the user (competence, training) and the process (stage) (Blili et al. 1998; Guimaraes et al. 1992; McKeen et al. 1994).

Taking the insight from the theory of technology acceptance, we can conclude that perceived usefulness and perceived ease of use are predictors for technology acceptance. Acceptance leads to system usage and user satisfaction, which in turn will positively influence the obtained performance. Many variables can be influenced by organizational and technical design.

Finally, we propose that perceived control is a decisive factor for perceived usefulness, attitudes, and behavior. Perceived control is determined by the level of participation, especially within the design process of a technical system. Participation within the design process also influences system characteristics, which in turn are a factor influencing perceived usefulness and perceived ease of use.

In Fig. 11.10, our research model has been summarized. We build on previous work in the core of the model, where we relate participation, involvement, perceived control, perceived usefulness, and behavioral intentions. Note the dynamic

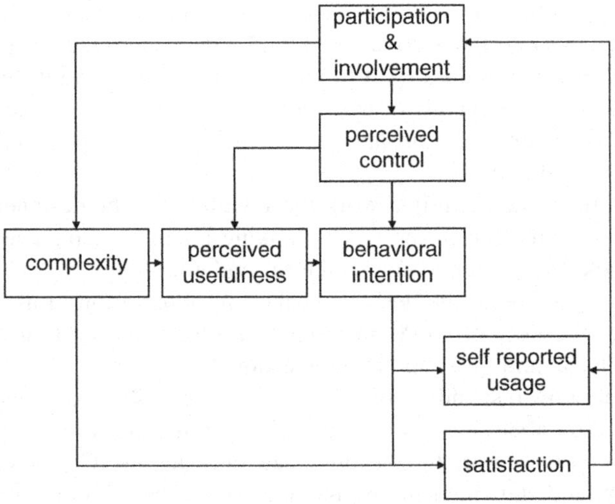

Fig. 11.10 Conceptual model

relationship in which satisfaction leads to participation and involvement, and participation and involvement lead to more complexity in the solution. The reasoning behind this part of the model is that extensive involvement and participation of the user make it difficult for the modeler to abstract from the scheduling problem, and this would lead to the inclusion of excessive detail in the model.

The main question of the research is "How does user participation in the modeling process influence model complexity and how might the model complexity change over time within a dynamic process?"

The hypotheses that we propose based on the developed theoretical model are:

1. Linking user participation and the Technology Acceptance Model (TAM)
   (a) The TAM is able to explain behavior also in situations with user participation.
   (b) Different levels of user participation affect the behavioral intention to use a special system. This is due to perceived control.
2. User participation influences model complexity and vice versa.
   (a) Higher degree of complexity leads to higher perceived usefulness but to poorer perceived ease of use and in the end to lower satisfaction.
   (b) If satisfaction is low the level of participation will be increased (decreased) in order to increase (decrease) the level of model complexity, dependent if the initial model complexity was low (high) $\rightarrow$ this should be an observable dynamic process.

11.6.4 Methodology

Case studies are the preferred strategy when there are phenomena to be explored, explained or described, preferably by "why" and "how" questions", when the investigator has little control over events, and when the focus is on a contemporary phenomenon within some real life context (Yin 1989). The objective of this study is to investigate, based on a developed theoretical model, the relationship between user participation, model complexity and user satisfaction when implementing DSS for industrial scheduling and how this changes over time until it reaches a certain "accepted stage". Developing and implementing a DSS is a complex project. Its success depends on a variety of variables. An analysis of quantitative data is not intended here.

A system implementation process occurs in many different companies with many different systems, users, and software specialists. The process, however, can be almost every time described by some characteristics also used for building the research model, i.e., level of participation and involvement, acceptance, resulting model complexity, satisfaction. So a multiple case study should be appropriate. The data collected in the case study should meet the propositions of the theoretical research model developed in advance. If the data from all case studies seem to fulfill this criterion then the research model is supported. The units of analysis are specific design and implementation processes for Decision Support Systems in an industrial setting.

For the data collection, multiple sources were used:

- Design specifications of the software
- Questionnaires to capture involvement, perceived usefulness, perceived ease of use, perceived control and satisfaction
- Characteristics of the environment (organization charts, equipment list, objective statements, etc.)
- Semi-structured interviews using a developed checklist with management, users, project leaders/ members of the project team, people from IS department; the statements were specified and supplemented with some documents and examples

For the following factors questionnaires and a checklist for the (semi-structured) interview were developed:

- Perceived usefulness and perceived ease of use (Davis 1989; Davis and Venkatesh 2004; Venkatesh and Davis 2000; Venkatesh 2000).
- System acceptance/ attitudes (Davis 1989; Dishaw and strong 1999; Taylor et al. 1995).
- Behavioural intention (Davis 1989; Davis et al. 2004; Dishaw and Strong 1999; Taylor and Todd 1995; Venkatesh 2000; Venkatesh et al. 2000).
- Perceived control (external and internal) (Hardgrave and Johnson 2003; Venkatesh 2000).
- User satisfaction (Kim et al. 1998).
- The extent of participation (Barki and Hartwick 1994; Doll and Torkzadeh 1991; Hawk and Dos Santos 1991; Kappelman 1995). For user participation, a distinction was made between several stages: project initiating and leading, system definition, system (physical) design, system implementation. For each stage, the relative activities are outlined. During the interview the user should indicate to what extend he/she participated in the mentioned activities (e.g., no participation | very little | little | moderately | much | very much).
- Involvement is considered as a psychological stage (Barki et al. 1994). For determining this stage several scales consisting of seven points between two semantic expressions (Kappelman 1995) were used.
- For actual system use the self-reported usage of the system for completing assigned tasks seemed to be appropriate.
- For assessing model complexity, the entities and relations of the real system were compared to the entities, objectives, and constraints in the model.
- Model/system performance (Blili et al. 1998).

Furthermore, some situational factors were captured in order to serve the comparability of the different cases and also to exclude or possibly find some alternative explanations. The following controlling variables were captured: employee qualification, prior use of IT in general and of DSS in particular, former software implementation projects and their characteristics, management support, problems during the development and implementation process, significant changes in the company's situation, developer responsiveness, collaboration with the software company, analysts, consultants in general, user training, general attitudes and beliefs regarding IT and DSS as well as their benefits.

11.6.5 Description and analysis of cases

11.6.5.1 Case #1

Company 1 produces moulds for the automotive industry and it is a leading company in Germany in this field. It employs nearly 350 people and it covers the whole value adding chain from development, design/ engineering, NC programming, metal working (milling, drilling, eroding), assembly as well as injection molding for testing purposes. Planning functions are distributed among the company. That means, people from sales department are responsible for basic data of customer's orders, another one is responsible for the processing of the orders, a special planning and scheduling department is responsible for capacity planning and for the allocation of jobs and resources at the job floor level. Beyond that, also the foremen in metal working and assembly make decisions regarding the assignment and sequence of jobs.

A Decision Support System was introduced to support the assignment and sequencing of jobs in the manufacturing department. This department consists of approximately 60 machines. Most of them could be grouped by technology and by geometrical and performance capabilities. The main objectives function covers the utilization of the machines and the compliance with due dates. A lot of jobs can be processed on different machines, but there can be differences in processing time and quality. Restrictions that have to be considered when optimizing the schedules are the availability of the machines, the optimal work sequence and also alternative sequences for the parts, the different outcomes when choosing different machines and the due dates set by predecessors (engineering, NC programming, availability of material) and successors.

The DSS was developed especially for companies with one-of-a-kind production. It is a so-called "Leitstand" that provides different views (order based or resource-based view) to the situation at the job floor. Results can be displayed as a Gantt chart, as a list and as a utilization diagram for resources and resource groups. The data in the system cover nearly all machines but only the important (or critical) parts of the products. The work sequence of the parts usually covers between three and five process steps (rough milling, fine milling, eroding, drilling). That is why the complexity can be described as moderate.

The system was introduced in 2005. During the conception and design phase, participation was quite low. The participation in the implementation phase was different. For one person the participation level was high. She tested the system very extensively and gave direct feedback to the software experts. The level of involvement and the level of perceived control were quite high. The usefulness and ease of use were rated moderate to high. The person showed a high acceptance of the system. The observed and the self-reported level of usage were also quite high. The other person in the planning department showed a very low level of participation in the implementation phase. The system was more or less refused; there was only little collaboration with the software experts. He continued using his former spreadsheets and project plans for planning. The level of involvement and the acceptance level were also low. The perceived performance was rated by the first person as very good and by the second person as not good.

Both persons do not have any academic education. The first comes from an administrative the second comes from a technical background. The first person had former experience with planning software, especially with software for order processing. The second person had experience with "normal" office software but no experience with DSS or planning software in general.

Interpretations of those findings are that:

1. Former experience may drive behavioral intention and therefore system usage.
2. Higher participation in later stages may compensate lower participation in earlier stages so that acceptance and the level of actual usage are not negatively affected by a lack of participation in earlier stages (as long as it is compensated).
3. Perceived usefulness and a high level of perceived control influence positively the level of actual usage, as proposed in the TAM and in our model.

11.6.5.2 Case #2

Company 2 employs nearly 300 people and produces moulds as well as molded parts for customers in automotive, electronics, healthcare, and biotechnological industry. The company is divided into two main departments. One department is responsible the engineering and production of moulds for external customers as well as for the own parts production. It is organized as a one-of-a-kind production and covers sales, project management, engineering, process engineering, production planning and scheduling, manufacturing, assembly, service, and a lab for testing. The other department is responsible for parts production. The production is a serial production with different lot sizes. Usually the customer indicates a certain number of parts that he needs in a certain period, e.g. a year. This framework is then broken down in several batches that are ordered separately. The department consists then again of several departments: sales and order processing, production planning and scheduling, materials, production, assembly and shipment.

A standard ERP System is used in both departments containing common functionalities for MRP and order processing but also APS/ DSS functionalities. In the mould production department the system was specifically customized to the needs of the users. Like in company 1 the system covers the machines of the manufacturing department. Objectives are mainly capacity utilization and due date adherence. There are fewer machines than in company 1 (appr. 20–25). It is only possible for certain technologies (esp. eroding) to group machines. Restrictions are the capacity of the machines and the process sequence. Major challenges are unforeseeable jobs caused by brake downs and following repairs in the parts production departments. Such jobs are usually prioritized. The complexity can be judged as moderate. The level of participation during system design and implementation was quite high in all stages. Perceived usefulness, perceived ease of use and perceived control were quite high. The level of acceptance and the observed level of usage were high. The users reported to be satisfied with the system and its performance.

In the parts production department, the system with its standard functionalities was used. Participation during system design and implementation was quite low in all stages. The production itself covers nearly 70 injection molding machines. The decisions to be supported by the system cover mainly the job sequences for each machine. Objectives are also capacity utilization and due date adherence. Although it would be possible to extend the optimization to alternative lot sizes and alternative machines for a certain job such considerations are not taken into account. Therefore, the complexity can be judged as low to moderate. The perceived usefulness was at a moderate and the perceived ease of use was at a quite good level. Perceived control was judged by the employees to be low. Satisfaction and perceived performance were at a moderate to level. The observed level of usage was low.

The insights of this special case can be summarized as follows:

1. The assumptions of the TAM and of our model that state a relationship between perceived ease of use, perceived usefulness and actual usage can be supported by the experiences. The relationships seem to be moderated by perceived control (especially in the parts production department).
2. In the department with high participation the complexity of the system was moderate (compared to the ratio of possible objects and actual objects a high percentage of the real entities were modeled). In the department with low complexity of the system was low. Therefore, there may be a relationship between participation level and complexity as proposed in the research model.
3. When complexity is low, especially compared to the real world, the perceived performance is also low. The low complexity may also influence perceived usefulness and therefore acceptance.

11.6.5.3 Case #3

Company 3 produces forming dies for the automotive industries. It employs approximately 150 people. Production and order processing can also be characterized by the one-of-a-kind principle. The functions in the company cover Engineering, programming, manufacturing/ machining, assembly and tryout. Only the production of Styrofoam models, steel casting, and heat treatment is done by external cooperation partners. The company introduced a new ERP system in 2002. In this context, also some decision support mainly in order to improve capacity utilization was asked for. Some decision support was designed and implemented using data provided by the ERP-system. The objects in the optimization model were approximately 20 machines covering milling, drilling and grinding technology. The jobs comprised only important parts of the final product whereby the importance was determined by the critical path in the process network. That means if the processing time of a part in the manufacturing department was decisive for due date compliance then it was included in the optimization. Due to the limited amount of objects, restrictions, and objectives, the complexity can be classified as moderate to low.

The participation during system conception, design, and implementation was different in the particular stages: moderate in the conception phase, low in the design phase and moderate in the implementation phase. One of the employees indicated a higher participation level than his colleagues in the design phase. Perceived usefulness and perceived ease of use as well as perceived control were quite high. Involvement (in the system as well as in the process) was also high. The persons who responded to the questions reported a quite high satisfaction and a quite high level of acceptance but a rather moderate level of usage. The last statement could be verified by own observations. When being asked for the performance of the system almost all judged the performance of the system as moderate to poor. Only the one who reported more participation in the design process rated the performance higher.

Some conclusions that could be drawn from this case are:

1. The perception of low complexity may lead to a perceived low performance. This perception may drive behavioral intentions so that the system is used less than being expected due to the high degree of perceived usefulness and perceived ease of use.
2. A high level of involvement and a high level of perceived control may lead to higher level of satisfaction even though the performance of the system is rated rather poor.
3. Interestingly there is no direct relation between perceived ease of use, perceived usefulness, and actual usage. This may be moderated by the level of complexity.

11.6.5.4 Case #4

Company 4 produces consumer packaged goods in two stages: in the first stage, the semi-finished product is made from its raw material, and in the second stage, the product is packed. The APS implementation focused on the first stage of production. The main challenge for the planners in this stage is to make sure that there is enough inventory of semi-finished product for the second stage of production to run uninterrupted. The first production stage can be characterized as semi-batch production with the following characteristics (Fransoo 1994):

- Materials involved are process oriented (measured in kilograms)
- Resources can be physically linked together temporarily, and when one link is made, some other links are not possible at the same time
- Large number of process steps from the raw material to the semi-finished material
- There is limited buffer capacity – work in process is stored in silo's with a limited capacity
- Change-over times are sequence dependent

- Both convergent and divergent material flows
- Production involves manual labor that has to be shared by different operations

The need for an APS was identified to improve the efficiency of the first production stage. The objective was to improve efficiency and predictability of the production process in such a way that maintenance activities could be carried out during the week instead of in the weekend, thereby making large cost savings in maintenance possible. At the same time, the availability of semi-finished material to the second stage, which was already on a high level (above 95%), should not be compromised.

The selection of the APS started in 2001 and the implementation started in 2003. The requirements for the implementation were initially based on a document that was also used during the selection process – there had been little involvement of the schedulers so far. During the implementation process, the schedulers were requested to test parts of the system that had been modeled. Starting up the testing process was hampered by the lack of experience of the schedulers with APS implementations and the fact that until then they had little involvement in the implementation.

During the testing process, the schedulers started to indicate that some of the design choices that had been made in the APS were wrong: in the APS, the production steps of the first production stage were coupled with each other. This meant that when a batch was rescheduled at the end of the process, there would be consequences upstream to the start of the process. The schedulers believed that such a concept was too nervous and they advised on a more decoupled model. However, the IT dominated project group did not consider this input according to the schedulers. The schedulers felt that they were ignored more and more by the project team and that their input was regarded as a nuisance.

The system went live in April 2004 but after two months of trying to get the system to work, the system was shut off again. The schedulers complained about the nervousness of the system and the large amount of time it needed to keep the schedule up to date. Because of this initial failure, the schedulers were given more control over the APS model. They specified how the model should be changed, introducing a number of decoupling points to make the schedule less nervous. Various parts of the production process were decoupled and could be scheduled relatively independently. So, a change at one point in the process did not immediately lead to many changes in other parts of the process.

After a few months the system went live again and today is still used successfully. The objectives of the system implementation had been formulated as follows: Increase output by 20%, and save maintenance costs by moving maintenance activities from the weekend to the week. From this point of view, the system implementation has been successful. In the first year, the system has returned its investment by a decrease in inventory costs. The output figures have not risen as originally expected; however, this is also due to the order book. Instead, many savings have been made in the manning in the production process. The number of planners has also decreased, from 6 to 5. However, in the last year of its usage, the maintenance costs are rising again, without a clearly understood reason.

The current system is used fully – every modeled functionality is used by the schedulers. The manual planning is supported by the system and there are macro's that contain algorithms to carry out some of the routine scheduling tasks.

The model contains 211 resources, 422 routing constraints (e.g. resource A and B cannot be used simultaneously), 30 end products. Demand is received by the system by another APS (the one that is used to schedule the second stage of the factory). The demand is specified by a stock consumption per product per silo. The task of the schedulers is to make sure that the projected stock does not get negative.

The planners are still initiating changes to the system; for example, one planner has created a special view to visualize routing conflicts. Another recent change was related to improving the speed of the interface to the shop floor.

The planners did not have any previous experience with APS systems. All planners had a background in production. Most planners had experience with using parts of ERP systems and all planners were using spreadsheets to generate a schedule before the APS was implemented. In addition, the shop floor had much more autonomy in the old situation where XL was used.

It can be concluded that the limited participation of the schedulers in the first attempt of the implementation has had a large influence on the initial failure of the system. Part of this is due to the wrong design choices in the model, but trust will also have played a role. The schedulers were very hesitant to test the system and to use something that they had not designed themselves. The intensive involvement of the planners in the re-implementation has led to an APS model that one the one hand better suits their way of working, but also to an APS model that they trust, because they have created the specifications themselves.

11.6.6 Conclusion

In our field study, we have partly analyzed relationships that have been studied earlier, but not in a context of advanced planning systems. Furthermore, we have explored the impact and dynamic relationships caused by complexity. Complexity of Advanced Planning Systems solutions is a dominant problem in many implementations.

Looking at our hypothesized relationships, we can observe the following.

For hypothesis 1a, we found supportive evidence that the earlier results from the TAM literature holds in settings of advanced planning systems in Case #1. Specifically, former experiences (Taylor and Todd 1995) and perceived ease of use (Davis 1989) were found to have substantial influence on the actual usage of the system. Conversely, the findings in Case #3 are ambiguous, since the relationships sought after could not be found. Possibly, this is due to the moderating effect of complexity.

The impact of perceived control in developing and implementing APS systems with user involvement is clearly present in all four cases. These findings provide strong support for earlier developed theories (Davis and Kottemann

1994) based on an experiment with MBA students. It therefore addresses one of the limitation suggested (Davis et al. 1994).

For the dynamic relationship between participation, complexity and usage (as mentioned in hypotheses 2a and 2b), we find initial support from the case studies. In Case #2 and Case #3, the proposed relationship appears to be present. Participation negatively affects complexity (i.e., more participation leads to increased complexity of the model) and hence user satisfaction and actual usage. In the two other cases (Case #1 and Case #4), especially the dynamic development over time was clear. In both studies, it was demonstrated that a low level of user involvement in the initial stages of the project could be compensated by increased involvement at later stages. It is interesting to compare this to and earlier result (Davis et al. 2004), which shows that having an early prototype positively affects usage of the system. Our results suggest that not having such an involvement at early stages could be compensated by closely involving the user at later stages.

Our results extend the current knowledge in this area in two ways. First, we have explicitly focused on Advanced Planning Systems. There are only a few studies around that have empirically studied implementation processes of these systems (e.g. Zoryk-Schalla et al. 2004). Second, we have been able to show the validity of earlier results of laboratory experiments in a field setting with actual projects. Furthermore, we have introduced the mechanism by which complexity of the model, user participation and actual usage are related. We have theorized on this relationship and found some supporting evidence in our case studies.

While our results are interesting and extend the current knowledge in this area, further research is needed. Obviously, the number of observations in our studies is limited. Further research is needed to find a broader empirical base for the theories that have been suggested.

11.7 Human Factors when Designing and Implementing DSS: The Role of Trust

11.7.1 Introduction

Because machines can be designed to perform more and more scheduling tasks, the role of the human is constantly reduced. However, neglecting the human role may lead to failures in the design of human-machine cooperation (see, e.g., Kerr 1992). Four main difficulties have been previously stressed (Parasuraman and Riley 1997; Hoc 2001): *loss of expertise*, *loss of adaptivity*, *complacency* and *trust miscalibration*.

Having a machine autonomously fulfilling the tasks leads the human to be in a monitoring role. This role might lead the schedulers to become passive, poorly practicing, thus not maintaining their scheduling skills (*loss of expertise*). This might have severe consequences when the human must perform actions in case of exceptions or automation failures. Moreover, being in a monitoring role reduces the opportunity for the human to learn from experience, to anticipate events and to analysis feedbacks (*loss of adaptability*).

The *complacency* phenomenon has been described as an unjustified assumption of satisfaction in the situation faced, although some improvements could apply (Parasuraman et al. 1993). In the scheduling domain, Cegarra and Hoc (2008) noted that complacency could result from the high cost of interacting with the automation, implying a high cognitive workload.

This section focuses on the remaining difficulty: *trust miscalibration.*

During the last two decades the concept of trust – coming originally from social sciences – has been adapted to the usage of automation. So far, the main focus was for instance on aircraft devices, military and routing problems but not on planning and control in manufacturing. The section transfers and adapts the concept of trust in automation to the usage of DSS for planning. A research model was developed that covers the relations of selected system parameters, user variables, and situational factors and their impact on trust in the system as well as the actual usage of the system.

For the test of the research model a survey was developed where people indicated their trust in a system being given a certain set of outcomes (with different performance levels and different consistency) in different situations.

This Section is organized as follows. First some theoretical considerations are presented in order to describe the field of Decision Support Systems and their application in planning and scheduling. This is followed by a review of the concept of trust, its application to the use of automation as well as a discussion of prior studies in this field. Based on the theory a research model that describes the relations between the characteristics of a DSS and trust and usage is developed. From this certain research hypotheses are derived. The hypotheses are to be tested by an empirical investigation; the methodological approach for this survey is also presented. The Section ends with some concluding remarks regarding the expected outcome of the empirical research and its implications.

11.7.2 Literature Review

11.7.2.1 The Concept of Trust

Trust in general is the attitude that an agent will help to achieve an individual's goals in a situation characterized by uncertainty and vulnerability. In this definition an agent can be automation or another person that actively interacts with the environment on behalf of the person. In our context the agent would be a Decision Support System. Trust is an attitude toward automation (or toward the system) that affects reliance and that can be measured consistently. Trust sometimes is also characterized as an expectation related to the subjective probability an individual assigns to the occurrence of some set of future events, (Muir 1987, 1988; Lee and See 2004).

Trust is a concept with many facets. Most of the concepts of (personal) trust share the following common elements:

- A certain degree of interdependence between trustor and trustee
- The assumption that trust provides a way to cope with risk or uncertainty in exchange relationships
- A belief or expectation that the vulnerability, resulting from the acceptance of risk, will not be taken advantage of by the other party of the relationship

According to Luhmann (1979) trust is a mechanism to reduce internal complexity of a system of interaction through the adoption of specific expectations about the future behavior of the other by selecting amongst a range of possibilities. Trust absorbs complexity insofar as someone who trusts acts if the trustee's acts are – at least to some degree – predictable.

11.7.2.2 A Psychological Perspective on Trust

The concept of trust has been widely studied by researchers in many areas such as psychology, sociology, history and political science. However, it remains a difficult concept to define because of its dynamic, evolving and multi-facet nature (Lewicki and Bunker 1996). In the social psychology literature, trust is traditionally defined on an interpersonal-level (e.g., "I do or do not trust a person or group"). According to Mayer et al. (1995), trust is "the willingness of a party to be vulnerable to the actions of another party based on the expectation that the other will perform a particular action important to the trustor, irrespective of the ability to monitor or control that other party" (p. 712).

Many researchers perceive trust in terms of individuals' expression of confidence in others' intention and motives (e.g. Deutsch 1958). This viewpoint attributes trust to an interpersonal relationship, as Friedman et al. claim: "People trust people, not technology" (Friedman et al. 2000, p. 36). On the other hand, several researchers have found that human-automation teams and human-human teams function similarly Bowers et al. 1996, 1998. According to Madhavan and Wiegmann (2004), people enter into "relationships" with computers, robots, and interactive machines which are similar to their relationships with other humans (Nass et al. 1996a, b; Reeves and Nass 1996). Several studies (Nass and Moon 2000; Nass et al. 1995) have demonstrated that social rules guiding human-human interaction may apply equally to human–computer interaction, with users responding to machines as independent entities rather than as a manifestation of their human creators (Madhavan and Wiegmann 2004; Sundar and Nass 2000).

Zuboff (1988) discussed three components of trust as reported by the participants in her study: understanding of the technology, trial-and-error experience, and faith. In her research, subjects reporting higher trust also reported greater understanding of the system (the amount to which they felt they understood the automated system, not necessarily the amount to which they actually understood the automation). As trial and error experience increased, those perceiving higher system reliability reported higher trust. Lastly, some subjects reported an almost "blind faith" that automation would perform as advertised and subsequently reported higher trust scores (Johnson et al. 2004). Much like Zuboff (1988), Lee and Moray (1992) provided a multi-component description of trust. They, however, described the development of trust as a progression across dimensions of trust: foundation, performance, process, and purpose (Johnson et al. 2004).

11.7.2.3 Trust and Credibility

Trust and credibility are not the same concept. Although these two terms are related, trust and credibility are not synonyms. Trust indicates a positive belief about the perceived reliability of, dependability of, and confidence in a person, object, or process (Rempel et al. 1985; Rotter 1980). For example, users may trust in a computer system designed to keep financial transactions secure. On the other hand, credibility refers to the concept of believability. Credibility is a perceived quality; it does not reside in an object, a person, or a piece of information. Therefore, in discussing the credibility of a computer product, one is always discussing the perception of credibility (Fogg and Tseng 1999).

11.7.2.4 Trust in Automation, Trust in Systems

Trust in automation guides reliance when the complexity of the automation makes a complete understanding impractical and when the situation demands adaptive behavior that procedures cannot guide. System trust plays a role when a trustor, the user, interacts with a trustee, the system; the necessary precondition is that this user-system interaction takes place voluntarily. In addition, the incentive for the user to engage in interaction with the system is the expected benefit from the outcomes of the task. The actual outcome, however, is uncertain, due to a lack of sufficient evidence. The degree to which the actual amount of evidence available can be called sufficient depends on the risk involved; compared to situations of low risk; high-risk situations may require more evidence for uncertainty to be reduced.

System trust is considered to be an expectation a user has about the system, that, when the system is activated, it will perform a certain task that is beneficial for the user, in a situation in which a lack of sufficient evidence causes the actual outcome of that task to be uncertain, in that using the system can have both positive and negative consequences.

Some research has been done so far to determine the factors that influence trust in automation. The most helpful approach for the research question discussed in here is the concept of Lee and Moray (1992) which identified performance, process, and purpose as the general bases of trust.

Performance refers to the current and historical operation of the automation and includes characteristics such as reliability, predictability, and ability. Performance information describes what the automation does. More specifically, performance refers to the competency or expertise as demonstrated by its ability to achieve the operator's goals.

Process is the degree to which the automation's algorithms are appropriate for the situation and able to achieve the operator's goals. Process information describes how the automation operates. In interpersonal relationships, this corresponds to the consistency of actions associated with adherence to a set of acceptable principles. Process as a basis for trust reflects a shift away from focus on specific behaviors and toward qualities and characteristics attributed to an agent. With process, trust is in the agent and not in the specific actions of the agent. In the context of automation, the process basis of trust refers to the algorithms and operations that govern the behavior of the automation. The operator will tend to trust the automation if its algorithms can be understood and seem capable of achieving the operator's goals in the current situation.

Purpose refers to the degree to which the automation is used within the realm of the designer's intent. Purpose describes why the automation was developed. With interpersonal relationships, the perception of such a positive orientation depends on the intentions and motives of the trustee. The purpose basis of trust reflects the attribution of these characteristics to the automation. Frequently, whether or not this attribution takes place will depend on whether the designer's intent has been communicated to the operator. If so, the operator will tend to trust automation to achieve the goals it was designed to achieve.

11.7.2.5 Psychological Factors Affecting Trust

There are several factors influencing user's trust in a system or in a computer. These are features of the situation in which the system is used, characteristics of the user himself, as well as system variables:

- Situational factors that affect trust
  - Familiarity of the situation
  - Dynamics
- User variables that affect trust
  - User familiarity and user understanding
  - Experience and expertise
- System variables that affect trust
  - System reliability
  - Types of system errors
  - Usefulness and usability

11.7.2.6 Situational Factors that Affect Trust

Familiarity of the Situation

Trust in a system does not always depend on the system itself. The context of system use can affect trust. Following Fogg and Tseng (1999), three related situations increase trust in a system:

- In unfamiliar situations people give more credence to a system that orients them (Muir 1988).
- Computer systems have more credibility after people have failed to solve a problem on their own (Waern and Ramberg 1996).
- Computer systems seem more credible when people have a strong need for information (Hanowski et al. 1994; Pancer et al. 1992).

Indeed, other situations are likely to affect trust, such as user's familiarity with the domain in which the system is used and user's understanding of how the system arrives at its conclusions (see below).

Dynamics

Another cluster of research examines the dynamics of computer credibility – how it is gained, how it is lost, and how it can be regained. Some studies demonstrate what is highly intuitive: Computers gain credibility when they provide information that users find accurate or correct (Hanowski et al. 1994; Kantowitz et al. 1997; Muir and Moray 1996); conversely, computers lose credibility when they provide information users find erroneous (Kantowitz et al. 1997; Lee 1991; Muir and Moray 1996).

Researchers have also examined how computer products can regain credibility (Lee and Moray 1992) Two paths are documented in the literature:

- The computer product regains credibility by providing good information over a period of time (Hanowski et al. 1994; Kantowitz et al. 1997)
- The computer product regains some credibility by continuing to make the identical error; users then learn to anticipate and compensate for the persistent error (Muir and Moray 1996).

However, regaining credibility is difficult. Once users perceive that a computer product lacks credibility, they are likely to stop using it, which provides no opportunity for the product to regain credibility (Muir and Moray 1996).

11.7.2.7 User Variables that Affect Trust

User Familiarity and User Understanding

Users who are familiar with the content in which a system is used will evaluate the system more stringently and likely perceive it to be less credible (Honaker et al. 1986; Kantowitz et al. 1997; Lee and Moray 1992). Conversely, those not familiar with the subject matter are more likely to view the system as more credible (Waern and Ramberg 1996). These findings agree with credibility research outside of HCI (Gatignon and Robertson 1991; Self 1996; Zajonc 1980).

Users who understand how the computer arrives at a conclusion are more inclined to view a computer as credible (Lee 1991; Lerch and Prietula 1989;

Miller and Larson 1992; Muir 1988). In this line of research users either learned about the computer product before using it (Muir 1988; Waern and Ramberg 1996), or the computer justified its decisions in real time through dialog boxes (Miller and Larson 1992).

Experience and Expertise

Trust is an important aspect of human interaction with automation (Lee and Moray 1992, 1994). Operators may not use a well-designed, reliable automated system if they believe it to be untrustworthy. Conversely, they may continue to rely on automation even when it malfunctions and may not monitor it effectively. Both phenomena have been observed (Parasuraman and Riley 1997). Several studies deal with the role of expertise in the occurrence of excessive trust and complacency in automation (Miller et al. 2005). Metzger and Parasuraman (2001) report evidence of increased complacency among experienced air traffic controllers using decision aiding automation. In this study, users effectively grant a higher level of automation to a system than it was designed to support by virtue of coming to accept automatically the system's recommendations or processed information even though the system sometimes fails. Riley (1994b) documents a similar phenomenon, overreliance on automation, by trained pilots. In an experiment where the automation could perform one of a pair of tasks for the operator, but would occasionally fail, almost all of a group of students detected the failure and turned the automation off, while nearly 50% of the pilots failed to do so. While it is impossible to conclude that pilots' increased experience with (albeit, reliable) automation is the cause for this overreliance, it is tempting to do so (Miller et al. 2005).

11.7.2.8 System Variables that Affect Trust

System Reliability

Trust can be described as a subjective measure of one's confidence in something or someone else. In the context of automation, research has begun to distinguish this subjective rating from its closely related counterpart reliability. Trust refers to the subjective reports of the operators or their feelings about automation; whereas, reliability is the objective measure of performance, such as automation utilization or task efficiency (Wiegmann et al. 2001). From a system perspective, reliability is at the heart of trust. After all, a system that unreliable will not be trusted, relied upon or even used. Unreliability, real or perceived, doesn't simply lead to distrust, but adds demands to working memory and as a result leads to increased workload (Ho et al. 2005). In several studies participants' subjective ratings of trust in automation may be typically lower than their usage of the automation (Sanchez et al. 2004; Wiegmann et al. 2001). This was most notable when trust ratings were less than 100% despite utilization rates of 100% (Sanchez et al. 2004;

Wiegmann et al. 2001). This could quite possibly reflect the human feeling that nothing is ever perfect. Wiegmann et al. also showed that as the reliability of the automated system decreased the participants' subjective ratings of trust in the system in general decreased and vice versa. Muir (1994) found that operators' perception of trust was only changed by the performance of the machine and people quickly reverted to manual operations when they felt technology was unreliable.

Types of System Errors

A few studies have investigated the effects of computer errors on perceptions of computer credibility. Although researchers acknowledge that a single error may severely damage computer credibility in certain situations (Kantowitz et al. 1997), no study has clearly documented this effect. In fact, in one study, error rates as high as 30% did not cause users to dismiss an onboard automobile navigation system (Hanowski et al. 1994; Kantowitz et al. 1997).

As described in classic signal detection theory, when an automated engine status indicator malfunctions, one of two types of errors will occur. First, if a system malfunctions and the automation does not indicate a malfunction (i.e., no signal detected), a miss has occurred. Second, if the automation erroneously indicates a malfunction when the system is working properly (a non-existent signal is detected), a false alarm has occurred. Research to date has primarily focused on how false alarms affect trust and reliance in the automation (Bliss 1997; Bliss and Dunn 2000; Xu et al. 2004). Many researchers (e.g., Breznitz 1984; Bliss and Dinn 2000) have supported the notion that persistent or pervasive false alarms negatively affect operator trust in automated systems. On the other hand, very little has been done to examine how, if at all, misses affect trust (Johnson et al. 2004).

Usefulness and Usability

Davis (1993) found that perceived usefulness of a system (i.e. does it perform the task) was 50% more influential than the ease of use of the system in determining how much the system was actually used. This research emphasizes the importance of designing new systems with appropriate functional capabilities to suit user expectations and how operators will adjust to the different functions of future technology.

Usability refers to the extent to which users can exploit the utility of a system. Thus, systems with equivalent utility may result in different levels of usability depending on how the design is implemented. Usability is operationally defined as effectiveness, efficiency and satisfaction with which specified users can perform particular tasks in a given environment (Nielsen 1993).

The relationship between usability and trust is a very complex relationship. Different schools of thought project the relationship between trust and usability in a different manner. Scientists like Egger (Egger and de Groot 2000) are of the opinion that usability is a component affecting trust whereas Fogg et al. (2001) suggests that trust and usability both are components of credibility (Banati et al. 2006). Coming back to the definition of trust as the "*expectancy* held by an individual or a group that the word, promise, verbal or written statement of another individual or group can be *relied* upon" (Rotter 1967) than *predictability* is implied in this concept. However, predictability is at least a part of the concept of usability. Systems with good usability are consistent, controllable and *predictable.* In this sense usability and trust are closely related to each other.

Several studies address the role of usability for trust in Web sites. These studies suggest that interface elements, such as ease of navigation and feed-back mechanisms, are important to establish trust (Cheskin Research 2000; Cheskin Research and Studio Archetype 1999; Rhodes 1998).

11.7.3 Prior Research and Results

Many studies have demonstrated that trust is a meaningful concept to describe human automation interaction in both naturalistic (Zuboff 1988) and laboratory settings (Halprin et al. 1973; Lee and Moray 1992; Lewandowsky et al. 2000; Muir 1989; Muir and Moray 1996). From special interest are two studies which are tangent to the research question discussed in this chapter.

The first one is a study by de Vries et al. (2003) which researched trust in the domain of route planning. Their work was focused on the way people deal with complex systems. The authors consider trust as a mental state. The underlying assumption of the study is "the expectation of a user about the system, that the system will perform a certain task for him or her, while the outcome of that task is uncertain, in that it can have both positive and negative consequences." The participants had to perform the task of determining an optimal route compared to a database, that contained route information based on experiences of ambulance personnel and police officers. They had to find the quickest possible route, either automatically or manually. After some practice trials the participants had to choose between the two modes (automatic and manual). During the practice trials errors were introduced: routes turned out, after comparison to the database, to be "slow" or "fast". The authors of the study expected the relation between error rates and control allocation to be mediated by trust. That means that the decrease of trust in the system would lead to a less often use of the automatic mode whereas a decrease of trust in own abilities (self confidence) would lead to a more often use of the automatic mode. A certain risk was associated with control allocation, that means participants received ten credits per trial, which could be put at stake (slower route = lose money; faster route = double amount). They were told that they would receive the sum of money at the end of the experiment. The authors assumed that staked credits would be an accurate measure of trust. After all trials, participants were asked to rate their trust in the system and themselves in 7-point-scales, ranging from "very little" to "very much". The results were that

- Automatic mode was selected more often in (1) low automation error rate and a high manual error rate condition than in (2) a high automation error rate and a low manual error rate condition
- The number of credits staked in automatic mode was higher in (1) than in (2)
- The number of credits staked in manual mode in (1) did not differ from (2)
- Ratings of system trust were higher in (1) than in (2)
- The difference between trust in the system and self confidence was higher in (1) than in (2)

The second set of studies was conducted by Dzindolet et al. (2002, 2003). The experiments were conducted in a military context. Participants viewed 200 slides displaying a terrain on a computer screen. After each slide they should indicate if a soldier was present or not. They should also indicate the confidence in their decision on a 5-point scale. Next, a contrast detector decided. After four practice trials (aid supplied correct decision) participants were asked to estimate their and their aid's performance during 200 trials on a 9-point scale. The authors experimented with different error rates, with different kind of feedback (continuous, cumulative), and with different explanations for the errors the system made. The experiment led to the following results:

- The more the participants trusted the aid, the better they expected it to perform and the fewer errors they expected.
- Participants believed they lost trust in the automatic aid when it made an error, and they did not deem the aid as trustworthy.
- Trust mediated the effect that providing a rationale for why the aid might error increased reliance on the aid.

11.7.4 Research Model

On a general level, the derived research question addresses distinct factors that affect the actual usage of a software system. The factors influencing the decision to use the system could be inherent to the software itself, the human user or could be found in the presence of another human who is qualified to be an expert in the particular planning topic. On the part of the software system, the parameters performance and performance variability are identified to have a crucial influence on trust and the decision to utilize the system. On the part of the human user, typical moderating variables like age and gender need to be considered as well as more specific factors like experience in the field of planning, the judgment of own planning abilities, and general attitudes towards software systems. The third major impact could be due to the human expert who might have an impact on the user's decision to utilize the software. Figure 11.11 summarizes the considered factors and provides an overview of how these factors are thought to influence trust and the system usage.

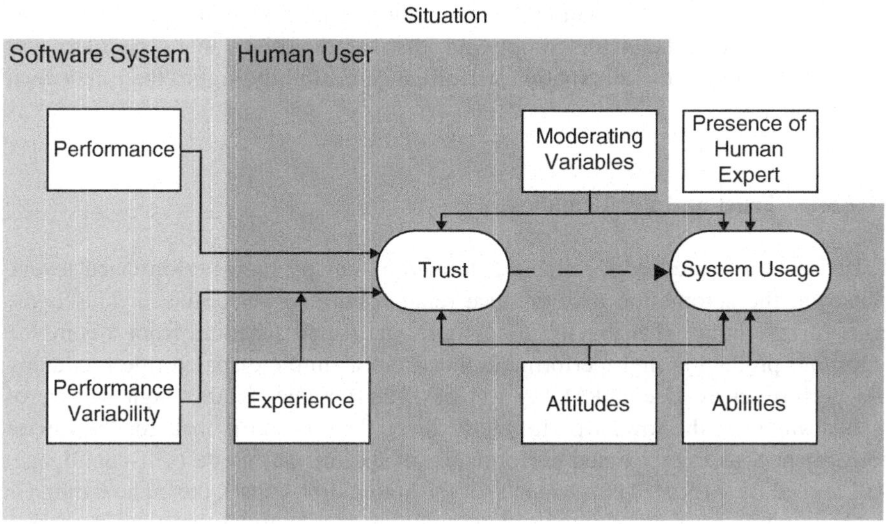

Fig. 11.11 Research model

11.7.5 Hypotheses

According to the research model displayed in Fig. 11.11, the following five hypotheses were derived.

Hypothesis 1:	Higher performance leads to higher trust
Hypothesis 2:	The higher the performance variability, the lower the trust
Hypothesis 3:	The lower the experience, the larger the effect of variability
Hypothesis 4:	Interrelations exist between general attitudes towards DSS and trust/ intention to use in a specific situation
Hypothesis 5:	Interrelations exist between the judgment of own abilities and intention to ask an expert

Several authors have suggested that human operators' subjective trust towards automation governs the effective use of automation (Lee and Moray 1992, 1994; Lee and See 2006). As for other automated systems, to benefit from a computer-generated schedule, the human scheduler must learn, or be trained to anticipate, identify and act depending of the performance of the computer-generated schedule. How does this trust evolve from interactions with the computer-generated schedule?

When the automation is reliable (i.e. performing as expected), the trust towards automation increases with exposure time, whereas if the automation is not reliable, the trust decreases (Lee and Moray 1992). This level of trust also governs the allocation of functions to the automation or to a manual mode. In addition, Lee and Moray (1994) pointed out the role of self-confidence in this allocation decision. They noted that trust towards automation must be slightly greater than operator's self-confidence in order to favor the automation over the manual mode. A similar bias was also noted in scheduling by Liu et al. (1993): Participants had greater confidence in their ability than in the automation, even if automation was replicating their exact own behavior. Simply put, these studies show that schedulers will favor automation if the automation performance is slightly higher than their own performance.

11.7.5.1 Trust and Performance

Trust is often related with reliable automations having high performance levels. However, the automation may produce reliable but low performance. This is the case, for example, when having a computer-generated schedule from a complex algorithm producing high performances or from a simple dispatch rule producing low performances. Lee and Moray's (1992, 1994) studies predict higher usage of manual mode in the lowest performance level. The reason is that, for the lowest performance, as the expected performance of the manual mode (self-confidence) will exceed the expected performance of the automation (trust), the manual mode is likely to be performed. Moreover, Lee and See (2004) noted that trust decreases when reliability decreases, but, below a certain level of performance trust declines quite rapidly.

11.7.5.2 Trust and Performance Variability

Riley (1994a) suggested factors outside performance evaluation determining trust towards automation, such as the risk of using the automation. This risk may result from a performance variability of automation in contexts that are almost identical from the human point of view. The importance of the risk evaluation has also been noted by Cohen et al. (1999): They consider that trust has to be defined, not only as a measure of expected performance, but has to be considered as a measure of uncertainty regarding automation performance.

Previous studies predict higher usage of automation for the high consistency level than for the low consistency level. For example, Lee and Moray (1994) noted, after a severe lack of consistency in automation performance, that recovery of trust in the automation is not instantaneous. There is inertia to switch allocation strategies related to previous uses of automation. In the same way, Parasuraman et al. (1993) demonstrated that inconsistent reliability in automation leads to lower level of trust than consistent reliability.

11.7.5.3 Trust, Performance and Performance Variability

When crossing the two dimensions (consistency and performance), other predictions arise, see Fig. 11.12.

The condition A (high performance and low performance variability) will produce the highest usage of the automated mode, due to the highest level of

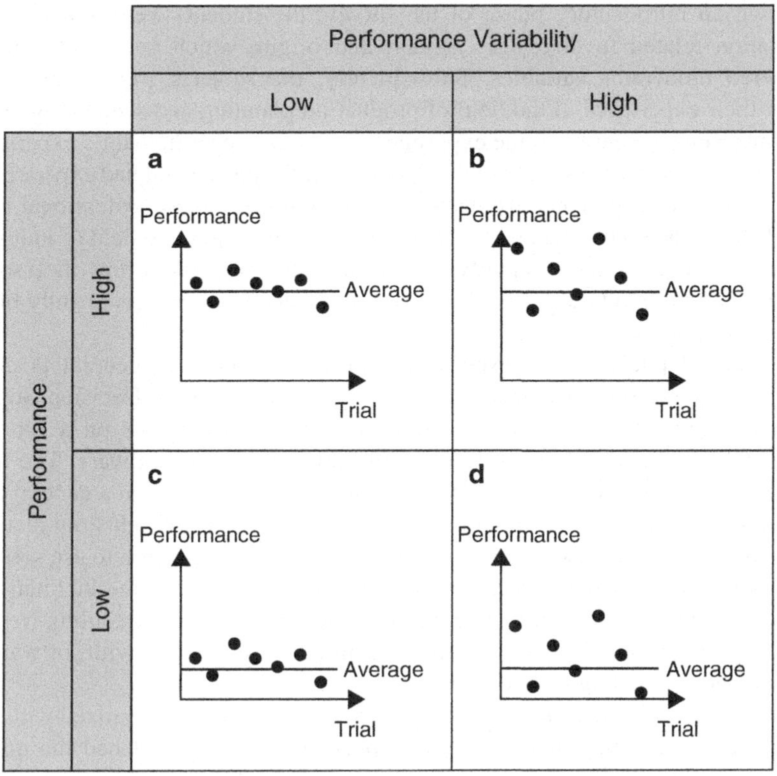

Fig. 11.12 Combinations of system parameters

trust. Conversely, the condition D will produce the lowest usage of the automated mode due to the lowest level of trust (with the requirement that self-confidence is higher).

Then, we predict that the B-condition will produce higher usage of automation than the C- condition. In condition B, participants will experience, in some trials, high levels of automation performance. Then, after a lack of consistency trust will progressively decrease. Whereas in the C condition, operators will, from the start, totally shift away from using the automation as the performance is too low to be useful (self-confidence exceeding trust towards automation).

11.7.6 Research Methodology

An online survey was developed in order to investigate the proposed influence of factors inherent to the software system, the human user and the presence of an expert. Participants of the survey were students from German, Swiss, and Dutch universities as well as universities of applied science.

During an introductory phase of the survey, the students were asked to give information related to their age, gender, and origin, which covers the area of considered moderator variables. Subsequently, the subjects were requested to specify their experience in the field of production planning and scheduling. Three major measures were used in the experiment to determine participant's experience. The first measure stressed the kind of experience if the participant had experience at all. Two kinds of experience were discerned with internship or professional background. As to the second measure, the duration of working in the field of industrial scheduling was considered. Finally, the participants were asked how they solved particular scheduling tasks, which relates to tools like spreadsheets or fully blown scheduling systems.

The second phase of the experiment shifted the focus to general positions participants possessed concerning software systems, especially Decision Support Systems (DSS). Five items of the survey asked the participants on a six-point Likert-scale for their wide-ranging point of view concerning software. The items were the degree to which software can support human decisions on a general basis (support of decisions), the extent that software sustains tasks of industrial planning and control (sustain industrial tasks), the intention of the participant to use software as a production planner for decision support (intension to use), the inclination to rely on computer-based decisions in the field of planning and scheduling (rely on computer), and the tendency to rely on own abilities when dealing with software for planning and scheduling (rely on oneself).

During the third phase of the survey, the participants were familiarized with their specific task. The assignment was a transportation problem that had the goal to optimize the distribution of customer orders at lowest possible costs given that warehouse capacities were finite, customers had a defined demand with specific delivery dates, and the limited number of transportation vehicles owned finite transportation capacities. The task could be solved either manually or with the help of an imaginative planning system called R-PLAN+. The results of the last ten trials to create a transportation plan using R-PLAN+ were presented to the participants who were randomly assigned to three different groups. These three groups were determined by the system parameters performance and performance variability as it is outlined in Fig. 11.12 omitting the condition A, which was considered to posses too little valence for further investigation. To facilitate a frame of reference for the participants, the test persons were introduced to the results of the previous ten trials using two other software systems. These two additional results represented the missing two conditions of the two other groups (e.g. if the participant was assigned to the condition B, the results of C and D were provided to the test person). The heuristic calculation results of R-PLAN+ for each trial were compared with the theoretical optimum based on total enumeration. The optimal solution guaranteed to meet all given constrains at the lowest possible costs which is contrary to the sub-optimal solution provided by the fictive planning system R-PLAN+ that might entail higher costs. Hence, the relative comparison of the heuristic and the optimum yields an average level of performance, which was visualized for the participants with simple graphs that henceforth contained the spread of each heuristic calculation result around the mean performance (i.e., performance variability). Therefore, the test persons obtained the system parameters performance and performance variability as graphical representations comparable to the graphs shown in Fig. 11.12. Furthermore, the values for mean performance and standard deviation were provided to summarize both system parameters.

As confirmation of the intended manipulation due to the assignment to one of the three groups, the participants were asked to evaluate performance and performance variability on a six-point Likert-scale. Furthermore, this manipulation check marked the transition to the fourth phase of the survey. During this stage, the participants were requested to imagine completing a new planning task and to provide their evaluation on the five subsequently pictured measures of trust[1] on a six-point scale.

I. *E*xpected *S*ystem *P*erformance during the next planning task (ESP)
II. *A*nticipated *O*wn *P*erformance during the following planning task (AOP)
III. Relative *C*omparison of *S*ystem and *O*wn *P*erformance (CSOP)
IV. Tendency to *R*ely on the decisions of *R*-PLAN+ (RR)
V. Preference to *R*ely on *O*wn decisions (RO)

During the fifth and final phase, the subjects had to decide first if they use the software system (yes/no answer). The following item appeared possibly as a surprising turn since the participants had now the option to neglect the computer system by asking a human expert to solve the given task instead. Consequently, the associated answer option for this question were asking the expert, solve the task by oneself, or use the planning system. The last item of the online survey gave room to the participants to state with an open ended qualitative answer their grounds to use the system.

11.7.7 Results

170 people initially participated in the online experiment and 97 completed all survey parts successfully. These 97 cases were the foundation for all further investigation and analysis. The participants consisted of 17 female and 80 male students who spent on average 9.14 min ($SD = 5.72$ min) to complete the web-based survey. The average age of the international participants was 24.47 years ($SD = 3.81$ years).

[1]Note that the trust related measures are coded with two up to four letters. These abbreviations will appear frequently throughout the following text.

11.7.7.1 Performance and Performance Variability

Both system parameters were investigated by conducting hierarchical regression analysis, which included two steps. The first step controlled for possible moderators and the second step was performed to determine the effects of both system parameters on the five previously pictured measures of trust. Three trust related measures were found to be significantly influenced by the chosen predictor variables. These were the expected performance of the system (ESP), the comparison of individual and system performance (CSOP), and the participants' tendency to rely on the decisions of the planning system (RR). Related results of the significant regression model are shown in Table 11.3. Two criterion variables concerning the anticipation of individual performance (AOP) and the tendency to rely on own decisions (RO) seem to evade being predicted with the selected dependents, which means that neither the moderating variable nor the measures of trust showed a significant relation to these two measures.

As to the significant linear regressions, the moderator variables revealed that they seemed of no predictive relevance, which was supported by the t tests of the standardized coefficients shown in Table 11.3 as well as by – not displayed – correlation results.

The system parameters performance and performance variability were both of significant importance for predicting trust in the planning system. Moreover, and in accordance with hypotheses 1 and 2, the signs of the standardized coefficients gave further evidence that high performance was positively related to trust and high variability had negative influence on trusting the system. Hence, both hypotheses 1 and 2 were supported without limitations.

Regarding the combined outcome of both system parameters, the interaction effect of performance and performance variability was taken into further consideration based on a post-experimental grouping of the participants based on their

Table 11.3 Coefficients of significant linear regressions

	Predictor	β	$t(91)$	p
ESP	Age	0.00	0.06	0.95
	Gender	−0.11	−1.50	0.14
	Origin	0.04	0.58	0.56
	Performance	0.63	8.48	<0.01
	Performance variability	−0.21	−2.77	0.01
CSOP	Age	0.04	0.41	0.68
	Gender	0.08	0.84	0.40
	Origin	−0.02	−0.25	0.80
	Performance	0.38	4.13	< 0.01
	Performance variability	−0.25	−2.61	0.01
RR	Age	0.06	0.71	0.48
	Gender	0.01	0.05	0.96
	Origin	0.07	0.84	0.40
	Performance	0.45	4.95	<0.01
	Performance Variability	−0.22	−2.46	0.02

Table 11.4 Results of Mann-Whitney U-tests

	U	*p*
ESP	49.50	<0.01
AOP	82.50	0.11
CSOP	81.00	0.09
RR	95.00	0.27
RO	110.50	0.68

evaluation of the system parameters. Employing the newly formed groups, Mann-Whitney U-tests were performed for each dimension of trust. The obtained test outcomes are shown in Table 11.4.

Concerning the expected system performance (ESP), a significant relation of the interaction was found which indicated that low performance/ low variability will yield lower trust (mean rank = 10.63) than the combination high performance/ high variability (mean rank = 20.03). Consequently, it may be concluded that the particularly expectancy related dimension of trust is more absorbed by potentially high performance on the costs of higher risk due to greater variability than by low performance combined with low variability.

11.7.7.2 Experience and the Effect of Performance Variability

The analysis of hypothesis 3 took only the dimensions of trust into account that were significantly influenced by performance variability. These were the expected system performance (ESP), the relative evaluation of performance (CSOP), and the tendency to rely on the system (RR).

Three multiple linear regressions were conducted taking the measures of experience as predictors and the variability-influenced dimensions of trust as criterion variables. Moderators were not being controlled for since it was already shown in the previous section that they were of insignificant influence on the trust related measures. All three linear regression equations were found to be insignificant.

In the light these findings, it may be concluded that experience had no influence on the effect of varying calculation results (i.e., performance variability) refuting hypothesis 3. Nonetheless, the participants had no explicit experience with the given task which might limit the generalizability of the results since it could be argued that familiarity with the particular task elicits different effect of variability than experience on a more general level of industrial scheduling, which was addressed with the selected predictors in the present survey.

11.7.7.3 General Attitudes

The analysis of hypothesis 4 was twofold. First, it was necessary to separate the inclination to trust the system from the actual decision to use the software. Building on this distinction, the predictors of actual system usage needed to be identified first. The second part of the analysis examined the relation of common thoughts on computer systems and the tendency to trust R-PLAN+.

11.7.7.4 Trust Related Predictors of System Usage

The statistical analysis of actual system usage was implemented with a binary logistic regression. Since the moderator variables and all measures of trust (ESP, AOP, CSOP, RR, and RO) appeared suitable as dependent variables, a stepwise forward selection of parameters had been considered which made use of the *Wald*-statistic to test for variable entry or removal. The selection algorithm included the tendency to rely on decisions of R-PLAN+ (RR), and the tendency to rely on own decisions (RO). Table 11.5 gives an overview of the resulting regression equation.

It can be concluded from the coefficients in Table 11.5 that participants who showed a higher tendency to rely on the decisions of the software system R-PLAN+ (RR) were more likely to state that they use the system (answer = 0) due to the negative sign of the coefficient. On the contrary, test persons who indicated to preferably rely on their own decisions (RO) had a higher inclination to solve the task without the software (answer = 1) due to the positive sign of the coefficient.

11.7.7.5 Relation of General Attitudes and Trust in a Specific Situation

The subsequent analysis considered only the measures of trust that were related to actual software usage, which were identified in the previous section. Building on a significant correlation, a multiple linear regression was conducted taking RR as dependent variable and all measures of general attitude as predictors since they are all highly correlated to each other. The resulting regression model was significant, $F(5, 91) = 3.08$, $p = 0.01$ accounting for 10% of the variance in the sample ($R^2 = 0.14$, adjusted $R^2 = 0.10$). The only regression coefficient which was significant according to t test was the inclination to rely on computer-based decisions on a general level (rely on computer) to predict the tendency to rely on R-PLAN+ in the specific context (RR). Nonetheless, the other predictors of general attitude should not be excluded from the model too easily since they were all correlated to each other. It can be summarized for hypothesis 4 that the predicted relationship between general attitude and trust – and furthermore – the intension to use the software exist. As a primary bases, the tendency to rely on computer systems seemed to have the greatest importance.

Table 11.5 Coefficients of Binary Logistic Regression

	Coefficient B	Wald(1, $N = 97$)	p
RR	−1.09	8.70	<0.01
RO	0.73	4.33	0.04
Constant	−0.20	0.01	0.90

11.7.7.6 Personal Abilities and the Human Expert

Figure 11.13 provides the first impression of how the test persons decided on using the system and on asking an expert. It should be noted that Fig. 11.13 contains the answers for two questions: The first question asked the participants if they use the planning software. Subsequent to this question, the test persons were introduced to the option of asking a human expert to solve the given task.

It appears as a rather striking finding that most of the participants ($N = 66$) decided first to use R-PLAN+ but changed their mind when being confronted with the option to consult an expert in the particular field ($N = 65$). A possible explanation of this behavior could be found in the analysis of the qualitative answers the participants gave at the end of the survey where they were asked to state their grounds to use the system or not. 55 test persons actually gave an explanation of their reasons, whereas 39 had decided to use the system and 16 had stated to not make use of the system. The given qualitative answers were categorized into 12 groups by two independent raters, who initially agreed on the classification of 83% of the cases. The remaining cases were discussed and agreed upon shortly afterwards.

The most frequently ($N = 21$) given qualitative answer for using the planning system was that the planning system provides a first impression of what could be expected from a solution and then compare this result with own calculations or the results acquired from the human expert. Hence, the participants indicated a critical approach towards the system and cared for multiple perspectives on the given task in order to make their decision on the basis of all information available. This may also serve as an explanation of the high share of participants choosing to consult the expert. On the other hand, major reasons for not using the system were rather straightforward: high variability and low performance of the system.

As to the analysis of hypothesis 5, a multinomial logistic regression was implemented with the reference category "ask the expert to solve the task".

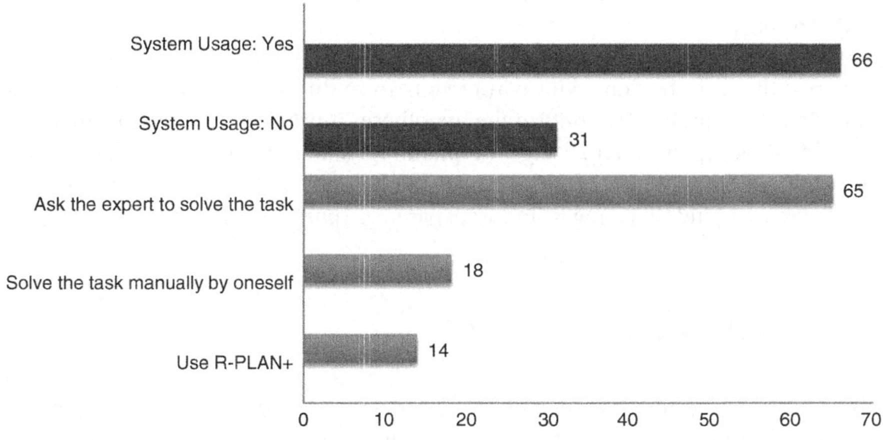

Fig. 11.13 Answer frequencies for system usage and the decision to ask an expert

The regression consisted of two parts. The first part controlled for the moderator variables age, gender, and origin. In the second part, four measures of participants own abilities were considered to be included into the regression equation. The measures of own abilities were the tendency to rely on own abilities when dealing with software for planning and scheduling (rely on oneself), the anticipated own performance during the following planning task (AOP), the relative comparison of system and own performance (CSOP), and the preference to rely on own decisions (RO). However, the only significant predictors to be found was related to the participant's origin and no measures of own abilities were capable to explain the tendency to ask the expert. Hence, hypothesis 5 was not supported.

11.7.8 Conclusion

Taking into account the great amount, complexity, and dynamics of information that has to be processed when executing a planning or scheduling task it becomes obvious that support by software is crucial for both the quality of results and the efficiency of the planning process. However, if a system is used by the employees that are supposed to do so depends on several factors. To determine those factors and to work on guidelines for their consideration and for their influence is important for the success of design and implementation projects of software for planning and scheduling.

It was shown that trust is important for the usage of a software system also in the field of production planning and scheduling. Existing studies can be found especially in the field of automation. There are only a few available for planning tasks. The concept of trust was applied to the field of DSS in planning and scheduling. Especially in this field the performance and stability, leading to predictability, is important for trust. Those characteristics of a system determine together with self confidence and other variables if trust in a system is being developed or not.

In order to validate the hypotheses derived from this model a survey was developed that is to be done with participants from different countries. The results of the survey – should they confirm the hypotheses – provide valuable insights into the field of acceptance and usage of planning and scheduling software. They may improve the process of DSS design and implementation by incorporating knowledge and guidelines regarding transparency (purpose, process) and training.

11.8 Conclusions and Outlook

Design and developing decision support systems for planning and scheduling is not an easy task. In general, building systems such that technology is accepted by the eventual user is a challenging process. Decision support systems for planning and scheduling are usually developed together with the user. While it appears that by doing this the probability of the user accepting the system is enhanced, we show in this chapter that this is not necessarily the case. This is due to the resulting dynamic nature of the development process, and the way in which the modeling details and system performance are related. Our results show that more user involvement is not necessarily better, and a balance needs to be found between leadership by the modeler and involvement of the user.

In the second study presented in this chapter, we specifically address the issues of trust in planning and scheduling systems. We show that building trust is more related to consistency in system performance that to absolute system performance. While trust is a very comprehensive concept, our results suggest that the use of algorithms that typically lead to a substantial variability in performance, as may the case with some methods from the field of deterministic optimization, trust in the system may be severely affected.

The insights from other studies reported in this chapter and the new studies presented show that both user acceptance and user trust can be significantly enhanced if the proper steps are taken in system development. This implies engineers, software developers, and operations researchers need to have more extensive knowledge and understanding of concepts such as those presented in this chapter.

References

Alter, S. (2004). A work system view of DSS in its fourth decade. *Decision Support Systems, 28*, 319–327.

Amoako-Gyampah, K., & Salam, A. F. (2004). An extension of the technology acceptance model in an ERP implementation environment. *Information & Management, 41*, 731–745.

Andriole, S. J. (1989). *Handbook of decision support systems*. Blue Ridge Summit, PA: TAB Professional and Reference Books.

Anthony, R. N. (1965). *Planning and control systems: A framework for analysis*. Boston: Harvard University Graduate School of Business Administration.

Baecker, R. M., Grudin, J., Buxton, W. A. S., & Greenberg, S. (Eds.). (1995). *Readings in human–computer interaction. Toward the Year 2000*. San Francisco, CA: Morgan Kaufmann.

Banati, H., Bedi, P., & Grover, P. S. (2006). Evaluating web usability from the user's perspective. *Journal of Computer Science, 2*(4), 314–317.

Barki, H., & Hartwick, J. (1994). Measuring user participation, user involvement, and user attitude. *MIS Quarterly, 18*(1), 59–82.

Baroudi, J. J., Olson, M. H., & Ives, B. (1986). An empirical study of the impact of user involvement on system usage and information satisfaction. *Communications of the ACM, 29* (3), 232–238.

Bartoelke, K., Eschweiler, W., Flechsenberger, D., & Tannenbaum, A. S. (1982). Workers' participation and the distribution of control as perceived by members of ten german companies. *Administrative Science Quarterly, 27*(3), 380–397.

Bennet, J. L. (Ed.). (1983). *Building decision support systems*. Reading, MA: Addison Wesley.

Beyer, H. R., & Holtzblatt, K. (1995). Apprenticing with the customer. *Communications of the ACM, 38*(5), 45–52.

Beyon, M., Rasmequan, S., & Russ, S. (2002). A new paradigm for computer-based decision support. *Decision Support Systems, 33*, 127–142.

Blili, S., Raymond, L., & Rivard, S. (1998). Impact of task uncertainty, end-user involvement, and competence on the success of end-user computing. *Information & Management, 33*, 137–153.

Bliss, J. P. (1997). Alarm reaction patterns by pilots as a function of reaction modality. *International Journal of Aviation Psychology, 7*(1), 1–14.

Bliss, J. P., & Dunn, M. C. (2000). Behavioral implications of alarm mistrust as a function of task workload. *Ergonomics, 43*, 1283–1300.

Bowers, C. A., Jentsch, F., Salas, E., & Braun, C. C. (1998). Studying communication patterns among aircrews: Implications for team training needs assessment. *Human Factors, 40*, 672–679.

Bowers, C. A., Oser, R. L., Salas, E., & Cannon-Bowers, J. A. (1996). Team performance in automated systems. In R. Parasuraman & M. Mouloua (Eds.), *Automation and human performance: Theory and applications* (pp. 243–263). Mahwah, NJ: LEA.

Breznitz, S. (1984). *Cry wolf: The psychology of false alarms*. Hillsdale, NJ: Lawrence Erlbaum Associates.

Carlson, C., & Turban, E. (2002). DSS: directions for the next decade. *Decision Support Systems, 33*, 105–110.

Cegarra, J., & Hoc, J.-M. (2008). The role of algorithm and result comprehensibility of automated scheduling on complacency. *Human Factors and Ergonomics in Manufacturing & Service Industries, 18*(6), 603–620.

Chau, P. Y. K. (1996). An empirical assessment of a modified technology acceptance model. *Journal of Management Information Systems, 13*(2), 185–204.

Chau, P. Y. K. (2001). Influence of computer attitude and self-efficacy on IT usage behaviour. *Journal of End User Computing, 13*(1), 26–33.

Cheskin Research (2000). Trust in the wired Americas. http://www.cheskin.com/think/studies/trust2.html.

Cheskin Research and Studio Archetype (1999). eCommerce trust study. http://www.studioarch-andype.com/cheskin/index.html.

Cohen, M. S., Parasuraman, R., & Freeman, J. (1999). *Trust in decision aids: A model and its training implications (Tech. Report USAATCOM TR 97-D-4)*. Arlington, VA: Cognitive Technologies.

Cooper, A., Reimann, R. M., & Cronin, D. (2007). *About Face 3: The essentials of interaction design*. New York: Wiley.

Courtney, J. F. (2001). Decision making and knowledge management in inquiring organizations: toward a new decision-making paradigm for DSS. *Decision Support Systems, 31*, 17–38.

Davis, F. D. (1993). User acceptance of information technology: system characteristics, user perception and behavioral impacts. *International Journal of Man-Machine Studies, 38*, 475–487.

Davis, F. D. (1986). *A technology acceptance model for empirically testing new end-user information systems: theory and results*, PhD Thesis, Sloan School of Management, MIT, Cambridge.

Davis, F. D. (1989). Perceived Usefulness, Perceived Ease of Use, and User Acceptance of Information Technology. *MIS Quarterly, 13*(3), 319–340.

Davis, F. D., Bagozzi, R. P., & Warshaw, P. R. (1989). User acceptance of computer technology. A comparison of two theoretical models. *Management Science, 35*(8), 982–1003.

Davis, F. D., & Kottemann, J. E. (1994). User perceptions of decision support effectiveness: Two production planning experiments. *Decision Sciences, 25*(1), 57–78.

Davis, F., & Kottemann, J. (1995). Determinants of decision rule use in a production planning task. *Organizational Behavior and Human Decision Processes, 63*, 145–157.

Davis, F. D., & Venkatesh, V. (2004). Toward preprototype user acceptance testing of new information systems: implications for software project management. *IEEE Transactions on Engineering Management, 51*(1), 31–36.

de Vries, P., Midden, C., & Bouwhuis, D. (2003). The effects of errors on system trust, self-confidence, and the allocation of control in route planning. *International Journal of Human-Computer Studies, 58*, 719–735.

De Kok, A. G., & Fransoo, J. C. (2003). Planning supply chain operations: Definition and comparison of planning concepts. In S. C. Graves & A.G. de Kok (Eds.), *Supply Chain Management: Design, Coordination and Operation* (pp. 597–675), (Handbooks in Operations Research and Management Science, vol. 11), Amsterdam: Elsevier.

Deutsch, M. (1958). Trust and suspicion. *Journal of Conflict Resolution, 2*, 265–279.

Dillon, A. (2000). Group dynamics meet cognition: combining socio-technical concepts and usability engineering in the design of information systems, http://dlist.sir.arizona.edu/1282/01/Ad2000.pdf

Dishaw, M. T., & Strong, D. M. (1999). Extending the technology acceptance model with task-technology fit constructs. *Information & Management, 36*, 9–21.

Doll, W. C., Hendrickson, A., & Deng, X. (1998). Using Davis's perceived usefulness and ease-of-use instruments for decision making: A confirmatory and multigroup invariance analysis. *Decision Sciences, 29*(4), 839–869.

Doll, W. J., & Torkzadeh, G. (1991). A congruence construct of user involvement. *Decision Sciences, 22*(2), 443–453.

Druzdzel, M. J., & Flynn, R. R. (1999). Decision support systems. In A. Kent (Ed.), *Encyclopedia of library and information science*. New York: Marcel Dekker.

Dzindolet, M. T., Peterson, S. A., Pomranky, R. A., Pierce, L. G., & Beck, H. P. (2003). The role of trust in automation reliance. *International Journal of Human–Computer Studies, 58*, 697–718.

Dzindolet, M. T., Pierce, L. G., Beck, H. P., & Dawe, L. A. (2002). The perceived utility of human and automated aids in a visual detection task. *Human Factors, 44*, 79–94.

Egger, F., & de Groot, B. (2000). Developing a model of trust for electronic commerce. An application to a permissive marketing web site. In *Proceedings of the 9th International World-Wide Web Conference*. Foretec Seminars.

Elbeltagi, I., McBride, N., & Hardaker, G. (2005). Evaluating the factors affecting DSS usage by senior managers in local authority in Egypt. *Journal of Global Information Management, 13* (2), 42–65.

Faranda, W. T. (2001). A scale to measure the cognitive control form of perceived control: Construction and preliminary assessment. *Psychology & Marketing, 18*(12), 1259–1281.

Franz, C. R., & Robey, D. (1986). Organizational Context, User Involvement, and the Usefulness of Information Systems. *Decision Sciences, 17*(3), 329–356.

Fogg, B. J., Marshall, J. Laraki, O., Osipovich, A., Varma, C., Fang, N., et al. (2001). *What Makes Web sites Credible? A Report on a Large Quantitative Study*, Seattle, WA: SIGCHI.

Fogg, B. J., & Tseng, H. (1999). The elements of computer credibility. In *Proceedings of the SIGCHI Conference on Human Factors in Computing Systems: The CHI is the Limit (Pittsburgh, Pennsylvania, United States, May 15–20, 1999). CHI '99* (pp. 80–87). New York, NY: ACM

Foster, S. T., & Franz, C. R. (1999). User involvement during information systems development: A comparison of analyst and user perceptions of system acceptance. *Journal of Engineering and Technology Management, 16*(3–4), 329.

Friedman, B., Kahn, P., & Howe, D. C. (2000). Trust online. *Communications of the ACM Press, 43*(12), 34–40.

Gatignon, H., & Robertson, T. S. (1991). Innovative decision processes. In T. S. Robertson & H. H. Kassarjian (Eds.), *Handbook of consumer behavior*. Englewood Cliffs, NJ: Prentice-Hall.

Gefen, D., & Keil, M. (1998). The impact of developer responsiveness on perceptions of usefulness and ease of use: An extension of the Technology Acceptance Model. *Database for Advances in Information Systems, 29*(2), 35–49.

Gerrity, T. P., Jr. (1971). Design of man–machine decision systems: An application to portfolio management. *Sloan Management Review, 12*(2), 59–75.

Gorry, G. A., & Morton, M. S. (1971). A framework for management information systems. *Sloan Management Review, 13*(1), 55–70.

Greenbaum, J., & Kyng, M. (Eds.). (1991). *Design at work: Cooperative design of computer systems*. Hillsdale, NJ: Lawrence Erlbaum Associates.

Guimaraes, T., Igbaria, M., & Lu, M.-T. (1992). The determinants of DSS success: An integrated model. *Decision Sciences, 23*(2), 409–430.

Halprin, S., Johnson, E., & Thornburry, J. (1973). Cognitive reliability in manned systems. *IEEE Transactions on Reliability, 22*, 165–169.

Hanowski, R. J., Kantowitz, S. C., & Kantowitz, B. H. (1994). Driver acceptance of unreliable route guidance information. *Proceedings of the Human Factors Society 381 h Annual Meeting* (pp. 1062–1066.

Hardgrave, B. C., & Johnson, R. A. (2003). Toward an information systems development acceptance model: the case of object-oriented systems development. *IEEE Transactions on Engineering Management, 50*(3), 322–336.

Hartwick, J., & Barki, H. (1994). Explaining the role of user participation in information system use. *Management Science, 40*(4), 440–465.

Hawk, S. R., & Dos Santos, B. L. (1991). Successful system development: The effect of situational factors on alternate user roles. *IEEE Transactions on Engineering Management, 38*(4), 316–325.

Ho, G., Kiff, L. M., Plocher, T., & Haigh, K. Z. (2005). A model of trust and reliance of automation technology for older users. *AAAI Fall Symposium on Caring Machines, November 2005*, Washington, DC

Hoc, J.-M. (2001). Towards a cognitive approach to human-machine cooperation in dynamic situations. *International Journal of Human-Computer Studies, 54*, 509–540.

Hodgson, L., & Aiken, P., (1998). Organizational change enabled by the mandated implementation of new information systems technology: A modified technology acceptance model. *Proceedings of the 1998 ACM SIGCPR conference on Computer personnel research*. Boston, MA

Hogue, J. T., & Watson, H. J. (1983). Management's Role in the Approval and Administration of Decision Support Systems. *MIS Quaterly, 7*(2), 15–26.

Honaker, L. M., Hector, V. S., & Harrell, T. H. (1986). Perceived validity of computer versus clinician-generated MMPI reports. *Computers in Human Behavior, 2*, 77–83.

Hubona, G. S., & Kennick, E. (1996). The Influence of External Variables on Information Technology Usage Behavior. *Proceedings of the 29th Annual Hawaii International Conference on System Sciences*

Hunton, J. E., & Beeler, J. D. (1997). Effects of user participation in systems development: A longitudinal field experiment. *MIS Quarterly, 21*(4), 359–388.

Hwang, M. I., & Thorn, R. G. (1999). The effect of user engagement on system success: A meta-analytical integration of research findings. *Information & Management, 35*, 229–236.

Igbaria, M., & Guimaraes, T. (1994). Empirically testing the outcomes of user involvement in DSS development. *Omega, 22*(2), 157–172.

Igbaria, M., & Iivari, J. (1995). The effects of self-efficacy on computer usage. *International Journal of Management Science, 23*(6), 587–605.

Igbaria, M., & Tan, M. (1997). The consequences of information technology acceptance on subsequent individual performance. *Information & Management, 32*, 113–121.

Ives, B., & Olson, M. H. (1984). User involvement and MIS success: A review of research. *Management Science, 30*(5), 586–603.

Jacko, J. A., & Sears, A. (Eds.). (2003). *Handbook for human computer interaction*. Mahwah, NJ: Lawrence Erlbaum Associates.

Jackson, C. M., Chow, S., & Leitch, R. A. (1997). Toward an understanding of the behavioral intention to use an information system. *Decision Sciences, 28*(2), 357–389.

Johnson, J. D., Sanchez, J., Fisk, A. D., & Rogers, W. A. (2004). Type of automation failure: The effects on trust and reliance in automation. *Proceedings of the Human Factors and*

Ergonomics Society 48th Annual Meeting. Santa Monica, CA: Human Factors and Ergonomics Society.

Kantowitz, B. H., Hanowski, R. J., & Kantowitz, S. C. (1997). Driver acceptance of unreliable traffic information in familiar and unfamiliar settings. *Human Factors, 39*(2), 164–176.

Kappelman, L. A. (1995). Measuring user involvement: A diffusion of innovation perspective. *ACM SIGMIS Database, 26*(2–3), 65–86.

Karahanna, E., & Straub, D. W. (1999). The psychological origins of perceived usefulness and ease-of-use. *Information & Management, 35*, 237–250.

Karwowski, W. (Ed.). (2006). *International encyclopedia of ergonomics and human factors*. Boca Roton, FL: CRC.

Katz, D., & Kahn, R. L. (1966). *The social psychology of organizations*. New York: Wiley.

Keen, P., & Morton, M. S. (1978). *Decision support systems: An organizational perspective*. Reading, MA: Addison-Wesley.

Kensing, F. (2003). *Methods and practices in participatory design*. Copenhagen, Denmark: ITU.

Kerr, R. M. (1992). Expert systems in production scheduling: Lessons from a failed implementation. *Journal of Systems Software, 19*, 123–130.

Kidwell, B., & Jewell, R. D. (2003). An examination of perceived behavioral control: Internal and external influences on intention. *Psychology & Marketing, 20*(7), 625–642.

Kim, C., Suh, K., & Lee, J. (1998). Utilization and user satisfaction in end-user computing: A task contingent model. *Information Resources Management Journal, 11*(4), 11–24.

Kleingeld, A., van Tuijl, H., & Algera, J. A. (2004). Participation in the design of performance management systems: a quasi-experimental field study. *Journal of Organizational Behavior, 25*(7), 831–851.

Kleintop, W. A., Blau, G., & Currall, S. C. (1994). Practice makes use: using information technologies before implementation and the effect on acceptance by end users. *Proceedings of the 1994 Computer Personnel Research Conference on Reinventing IS: Managing Information Technology in Changing Organizations*. ACM: New York, NY

Klopping, I. M., & McKinney, E. (2004). Extending the technology acceptance model and the task-technology fit model to consumer E-COMMERCE. *Information Technology, Learning, and Performance Journal, 22*(1), 35–48.

LaForge, R. L., & Craighead, C. W. (2000). Computer-Based Scheduling in Manufacturing Firms: Some Indicators of Successful Practice. *Production and Inventory Management Journal, 41* (1), 29–34.

Langer, E. J. (1975). The illusion of control. *Journal of Personality and Social Psychology, 32*, 311–328.

Landy, F. J., & Conte, J. M. (2007). *Work in the 21st century: An introduction to industrial and organizational psychology* (2nd ed.). Malden, MA: Blackwell Publishing.

Lawrence, M., Goodwin, P., & Fildes, R. (2002). Influence of user participation on DSS use and decision accuracy. *Omega, 30*, 381–392.

Lee, J. (1991). The dynamics of trust in a supervisory control simulation. *Proceedings of the Human Factors Society 3Sh Annual Meeting* (pp. 1228–1232).

Lee, J., & Moray, N. (1992). Trust, control strategies and allocation of function in human-machine systems. *Ergonomics, 35*(10), 1243–1270.

Lee, J. D., & Moray, N. (1994). Trust, self-confidence, and operators' adaptation to automation. *International Journal of Human-Computer Studies, 40*(1), 153–184.

Lee, J. D., & See, K. A. (2004). Trust in Automation: Designing for Appropriate Reliance. *Human Factors, 46*(1), 50–80.

Legris, P., Ingham, J., & Collerette, P. (2003). Why do people use information technology? A critical review of the technology acceptance model. *Information & Management, 40*, 191–204.

Lerch, F. J., & Prietula, M. J. (1989). How do we trust machine advice? Designing and using human-computer interfaces and knowledge based systems. *Proceedings of the Third Annual Conference on Human Computer Interaction* (pp. 411–419). Amsterdam: Elsevier.

Lewandowsky, S., Mundy, M., & Tan, G. P. A. (2000). The dynamics of trust: Comparing humans to automation. *Journal of Experimental Psychology, 6*(2), 104–123.

Lewicki, R. J., & Bunker, B. B. (1996). Developing and maintaining trust in work relationships. In R. M. Kramer & T. R. Tyler (Eds.), *Trust in organizations: Frontiers of theory and research* (pp. 114–139). Thousand Oaks, CA: Sage Publications.

Lin, W. T., & Shao, B. B. M. (2000). The relationship between user participation and system success: A simultaneous contingency approach. *Information & Management, 37*(6), 283–295.

Liu, Y., Fuld, R., & Wickens, C. D. (1993). Monitoring behavior in manual and automated scheduling systems. *International Journal of Man-machine Studies, 39*(6), 1015–1029.

Luhmann, N. (1979). *Trust and power: Two works by Niklas Luhmann.* Chichester: Wiley.

Ma, Q., & Liu, L. (2004). The technology acceptance model: A meta-analysis of empirical findings. *Journal of Organizational and End User Computing, 16*(1), 59–72.

Madhavan, P., & Wiegmann, D. A. (2004). A new look at the dynamics of human-automation trust: is trust in humans comparable to trust in machines? *Proceedings of the Human Factors and Ergonomics Society 48th Annual Meeting*. Santa Monica, CA: Human Factors and Ergonomics Society.

Mathieson, K., Peacock, E., & Chin, W. W. (2001). Extending the technology acceptance model: The influence of perceived user resources. *Database for Advances in Information Systems, 32* (3), 86–112.

Mayer, R. C., Davis, J. H., & Schoorman, F. D. (1995). An integrative model of organizational trust. *Academy of Management Review, 20*(3), 709–734.

McHaney, R., & Cronan, T. P. (2001). A comparison of surrogate success measures in on-going representational decision support systems: An extension to simulation technology. *Journal of End User Computing, 13*(2), 15–25.

McKay, K. N., & Black, G. W. (2007). The evolution of a production planning system: A 10-year case study. *Computers in Industry, 58*, 756–771.

McKay, K. N., & Buzacott, J. A. (2000). The application of computerized production control systems in job shop environments. *Computers in Industry, 42*, 79–97.

McKeen, J. D., Guimaraes, T., & Wetherbe, J. C. (1994). The relationship between user participation and user satisfaction: An Investigation of four contingency factors. *MIS Quarterly, 18*(4), 427–451.

McNurlin, B. C., & Sprague, R. H., Jr. (2002). *Information systems management in practice* (5th ed.). Upper Saddle River, NJ: Prentice-Hall International.

Metzger, U., & Parasuraman, R. (2001). The role of the air traffic controller in future air traffic management: An empirical study of active control versus passive monitoring. *Human Factors, 43*, 519–528.

Miller, C. A., Funk, H. B., Goldman, R. P., Meisner, J., & Wu, P. (2005). Implications of adaptive vs. adaptable UIs on decision making: Why "automated adaptiveness" is not always the right answer. *Proceedings of the 1st International Conference on Augmented Cognition*. Las Vegas, NV.

Miller, C. A., & Larson, R. (1992). An explanatory and "argumentative" interface for a model-based diagnostic system. *Proceedings of the ACM Symposium on User Interface Software and Technology* (pp. 43–52).

Mitroff, I. I., & Linstone, H. A. (1993). *The unbonded mind breaking the chains of traditional business thinking*. New York: Oxford University Press.

Morris, S. A., & Marshall, T. E. (2004). Perceived control in information systems. *Journal of Organizational and End User Computing, 16*(2), 38–56.

Muir, B. M. (1987). Trust between humans and machines. *International Journal of Man–Machine Studies, 27*, 327–339.

Muir, B. M. (1988). Trust between humans and machines, and the design of decision aids. In E. Hollnagel, G. Mancini, & D. D. Woods (Eds.), *Cognitive engineering in complex dynamic worlds* (pp. 71–84). London: Academic.

Muir, B. M. (1989). *Operators' trust in and use of automatic controllers in a supervisory process control task*. Ph.D. dissertation, University of Toronto, Toronto, Canada.
Muir, B. M. (1994). Trust in Automation: Part I. Theoretical issues in the study of trust and human intervention in automated systems. *Ergonomics, 37*(11), 1905–1922.
Muir, B. M., & Moray, N. (1996). Trust in automation: 2. Experimental studies of trust and human intervention in a process control simulation. *Ergonomics, 39*, 429–460.
Muller, M. J. (2007). Participatory design: The third space in HCI. In J. Jacko & A. Sears (Eds.), *Handbook of HCI*. Mahwah, NJ: Lawrence Erlbaum Associates.
Musselmann, K., & Uzgoy, R. (2001). Advanced planning and scheduling for manufacturing. In G. Salvendy (Ed.), *Handbook of industrial engineering* (3rd ed.). New York: Wiley.
Nass, C., Fogg, B. J., & Moon, Y. (1996a). Can computers be teammates? *International Journal of Human–Computer Studies, 45*, 669–678.
Nass, C., & Moon, Y. (2000). Machines and Mindlessness: Social Responses to Computers. *Journal of Social Issues, 56*(1), 81–103.
Nass, C., Moon, Y., Fogg, B. J., Reeves, B., & Dryer, D. C. (1995). Can computer personalities be human personalities? *International Journal of Human-Computer Studies, 43*, 223–239.
Nass, C., Reeves, B., & Leshner, G. (1996b). Technology and roles: A tale of two TVs. *Journal of Communication, 46*(2), 121–128.
Ndubisi, N. O., Gupta, O. K., & Ndubisi, G. C. (2005). The Moguls' model of computing: Integrating the moderating impact of users' persona into the technology acceptance model. *Journal of Global Information Technology Management, 8*(1), 27–47.
Nielsen, J. (1993). *Usability Engineering*, Academic Press, Boston, MA
Palanisamy, R. (2001). User involvement and flexibility in strategic MIS planning: A path analytic study. *Global Journal of Flexible Systems Management, 2*(4), 15–32.
Pancer, S. M., George, M., & Gebotys, R. J. (1992). Understanding and predicting attitudes toward computers. *Computers in Human Behavior, 8*(2–3), 211–222.
Parasuraman, R. M., Molloy, R., & Singh, I. L. (1993). Performance consequences of automation induced 'complacency'. *International Journal of Aviation Psychology, 3*, 1–23.
Parasuraman, R., & Riley, V. (1997). Humans and automation: Use, misuse, disuse, abuse. *Human Factors, 39*, 230–253.
Pearson, J. M., & Shim, J. P. (1995). An empirical investigation into DSS structures and environments. *Decision Support Systems, 13*, 141–158.
Ramesh, R., & Sekar, G. C. (1988). An Integrated Framework for Decision Support in Corporate Planning. *Decision Support Systems, 4*, 365–375.
Rawstorne, P., Jayasuriya, R., & Caputi, P. (2000). Issues in predicting and explaining usage behaviors with the technology acceptance model and the theory of planned behavior when usage is mandatory. *Proceedings of the 21st International Conference on Information Systems* (pp. 35–44). Atlanta, GA: Association for Information Systems
Reeves, B., & Nass, C. (1996). *The media equation. How people treat computers, television, and new media like real people and places*. New York: Cambridge University Press.
REFA. (1984). *Methodenlehre des Arbeitsstudiums -Teil 1 – Grundlagen*. München: Carl Hanser Verlag.
Rempel, J. K., Holmes, J. G., & Zanna, M. P. (1985). Trust in close relationships. *Journal of Personality and Social Psychology, 49*(1), 95–112.
Rhodes, J. S. (1998). How to gain the trust of your users. http://webword.com/moving/trust.html.
Riley, V. (1994a). A theory of operator reliance on automation. In M. Mouloua & R. Parasuraman (Eds.), *Human performance in automated systems: Recent research and trends* (pp. 8–14). Hillsdale, NJ: Lawerence Erlbaum Associates.
Riley, V. (1994b). *Human use of automation*. Unpublished doctoral dissertation, University of Minnesota, Minneapolis, MN.
Rittel, H. W. J., & Webber, M. M. (1973). Dilemmas in a general theory of planning. *Policy Sciences, 4*, 155–169.

Rotter, J. B. (1967). A new scale for the measurement of interpersonal trust. *Journal of Personality, 35*, 651–665.

Rotter, J. B. (1980). Interpersonal trust, trustworthiness, and gullibility. *American Psychologist, 35* (1), 1–7.

Saade, R., & Bahli, B. (2005). The impact of cognitive absorption on perceived usefulness and perceived ease of use in on-line learning: an extension of the technology acceptance model. *Information & Management, 42*, 317–327.

Saffer, D. (2006). *Designing for interaction.* New Riders.

Sage, A. P. (2001). Decision support systems. In G. Salvendy (Ed.), *Handbook of industrial engineering* (3rd ed.). New York: Wiley.

Sanchez, J., Fisk, A. D., & Rogers W. A. (2004). Age-related and reliability related effects on trust of a decision support aid. *Presented at the 2004 Human Performance, Situation Awareness and Automation Technology Conference.* Daytona Beach, FL.

Saxena, K. B. C. (1991). *Decision support engineering: A DSS development methodology.* Kowloon, Hong Kong: Department of Computing, Hong Kong Polytechnic Hung Hom.

Schuler, D., & Namioka, A. (1993). *Participatory design: Principles and practices.* Hillsdale, NJ: Lawrence Erlbaum Associates.

Self, C. S. (1996). Credibility. In M. Salwen & D. Stacks (Eds.), *An integrated approach to communication theory and research.* Mahway, NJ: Lawrence Erlbaum Associates.

Sharp, H., Rogers, Y., & Preece, J. (Eds.). (2002). *Interaction design – beyond human–computer interaction.* New York: Wiley.

Shim, J. P., Warkentin, M., Courtney, J. F., Power, D. J., Sharda, R., & Carlsson, C. (2002). Past, present, and future of decision support technology. *Decision Support Systems, 33*, 111–126.

Simon, H. A. (1960). *The new science of management decision.* New York: Harper Brothers.

Sprague, R. H., Jr., & Carlson, E. D. (1982). *Building effective decision support systems.* Upper Saddle River, NJ: Prentice Hall.

Stadler, H., & Kilger, C. (2000). *Supply chain management and advanced planning – concepts, models, software and case studies.* Berlin: Springer.

Straub, D., Keil, M., & Brenner, W. (1997). Testing the technology acceptance model across cultures: A three country study. *Information & Management, 33*, 1–11.

Strohm, O. (1996). Produktionsplanung und –steuerung im Industrieunternehmen aus arbeitspsychologischer Sicht Arbeits- versus technikorientierte Gestaltungskonzepte, vdf Hochschulverlag AG an der ETH Zürich

Sundar, S. S., & Nass, C. (2000). Source orientation in human-computer interaction. *Communication Research, 27*(6), 683–703.

Szajna, B. (1994). Software evaluation and choice: Predictive validation of the technology acceptance instrument. *MIS Quarterly, 18*(3), 319–324.

Szajna, B. (1996). Empirical evaluation of the revised technology acceptance model. *Management Science, 42*(1), 85–92.

Tait, P., & Vessey, I. (1988). The effect of user involvement on system success: A contingency approach. *MIS Quarterly, 12*(1), 91–108.

Taylor, S., & Todd, P. (1995). Assessing IT usage: The role of prior experience. *MIS Quarterly, 19* (4), 561–570.

Tsubone, H., Matsuura, H., & Kimura, K. (1995). Decision support system for production planning – concept and prototype. *Decision Support Systems, 13*, 207–215.

Turban, E. (1995). *Decision support and expert systems: management support systems.* Englewood Cliffs, NJ: Prentice Hall.

Venkatesh, V. (2000). Determinants of perceived ease of use: Integrating control, intrinsic motivation and emotion into the technology acceptance model. *Information Systems Research, 11*(4), 342–365.

Venkatesh, V., & Davis, F. D. (1996). A model of the antecedents of perceived ease of use: Development and test. *Decision Sciences, 27*(3), 451–481.

Venkatesh, V., & Davis, F. D. (2000). A theoretical extension of the technology acceptance model: four longitudinal field studies. *Management Science, 46*(2), 186–204.

Venkatesh, V., Speier, C., & Morris, M. G. (2002). User acceptance enablers in individual decision making about technology: Toward an integrated model. *Decision Sciences, 33*(2), 297–316.

Waern, Y., & Ramberg, R. (1996). People's perception of human and computer advice. *Computers in Human Behavior, 12*(l), 17–27.

Wiegmann, D., Rich, A., & Zhang, H. (2001). Automated diagnostic aids: the effects of aid reliability on users' trust and reliance. *Theoretical Issues in Ergonomics Science, 2*(4), 352–367.

Xie, H. (2003). Supporting ease-of-use and user control: desired features and structure of Web-based online IR systems. *Information Processing and Management, 39*, 899–922.

Xu, X., Wickens, C. D., & Rantanen, E. (2004). *Imperfect conflict alerting systems for the cockpit display of traffic information (Technical Report AHFD-04-8/NASA-04-02)*. Savoy, IL: University of Illinois at Urbana-Champaign, Institute of Aviation, Aviation Human Factors Division.

Yin, R.W. (1994). *Case Study Research: Design and Methods*, 2nd edn. Sage, Thousand Oaks, CA.

Zajonc, R. B. (1980). Feeling and Thinking: Preferences need no inferences. *American Psychologist, 35*, 151–175.

Zoryk-Schalla, A. J., Fransoo, J. C., & de Kok, T.G. (2004). Modeling the planning process in Advanced Planning Systems. *Information and Management, 42*, 75–87.

Zuboff, S. (1988). *In the age of smart machines: The future of work technology and power*. New York: Basic Books.

Part III
Design and Support of the Planning and Scheduling Task

Chapter 12
Design of Scheduling Algorithms

Jan Riezebos, Jean-Michel Hoc, Nasser Mebarki, Christos Dimopoulos, Wout van Wezel, and Guillaume Pinot

Abstract The accomplishment of a manufacturing company's objectives is strongly connected to the efficient solution of scheduling problems that are faced in the production environment. Numerous methods for the solution of these problems have been published. However, very few of them have been adopted by manufacturing companies. This chapter suggests that the basic reason behind this imbalance is the inadequate representation of the scheduling process when designing decision support systems. Hence, the algorithms that are designed and included in these systems might not reflect the problems that actually have to be solved. The relevance of algorithmic design can be improved by using a more complete representation of the scheduling process, which would be highly relevant for increasing the adoption rate of new support systems.

The main contribution of the chapter concerns the development of a theoretical framework for the design of scheduling decision support systems. This framework is based on an interdisciplinary approach that integrates insights from cognitive psychology, computer science, and operations management. The use of this framework implies that the design of a decision support system should start with an examination of the human, organizational, and technical characteristics of the

J. Riezebos (✉) and W. van Wezel
Faculty of Economics and Business, University of Groningen, Groningen, The Netherlands
e-mail: j.riezebos@rug.nl, w.m.c.van.wezel@rug.nl

J.-M. Hoc and G. Pinot
Centre National de la Recherche Scientifique, (CNRS: French National Research Centre), University of Technology of Compiègne, Compiègne, France
e-mail: Jean-Michel.Hoc@irccyn.ec-nantes.fr, guillaume.pinot@irccyn.ec-nantes.fr

N. Mebarki
Département Qualité, Logistique Industrielle et Organisation (QLIO), University of Nantes, Nantes, France
e-mail: nasser.mebarki@univ-nantes.fr

C. Dimopoulos
European University Cyprus, Nicosia, Cyprus
e-mail: c.dimopoulos@euc.ac.cy

J.C. Fransoo et al. (eds.), *Behavioral Operations in Planning and Scheduling*,
DOI 10.1007/978-3-642-13382-4_12, © Springer-Verlag Berlin Heidelberg 2011

scheduling situation that has to be supported. This information can be obtained and analyzed using appropriate methodologies such as hierarchical task analysis, cognitive task analysis and cognitive work analysis as well as other methodologies, such as interviews, observations, context diagrams, and data flow diagrams. The designer of the decision support system can then match the results of the analysis to the guidelines of the theoretical framework and proceed accordingly.

12.1 Introduction

The accomplishment of a manufacturing company's objectives is strongly connected to the efficient solution of complex scheduling problems that are faced in the production environment. The academic field of production research has been growing rapidly over the last decades with researchers proposing numerous analytical and heuristic optimization methodologies for the solution of scheduling problems (Slack et al. 2004). However, very few of them have been extensively adopted by manufacturing companies. The basic reason behind this imbalance is the inadequate representation of the very complex scheduling process, as this is implemented in practice.

The main aim of this chapter is to provide a theoretical discussion on the design of scheduling algorithms. In the first part of this discussion a detailed description of the general problem-solving process and its relation with planning and scheduling will be presented. The second part will concern a critical review of traditional production research algorithmic design approaches. This review will contrast the model of the scheduling environment as this is conceived by traditional production research approaches against models that specifically address human and organizational issues. The final part of this chapter builds on the discussion of the previous sections and proposes the development of a scheduling theoretical framework. This framework will help practitioners to design decision support tools that specifically address human and organizational considerations of the scheduling case considered.

12.2 Problem Solving Fundamentals

Modeling a planning/scheduling problem and designing an algorithm to support the process of solving the problem is rather complicated. This process is sensitive to errors, assumptions, and perceptions, which might result in unsatisfactory solutions. This section gives an overview of problem solving fundamentals in order to improve our understanding of the role of algorithms and the process of designing such algorithms. First, we will give attention to a systems view on problem solving. Next, a cognitive psychology perspective of problem solving will be presented.

From a systems perspective, problem solving can be described as a process consisting of several stages. A landmark paper in this respect is Mitroff et al. (1974).

They distinguish four elements and six stages (see Fig. 12.1) and argue that many problem-solving processes differ in terms of the starting point and the subset and sequencing of stages taken into account.

It might seem that the regular way of problem solving is to start in I and follow the stages 1, 2, 3 and 4, while giving attention to the validation steps 5 and 6. However, Mitroff et al. argue that – from a systems perspective – all elements can be a suitable starting point for a fruitful problem solving process. If, for example, a standard solution is being applied that seems not appropriate anymore, a problem solver might consider the scientific model (III) that was behind the standard solution or directly criticize the conceptual model (II) that has been used. Hence, different paths can be followed in order to solve a problem.

The next fundamental issue they raise is that each element and stage requires different skills of people involved. Activities in the right hand side of the figure are said to require formal, analytic skills and people that excel in these skills are, according to Jung, denoted as Thinking types. Activities at the left hand side require intuitive thinking and human relations skills. Jung denotes people that excel in these activities as Intuitive and Feeling types. There are not that many persons that excel in both set of skills (Jung 1976).

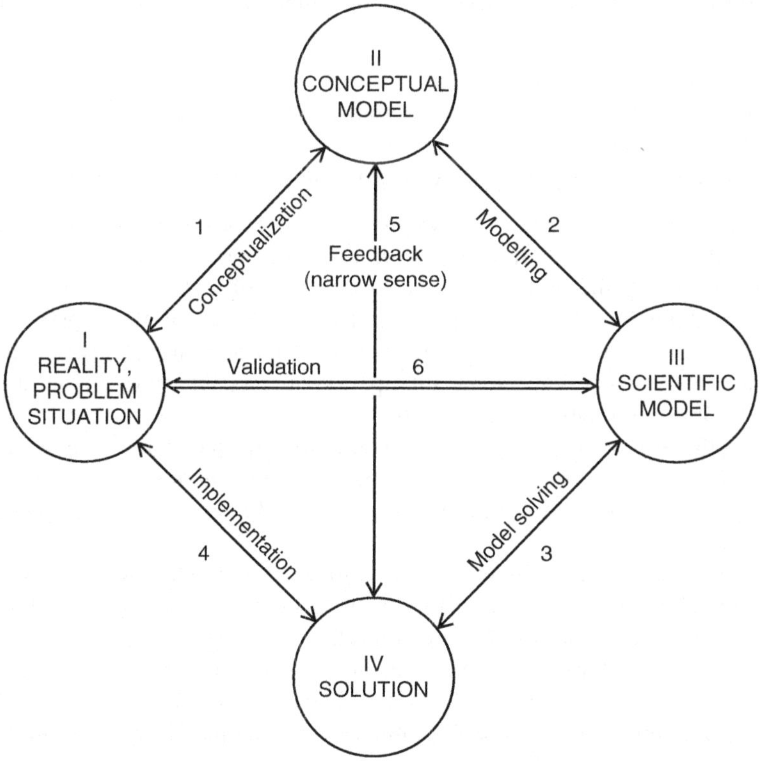

Fig. 12.1 Mitroff's view of problem solving

From a cognitive psychology perspective, a landmark in the field of problem solving is the work of Herbert A. Simon and Allen Newell, starting with their 1958 paper on human problem solving (Newell et al. 1958). Problem solving is seen as the interaction of a problem solver with the problem's task environment. The types of problems that they primarily consider are transformation problems consisting of an initial state, a goal state, and the rules of the game (available options and restrictions). They distinguish sub-goal setting and strategy selection as the main activities in problem solving, and focus on the differences in memory storage and usage in the solution process. Special attention is given to planning as a cognitive strategy. Newell & Simon distinguish a planning space and a basic space. Their first planning strategy was described as the resolution of the problem in abstract terms (within the planning space) before implementing it into the basic space with all the details. Further research on problem solving showed that such a purely top–down strategy was not practical. Newell and Simon see planning, in accordance to Miller et al. (1960), as a hierarchically controlled process in which several cognitive mechanisms, such as working memory, executive control, and knowledge representation co-operate. The working memory is used to store and retrieve plans when they are being generated or executed (1960: 207).

In order to solve a problem, Newell and Simon (1972) state that a representation of the transformation problem needs to be developed or selected. A representation includes:

1. A description of the current state.
2. Operators/actions in order to change/transform the current state, including the constraints/limitations.
3. Tests in order to verify that the achieved state corresponds to the goal.

The task environment constitutes of an objective structure of all states linked by operators. However, problem solvers have no complete view of the task environment when solving the problem. According to Newell and Simon, they base the decision making on a problem space, which constitutes of their instantaneous representation of a part of the task environment: the present state, some memory of previous visited states, and some anticipation of reachable states in the future, et cetera. Their representations may be false.

During the resolution of a problem, several problem spaces are being generated. The union of all these problem spaces is denoted as the search space. However, this search space is not known in advance, so it cannot be used during the process of problem solving.

Everyone that applies planning as a cognitive strategy when solving a problem can be considered to be a planner. However, this definition is not very limitative. Wezel (2001:36) concludes that there is a difference between "planning for yourself" and "planning for others". We therefore suggest using the term "planner" for people that have to solve problems in the "planning for others" category. When designing decision support for such problems solvers, one might aim to model the whole task environment. However, this goal is unattainable, as the number of possible states is generally very large and applying operators as well as testing for verification costs time. Therefore, it requires efficient and effective procedures to search and test solutions. The computer might provide support in the process of selecting operators, memorizing past states, anticipating future states, and testing (partial) solutions. Note that by not considering the whole task environment, the best solution might be overlooked, resulting in non-optimal solutions. Furthermore, the support provided by the computer has to deal with limitations as well, such as dynamic memory usage and conflict resolution if several non-congruent partial solutions have been found.

An alternative view of problem solving is developed by Hayes-Roth and Hayes-Roth (1979). Their main criticism is that a cognitive theory of problem solving should start with the actual behavior of problem solvers. They found that humans behave opportunistically when solving a problem. While dealing with a problem, they use information and anticipate on future situations and behave accordingly. Problem solving is seen as a process in which opportunities that emerge are as important for the decisions to be made as predetermined sub-goals and sequences of operations. This view of problem solving focuses more on the cognitive processes during the execution or implementation of a plan and seems to fit better in case of ill-structured problems, where at least one of the three parts of a problem representation is not completely available.

An important category of these ill-structured problems are so-called design problems (Simon 1981). Hoc (1988) defines a design problem as a task represented by the problem solver as a search for a precise representation of the goal. For design problems, there are several acceptable goals and several evaluation criteria. Design problems occur when creativity is required, i.e., use of pattern recognition and reuse of already developed solutions to similar problems are not sufficient to solve the problem. The number of states in the task environment is infinite in such problems. However, decomposition of such problems and applying search strategies may still be worthwhile, as empirical research has shown (Ormerod 2005).

12.2.1 Algorithms

Procedures for searching and testing solutions are denoted as algorithms. An algorithm for solving a problem determines how searching and testing is done.

Algorithms differ in the efficiency and effectiveness in which they operate.

Less efficient algorithms take more time or storage memory than efficient algorithms. In general, this is denoted as the computational complexity of an algorithm. As the runtime or storage requirement normally depends on the size of the problem, computational complexity is expressed as a function of a suitable input measure of the problem, e.g., the number of jobs n that need to be scheduled, or the number of machines m that have to be planned.

A problem is said to be polynomially solvable if an algorithm exists for which the computational complexity can be expressed as a polynomial in a suitable input measure. The idea is that if problem sizes increase, time or storage requirements will also increase, but according to a polynomial function, e.g., $c \cdot b^n$, where b and c are constants and n is an input measure. Although an algorithm is only more efficient than another one if the outcome of the function is smaller than for the alternative algorithm, the degree of the polynomial (b) is often considered to be the relevant factor for comparing the efficiency of algorithms for a problem. Problems for which a polynomial algorithm is available belong to the class P and are considered to be easy, although the time or storage requirements needed in order to find a solution may still be huge.

For many well-structured problems, no polynomial algorithms are known. The algorithms that are known have an input measure in the exponent of the function, e.g., $c \cdot b^n$. A special subset of problems in this set belongs to the so-called NP-complete problems. If for any of these problems a polynomial algorithm will be found, this algorithm can be used to solve the other problems as well within polynomial time. There are not so many scientists who believe that for these problems once a polynomial algorithm will be found. Hence these problems are considered to be hard, although for small values of the input measure it might be possible to find a solution quickly. See for further details Garey and Johnson (1979).

Less effective algorithms perform less in finding or testing satisfying solutions. Some algorithms do not even find a solution; others are not able to guarantee that the solution that has been found is of sufficient quality. The latter are often denoted as heuristics. The effectiveness of heuristics can be assessed through a worst-case analysis. Note that the effectiveness of an algorithm depends on the formulation of the goal. If the goal is to find the optimal solution (i.e., no better solution is available in terms of the objective function), the algorithm should not only find a solution in the set of optimal solutions, but also prove that there is no better solution. If the goal is rephrased to finding a solution with an objective function value larger than z, any solution that satisfies this criterion qualifies, and the algorithm will be much more efficient as it does not have to proof uniqueness or optimality.

This brings us to the last remark on algorithms, a remark with respect to their *robustness*. Scheduling problems are characterized according to some characteristics of the problem. For example, a flow shop scheduling problems with two machines (i.e., all jobs visit the machines in the same sequence), n jobs and objective function "minimize the make span". Johnson (1954) found a polynomially solvable algorithm for this flow shop problem. However, if this problem is slightly changed to a three-machine flow shop scheduling problem with the same objective, Johnson's algorithm no longer solves this problem optimally. Garey et al. (1976) even proved that problem to be NP complete. Some algorithms are therefore strongly connected to a single scientific model (III in Fig. 12.1) and often not optimal for a slightly different conceptual model of the problem (II). Applying it to such a problem might result in near-optimal results, but a very disappointing result might also be possible. They may even be inapplicable to such a problem, hence providing no solution at all. Other algorithms are more robust if changes in the characteristics of the problem occur. For example, simple heuristics, such as hill-climbing or an Earliest Due Date dispatching rule, can be applied to a large set of problems, although not always generating good quality solutions, and hence are much more flexible to changes in problem characteristics than specific tailor-made algorithms. The latter might be quicker or better in the process of finding good solutions. We see a similar design problem in nature. Some insects are very specialized on a type of plant for their food consumption, while others are not sensitive for changes in the availability of specific natural resources. The design of algorithms is hence the result of a decision process in which many factors have to be taken into account.

12.2.2 Improving Problem Solving

Many attempts have been made in the past in order to improve the result of the process of problem solving. These attempts range from

- Problem structuring improvements (what aspects of the problem situation are relevant).
- Modeling improvements (multi-objective models, constraint programming, combinatorial models).
- Technical improvements (the technique used by the algorithm, i.e., the programming language or the hardware platform).
- Skills improvements (better training of planners in using the tools).
- Social improvements (improve acceptance of outcome, reduce resistance).

These attempts can be characterized according to the scheme of Mitroff et al. (1974) (see Fig. 12.1). Problem structuring improvements focus on stage 1. Modeling improvements focus on stage 2. Technical improvements focus on stage 3. Skills and Social improvements focus on stage 4.

12.2.2.1 Problem Structuring

Many attempts to improve problem structuring were initiated after the criticism of Ackoff (1978, 1979) on the developments in the Operations Research community appeared. Flood and Jackson (1991) constructed a system of systems methodologies, that gives much attention to problem identification in terms of systems and participants, and describe various approaches, such as Operations Research, Interactive planning, Soft Systems Methodology, and Critical Systems Heuristics. Note that their grouping of these methodologies is based on the (often implicit) assumptions that the methodologies make about the problem context.

Examples of such assumptions might concern the:

- Moment that information will be available for problem solving.
- Moment that all relevant stakeholders are identified.
- Presence of ambiguity in the interpretation of problem and its relevant aspects.

- Role of ratio, politics, and power in defining a problem and accepting a solution.

From this list it will be clear that efforts to improve problem structuring are strongly related to reconsidering traditional assumptions with respect to the problem and its context.

12.2.2.2 Model Building

Improvements in model building during the last decade have shown a huge increase in new methods that better fit with other representations of a problem. Traditionally, Operations Research has focused on optimizing single objective deterministic problems. Heuristics were considered to be of limited value, at least from a scientific point of view. However, many engineers and computer scientists have started working at other solution approaches, such as Constraint Programming (Baptiste et al. 2001), Simulated Annealing (Laarhoven and Aarts 1987), Tabu Search, Neural Networks, Genetic Programming (Pham and Karaboga 1998, Gen and Cheng 2001), and Evolutionary Algorithms (Coello Coello 2006). Characteristics of the problems for which these approaches are suitable are:

- Satisfying goals instead of optimizing
- Large problem spaces suitable for mimicking search patterns that are found in nature

Other improvements have focused on multi-objective instead of single-objective problem solving. In a pluralistic or even coercive problem context, various stakeholders will have different goals and objectives. These objectives might even be contradictive. In order to take multiple objectives into account in a (mathematical) model, several possibilities have been explored. First, a weighted function of the various objectives could be used. However, this raises the problem of scaling, weight parameter setting, and interpretation of the solution. Next, the notion of dominance of solution was introduced. The concept of Pareto optimality as a dominance criterion has been proposed most frequently (Coello Coello 2006). The idea of this concept is that a solution is not dominated until another solution has been found that improves the performance on all objectives that have to be taken into account. The set of non-dominated solutions is therefore in practice considerably large.

Finally, many improvements have been realized in the field of non-deterministic problem solving, although still a huge number of papers that is being published mainly focuses on deterministic problems. An important characteristic of non-deterministic problems is that the situation at the moment of taking the decision is not assumed to be known in advance. The future is therefore modeled as uncertain. Based on this incomplete knowledge, still a good solution has to be proposed. The question is what will be considered a good solution. Literature has introduced the notion of risk preference of the decision maker in order to cope with this issue. The goodness of fit of a solution depends on the preference for risk of the decision maker/problem owner. Some try to avoid risks, others focus on management of risks, and some even seek risks. Dorfman (2007) distinguishes four basic positions: tolerate, treat, terminate, or transfer risks. Traditionally, models have mainly focused on risk avoidance. New directions are to handle other risk mitigation strategies as well.

Using a model should be possible not only because of the foreseeable future at the moment of decision making/selecting a solution, but also because of the time available for building a model and finding a solution. If the time available is too small to do both, one can either decrease the time for model building or for model solving. Recent developments in Advanced Planning Systems show that it is sometimes possible to reduce the time needed for model building by using intelligent software that enable the use of standard OR techniques (Pochet and Wolsey 2006).

12.2.2.3 Model Solving

Several attempts have been made in order to improve the process of solving a model (step 3 of Fig. 12.1). The efforts invested in finding algorithms with a lower time-complexity for some well-defined problems are still huge. The same holds true for the improvements in the speed of computers, programming languages, parallel processing, et cetera. These improvements are strongly related to the fields of Informatics and Computer Science.

From a cognitive perspective, Chap. 14 discusses possible roles of human in solving models for multi-actor problem situations, where humans co-operate with machines (i.e., combinations of hard and software) in order to make a schedule. Various tasks need to be accomplished by both human and machine in order to make a good schedule. Some of these tasks can be performed without interaction with another actor, some need to be performed in co-operation, because of strong interdependencies and/or a high impact on the quality or acceptability of the resulting plan. This issue will be explored further in Sect. 12.3.

12.2.2.4 Implementation

Finally, large investments have been made in improving skills of problem solvers/ decision makers. Planners, for example, have received more training in using computerized planning systems as well as the organizational context of planning problems. Due to the introduction of participative techniques in problem structuring, stakeholders are involved earlier, which might affect the acceptance of solutions. Nevertheless, many people think that a problem is solved at the moment the model has identified a solution. Such a limited view of problem solving will stay a stumbling block on the road to improving problem solving.

12.3 Planning and Scheduling from a Problem Solving Perspective

There are two approaches to planning from a problem solving perspective. First, any problem solving activity consists of a number of reasoning steps. The order of these steps is not always prescribed by a clear procedure, and planning is one way to order the activities within a problem solving cycle. Second, planning can be the problem to solve. As mentioned in the previous section, it is this kind of planning that we analyze in this chapter. This section will further define this second kind of planning.

An important feature of planning is that it is about choosing one alternative out of a huge number of alternatives that are structurally similar. Examples are routing trains, making a production schedule, making a staff schedule and determining the trajectory of an automatic vehicle. Each of these examples concerns choosing one out of a number of similar alternatives of future states. With this definition, planning problems can be modeled as follows:

A planning problem consists of groups of entities, whereby the entities from different groups must be assigned to each other. The assignments are subject to constraints, and alternatives can be compared on their level of goal realization.

For example, production scheduling is a problem where orders must be assigned to machines, in a shift schedule people are assigned to shifts, and in task planning tasks are assigned to time slots and resources. Now we can also specify what we mean by "structurally similar"; it means that plan alternatives have the same structure (e.g., orders are assigned to machines), but a different content (e.g., in plan alternative A, "order 1" is assigned to "machine 1", and in plan alternative B, "order 1" is assigned to "machine 2"). This definition also precludes some areas that are commonly regarded as planning, for example strategic planning and retirement planning. Although the boundaries are debatable, in such planning problems alternatives are not structurally similar.

An important aspect of planning is decomposition, either because of complexity or uncertainty. If decisions at hierarchical levels are distinctive, then there should also be planning models at multiple hierarchical levels. Most apparently, this is the case in organizational planning, where a planning department can be said to make a plan, but where individual human planners make the sub-plans.

Two types of sub-plans can be distinguished in a planning hierarchy: aggregation and decomposition. In aggregation, the dimensions that exist in the planning problem stay the same, but individual entities of a dimension are grouped. For example, a plan that contains the assignment of individual orders to production lines for a certain week can be aggregated to a plan that contains the assignment of orders per product type to production lines in that week. Aggregation can be used to establish boundaries or constraints for individual assignments of entities that fall within an aggregated group. For example, it is first decided how much caramel custard will be made next week. Then individual orders that fall in this product family can be assigned to a specific production time. In this way, several stages of aggregation can be sequentially followed, each stage creating boundaries for the next stage.

In the second type of sub-plan, decomposition, a subset of the entities that must be planned is considered as a separate planning problem. Decomposition can deal with all entities of a subset of the dimensions, all dimensions with a subset of the entities, or a combination of subsets of dimensions and entities. For example, if we attune orders, machines, and operators, we could first assign orders to machines and then operators to the chosen order/machine combinations. Or, we can first assign all customer orders, after which we assign all stock orders.

Planning problems as defined above can be modeled with object oriented techniques. Groups of similar entities are classes, and the entities themselves are object instances. An example of an object model of a nurse schedule is depicted in Fig. 12.2. Objects are either singular (nurse, starting time, and ending time) or composed of other objects, i.e., combination objects (shift, scheduled shift, and schedule). The nurse schedule example consists of a number of scheduled shifts. A scheduled shift links a particular shift to a nurse. Shifts themselves are composed of a starting time and an ending time. Assignments in the schedule are depicted by instances of objects. A constraint in the model in Fig. 12.2 could be that student nurses may not work in the night shift. This is a constraint at the level of the scheduled shift because it can be checked only when nurses are assigned to shifts. A goal could be the minimization of the deviation of worked hours and the number of hours in the contract of the nurse. This goal is at the level of the "schedule" object because it is about collections of scheduled shifts. Of course, a domain model can contain several constraints and goals.

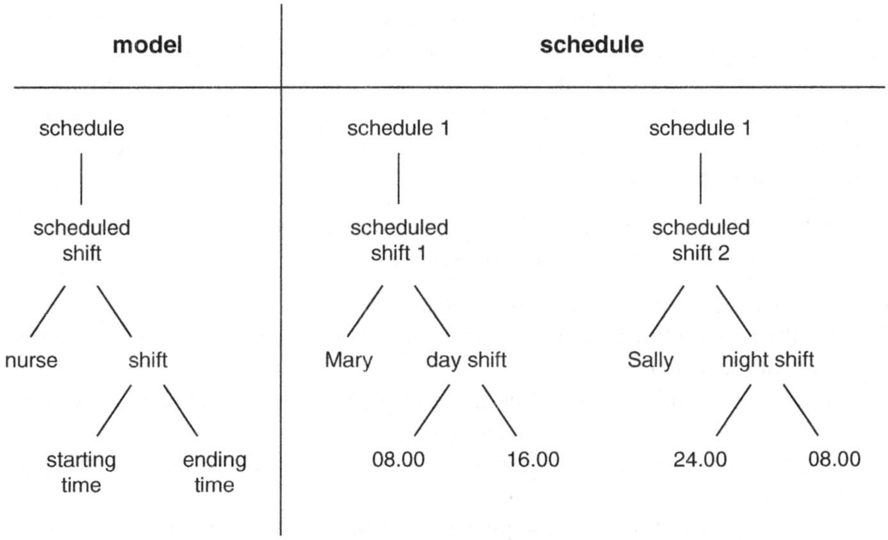

Fig. 12.2 Example of an object model of a nurse schedule

This modeling technique matches the traditional problem solving approach as described in the previous section. Both the cognitive perspective on planning and the algorithmic approach can be described by the way in which the objects are structured, aggregated, reasoned with, and assigned.

Further extending this way of looking upon planning and scheduling, Wezel and Jorna (2006) analyze recurring issues and elements in planning. A planning problem does not exist as such. It is embedded within its environment and it is subject to perception by the one who plans and the one who must execute the plan. Furthermore, there are characteristics that typify the kind of method used to solve the planning problem. The following overview specifies the characteristics of these elements (Wezel and Jorna 2006: 17):

Planning environment. Different planning approaches have different environments. Even within approaches different environments can be distinguished. Some characteristics are:

1. The predictability of the environment (is the plan executed in a closed world in which the presumptions will not change or is it executed in an open world?).
2. Kinds of events that trigger planning, for example:
   - Time based: e.g., a plan needs to be made each week.
   - Event based: a plan must be made after an event, for example a rush order in a factory.
   - Disturbance based: a plan must be adjusted because a disturbance occurs that renders the plan invalid, for example a shop that is closed when I do my shopping.
3. Goals and constraints. The goals and constraints are often determined by the planning environment. They can be about the plan itself, e.g., the time at which the actions that are specified in the plan must be finished or the amount of resources that are used. They can also be about the process of making the plan, e.g., when the plan must be available or how much people may be involved in making the plan. Goals and constraints can be about, e.g., time, materials, remaining life-span of the system, energy, the degree of fault-tolerance, money, capacity usage, etc. Making the constraints and goals explicit is often the hard part in planning.

Planned entities. The actions in the plan must be performed by someone or something. Several aspects are important, e.g.,

1. Is the planned entity the same as the planning entity?
2. Does the plan deal with actions of individuals or actions that are performed by groups of individuals?
3. Is the planned entity a natural entity (e.g., human) or an artificial entity (e.g., a machine, a robot, or a computer program)?
4. Does the planned entity possess intelligence itself, i.e., can it interpret the plan and change it if necessary?
5. What kind of constraints do the planned entities impose on the plan?
6. Does the planned entity use scarce resources that also have to be planned?

Planning entities. Someone or something must make the plan, i.e., search for alternative plans and choose one. Important aspects of the planning entity are:

1. Does the planning entity execute the plan itself or is the plan executed by others?
2. Is it a natural entity (human) or is it an artificial entity (computer program)?
3. What kind of planning methods does it use?
4. What is the planning strategy, i.e., how does it choose an appropriate planning method?
5. What kind of information processing mechanism does it have?
6. What are the architectural components?
7. What kinds of representations does it use?
8. How does it communicate?
9. How does it coordinate with other planning entities and with planned entities?

Plan. The plan itself is the specification of future actions. A plan that contains explicit temporal assignments on an interval or ratio level of measurement is called a schedule.

1. Horizon: what time span does the plan cover?
2. Frequency: how often is the plan created or adapted?
3. Level of detail: Does the plan need more detail in order to be executed? Does the executing entity have to fill in the details, or is the plan used as a template for another planner?
4. Structure: what is actually planned, e.g., human actions, machines, time, locations, vehicles, movements, etc.
5. (Re)presentation: how is the plan represented or depicted? Does it specify the end-state, or does it provide a process description that leads to the end state?

Planning methods. The planning method depicts the decision process of the planning entity. A planning entity can have multiple planning methods to choose from. Some generic issues with regard to planning methods are:

1. How does the planning entity deal with combinatorics, for example:
   - Plan partitioning: divide the plan in multiple sub-plans and treat the sub-plans independently.
   - Multi-resolutional planning: make a plan with less detail (and less complexity), and use that plan as a template for a plan with more detail (at a higher level of resolution).
   - Learning: use (and possibly adapt) a previously found solution for a problem that was equal or similar.
   - Opportunistic planning: apply the first feasible solution that is found without looking whether there are better solutions (e.g., when planning under strict time constraints).
2. How much does the use of a planning method cost? Methods can for example be costly in terms of the information processing capacity that is needed, or in the tools that are used, or in the throughput time that is needed.

3. What is the starting point? For example:

   – An empty plan
   – An existing plan that must be supplemented
   – An existing plan with errors that must be corrected
   – A previous plan that can be used as a template

4. How are conflicts during the solution process solved (i.e., when the planning entity gets stuck)?

   – Backtracking
   – Repair

5. Adjustment of constraints to make the plan valid
6. Does the method search for an optimal solution or state, or does it search for a satisfactory solution?

Many authors distinguish planning from scheduling. Often, the difference is only a matter of abstraction or aggregation level that is been taken into account. Planning is in that case considered to be an activity with a larger distance to the execution or implementation (both in terms of time horizon and level of detail) than scheduling, but from the problem solving perspective they are structurally the same.

The elaborate definition of planning in this section distinguishes planning and scheduling from other problem solving tasks. Linking to the above definition, we can state that algorithm development mainly focuses on developing planning methods to create a time-based plan within a closed world, having a limited model of the planned entities, and not taking into account the planner. In the following sections, this background will be used to critically review the paradigms used in designing scheduling algorithms, and to propose a framework in which these issues are taken into account.

12.4 Designing Algorithms for Scheduling Problems: A Critical Review

In the previous section, a rigorous description of theoretical issues related to the general problem-solving process was provided. This section takes a closer look on the procedure of designing algorithms for the solution of scheduling problems. As discussed in the previous section, the design of a scheduling algorithm is mainly related to stages 2 and 3 of Mitroff's problem-solving model (Fig. 12.1). As it has been observed, scheduling algorithms designed through the traditional production research approach are rarely favored to ad-hoc approaches in realistic industrial environments (Berglund and Karltun 2005). This section examines in detail the traditional approach provides a critical review of its shortcomings. It then proceeds to discuss the characteristics of the scheduling process as it happens in practice, and proposes the development of a new theoretical framework for the design of scheduling decision support systems that explicitly consider the human and organizational factors within a scheduling environment.

12.4.1 Scheduling Environment: The Traditional Production Research View

As it has been discussed by various researchers, the traditional production research approach to the design of scheduling algorithms has rarely made an impact in realistic industrial environments (Portougal and Robb 2000; Fransoo and Wiers 2005), (Berglund and Karltun 2005). In practice, the majority of schedulers assume full control of the scheduling process by employing ad-hoc solution approaches. While custom-built IT decision support systems are occasionally used to support the schedulers' work, only a limited number of fully automated scheduling systems exist. In order to examine the reasons for the lack of practical use for algorithms designed through the traditional production research approach, it is important to examine the nature of the scheduling process as it is assumed by such a problem-solving process. The conceptual model, illustrated in Fig. 12.3, provides a basis for this discussion.

In general, the traditional production research view of the scheduling environment assumes that a scheduling task defined by a rigorous mathematical model has to be implemented at a specific moment in time, or at well-defined intervals (Meredith 2001). This view magnifies the importance of the scheduling algorithm

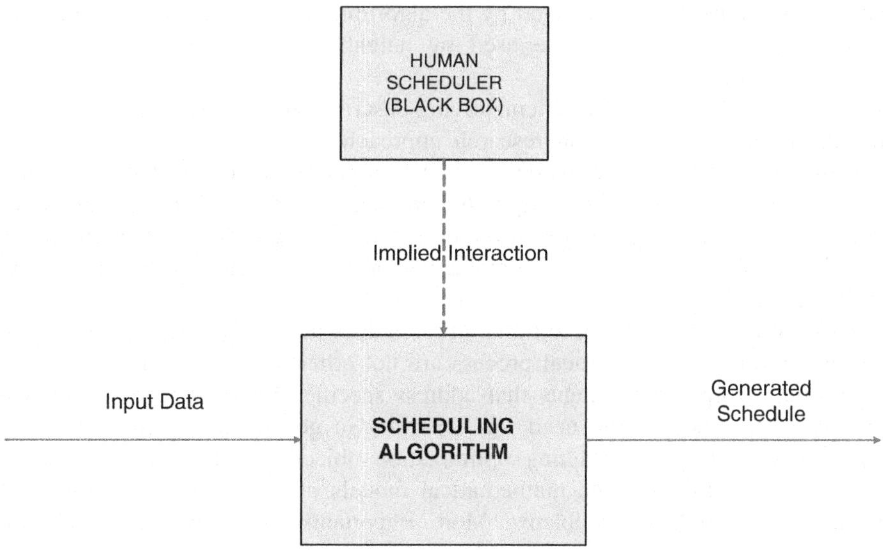

Fig. 12.3 The technological view of the scheduling environment

within the scheduling environment and degrades the implications of human and organizational considerations (as depicted in Fig. 12.3).

In terms of the cognitive aspects of the process, the human scheduler is considered a black box in the scheduling environment. Her/his participation in the overall process is implied, but is not explicitly considered in terms of the mathematical model. In principle, the presence of a human scheduler is not even required, since there is no reference to the source of the scheduling information that is used as input data by the scheduling algorithm. As a result, the interaction between the human scheduler and the scheduling algorithm is given no special consideration. The design of the Human Computer Interaction (HCI) environment for the algorithm is based on the skills and the intuition of programmers, rather than the use of an appropriate scientific methodology.

The organizational structure of the scheduling environment is modeled as an automated flow line process. The necessary scheduling information is provided to the scheduling algorithm by unknown sources, however, this information is always considered to be in the required format as well as timely and accurate. The algorithm generates schedules in a format that is assumed to be meaningful and understandable by all parties who receive them. Manual or automatic editing of generated schedules is not considered to be part of the decision-making process. The possible existence of organizational structures within the scheduling environment, such as a team of human schedulers, and their relationship with neighboring organizational structures such as the planning department and the shop-floor environment are also excluded from consideration.

The IT infrastructure that supports the scheduling process is considered to be simple and is based on the existence of an autonomous computing facility required for the execution of the algorithm. The technical specifications of the electronic data which are used and generated by the algorithm are assumed to be consistent with the specifications that are used by neighboring departments within the manufacturing environment.

Given Mitroff's view of problem-solving described in Sect. 12.2, it can be said that the traditional production research approach to scheduling starts with the construction of the scientific model (step III), implicitly assuming that a standard solution can be applied successfully to the problem considered. Thus, the problem-solving process reduces to the development of an algorithm that attempts to optimize an objective for the rigid mathematical representation of the scheduling environment.

This critical view of the traditional process does not imply that the algorithms designed through the traditional process are not efficient. In fact, many of these algorithms incorporate elements that address specific human and organizational considerations, such as the need for flexibility in generated schedules and the existence of multiple conflicting optimization objectives. However, it is very difficult to develop accurate mathematical models of the problem environment for practical scheduling problems. More importantly, there exists a trade-off between the accuracy of the developed mathematical model and the computational complexity of the solution algorithm. Since the development of mathematical models that explicitly consider all human and organizational factors in a scheduling environment is a difficult (if not impossible) task, the implemented algorithms are rarely used in practice, at least in the manner envisaged by their designers.

The question that naturally arises from the observations of the previous sections is the following: How can we improve the problem-solving process of scheduling problems in a way that will provide useful support to the human scheduler in realistic production environments? In order to answer this question it is necessary to examine the scheduling environment from a realistic perspective.

12.4.2 Scheduling Environment: The Realistic View

In contrast to the view of the traditional production research approach, in a realistic industrial environment the scheduling process is a complex interpersonal and interdepartmental process that takes place dynamically over time (Fig. 12.4).

The human scheduler resides at the heart of the scheduling environment, since s/he processes or communicates the majority of scheduling information needed for the implementation of the process. The human scheduler generates schedules through a cognitive process which can be assisted by the existence of decision support tools. These tools are not necessarily in a software form. If the implementation of the scheduling process requires the cooperation of a group of schedulers,

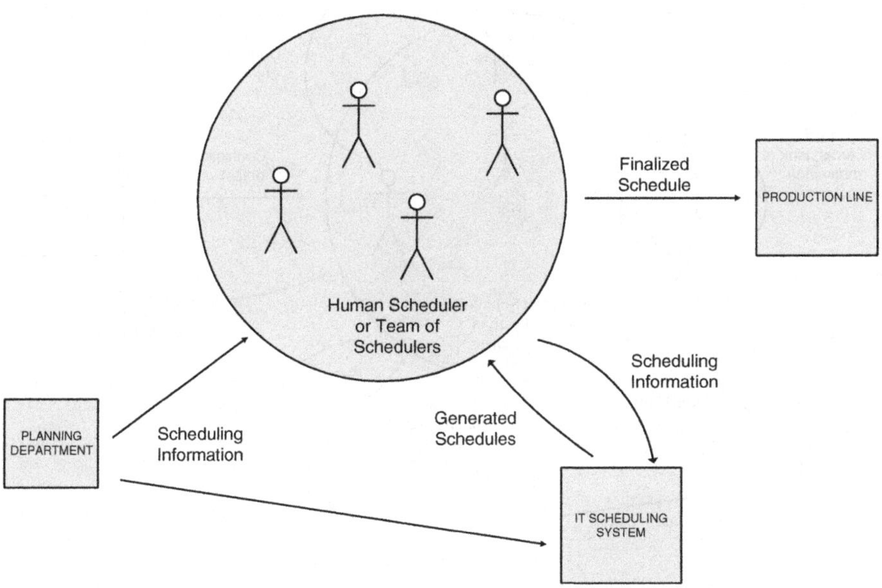

Fig. 12.4 The realistic view of the scheduling environment

rather than the skills of a single human operator, the complexity of the overall process increases further.

From an organizational point of view, scheduling information flows in the environment from various sources (employees, departments, other IT systems) and in various forms (verbal communication, paperwork, electronic data). The scheduling algorithm is a part of the software system that provides support in the decision making process. It receives input information either from the scheduler(s) or from other software systems and database facilities. The algorithm generates schedules based on the scheduling information provided; however, these schedules are not necessarily accepted by the human scheduler or group of schedulers. The finalized schedule might be the result of a manual editing process that takes place based on personal experience, group experience, departmental negotiations and unexpected events.

The conceptual model depicted in Fig. 12.4 provides a more realistic representation of the complexity of the scheduling process in relation to the traditional production research view. Still, it does not portray the dynamic human and organizational factors that can seriously disturb its implementation. Some of these factors are illustrated in Fig. 12.5.

Since the human scheduler plays the central role in the overall process, his/her personal skills as well as his/her physical and psychological condition are of great importance to the smooth implementation of the decision making process.

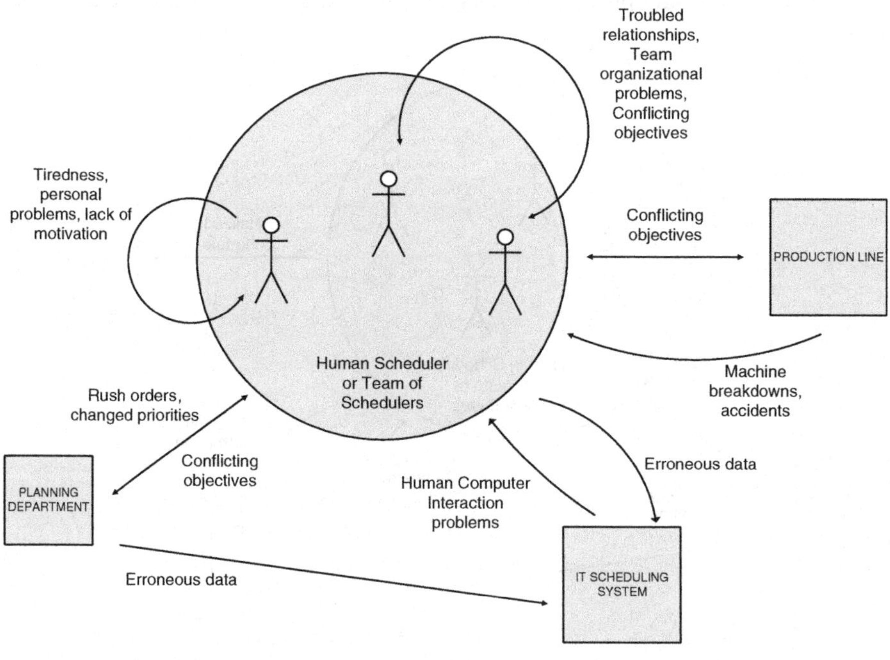

Fig. 12.5 Human and organizational "noise" in the scheduling environment

Inadequate training, tiredness, personal problems, poor communicational skills, limited IT skills, inability to handle pressure and lack of motivation are just some of the factors that might have on effect on the cognitive process of generating, editing and deciding on the suitability of schedules.

This cognitive process must also take in account the existence of various objectives within the scheduling environment, which may have conflicting nature. The term "objective" is used here not only as a description of a rigorously defined mathematical function that needs to be optimized, but in the more general sense. These objectives can be attributed to various sources (scheduler's objectives, departmental objectives, managerial objectives) and they can have different time horizons (short, medium and long-term objectives). As a result, it is extremely difficult to accommodate them all within the context of a typical algorithmic design. The process of generating schedules is heavily influenced by the existence and the prioritization of scheduling objectives within the production environment. Especially the existence of contradicting high-level objectives of the departments can lead to tensions and generate mistrust on the priority of jobs that need to be scheduled.

Another factor that plays a crucial role on the successful implementation of the cognitive process is the design of the Human–Computer Interaction (HCI) environment between the scheduler and the software-based decision support system. An HCI design that is not based on a rigid design methodology can lead to misunderstandings, erroneous decisions, and most importantly distrust by the human operators.

From an organizational point of view, the flow of scheduling information between all parties involved is not necessarily timely or accurate, since, as explained earlier, this information originates from various sources and can take different forms. The existence of scheduling data in both handwritten/printed and electronic format can generate significant problems since the human operators will have to provide this information to the decision support system manually. Erroneous data input will generate erroneous schedules and will subsequently lead to poor decision making.

These observations leads us back to the question asked in Sect. 12.4.1: how can we improve the problem-solving process of scheduling problems in a way that will provide useful support to the human scheduler in realistic production environments? Although there cannot be a simple or easy answer to this question, a potential approach to a realistic design methodology is described in the following section.

12.4.3 The Way Forward: A Framework for Designing Scheduling Algorithms with Human and Organizational Considerations

The complexity of the scheduling process as described in the previous sections does not imply that the design of a scheduling algorithm is an unnecessary or impossible task. It indicates though that if these algorithms are to be employed within the context of a decision support system in realistic industrial environments, their design should be based on a new approach.

The development of this new approach can benefit from the analysis of the scheduling process that was described in Sects. 12.4.1 and 12.4.2. This analysis established the following facts:

- Each scheduling problem is unique and has its own particular human and organizational characteristics
- The development of a mathematical model that incorporates all human and organizational characteristics of the scheduling process is most of the times not possible or even necessary (due to the mathematical complexity of the resulting model)
- A scheduling algorithm is only a part of an overall decision support system that will be used by a human scheduler

What has become evident from the analysis is that there is a need for a robust development process which will drive not only the design of the scheduling algorithm, but also the design of the overall decision support system that will function within the scheduling environment. We propose the development of a theoretical framework for the design of scheduling Decision Support Systems (DSS) that will explicitly consider human and organizational considerations of the scheduling case considered. The proposed framework (Fig. 12.6) contains the following information:

- A categorization of cognitive, organizational and technical attributes of production scheduling processes.
- A categorization of cognitive, organizational and technical characteristics explicitly considered by existing scheduling algorithms.

- o A categorization of cognitive, organizational and technical attributes of production scheduling processes.
- o A categorization of cognitive, organizational and technical characteristics explicitly considered by existing scheduling algorithms.
- o Recommendations on the type of core algorithms that should be employed for the design of a production scheduling decision support system based on the cognitive, organizational and technical characteristics of the scheduling case considered.
- o Guidelines on the design of new algorithms (or the modification of existing ones) that will address cognitive, organizational, and technical considerations not currently handled by existing scheduling algorithms
- o Guidelines on the design of interactive computing environments (graphical user interfaces) that will address the cognitive, organizational and technical considerations of the human operator while performing the required scheduling tasks.

Fig. 12.6 Framework for design of scheduling Decision Support System

- Recommendations on the type of core algorithms that should be employed for the design of a production scheduling decision support system based on the cognitive, organizational and technical characteristics of the scheduling case considered.
- Guidelines on the design of new algorithms (or the modification of existing ones) that will address cognitive, organizational, and technical considerations not currently handled by existing scheduling algorithms
- Guidelines on the design of interactive computing environments (graphical user interfaces) that will address the cognitive, organizational and technical considerations of the human operator while performing the required scheduling tasks.

The use of this framework implies that the design of a decision support system should start with an examination of the human, organizational, and technical characteristics of the scheduling case considered. This information can be obtained and analyzed using appropriate methodologies such as interviews, observations, context diagrams, data flow diagrams, as well as hierarchical and cognitive task analysis. The designer of the decision support system can then match the results of the analysis to the guidelines of the theoretical framework and proceed accordingly. This process might lead to the use of an existing general-purpose algorithm with a very simple user interface, but may also lead to the design of a new complex case-based algorithm with high-end graphical user interface provisions, depending on the analysis of the scheduling case considered.

12.5 Conclusions

The design of a scheduling algorithm is a complex process that should not be solely based on the development of a scientific model for the scheduling problem considered. Human and organizational characteristics of the scheduling environment play a significant role on the practical adoption of a scheduling decision support tool.

This chapter discussed in detail the process of designing algorithms for the solution of scheduling problems. A rigid description of the general problem-solving process was initially provided with references to Mitroff's model as well to models from the field of psychology. This was followed by a description of the alternative types of algorithms that exist and their corresponding characteristics. It concluded with an analysis of the possible improvements that can be applied to each step of the problem-solving process, as this is conceived by Mitroff. Next, attention has been given to differences between planning and scheduling. Using an object-oriented description, we have been able to distinguish between both types of activities.

The second part of the chapter focused on the scheduling problem-solving process. It discussed and reviewed the traditional production research algorithmic design process using as a tool the model of the scheduling environment employed by such processes. It then contrasted this model with the realistic view of the scheduling environment as it happens in practice pointing out the important human and organizational considerations which are not addressed by the traditional production research design approach.

Based on the findings of this analysis, a possible development of a theoretical framework which can be used on a case-by-case basis for the design of scheduling decision support tools has been discussed. This framework will provide guidelines on both the design of scheduling algorithms and their associated graphical user interface. The aim is to develop not just scheduling algorithms, but integrated support tools that will address human, organizational, as well as technological considerations of the scheduling environment considered.

As it is obvious, the development of such a framework requires the completion of large-scale psychological, organizational and technological studies that will provide the respective framework information. The authors of this chapter aim to work towards the development of such a framework.

References

Ackoff, R. L. (1978). *The art of problem solving*. New York: Wiley.

Ackoff, R. L. (1979). The future of operational research is past. *Journal of the Operational Research Society, 30*(2), 93–104.

Baptiste, P., Le Pape, C., & Nuijten, W. (2001). *Constraint-based scheduling: Applying constraint programming to scheduling problems* (International Series in Operations Research & Management Science, Vol. 39). Heidelberg: Springer.

Berglund, M., & Karltun, J. (2005). Human, technological and organizational aspects influencing the production scheduling process. *Proceedings of the International Conference of Production Research (ICPR '05)*, Salerno, Italy.

Coello Coello, C. A. (2006). Evolutionary multi-objective optimization: A historical view of the field. *IEEE Computational Intelligence Magazine, 2006*, 28–36.

Dorfman, M. S. (2007). *Introduction to risk management and insurance* (9th ed.). Englewood Cliffs, NJ: Prentice Hall.

Flood, R. L., & Jackson, M. C. (1991). *Creative problem solving: Total systems intervention*. Chichester: Wiley.

Fransoo, J.C., & Wiers, V.C.S. (2005). Production planning and actual decisions: an empirical study. *Proceedings of the International Conference of Production Research (ICPR '05)*, Salerno, Italy

Garey, M., & Johnson, D. (1979). *Computers and intractability: A guide to the theory of NP-completeness*. San Francisco: W.H.Freeman.

Garey, M. R., Johnson, D. S., & Sethi, R. (1976). The complexity of flowshop and jobshop scheduling. *Mathematics of Operations Research, 1*(2), 117–129.

Gen, M., & Cheng, R. (2001). *Genetic algorithms and engineering optimization (Engineering design and automation)*. New York: Wiley.

Hayes-Roth, B., & Hayes-Roth, F. (1979). A cognitive model of planning. *Cognitive Science, 3*, 275–310.

Hoc, J.-M. (1988). *Cognitive psychology of planning*. London: Academic.

Johnson, S. M. (1954). Optimal two-and-three-stage production schedules with set-up times included. *Naval Research Logistics Quarterly, 1*, 61–68.

Jung, C. G. (1976). Psychological types. In A. Gerhard & R. F. C. Hull (Eds.), *Collected works of C.G. Jung* (Vol. 6). NJ: Princeton University Press.

Laarhoven, P. J. M., & Aarts, E. J. L. (1987). *Simulated annealing: Theory and applications*. Norwell, MA: Kluwer Academic.

Meredith, J. R. (2001). Reconsidering the philosophical basis of OR/MS. *Operations Research, 49* (3), 325–333.

Miller, G. A., Galanter, E., & Pribram, K. H. (1960). *Plans and the structure of behavior*. New York: Holt, Rinehart and Winston.

Mitroff, I. I., Betz, F., Pondy, L. R., & Sagasti, F. (1974). On managing science in the systems age: Two schemas for the study of science as a whole systems phenomenon. *Interfaces, 4*(3), 46–58.

Newell, A., Shaw, J. C., & Simon, H. A. (1958). Elements of a theory of human problem solving. *Psychological Review, 65*, 151–166.

Newell, A., & Simon, H. A. (1972). *Human problem solving*. Englewood Cliffs, NJ: Prentice Hall.

Ormerod, T. C. (2005). Planning and ill-defined problems. In R. Morris & G. Ward (Eds.), *The cognitive psychology of planning* (pp. 53–70). Hove: Psychology Press.

Pham, D. T., & Karaboga, D. (1998). *Intelligent optimisation techniques: Genetic algorithms, Tabu search, simulated annealing and neural networks*. New York: Springer.

Pochet, Y., & Wolsey, L. A. (2006). *Production planning by mixed integer programming* (Springer series in Operations Research and Financial Engineering). Heidelberg: Springer.

Portougal, V., & Robb, D. J. (2000). Production scheduling theory: just where is it applicable? *Interfaces, 30*(6), 64–76.

Simon, H. A. (1981). *The sciences of the artificial*. Cambridge: MIT.

Slack, N., Chambers, S., & Johnston, R. (2004). *Operations management*. London: Prentice Hall.

Van Wezel, W. M. C., & Jorna, R. (2006). Chapter 1 Introduction. In W. M. C. van Wezel, R. Jorna, & A. Meystel (Eds.), *Planning in intelligent systems: Aspects, motivations and methods* (Wiley Series on Intelligent Systems, pp. 1–22). New York: Wiley.

Van Wezel, W. M. C. (2001). *Tasks, hierarchies, and flexibility; planning in food processing industries*, PhD Thesis, University of Groningen, The Netherlands.

Chapter 13
A Comparison of Task Analysis Methods for Planning and Scheduling

Julien Cegarra and Wout van Wezel

Abstract Planning and scheduling experts in practice are often faced with the question of how a company can improve its planning performance. Such improvements can be related to, for example, computer support, organizational task division, performance analysis, etc. The multitude of planning and scheduling factors and their interrelatedness makes it difficult to integrally explain current performance and assess the consequences of changes. We analyze how different perspective on task analysis methods complement each other for the various questions that planning and scheduling experts encounter in practice. There are two main findings. On the one hand, a combination of methods is often necessary in order to avoid myopia and biased results. On the other hand, however, the analysis shows that not all questions require a full-scale analysis of the situation.

13.1 Introduction

Task analysis is the keystone in the practice of various experts such as ergonomists, psychologists, organizational designers, engineers, etc. Different variants of task analysis methods exist (for a review see Diaper and Stanton 2004). Some of the most well-known are *Hierarchical Task Analysis* (Annett and Duncan 1967; Annett 2000), *Cognitive Task Analysis* (Schraagen et al. 2000a), *Cognitive Work Analysis* and its *Work Domain Analysis* (Rasmussen et al. 1994; Vicente 1999a). Other methods of interest are *Cognitive Function Analysis* (Boy 1998) and *Scenario-Based Task Analysis* (Carroll 1995). All these methods allow experts to undertake

J. Cegarra (✉)
Université de Toulouse, Toulousse, France
e-mail: julien.cegarra@univ-jfc.fr

W. van Wezel
Faculty of Economics and Business, University of Groningen, Groningen, The Netherlands
e-mail: w.m.c.van.wezel@rug.nl

J.C. Fransoo et al. (eds.), *Behavioral Operations in Planning and Scheduling*,
DOI 10.1007/978-3-642-13382-4_13,

detailed studies in order to put together a model of work situations, which allows for improvements in an efficient, economical, and safe direction.

The various experts focus on different objectives, for example, designing the organizational structure, designing individual tasks, training employees, designing support tools, evaluating performance of human individuals and so on. Therefore, the appropriate method for the task analysis may depend on the objective and of the situation under consideration. Our question is, how do these methods cope with the different objectives in planning and scheduling such as evaluating human scheduling performance, training human schedulers, or determining task requirements?

This chapter will assess this question in order to facilitate interventions of experts in planning and scheduling situations. Several methods relevant to perform tasks analyses of scheduling are described in literature (Crawford et al. 1999; Higgins 1999; Usher and Kaber 2000), but comparisons of task analysis methods are not abundant. When they are done, the focus is on their respective characteristics and not on their relevance to a specific domain of application (see for instance, Miller and Vicente 2001). In order to relate the different objectives of the experts to the characteristics of the task analysis methods, we will refer to two main categorizations:

- Rasmussen's (1997) distinction of normative, descriptive and formative perspectives for modeling the scheduling situation. The selected perspective defines several properties of the produced model for the work situation under consideration.
- Vicente's (1999b) characterization of the models of scheduling situation, referring to three dimensions: device dependency, event dependency, and psychological relevance. The goals of the experts intervening in planning and scheduling situations may also be discussed in relation with these three dimensions.

We will refer to these two categorizations in the next sections to map the suitability of different task analysis frameworks to the various questions that scheduling experts face in practice.

13.2 Normative, Descriptive and Formative Perspectives of Scheduling Situations

As previously noted, there are different methods to handle task analyses. These methods may be used in different ways, with different goals or data collection techniques. Various methods may also share the same goal or a common point of view. Therefore, in order to compare task analysis methods, we refer to Rasmussen's (1997) distinction of modeling methods on the basis of three different perspectives:

- A normative perspective prescribes how the task should be done. It typically produces a list of sub-tasks the user must perform to complete the main task.

- A descriptive perspective describes how the task is currently performed in practice. It generally focuses on modeling the knowledge, thought processes and goals that underlie the task effectively done by the human.
- A formative perspective specifies an exhaustive description of the work domain and its physical, functional interrelations. As we will mention in the next section, this perspective facilitates the identification of crucial tasks in a novel situation.

We will present examples of these three perspectives in a scheduling context. We will focus on the perspective and not on the task analysis method as a task analysis may be performed with different perspectives.

13.2.1 A Normative Perspective: Usher and Kaber's (2000) Study

Most literature on planning and scheduling is in Operations Research.l In this field, many algorithms and heuristics have been developed that will automatically assign orders to machines and determine the starting and ending times of operations. These techniques create schedules by making detailed analyses of the objects to be scheduled and their constraints. And although algorithms have their use in the planning and scheduling processes in organizations, their scope is too limited to provide comprehensive solutions for scheduling. They represent a normative perspective because they focus on the prescribed tasks. Higgins (1999) introduces the word perplexity to describe that scheduling is more than a complex combinatorial assignment problem. All kinds of aspects beyond the formal model of the planned objects disturb the idea that scheduling is solving a neatly defined problem. Examples are uncertainty, confusion, politics, unexpected events, organizational conflicts, etc. Several authors stress this notion as the gap between theory and practice in scheduling (e.g., Mckay and Wiers 1999; McKay et al. 1988, Halsall et al. 1994; Dessouky et al. 1995; Buxey 1989).

Usher and Kaber (2000)'s study illustrates this normative perspective in a production scheduling problem. Starting from Martensson's study (1996), they tried to determine the overall goal, subgoals and objectives (i.e., sub-subgoal) of human scheduling. They identified three levels: the main goal is to meet production demand (customers' orders). On the second level, this goal could be accomplished by scheduling jobs to meet due dates, to avoid bottlenecks in the production, to start the most critical orders first, and finally to deal with disturbances to facilitate the schedule. On the third level, another decomposition level is specified, for example to avoid bottlenecks requires to both identify and resolve capacity bottlenecks and so on. Usually such analysis is presented using a graph that emphasizes the hierarchical relation between the goals and subgoals (Fig. 13.1).[1]

[1]The reader could refer to Usher and Kaber's (2000) article to access the complete task decomposition that lists the 37 end nodes of this tree diagram, listing all the individual sub-objectives (4th level) the scheduler has to complete.

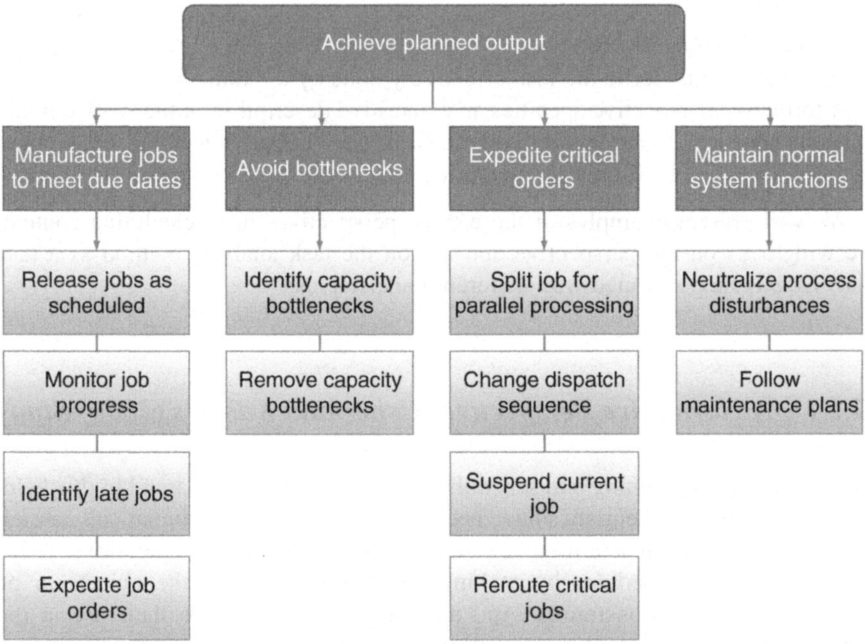

Fig. 13.1 A tree diagram from task decomposition in the industrial scheduling domain (from Usher and Kaber 2000). Meaning of the boxes and arrows is discussed in the text

To analyze a situation an expert will produce some "intuitive" representations, which the hierarchical representation of task decomposition is claimed to be (Paternò 2004). This representation allows the designer both to consider the hierarchical relationships between the components and to make a good representation of the complexity under consideration.

When one has to contribute to the design of support tools, such decomposition may be not detailed enough. As stated by Diaper (2004, p.11): "it is far less clear that the natural world and our social worlds are arranged hierarchically". Indeed, in Usher and Kaber's (2000) decomposition, the organization of units could convey different meanings. For example the top down order of objectives could indicate concurrent tasks ("release jobs as scheduled", "monitor job progress"); they could also illustrate precedence relations ("identify capacity bottlenecks", "resolve capacity bottlenecks"). Then, if the existing links are not sufficiently detailed, several missing links also need attention. For example, the authors consider that the scheduler has to resolve capacity bottlenecks (row 2) in order to operate at full capacity. But this could be in contradiction with the goal to neutralize process disturbances (row 4) because operating at full capacity does not allow many degrees of freedom to manage a disturbance. So this method imposes a hierarchy structure that could be inadequate to precisely understand the task as really performed.

To overcome the problem that the actual task is not fully taken into account during the task analysis, descriptive perspectives have since been applied.

13.2.2 A Descriptive Perspective: Crawford et al.'s (1999) Study

To overcome the gap between theory and practice of scheduling, work has been done to identify specific organizational and individual characteristics of scheduling (Sanderson 1989; Van Wezel and Jorna 2001; Crawford and Wiers 2001; Hoc et al. 2004). Not many authors empirically look at planning as an organizational process. Still, some work has been reported in this area. For example, Jackson et al. (2004) state that there are five interdependent factors that determine the context of human planners (the manufacturing process, the organizational structure, information systems, people, and performance measures), McKay and Wiers (2003a) and Van Wezel et al. (2006) describe the different phases in the planning process.

In most traditional planning and scheduling literature, the human individual is disregarded. Empirically looking at human planners to analyze planning and design support seems to be a research niche. The analyses that are available show a multitude of tasks that human planners must perform (e.g., Mietus 1994; McKay and Buzacott 2000; Jackson et al. 2004). Next to seeing structure in planning task performance, several authors have noted individual differences in scheduling strategies within the same situation as well (e.g., Mietus 1994; Wiers 1996; Taatgen 1999; Cowling 2001). Such differences can be attributed to, for example, differences in experience (Bainbridge 1974; Bi and Salvendy 1994; Mietus 1994), the cognitive dimensions of the situation, such as the level of complexity or uncertainty (Cegarra 2008), the department a planner works in (Kiewiet et al. 2005), the problem domain, and culture (Jorna 2006).

Differently from the Usher and Kaber consideration of the hypothetical behavior of the human, the descriptive perspective resorts to detailed studies of human activities to gain insight into the effective tasks. This is the case for example in studies on nurse scheduling (van Wezel et al. 1996), industrial scheduling (Crawford et al. 1999), train shunting scheduling (Van Wezel and Jorna 2004), and so on. Because this perspective is strongly related to the analysis of how experts perform a cognitive task, the designer can select from the many existing techniques to identify knowledge and representations. Schraagen et al. (2000b), for example, identify interviews, process tracking (e.g., verbal protocols) and modeling approaches (e.g., using the GOMS framework by Card et al. 1983). Each technique provides a different point of view on the implied cognitive processes.

Crawford et al. (1999) studied the work of one scheduler using different techniques: interviews, direct observation, verbal protocols and so on. From this data and after several debriefing they were able to describe a representation of the scheduler's activities (cf. Fig. 13.2). This descriptive perspective indicates that the starting point for the scheduler is to collect the planned orders in the mailbox (the top left unit). Then the scheduler has to translate these orders into a work-to-do list for the shop floor (the bottom left unit). The formal task structure specifies activities such as "Check mailbox", "Firm and cut the planned order", "order materials", "book the orders", and "allocate the orders to shop floor". But the analysis also stresses the importance of informal activities, especially in terms of information collection and validation; this is accounted by many units such as "phone the person [...] who changed the programme to find the reason".

The more cognitively-oriented nature of this descriptive perspective allows the resulting decomposition to focus on real practice (i.e., effective tasks). Indeed, the decomposition resulting from a cognitive point of view highlights relations between activities and the format of presentation is not imposed by the method. In the analysis done by Crawford et al. (1999), the different effective tasks are linked together and the designer could possibly identify the relation between those done by the scheduler (cf. Fig. 13.2).

Using the output of this method for designing a generic support tool is a weak point of existing literature as noted Chipman et al. (2000). In fact, for any analyst, the first step of analysis is to gain familiarity with the task, to acquire knowledge and specialized vocabulary (Chipman et al. 2000). This will allow the analyst with more data such as interviews or technical documents to build a more detailed representation of the situation, i.e., developing expertise then proceeds with the elaboration of the decomposition. In the Crawford et al. (1999) study, the decomposition focuses on only one expert scheduler. This could be insufficient to tackle the variety of practices; it will not necessarily be identical to the decomposition resulting from the study of another scheduler in the same situation. For example, Cowling (2001) noted in a case study that one scheduler prefers to stretch the technical capacity of the machines to their limits, whereas the other scheduler was unwilling to approach

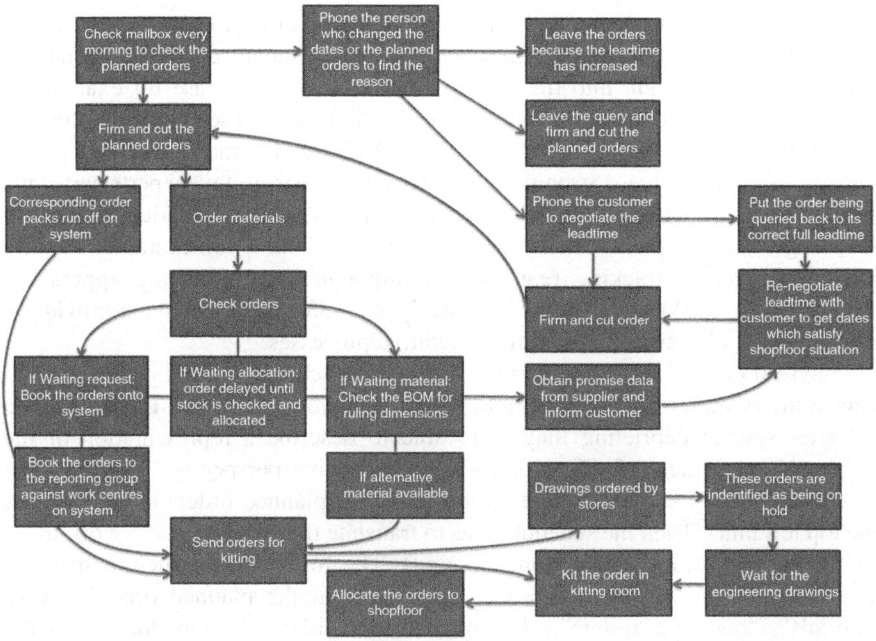

Fig. 13.2 A task analysis done by Crawford et al. (1999) in the scheduling industrial domain

these limits. In terms of task decomposition, the two schedulers provide two ways to schedule, resorting to different goals. So, a decomposition operated from the analysis of only one scheduler could neither be applied generally to the practice of another scheduler nor to produce a valid basis for the design of a support tool. These problems are dealt with by the formative perspective.

13.2.3 A Formative Perspective: Higgins's (1999) Study

In contrast with the two previous perspectives, the Higgins's (1999) study does not focus on the human's hypothesized or effective tasks but on the work domain, as was stressed by Vicente (2002, p.63): "A task can be defined as the set of actions that can or should be performed by one or more actors to achieve a particular goal. In contrast a work domain is the system being controlled, independent of any particular worker, automation, event, task, goal, or interface".

The decomposition is supported by an abstraction hierarchy: the higher levels of the hierarchy describe functional information, whereas lower levels describe physical information. Besides this physical to functional decomposition, there is a part–whole dimension, which takes into account several levels of details (e.g., system, subsystem, components). Five different levels of abstraction are under consideration: functional purposes, abstract functions, generalized functions, physical functions and the physical form.

The most complete decomposition following this method in the scheduling domain was done by Higgins (1999, 2001). At the top of the decomposition, there is the maximization of a long-term financial return. It depends on the capability to maximize short-term financial viability and customer patronage. When going to lower levels, the analysis focuses more and more on physical aspects such as the functions fulfilled (to print, to cut and so on) by the machines, and finally by the machines themselves (cf. Fig. 13.3).

Whereas this method shares Usher and Kaber's goal of decomposition, their focus differs. The Usher and Kaber's approach is a goal-oriented decomposition listing the different goals required to complete the upper goal; the Higgins's approach focuses on the possible actions in order to complete the upper goal (Miller and Vicente 2001). Vicente (2000) notes that this method leads the final product to present a list of possible actions the human has to choose from, whereas other perspectives produce an aid that is much more directing the human towards a specific goal. Because this perspective is independent of the current organizational structure and task performance it is a viable tool to design computers that will correctly function in novel situations. It is also a viable tool to take into account the diversity of strategies by allowing any of these strategies to be applied. More recently a formative perspective has been favored to facilitate the identification of the structural boundaries of a routing problem and to frame a Constraint Programming algorithm (Gacias et al. 2009).

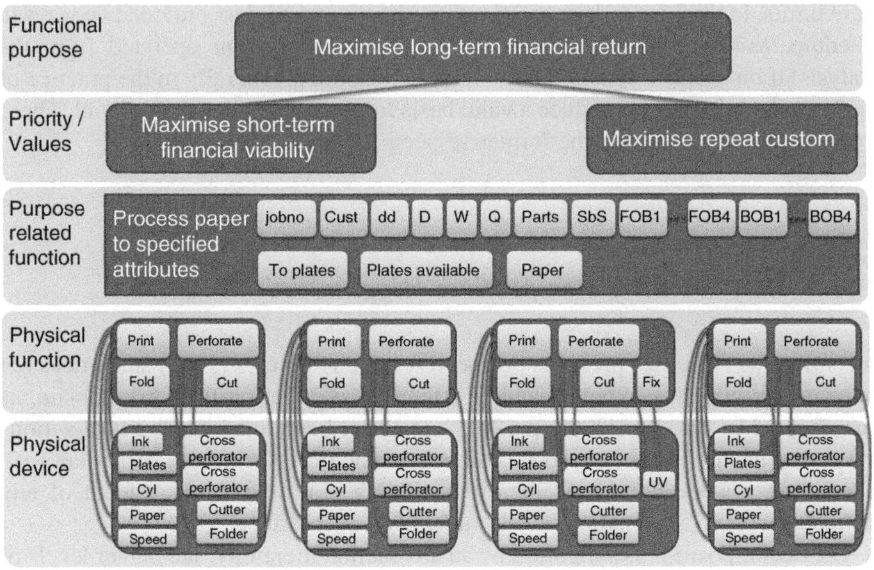

Fig. 13.3 An analysis done by Higgins (1999) in industrial scheduling domain

Table 13.1 Objective, unit and type of decomposition that relates to normative, descriptive and formative perspectives

	Normative	Descriptive	Formative
Objective of the method	To model the tasks the human has to accomplish	To model the knowledge, thought processes and goals that underlie the task	To model the work domain and its physical, functional interrelation
Unit of analysis	Task (designer's point of view)	Effective task (human scheduler's point of view)	Domain (independent of the human scheduler)
Type of decomposition	Hierarchical (task to subtasks)	Network of human activities	Means-ends and part-whole decomposition of the system functioning

In Table 13.1, we show the overall characteristics of three perspectives that have previously been applied to planning and scheduling. As can be seen, they have different objectives and hence focus on different aspects of the situation. In the next section, we will consider the capability of the normative, descriptive and formative perspectives to deal with the kinds of questions that we raised in the introduction: designing individual tasks, evaluating the task performance of a human planner, organizational design, design computer support, etc.

13.3 Relevance of Task Analysis Methods for the Various Goals of the Experts

As with every complex discipline, scheduling is a multi-faceted organizational phenomenon. Because it determines the performance of the primary process, all aspects have to be carefully attuned. The performance goals should be high but realistic. Human planners should have detailed knowledge about the planned processes but they also need numerical expertise and social abilities. Computer support for planning should combine sales forecasts, the order portfolio, stock positions, and production progress.

In the previous section, we presented multiple perspectives available to analyze planning and scheduling. However, looking at the list with aspects that should be taken into account, we draw the following conclusions. First, the goals of the methods are not the same. Second, none of the methods provides all the necessary tools and techniques to answer all questions. In this section, we analyze appropriateness of the perspectives for the various goals that experts have when looking at planning and scheduling.

Vicente (1999b) describes three dimensions on which models resulting from task analysis methods can be compared and with which they can be classified: device dependency, event dependency, and psychological relevance.

- The device dependency expresses to what extent the model depends on the devices that are used to perform the scheduler's work. In normative (Usher and Kaber 2000) as well as in descriptive (Crawford et al. 1999) perspectives, the authors focus on what they consider the usual information flows and existing actors of the scheduling process. Analyzing a system from the perspective of currently used devices (such as humans, computers, and organizational units) only will limit the scope and thereby the options for objective evaluation and redesign. The formative perspective (Higgins 1999) is not related to specific actors and the subtask may be done by human or computer indistinctly.
- The event dependency specifies whether the technique can be used to identify information requirements that are outside the normal working context of the system. Event independency allows novel situations to be taken into account. Normative and descriptive perspectives focus on known cases, hypothetical or encountered by the authors of the analysis. This decomposition may be inadequate to include subtasks for other situations not anticipated. In a formative perspective, all the constraints of the domain are stressed, which may allow for having constraints relevant for all situations, even those not yet encountered by the scheduler or the computer designer.
- The psychological relevancy indicates whether the method is able to take into account the point of view of the humans that schedule. In a normative or a formative perspective, the authors do not focus on the scheduler's point of view, which may imply they produce some task decomposition not necessarily relevant or sufficient for the human that has to perform the task. In a descriptive

Table 13.2 Focus of the normative, descriptive and formative perspectives

	Device independent	Event independent	Psychologically relevant
Normative	No	No	No
Descriptive	No	No	Yes
Formative	Yes	Yes	No

perspective, the effective task is the topic of the analysis, which definitely produces psychologically-relevant task decomposition.

If it is possible to categorize the perspectives of task analysis in relation to device, event independence and psychological relevance (see Table 13.2), the experts' goals may be also be characterized according to these dimensions.

13.3.1 Organizational Structure of the Planning Function

There are two main issues related to the organizational structure: (1) analyzing and (re)designing the structure, and (2) analyzing and improving the mutual awareness of organizational members.

Questions related to the design of the organizational structure and the specification of work jobs are device independent. Although some job requirements are related to psychological factors such as well-being at work, stress, etc. the psychological relevance at this stage is low, because the psychological factors are not related to specific individuals. Organizational design needs an event independent technique because it is about structure and uncertainty makes that not all circumstances can be modeled by looking at current practice. The combination of these characteristics makes a combination of formative and normative perspectives apparent. The outcome of a normative approach is an overview of the way in which the overall task is divided into subtasks and for each subtask it specifies what the goal is, who performs the task, what skills are necessary, what kind of temporal relation there is with other tasks, etc. Generic normative techniques and tools can be used. For example, planning specific subgoal templates can be designed in order to facilitate the analyst's task (see Ormerod and Shepherd 1998). The normative perspective can be linked to a formative one by specifying for each task what part of the planning domain is under consideration. For example, the goal of one subtask could be to decide what customer order will be in what batch, and the goal of another subtask is to specify what batch will be made on what machine. From such dependencies in the domain and the subtask allocation to planners, the need for interaction and coordination of subtasks can be inferred. Taking the goal of redesign, the analysis of dependencies can be used to assess the viability of the task division with, for example, techniques from Group Technology.

Analyzing mutual awareness is device dependent, psychologically relevant, and event dependent. This makes descriptive approach the obvious choice. However, this only provides a picture of how planners interpret their own and each other's work. To be able to analyze whether this is congruent with the organizational philosophy, a combination with normative is necessary.

13.3.2 Individual Task Performance

The descriptive approach is the obvious method to apply if individuals are analyzed. However, if we are to make a judgment of task performance, it might be necessary to compare the knowledge of the planner to objective knowledge of the domain and organizational structure. There are four reasons to analyze individual task performance.

First, an analysis might be needed to determine the requirements a specific job poses to the human. For example, in some planning tasks communication might be important whereas in other tasks problem solving would be dominant. Knowing the job requirements for planners is a problem that all companies have. An analysis of job requirements ought to be device and event independent because it is not based on the current task performance of a planner. However, the psychological relevance is high because we are talking about a human that has to perform the task. A combination of formative and normative perspectives is obvious, but we keep the problem that we need to deduce psychological characteristics of a job without having a current planner to analyze.

Second, an analysis of the task performance of a planner might be needed for two reasons: evaluate the quality of the planner's work to be able to determine opportunities for improvement, and design computer support. We will discuss the latter in the next section. For the first, the psychological relevance makes descriptive perspective the obvious method to use. However, merely modeling the task performance is device and event dependent, whereas assessing the performance is to a certain extent device and event independent. In other words, the quality of task performance cannot be determined without looking at the domain and organizational context. Therefore, a combination with normative and formative perspectives is necessary, but only to a limited extent.

The third reason to look at the individual's task performance is to design training for novice schedulers. Although in practice novices mostly learn by doing, in the past years we see initiatives in industry to design training programs specifically for planners and their managers. The important question here is to what extent planning and scheduling tasks are device and event dependent or independent. In other words, to what extent can we train planners without taking into account the domain characteristics of their job and the organizational setting they are in? The training program of the Planners Academy of Dehora Consultancy Group in The Netherlands that is offered to planners shows that generic training is certainly possible, although knowledge of specific characteristics of the job of individual planners is necessary as well. To design this latter kind of training, formative and normative perspectives are needed, but like with assessing the task performance of human individuals, they only need to be done to a limited extent.

13.3.3 Computer Support

We previously mentioned the gap between theory and practice in the planning and scheduling domain. This gap is strongly related to computer support. In the past, many systems were created and implemented by a device independent and event independent approach without taking into account psychological factors. Often, the formative and normative perspectives were not used but it was done the other way around: a system was implemented and the organization organized itself around the system. The introduction of configurable ERP and APS systems has improved this considerably in the past decade, and formative, normative perspectives could be applied nowadays before implementation. However, it is still not common practice to apply descriptive perspectives when designing the system, whereas the use of a computer program is device dependent and the psychological relevance is high. However, limiting the design of a computer system to a descriptive perspective alone is not wise as well, because of the event independence of a computer program. A combination of methods is necessary (see e.g., Van Wezel 2006).

Algorithms are an important part of scheduling systems. Although they are integrated in computer programs, they have different attributes. A scheduling algorithm implemented into a computer is generally designed as device dependent (the device being the computer) and event dependent. It is generally designed independently of the human and psychological relevance is considered to be low. A formative perspective provides information on the objects and their characteristics and constraints. However, this is not enough because algorithms not only try to find solutions which do not violate constraints; they also try to find the best solution possible. A normative perspective is needed to specify organizational context and the goal structure for the algorithms. One disadvantage of considering algorithms psychologically irrelevant is that the human that must use the algorithm is disregarded. Riezebos and van Wezel (2006) provide an example where a cognitive analysis is combined with domain knowledge and the organizational task structure to improve on this. Gacias et al. (2009) provide a method for designing support systems taking the most of algorithms designed in a formative perspective and human abilities to redefine the problem. Table 13.3 summarizes the applicability of the task analysis methods for the different issues in planning and scheduling. Further research has also to be done in order to precisely identify the benefits of producing algorithms that are psychologically relevant (Cegarra and Hoc 2008).

13.4 Conclusions and Further Research

Planning and scheduling experts in practice are often faced with the following question: how a company can improve its planning performance? The effect that planning and scheduling have on the performance of the primary process makes it

Table 13.3 The use of task analysis frameworks for planning and scheduling

Question	Characteristics	Applicable methods
Analyze/design the organizational structure	Device independent Not psychologically relevant Event independent	Formative & normative
Analyze/improve mutual awareness	Device dependent Psychologically relevant Event dependent	Descriptive & normative
Analyze the requirements for a specific job	Device dependent Psychologically relevant Event independent	Formative & normative
Analyze/design the individual task performance	Device dependent Psychologically relevant Event dependent	Descriptive, limited formative & normative
Design training for novices	Device dependent Psychologically relevant Event independent	Formative & normative
Create computer support	Device dependent Psychologically relevant Event independent	Formative, normative & descriptive
Design scheduling algorithms	Device dependent Not psychologically relevant Event dependent	Formative & normative

important that the scope in which this question is answered is not too narrow. Looking at only one link of the decision chain will almost certainly result in suboptimal performance. For example, the performance of a single human planner cannot be improved without looking at the planned domain, the organizational context, and the available support as well.

We have confronted three perspectives for performing task analysis with the various questions that planning experts face in practice. There are two main findings. On the one hand, a combination of task analysis methods is often necessary in order to avoid myopia and biased results. On the other hand, however, the analysis shows that not all questions require a full-scale analysis of the planning situation. The insights gained might help professionals in the planning and scheduling domain to both expand their horizon and at the same time guide the choice of appropriate methods to analyze the planners' work environment.

The similarities that many planning and scheduling situations show will be used to extend these ideas further. We hope it will be possible to provide more details about what parts of the different task analysis methods can be used to answer specific questions. Examples are a model to assess the fit of the organizational planning structure with the domain and the business environment, a method to determine the appropriate kind of algorithm for supporting a subtask, and a method to determine the suitability of a person for a specific planning task. With such projects we hope to contribute to improving planning and scheduling, not only by looking at and optimizing the individual aspects but also integrating them into a more comprehensive picture.

References

Annett, J. (2000). Theoretical and pragmatic influences on task analysis methods. In J. M. Schraagen, S. F. Chipman, & V. L. Shalin (Eds.), *Cognitive task analysis* (pp. 25–37). Mahwah, NJ: Lawrence Erlbaum Associates.

Annett, J., & Duncan, K. D. (1967). Task analysis and training design. *Occupational Psychology, 41*, 211–221.

Bainbridge, L. (1974). Analysis of verbal protocols from a process control task. In E. Edwards & F. P. Lees (Eds.), *The human operator in process control* (pp. 146–158). London: Taylor and Francis.

Bi, S., & Salvendy, G. (1994). Analytical modeling and experimental study of human workload in scheduling of advanced manufacturing systems. *The international Journal of Human Factors in Manufacturing, 4*(2), 205–234.

Boy, G. (1998). *Cognitive functions analysis*. London, UK: Ablex Publishing.

Buxey, G. (1989). Production scheduling: Practice and theory. *European Journal of Operational Research, 39*, 17–31.

Card, S., Moran, T., & Newell, A. (1983). *The psychology of human-computer interaction*. Hillsdale, NJ: Erlbaum.

Carroll, J. M. (1995). *Scenario-based design: Envisioning work and technology in system development*. New York: Wiley.

Cegarra, J. (2008). A cognitive typology of scheduling situations: A contribution to laboratory and field studies. *Theoretical Issues in Ergonomics Science, 9*(3), 201–222.

Cegarra, J., & Hoc, J. M. (2008). The role of algorithm and result comprehensibility of automated scheduling on complacency. *Human Factors and Ergonomics in Manufacturing, 28*(6), 603–620.

Chipman, S. F., Schraagen, J. M., & Shalin, V. L. (2000). Introduction to cognitive task analysis. In J. M. Schraagen, S. F. Chipman, & V. L. Shalin (Eds.), *Cognitive task analysis* (pp. 3–23). Mahwah, NJ: Lawrence Erlbaum Associates.

Cowling, P. (2001). Design and implementation of an effective decision support system: A case study in steel hot rolling mill scheduling. In B. L. MacCarthy & J. R. Wilson (Eds.), *Human performance in planning and scheduling: fieldwork studies, methodologies and research issues* (pp. 217–230). London: Taylor & Francis.

Crawford, S., MacCarthy, B. L., Wilson, J. R., & Vernon, C. (1999). Investigating the work of industrial schedulers through field study. *Cognition, Technology & Work, 1*, 63–77.

Crawford, S., & Wiers, V. C. S. (2001). From anecdotes to theory: A review of existing knowledge on human factors of planning and scheduling. In B. L. MacCarthy & J. R. Wilson (Eds.), *Human performance in planning and scheduling: fieldwork studies, methodologies and research issues* (pp. 15–43). London: Taylor & Francis.

Dessouky, M. I., Moray, N., & Kijowski, B. (1995). Taxonomy of scheduling systems as a basis for the study of strategic behavior. *Human Factors, 37*, 443–472.

Diaper, D. (2004). Understanding task analysis for human-computer interaction. In D. Diaper & N. Stanton (Eds.), *The handbook of task analysis for human-computer interaction* (pp. 5–47). Mahwah, NJ: Lawrence Erlbaum Associates.

Diaper, D., & Stanton, N. (Eds.) (2004). The handbook of task analysis for human-computer interaction. Mahwak, NJ: Lawrence Erlbaum Associates.

Gacias, B., Cegarra, J., & Lopez, P. (2009). *An interdisciplinary method for a generic VRP decision support system*. Paper presented at the International Conference on Industrial Engineering and Systems Management (IESM 2009), Montréal, Canada.

Halsall, D., Muhlemann, A., & Price, D. (1994). A review of production planning and scheduling in smaller manufacturing companies in the UK. *Production Planning & Control, 5*(5), 485–493.

Higgins, P. G. (1999). *Job shop scheduling: hybrid intelligent human-computer paradigm*. PhD Thesis, University of Melbourne, Australia.

Higgins, P. G. (2001). Architecture and interface aspects of scheduling decision support. In B. L. MacCarthy & J. R. Wilson (Eds.), *Human performance in planning and scheduling: fieldwork studies, methodologies and research issues* (pp. 245–279). London: Taylor & Francis.

Hoc, J. M., Mebarki, N., & Cegarra, J. (2004). L'assistance à l'opérateur humain pour l'ordonnancement dans les ateliers manufacturiers. *Le Travail Humain, 67*, 181–208.

Jackson, S., Wilson, J. R., & MacCarthy, B. L. (2004). A new model of scheduling in manufacturing: Tasks, roles, and monitoring. *Human Factors, 46*, 533–550.

Jorna, R. J. (2006). Cognition, planning and domains: An empirical study into the planning processes of planners. In W. M. C. van Wezel, R. J. Jorna, & A. M. Meystel (Eds.), *Planning in intelligent systems: Aspects, motivations, and methods* (pp. 101–136). Hoboken, NJ: Wiley.

Kiewiet, D. J., Jorna, R. J., & van Wezel, W. M. C. (2005). Planners and their cognitive maps: An analysis of domain representations using multi dimensional scaling. *Applied Ergonomics, 36* (6), 695–708.

Martensson, L. (1996). The operator's requirements for working with automated systems. *The International Journal of Human Factors in Manufacturing, 6*, 29–39.

McKay, K. N., & Buzacott, J. A. (2000). The application of computerized production control systems in job shop environments. *Computers in Industry, 42*, 79–97.

McKay, K. N., Safayeni, F. R., & Buzacott, J. A. (1988). Job-shop scheduling theory: What is relevant? *Interfaces, 18*(4), 84–90.

McKay, K. N., & Wiers, V. C. S. (1999). Unifying the theory and practice of production scheduling. *Journal of Manufacturing Systems, 18*(4), 241–255.

McKay, K. N., & Wiers, V. C. S. (2003). Integrated decision support for planning, scheduling, and dispatching tasks in a focused factory. *Computers in Industry, 50*, 5–14.

Mietus, D. M. (1994). *Understanding planning for effective decision support*. PhD Thesis, University of Groningen, Groningen, The Netherlands.

Miller, C. A., & Vicente, K. J. (2001). Comparison of display requirements generated via hierarchical task and abstraction-decomposition space analysis techniques. *International Journal of Cognitive Ergonomics, 5*(3), 335–355.

Ormerod, T. C., & Shepherd, A. (1998). Using task analysis for information requirements specification: The sub-goal template (SGT) method. In D. Diaper & N. Stanton (Eds.), *The handbook of task analysis for human-computer interaction* (pp. 347–366). London: Lawrence Erlbaum Associates.

Paternò, F. (2004). Concurtasktrees: An engineered notation for task models. In D. Diaper & N. Stanton (Eds.), *The handbook of task analysis for human-computer interaction* (pp. 483–502). Mahwah, NJ: Lawrence Erlbaum Associates.

Rasmussen, J. (1997). Merging paradigms: Decision making, management, and cognitive control. In R. Flin, E. Salas, M. E. Strub, & L. Marting (Eds.), *Decision making under stress: Emerging paradigms and applications* (pp. 67–85). Aldershot: Ashgate.

Rasmussen, J., Pejtersen, A. M., & Goodstein, L. P. (1994). *Cognitive systems engineering*. London: Wiley.

Riezebos, J., & van Wezel, W. M. C. (2006). Planner-oriented design of algorithms for train shunting scheduling. In W. M. C. van Wezel, R. J. Jorna, & A. M. Meystel (Eds.), *Planning in intelligent systems: aspects, motivations, and methods* (pp. 477–496). Hoboken, NJ: Wiley.

Sanderson, P. M. (1989). The human planning and scheduling role in advanced manufacturing systems: an emerging human factors domain. *Human Factors, 31*, 635–666.

Schraagen, J. M., Chipman, S. F., & Shalin, V. L. (Eds.). (2000a). *Cognitive task analysis*. Mahwah, NJ: Lawrence Erlbaum Associates.

Schraagen, J. M., Chipman, S. F., & Shute, V. J. (2000b). State-of-the-art review of cognitive task analysis techniques. In J. M. Schraagen, S. F. Chipman, & V. L. Shalin (Eds.), *Cognitive task analysis* (pp. 467–487). Mahwah, NJ: Lawrence Erlbaum Associates.

Stanton, N. A. (2005). Hierarchical task analysis: Developments, applications, and extensions. *Applied Ergonomics, 37*, 55–79.

Taatgen, N. (1999). Learning without limits: From problem solving towards a unified theory of learning. PhD Thesis, University of Groningen, Groningen The Netherlands.

Usher, J. M., & Kaber, D. B. (2000). Establishing information requirements for supervisory controllers in a flexible manufacturing system using GTA. *Human Factors and Ergonomics in Manufacturing, 10*, 431–452.

Van Wezel, W. M. C. (2006). Interactive scheduling systems. In W. M. C. Van Wezel, R. J. Jorna, & A. M. Meystel (Eds.), *Planning in intelligent systems: Aspects, motivations, and methods* (pp. 205–242). Hoboken, NJ: Wiley.

Van Wezel, W. M. C., & Jorna, R. J. (2001). Paradoxes in planning. *Engineering Applications of Artificial Intelligence, 14*, 269–286.

Van Wezel, W. M. C., & Jorna, R. (2004). *Tracing cognition, tasks and support: Shunting support in the Netherlands railways*. Technical Report accepted for publication in the Journal of Cognition, Technology and Work, The Netherlands: University of Groningen.

van Wezel, W., Jorna, R., & Mietus, D. (1996). Scheduling in a generic perspective, knowledge-based decision support by domain analysis and cognitive task analysis. *International Journal of Expert Systems, 9*(3), 357–381.

Van Wezel, W. M. C., van Donk, D. P., & Gaalman, G. J. C. (2006). The planning flexibility bottleneck in food processing industries. *Journal of Operations Management, 24*(3), 287–300.

Vicente, K. J. (1999a). *Cognitive work analysis: Towards safe, productive, and healthy computer-based work*. Mahwah, NJ: Lawrence Erlbaum Associates.

Vicente, K. J. (1999b). Wanted: psychologically relevant, device- and event-independent work analysis techniques. *Interacting with Computers, 11*, 237–254.

Vicente, K. J. (2000). Work domain analysis and task analysis: A difference that matters. In J. M. Schraagen, S. F. Chipman, & V. L. Shalin (Eds.), *Cognitive task analysis* (pp. 101–118). Mahwah, NJ: Lawrence Erlbaum Associates.

Vicente, K. J. (2002). Ecological interface design: progress and challenges. *Human Factors, 44*(1), 62–78.

Wiers, V. C. S. (1996). A quantitative field study of the decision behavior of four shop floor schedulers. *Production Planning & Control, 7*(4), 381–390.

Chapter 14
Allocating Functions to Human and Algorithm in Scheduling

Wout van Wezel, Julien Cegarra and Jean-Michel Hoc

Abstract An important part of Advanced Planning Systems (APS) are algorithms. When algorithms are applied, the task is automated as much as possible. However, the human that is supposed to use the algorithm is generally ignored during the development process. As a consequence, a prior investigation whether and how an algorithm can or will be used in practice is not integrated in the development process. In contrast, in the field of cognitive ergonomics, function allocation methods explicitly take into account human factors in the design of human/computer systems. The function allocation literature, however, is mainly focused on dynamic systems where humans must make decisions in situations with time pressure and important safety risks, e.g., nuclear plants and air traffic control. We analyze the differences between such dynamic systems and planning and scheduling, and we propose a model for function allocation in planning and scheduling taking into account cognitive and human–machine cooperation aspects.

14.1 Introduction

A scheduling task consists of elaborating a plan for resources (e.g., machines and human operators) within a specified period of time, taking into account temporal constraints (waiting periods, precedence constraints, etc.) and constraints related to

W. van Wezel (✉)
Faculty of Economics and Business, University of Groningen, Groningen, The Netherlands
e-mail: w.m.c.van.wezel@rug.nl

J. Cegarra
Université de Toulouse, Toulousse, France

J.-M. Hoc
Centre National de la Recherche Scientifique, (CNRS: French National Research Centre), University of Technology of Compiègne, Compiègne, France
e-mail: Jean-Michel.Hoc@irccyn.ec-nantes.fr

J.C. Fransoo et al. (eds.), *Behavioral Operations in Planning and Scheduling*,
DOI 10.1007/978-3-642-13382-4_14, © Springer-Verlag Berlin Heidelberg 2011

the use and availability of the necessary resources. There are scheduling tasks in domains as diverse as manufacturing (production scheduling), personnel management (e.g., nurses scheduling) and transportation (e.g., truck and train scheduling). In practice, schedules are usually created manually by experienced humans and can require up to several days of work. In the many companies we visited for our research and educational projects, we noted that in a majority of companies planning systems are only used to perform calculations, edit the schedule, check constraints, and apply simple heuristics. This is in sharp contrast with the standpoints taken in scientific research, where the presumption is that many algorithms are available, and hence, they will probably be used in practice.

More precisely, in scientific research, the general paradigm in Operations Research is to utilize the possibilities that algorithms offer to the fullest extent, and allocating the tasks that remain to the human planner. However, in Cognitive Ergonomics, full automation of a task in the design of human–machine cooperation is often criticized (Parasuraman and Riley 1997; Hoc 2000). In general, criteria such as cognitive workload, situation awareness, complacency, skill degradation, risk of automation failure, trust, and cost of incorrect decisions are mentioned in the literature. Noting this discrepancy, several authors have suggested bringing together technical and human approaches in scheduling to improve the use of scheduling algorithms in practice (Sanderson 1989; Hoc et al. 2004).

In doing so, planning and scheduling experts need to undertake a number of detailed analyses in order to precisely identify the functions to allocate. In the function allocation theory, a function is defined as a unit coming from decomposition that the integrated human–machine is required to be capable of performing. In order to identify these functions, several methods have been developed such as *Hierarchical Task Analysis* (Annett 2000; Annett and Duncan 1967) and *Cognitive Task Analysis* (Schraagen et al. 2000). Their respective advantages and disadvantages have been already compared for planning and scheduling situations (see Cegarra and van Wezel, Chap. 13). When the functions have been identified, the designers have then to proceed to the identification of mandatory allocations and (re)define the human operator's task (Price et al. 1982; Sharit 1997). Surprisingly, very few researchers developing the methods of task analysis are also suggesting methods or rules for allocating functions (but see for instance in the case of HTA: Marsden and Kirby 2005; in the case of CTA: Neerincx and Griffioen 1996). However, function allocation methods obviously integrate the identification of functions to allocate (Dearden, Harrison, and Wright 2000). The main research question that we address in this chapter is, how can we determine on the function allocation between human planner and algorithm?

Function allocation decisions not only have cognitive implications at the individual level, but also at the human–machine cooperation level. After Hoc (2001), we approach cooperation as an activity of interference management between distinct agents in order to facilitate individual or collective tasks. Interference means that an agent's activity (goal, procedure, etc.) has something to do with another agent's activity. Function allocation is not only decomposition and allocation. More often than not, decomposition of a global function into several elementary functions does not result in independent functions. For example, if a human scheduler is in charge of the function of improving a schedule generated by a machine, the nature of the machine's strategy can influence the understanding of the schedule and thus its improving (Cegarra and Hoc 2008). The consideration of the human–machine cooperation implications of function allocation is crucial.

The objective of this article is to present a comprehensive model to determine the division of functions between human and computer within a scheduling task. This article favors a multidisciplinary approach integrating both a technical and a human point of view. In the next section, we will provide a general description of function allocation. In the subsequent section, the differences between dynamic situations and planning and scheduling will be investigated. Then we will use this to devise a function allocation framework for planning and scheduling. The framework will be demonstrated with the example of shunting scheduling. We end with conclusions and topics for further research.

14.2 Function Allocation in Dynamic Situations

Much of the research in the area of function allocation has been performed in the context of dynamic situations. Dynamic situations are situations which are only partly controlled by the human. He/she has to keep the situation between acceptable limits while managing high risk and/or time pressure. Examples are air-traffic control, aircraft piloting, and partly flexible manufacturing systems. Since the resulting theoretical and methodological backgrounds from dynamic situations do not necessarily apply to planning and scheduling, we will investigate what decision criteria for allocation can be used. Table 14.1 shows an overview of such criteria.

In the function allocation literature, two ways of allocating functions are recognized: static and dynamic allocation. At first, function allocation focused on a static allocation of functions between human and machine on the basis of competency with respect to the function at hand. Each function should be allocated to the actor that would achieve the supposed best performance. Lists with criteria to determine the allocation are called "Fitts' Lists", referring to one of the first of such lists (Fitts 1951). However, many authors have criticized this simplistic approach. Because machine intelligence is improving permanently, and because machines can be designed to perform specific tasks, the role of the human is constantly reduced (also called the "leftover" principle), ultimately resulting in a monitoring role. For example, Kantowitz and Sorkin (1987, p.359) noted: "Any table that can compare human and machine, especially if numerical indexes of relative performance or equations can be listed, as any good engineer would attempt, is bound to favor the machine". Such a role might lead to lower job satisfaction, complacency, and performance degradation, which can have severe consequences when the human must perform actions in case of exceptions or automation failures (Bainbridge 1983). Furthermore, allocation decisions might not be appropriate under all circumstances but situational factors are not taken into account in static allocation.

Table 14.1 Decision criteria for the allocation of tasks (Older et al. 1997, p. 158)

1	Task criticality
2	Environmental constraints, e.g., finances, use of specific technology, use of existing organizational structure
3	Workload: both underload and overload
4	Simultaneous demands on human or machine resources, e.g., task interaction, and the demands from other systems the operator is involved with; some tasks may enhance or degrade performance when carried out by the same operator
5	Number of personnel/amount of power required
6	Human well-being, e.g., safety, health, physical and mental requirements, psychological needs (e.g., job satisfaction, motivation, role importance, participation in decision making, responsibility, opportunity for skill development, task variety, range of mental demands)
7	Human and machine adaptability
8	Human and machine effectiveness, e.g., accuracy, speed, reliability, errors
9	Human and machine unpredictability
10	Technical feasibility
11	System efficiency, e.g., cost
12	System maintainability
13	Human and machine information processing requirements
14	Communication requirements (especially for dynamic allocation)
15	Training requirements
16	Notion of redundancy (i.e., how to cope with machine malfunction)
17	User-acceptance (particularly relevant to dynamic allocation), e.g., system familiarity, trust, self-confidence; the allocation should take into account the end-users' perception of the system, and enable the machine to provide explanations to the human for its actions

Most notably, time pressure, cognitive workload, and the current status of related tasks interact with each other and influence the appropriate allocation. Therefore, for specific subtasks, in order to overcome the disadvantages of static allocation, dynamic allocation or adaptive automation has been proposed (Hoc and Debernard 2002; Kaber et al. 2001; Rouse 1976; Scerbo 1996). In this approach, the allocation is not determined beforehand but during system functioning taking into account situational factors. An additional advantage is that the loss of human skills over time due to automation can be prevented because the function allocation decision can consider this explicitly by letting the human perform certain tasks at specified times.

Although function allocation theories and methods are generic, and therefore can be applied in any human–machine interaction situation, the context of most research in function allocation research has characteristics that do not apply in planning and scheduling. For its major part, research in planning and scheduling is dealing with algorithms. Typically, these algorithms are developed using the "leftover" principle. The algorithm is designed from the perspective of what technology has to offer and takes over as much decisions as possible. The tasks that remain are performed by the human planner. Human–computer interaction in scheduling literature is formulated from the perspective of the algorithm, for example:

- The human operator can choose from a number of alternative solutions that are generated by algorithms (Lauer et al. 1994; Ulusoy and Özdamar 1996)
- The human operator may specify weights on goal functions after which the algorithm generates a schedule (Gabrel and Vanderpooten 2002; Smed et al. 2000)
- The human operator can steer the backtracking process of the algorithm (Bernardo and Lin 1994)
- The human operator can specify parameters for the algorithm (Dockx et al. 1997; Oddi and Cesta 2000; Myers et al. 2002; Ulusoy and Özdamar 1996)

Criteria regarding the usage of algorithms such as cognitive workload, complacency, and trust are not taken into account. However, these factors might very well make that the predetermined goal of the human–computer interaction fails. For example, Endsley et al. (1999) empirically show that full automation does not necessarily lead to the highest system performance. There are several examples in the scheduling domain as well. Comparing solutions of complex scheduling problems requires the human scheduler to memorize and compare a large set of data. This leads to a very high cognitive workload and possibly to poor performance (Schakel 1976; Sanderson 1989; Cegarra and Hoc 2008). A real world example is described by Mietus (1994), who noted that nurse schedulers who were offered multiple solutions often focused on only one solution. The nurse scheduling system she described was developed to endorse the schedulers' individual differences in decision criteria, but the cognitive workload imposed by the system might have been too high to attain this goal.

This problem is amplified by changes in the task that are caused by the introduction of algorithms. Often, the role of the human changes from making a schedule to correcting a schedule that is created by the computer. This is a completely different task which can result in the need for learning and demotivation because of devaluation of the task.

The double predicament of ignoring cognitive aspects and changes in the task makes the effects of the introduction of algorithms for the joint human–computer system unpredictable. Consequently, there is a need for criteria, relevant in terms of human–machine systems, in order to decide under what circumstances what kinds of algorithms and interaction mode can be used. For example, suggesting multiple solutions in scheduling has to be avoided in complex scheduling problems due to the cognitive cost, but could be an interesting suggestion in simpler scheduling problems.

The Level of Automation (LOA) framework (Table 14.2) proposed by Sheridan and Verplanck (1978) and used by Parasuraman et al. (2000) is a starting point to analyze some consequences of the function allocation decisions. This framework specifies ten levels of automation from manual control to automatic control. However, the major drawback of the LOA framework is that it is clearly machine centered. LOA has mainly been defined for machine design purpose. It defines what the machine designer must do in each case, but it does not describe the consequence this could have on the human operator in terms of cooperation activity. In addition,

Table 14.2 Levels of automation of decision and action selection (Parasuraman et al. 2000, after Sheridan and Verplanck 1978)

10	The computer decides everything, acts autonomously, ignoring the human
9	Informs the human only if it, the computer, decides to
8	Informs the human only if asked, or
7	Executes automatically, then necessarily informs the human, and
6	Allows the human a restricted time to veto before automatic execution, or
5	Executes that suggestion if the human approves, or
4	Suggests one alternative
3	Narrows the selection down to a few, or
2	The computer offers a complete set of decision/action alternatives, or
1	The computer offers no assistance: the human must take all decisions and actions

some forms of automation based on human–machine cooperation principles are not considered. For example, within the problem-solving area, some redundancy between the human operator and the machine has been proved to be powerful, especially in planning and scheduling. Smith and Crabtree (1975) have shown that the human operators used computer support to planning more as a way to evaluate their own plans than as a source of plans. This activity is considered as "mutual control" in terms of human–machine cooperation.

Besides, the LOA levels need also adaptation because of the differences between dynamic situations, for which they have been described, and static planning and scheduling tasks. In the next section, we will first summarize the differences, after which we will propose a framework that can be used to gain insight in the appropriateness of algorithms for planning and scheduling subtasks.

14.3 Differences Between Dynamic and Static Scheduling

As previously noted, most of the literature on function allocation focuses on a human that must control a system in real time. One perspective that could be taken is that planners control the planned system, e.g., a production line. However, several authors state that the work of planners differs from the work of operators (Van Wezel and Jorna 2006; McKay and Wiers 2003; Wäfler 2001). Operators execute the schedule and are faced with the real-time behavior of the production system, whereas planners are only faced with real-time issues in specific circumstances, e.g., machine breakdowns and rush orders (Van Wezel et al. 2006a, b). Consequently, another perspective is that planners control the planning process rather than the process of plan execution. In this perspective, we can analyze the characteristics of the planning process in order to be able to determine the suitability of algorithms. We will discuss nine interrelated issues in which the planning process differs from processes that are typically found in the function allocation literature.

First, the function allocation literature often investigates systems where lives are at risk. As a consequence, the cost of the development of human–machine interaction and the human's job satisfaction in the joint cognitive system are of secondary importance. In planning and scheduling, performance is expressible as economic profits, which implies that the cost of developing the joint cognitive system is an important factor.

Second, in most dynamic situations, poor performance can lead to complete failure of the system. Obviously, this must be avoided to all costs in air-traffic control and aircraft cockpits. In static planning, however, the risk of a bad plan leading to failure of the planned system is negligible. Although poor performance of a planner can lead to "failure" of the planning process in the sense that the plans cannot be executed, there is always somebody who has to interpret the plan in order to be able to execute it. Plans that evidently lead to failure of the production system will not be executed but they will be discussed or disregarded. So, the harm risk in planning is much less than in other situations, although the economic risk remains.

Third, there is a time lapse between creating the schedule and executing it. The planning process usually starts several weeks before execution. Only in the very last phase is time pressure related to plan execution (Van Wezel et al. 2006a, b). Often, in many dynamic situations the time lapse is short; the decision is immediately executed.

Fourth, because of the time lapse, tasks can be serialized in planning, whereas in function allocation literature, one of the recurring questions is how to allocate primary and secondary tasks that must be performed in parallel under time pressure.

Fifth, again because of the time lapse, errors can be recovered before they have consequences. For example, a wrong estimate of the amount of raw material that is needed can be corrected by sending a rush order to the supplier.

Sixth, in order to be able to design a scheduling algorithm, constraints often have to be neglected for the algorithm to work. As a consequence, it is generally accepted that the outcome of a scheduling algorithm must be checked by the human planner as it might still contain constraint violations. In situations under time pressure where errors of a task allocated to the machine might lead to system failure, this is of course not acceptable.

Seventh, the function allocation literature pays much attention to failure of automation. For example in the case of computer failure in an aircraft, the human pilot must be able to land the aircraft manually. Therefore, function allocation must explicitly reckon with skill degradation. This does not play a role in static planning and scheduling. Scheduling algorithms run on desktop PC's that are not likely to fail and are easily replaced, and the time lapse between planning and execution reduces the need to immediately take over the planning tasks manually.

Eighth, the cycle duration of information acquisition, information analysis, decision and action selection, and action implementation in dynamic situations is typically very short, for example, seconds or minutes. In contrast, the same cycle in planning and scheduling can be large, for example, hours or days.

Ninth, as a consequence of the previous point, the underlying decision problem in dynamic situations must be small, otherwise a decision could never be made

Table 14.3 Overview of differences between dynamic situations and planning tasks

	Function allocation literature in dynamic situations	Static planning and scheduling tasks
Harm risk	High; lives	Low; profits
Possibility of system failure	Yes	No
Time lapse	Low	High
Task serialization	No	Yes
Time for error recovery	No	Yes
Acceptable error rate	Low	High
Failure of automation	Possible	Not an issue
Decision cycle duration	Short	Long
Information volume	Low	High

within the available time. Conversely, the underlying problem of decision and action selection in static planning and scheduling can be large. This implies that the human will draw on other cognitive resources during decision making.

Table 14.3 shows an overview of the differences between typical dynamic situations and planning and scheduling.

The task demands experienced by the human in dynamic situations and in static planning and scheduling tasks greatly differs. The differences in task demands induce different cognitive strategies and, consequently, different support requirements. The situations typically studied in dynamic situations are characterized by high risk and time pressure. The human could be in danger and must complete tasks in parallel while making decisions in a short time. A strategy to deal with the cognitive workload is to use reactive strategies that are less costly from a cognitive point of view because they imply a reduction of the anticipation span (Moray et al. 1991). However, in dynamic situations, the reduction of the anticipation span has the adverse effect of increasing the risk. Therefore, the designer of the supporting system has to prevent such heuristics and support the human in fully managing the task.

Compared to dynamic situations, static scheduling problems are large and complex. There is much information to process and multiple goals must be weighed. However, the human has much time to make a decision. An example to reduce the cognitive workload in such circumstances is to support abstraction, e.g., the categorization of situations or the detection of perceptual configurations (Cegarra 2008). Dutton's study (1964) offers an example of a categorization. In this study, schedulers reduced complexity (having more than one billion possible schedules) by categorizing customer orders into eight main categories. Other strategies that can be applied are, for example, plan partitioning, multi-resolutional planning, learning, and opportunistic planning (Van Wezel and Jorna 2006).

The task demands greatly differ between dynamic situations and planning tasks. The resulting strategies require a different function allocation between the human and the computer. So, we can gain insights from function allocation literature, but we cannot directly apply the factors used for the actual functional allocation decision by the designer as they presume different cognitive emphases. In the next section, we provide factors that can be considered when deciding on the function allocation for planning and scheduling.

14.4 Decomposition and Modes of Automation

Given the differences between dynamic situations and planning and scheduling, we have to analyze what planning and scheduling specific factors need to be taken into account in the function allocation decisions. Interestingly, the planning process is much more malleable than, say, a nuclear power plant or the cockpit of an aircraft. This means that decomposition not necessarily has to be derived from the planning process; if there are sensible reasons for a specific decomposition (e.g., because a cheap well performing algorithm is readily available), then the decomposition of the planning process could be derived from the use of the algorithm as well. In this light, the mode of automation depends on the characteristics of a function to allocate (solving a scheduling sub problem), the possibilities that algorithms can offer, and the constraints that the human imposes on the decision making process. We will discuss these three from the perspective of function allocation characteristics that were mentioned in the previous section, after which we propose a model in which the factors are related to each other and to the appropriate levels of automation.

14.4.1 Scheduling Sub-problems as the Unit of Function Allocation

The first step to do in function allocation is to determine the functions to be allocated. Planning and scheduling decisions can always be expressed in terms of assignments. Two or more objects of two or more types have to be assigned, adhering to constraints and trying to reach goals (Van Wezel and Jorna 2006). For example, production operations are assigned to machines, nurses to shifts and wards, etc. In decomposition, a scheduling problem is split in multiple smaller problems of which the solution processes can be serialized. For example, production orders can be grouped to product families, where each product family is treated as a separate scheduling problem. Each of such scheduling problems can be regarded as a function to allocate to the human, the computer, or both. The possibility to decompose scheduling problems into several problems of smaller size is important for function allocation, because characteristics of smaller problems differ from characteristics of larger problems.

As said, the unit of allocation is a planning sub-problem. In determining the allocation, we first need to know the goals and constraints of each sub-problem. We define four aggregate aspects. First, the *harm risk* for a function or planning sub-problem can be defined as the cost of a suboptimal decision. In this light, the error rate and failure of automation can be regarded as suboptimal or erroneous solutions of an algorithm. The harm risk determines whether such solutions are acceptable, and what role the human should have in improving or correcting the solution. Second, there might be the requirement of *situation awareness*; the planner needs to understand the situation, how the solution will affect the rest of the planning process, and what the consequences will be of executing the plan (Endsley 1995). The level of situation awareness attained by the human during problem solving depends on the level of automation. Endsley and Kiris (1995) show that full automation leads to lower situation awareness. Third, the *time lapse* and *time for error recovery* are important parameters of a scheduling sub-problem. In decomposition it has to be taken into account that the available time is decomposed as well. Because decomposition implies that multiple sub-problems must be solved, there is a maximum decision cycle duration even if execution of the plan is still some time away. Additionally, there are often deadlines for error recovery. For example, once raw material has been ordered, changing the amount might be very expensive or even impossible. Fourth, the *complexity* of the problem is important for function allocation. The complexity not only relates to the size of the problem, but also to the internal structure: "(a) the presence of multiple potential ways (i.e., paths) to arrive at a desired end-state, (b) the presence of multiple desired outcomes (i.e., end-states) to be attained, (c) the presence of conflicting interdependence among paths to multiple outcomes, and (d) the presence of uncertain or probabilistic links among paths and outcomes" (Campbell 1988). This is strongly related to decomposition. If the problem is too complex to be adequately solved by either the human or the algorithm, it can be decomposed in less complex problems, usually incurring degradation of performance.

14.4.2 Characteristics of Algorithms

Function allocation decides on the role of the algorithm and the role of the human in solving the sub-problem. The first issue with respect to algorithms is the *technical feasibility*. Pulliam and Price (1984) and Dearden et al. (2000) considered the state of automation design to decide about which functions to automate or not. Taking into account the state of algorithmic research is also relevant in the case of scheduling situations. Dearden et al. (2000) suggest that there is probably a feasibility continuum from the availability of an existing algorithm to the impossibility to develop one (Fig. 14.1).

The feasibility can not be evaluated without considering performance. There always exists some heuristics that can be applied in a particular function, although performance might be poor in complex problems. In order to decide if the state of algorithmic research is sufficient, it is necessary to take into account algorithms that adhere to the harm risk criterion. Furthermore, the cost of developing an algorithm

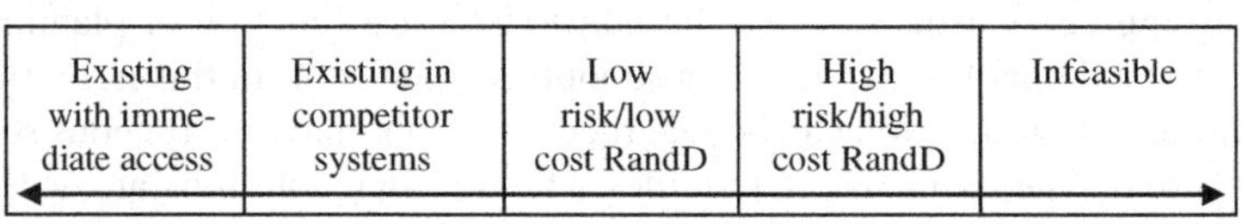

Fig. 14.1 Continuum of the feasibility of automation (from Dearden et al. 2000)

must be weighed against the improvement in performance, especially in cases where changes in the circumstances are likely to invalidate the algorithm. The quality of the solutions may deteriorate over time due to the fact that the optimization model is not a perfect representation of the planning problem faced by the planner. Constraints that were neglected in the optimization model can at some moment in time become important for the planners to include in their solution approaches. Two other factors are related to the use of the algorithm. First, the algorithm might not create complete correct solutions (the *error rate*), hence requiring adjustments by the human. Second, the algorithm might need *interaction during the solution process*, for example, to get feedback on partial solutions.

14.4.3 Characteristics of Human

The characteristics of the sub problem, together with the characteristics of the algorithm, determine the activities that must be performed by the human planner. If the human planner requires some support in order to fulfill a function, the algorithm may provide relevant support. However, some characteristics of the algorithm have to be avoided in order not to involuntarily hamper the human abilities. In order to design an efficient human–machine interaction, there is a need to take into accounts at least four characteristics:

Level of human involvement in the subtask. When a function is delegated or allocated to a machine in a multi-task situation, attention is shifted to those tasks which are not automated. This has two adverse consequences, related to human–machine cooperation:

- Loss of skills. Having an algorithm autonomously fulfilling the tasks leads the human to be in a monitoring role. This role might lead the scheduler to become passive, which can have severe consequences when the human must perform actions in case of exceptions or automation failures (Hoc 2000). Indeed, being in a monitoring role reduces the opportunity for the human to learn from experience and to maintain his/her skills. Bainbridge (1983) noted the "ironies of automation", which lead to a progressive loss of skills. Therefore the human has to be involved in the decision loop and cannot stay in a monitoring role.
- Complacency phenomenon. Moreover, after fully allocating a function to a machine, the human operator neglects the information necessary to perform the automated function, does not supervise the function and does not try to improve its results, even if it is possible. This has been termed the *complacency* phenomenon (Hoc 2000; Mosier and Skitka 1996; Parasuraman et al. 1993), which leads even those operators who are aware of the limits of the scheduling algorithm to accept the schedule because of the workload associated with the modification of this computer-generated schedule (Cegarra and Hoc 2008).

The algorithm design options then have to involve the human in a mutual control relationship. The tasks should be performed by the human to get the algorithm started, to interact with it, to modify the outcome, or to fix what is incomplete or not correct. And this interaction should be acceptable with respect to cognitive workload, necessary situation awareness, and job satisfaction. For example, in order to prevent complacency, Smith et al. (1997) suggested providing several alternatives from among which the humans had to choose. However, this suggestion has to be put in parallel with the complexity of the subtask. If the human has to compare very complex alternatives, the human may face cognitive limits (see Cegarra and Hoc 2008).

Performance level and performance variability. The human scheduler must trust the algorithm and has to perceive the algorithm's usefulness (Davis and Kottemann 1995; Riedel et al. 2007). Studies of trust by Lee and Moray (1992, 1994) indicate that the use of automation is determined by the ratio between trust in the machine and self-confidence. If the human considers his/her performance to be better than the performance of the machine, s/he would not use it. Therefore the performance level has to be sufficiently high for the algorithm to be used.

Moreover, Riedel et al. (2007) identified that a large variability in performance of the algorithm leads to a decrease in trust and finally to reduced use of the algorithm. However, Hoc (2001) and Lee and See (2004) have stressed the fact that the human operator can accept the machine's variability in performance if this variability can be understandable in relation to situational features. In other words, trust is not only related to a machine, but also to a situation. The human operator can trust the machine in a certain class of situations and distrust it outside this class. The comprehensibility of the machine performance in relation to differences between situations is a key factor of trust calibration.

Providing relevant feedback. An algorithm has also to be considered as reducing the adaptability of the human. Hoc (2000) stressed that the main reason comes from the lack of relevant feedback. In human–machine cooperation terms, there is a need for a common frame of reference (COFOR) between the agents. Such a COFOR integrates shared representations of the external situation, but also and above all shared understanding of the other agent's activity (goals, procedures, etc.). Not having relevant feedback, the human lacks the necessary situation awareness and the human adaptive processes cannot be brought into play.

In this way, Cegarra and Hoc (2008) did study two main levels of comprehensibility in scheduling: result comprehensibility (the schedule itself, especially through its graphical representation) and algorithm comprehensibility (the way the algorithm elaborates the schedule). The results of this study demonstrate that understanding the algorithm was not required for the best performance. This study also demonstrates that an algorithm with low result comprehensibility will lead to a lower level of performance. This was due to the fact that the graphical representation did not highlight some important information for the evaluation of the schedule, such as precedence constraints. As human attention is limited and because humans favor the achievement of satisfying performance rather than optimal performance (Hoc and Amalberti 2007), participants invested less attention in satisfying other constraints (such as those related to due date). This led to a poorer performance when the algorithm result was not comprehensible.

Flexibility with constraints. The tool must support the human view about the flexibility of the constraints. Different studies identified unsatisfied constraints in the schedule that are carried out deliberately; for example, a scheduler could use resources (especially machines) in a non-standard way to increase short-term capacity or put an order in late to finish several others in time. This stresses that an interaction mode has to let the scheduler decide on the flexibility of the constraints. Whilst the tool could inform the scheduler about constraints that are not satisfied, the scheduler should make the final decision. McKay and Buzacott (2000) also noted that a tool must be based on the human point of view by matching the terminology of the human planner, matching the information flow of the task, supporting the meta-objects (e.g., calendar time units of shifts) and meta-operations (e.g., delaying an operation to the next shift) of the planner.

14.4.4 Modes of Automation

The decision about function allocation should be based on the requirements of involvement and the possibilities of involvement with respect to the functions to allocate. From Sect. 4.1, we deduce the requirement for involvement of respectively the human and the algorithm. First, involvement might be necessary. As mentioned, the human is needed in cases where situation awareness or prevention of skill loss is required. An example where the algorithm is necessary is when errors in calculations must be prevented. Second, involvement could be impossible. For example, it might be too expensive to design an algorithm that finds acceptable solutions within the time available, or the problem could be too complex for the human to track. Third, involvement of either human or algorithm could be possible but not required.

In Sect. 14.2, we discussed that levels automation are related to the function allocation between the human and the computer. In the use of scheduling algorithms, we can discern several levels as well, depending on the involvement of the human and the algorithm in performing the subtask under consideration. Although a finer categorization is possible, for the purpose of this chapter we use the following levels of automation, formulated from the point of view of the human planner (Fig. 14.2):

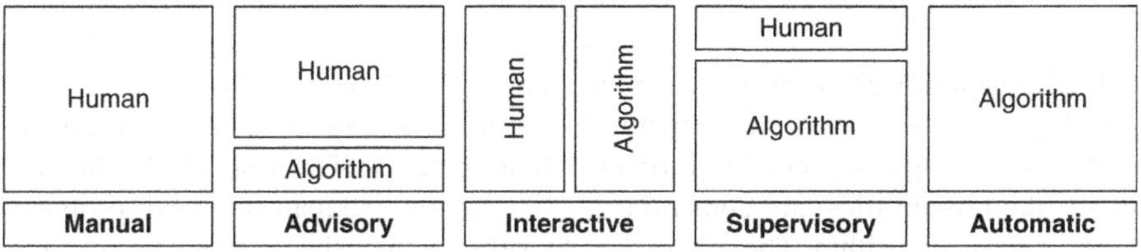

Fig. 14.2 Level of involvement

1. *Manual*. The user performs all activities in the subtask.
2. *Mutual control*. The user and algorithm both perform a part of the tasks, and they both have to consider the outcome of the other. We consider two kinds of mutual control:
   - *Advisory control*. The human proposes solutions or partial solutions, and the algorithm checks and evaluates them.
   - *Supervisory control*. The algorithm proposes solutions or partial solutions, and the human checks and evaluates them.
3. *Interactive*. The human and algorithm are both involved in each activity. Both may propose solutions or partial solutions, and each solution or partial solution is checked and evaluated by both.
4. *Fully automatic*. The algorithm performs all activities in the subtask.

Combining the levels for involvement with the need for involvement we deduce the following model to allocate functions for scheduling to human and algorithm:

- When it is impossible for both human and algorithm to be involved in solving the sub-problem, the decomposition is deficient.
- When an algorithm is not available, the human will solve the sub-problem manually.
- When the human cannot be involved but an adequate, either optional or necessary, algorithm is available, the sub-problem will be solved automatically.
- When both the human's involvement and the algorithm's involvement are optional, all levels of automation are possible, and the choice can be based on weighing the disadvantages such as development cost to the advantages such as reduction in the time the human planner needs to make a schedule.
- When the involvement of the human is optional, and the involvement of the algorithm is necessary, interaction, supervisory control, or automatic scheduling can be used.
- If the human's participation is required and an algorithm is optional, the function allocation depends on the reason why the human must participate. If skill loss or complacency must be prevented, all modes are possible if dynamic allocation is used. With dynamic allocation, sometimes the human, and sometimes the algorithm, solves the whole problem. If situation awareness is important, then the human must always be involved in the solution process, which implies advisory control or interactive scheduling. For reasons of efficiency, it is usually not sensible to resort to manual control if an algorithm is available.

The control modes are depicted in the matrix in Fig. 14.3.

In Fig. 14.4 we summarize our model for function allocation. A sub-problem defines the complexity and performance requirements. The role of the human defines the possibilities and constraints of human involvement for a sub-problem using a certain algorithm. The role of the algorithm defines the technical constraints and need for interaction with the human. Defining sub-problems, algorithms, and

		Human		
		Impossible	**Optional**	**Necessary**
Algorithm	**Impossible**	Infeasible	• Manual	• Manual
	Optional	• Automatic	Weigh cost and benefits: • Supervisory control • Mutual control • Advisory control • Interaction • Automatic	Prevent complacency: • dynamic allocation Prevent skill loss: • dynamic allocation Need for situation awareness: • advisory control • interaction
	Necessary	• Automatic	• Interaction • Supervisory control • Automatic	• Interaction

Fig. 14.3 Choice of control modes

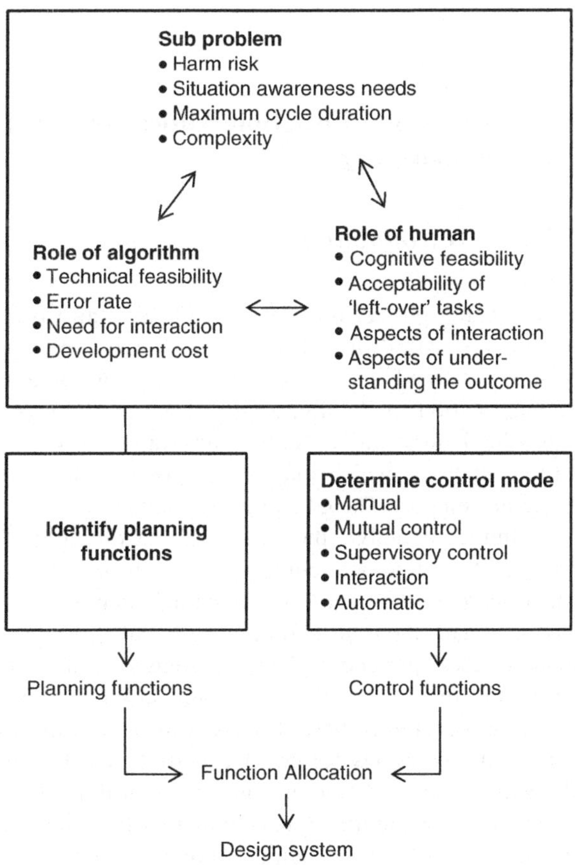

Fig. 14.4 Steps in function allocation

the human role is an iterative process. The starting point of this process can be any of the three comprising elements. For example, if a good algorithm exists, the problem decomposition (determining the sub-problems) and the effects on the role of the human can be based on that algorithm, but if the human is very important in the planning process, algorithms can be designed around the human's expertise.

Together, the characteristics of the sub-problem, the opportunities for algorithmic support, and the characteristics of the resulting functions that must be performed by the human planner identify the functions that must be performed to solve the sub-problem. Furthermore, the control mode, which defines the level of involvement of both the human and the algorithm for the identified functions, can be determined. The chosen control mode results in a number of control functions such as delegation and communication. The planning functions, control functions, and algorithms are functional requirements for the system that is being designed.

The model in Fig. 14.4 applies to a given sub-problem. An additional requirement is that the sequence of sub-problems to be solved and their control modes must still be feasible from a cognitive point of view. Furthermore, more, less, or a different decomposition of the total scheduling problem will result in different control modes.

14.5 Example: Function Allocation for Train Shunting Scheduling

14.5.1 *Introduction*

In a research project for The Netherlands Railways, called Rintel, five groups of researchers with different backgrounds and experience in various techniques and methods worked on the shunting planning for a representative medium sized railway station in Zwolle, a city in the northern part of The Netherlands. More than 130 planners in The Netherlands Railways work on shunting scheduling for the various stations, which makes it important and worthwhile to provide dedicated computer support. The goal of the research project was to compare different ways to algorithmically solve the shunting scheduling problem. Each research group received the same data containing a 24 h period of arriving and departing trains. The researchers could ask for all the information they needed (e.g., length of trains, technical data about the shunting yards, safety regulations, etc.), they were allowed to interview experienced planners extensively, they could observe how the planners work, and they were allowed to publish the results. The various approaches and resulting algorithms were discussed in several meetings with management, planners, and Operations Research experts of the logistical staff of The Netherlands Railways. In this section, we use the algorithms that were developed in the Rintel project to show the line of reasoning to follow in order to decide on the usability of the algorithms from a cognitive point of view. We would like to stress that the analysis we make is not meant to criticize the algorithms that were developed, as the goal of the research projects was to evaluate the usefulness of the various techniques rather than to create practical decision support. We use the algorithms because they give us an excellent opportunity for comparison; the algorithms were developed for the same complex practical situation using real world data and because the details about the algorithms are published. During the project, three algorithms were made without considering cognitive aspects of the prospective users (Abbink 2006; Lentink et al. 2006; Haijema et al. 2006), and two algorithms were created on the basis of extensive task analyses and incorporated in a prototype user interface to allow mixed initiative scheduling (Riezebos and Van Wezel 2009; Riezebos and Van Wezel 2006). The process of shunting scheduling and the algorithms we use in the example are described in detail in the book Planning in Intelligent Systems (Van Wezel et al. 2006a, b). We included one other algorithm in our analyses because it handles a subtask not dealt with by the other algorithms (Lamberts 2007).

14.5.2 Tasks and Subtasks of Making Shunting Schedules

During the night, passenger trains stay at the station. Because the main tracks are used during the night for freight trains, the passenger trains must be parked somewhere at the shunting yards. The parking space, however, is limited, which makes it a complex problem. Furthermore, the configuration of many trains that leave in the morning is different than the configuration in which they arrived in the preceding evening. Therefore, individual coaches must be disconnected, shunted, and connected. Additionally, several operations must be performed during the night such as external washing, internal cleaning, and fuelling. Roughly speaking, the task of the planner is to decide on what parking track the trains stay during the night and to schedule the movements of the trains. The arrival and departure times of the trains are derived from the national train timetable. As the timetable shows only slight changes during the year, the shunting planners never start from scratch; they always adapt the existing plan to include new events. Examples of such events are an additional train that should be shunted, a shunting track that can not be used because of maintenance, a train that gets an extra coach, and changes in the arrival or departure time of a train. A task analysis with shunting planners in practice shows there is a number of recurring activities that have to be performed (Van Wezel and Kiewiet 2006):

1. *Unit matching*. Incoming coaches must be matched to outgoing coaches. Because individual coaches of the same type are exchangeable, the planners have considerable flexibility.
2. *Track assignment*. Given the matching of individual units, the steps to realize the new configuration of the trains must be determined. This consists of a series of splitting trains into coaches, moving coaches, and connecting coaches into trains. At each moment, a train must be parked somewhere. Multiple trains

can be parked on the same track as long as the length of the combined trains does not exceed the length of the track, but the planner must make sure that a train is not blocked by another train when it must be moved. Three important requirements that have to be taken into account with track assignments are:

(a) *Washing*. Trains need to be washed frequently. In our reference case Zwolle, there is one washing track with automated washing equipment. All train units must be shunted to this track sometime during the night. Because the washing time is independent of the length of the train, it is advantageous to combine as much coaches as possible before washing. Of course, combining coaches requires additional shunting train movements.
(b) *Internal cleaning*. Trains must be cleaned internally every day. For safety reasons, there are a limited number of tracks that can be used by the cleaning staff, which means that shunting is required for internal cleaning as well.
(c) *Refueling*. There are both electrical and diesel trains. The diesel trains need to be refueled during the night at one of the refueling tracks, again needing shunting movements.

3. *Routing*. A movement from one track to another requires a route through the shunting yard. Because a train can not turn around like a car, it might have to change its driving direction (see Fig. 14.5 for a direction change in the route for a train that drives from track 3B to track 4B). Drivers realize the direction change by driving to a track where the train is allowed to stop, getting off the train, walking to the other side of the train, and starting it again. As the trains can be controlled from both the front side and the rear side, either the driver has to walk a distance equal to the total length of the train, or two drivers are used simultaneously. Both solutions are costly, making routes involving direction changes inefficient. Moreover, direction changes are only allowed at tracks that are long enough, as for safety reasons a train may only be halted at tracks that are surrounded by official rail signals. A planner will therefore try to minimize the

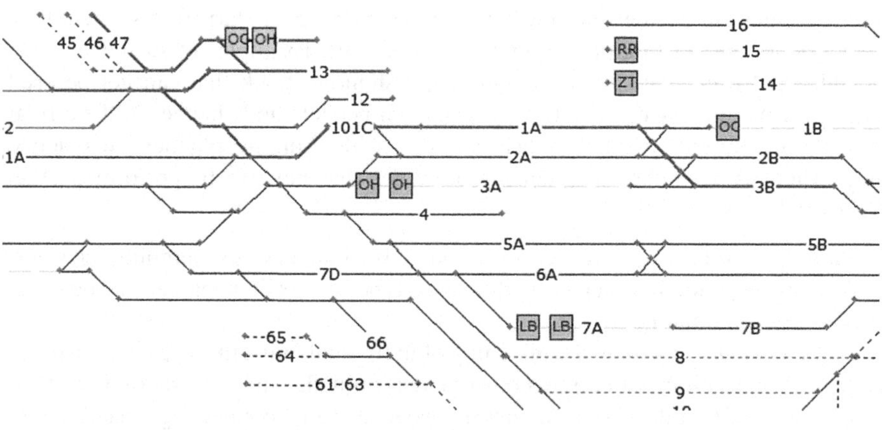

Fig. 14.5 Example of direction change (thick lines indicate the tracks used on the route)

number of direction changes in a route. The actual driving distance of the train on a route is of minor importance.

4. *Shunting staff assignment*. Connecting and disconnecting trains requires specialized staff, as does driving the train. The main trade-off for shunting staff assignment is walking time versus waiting time. For example, a shunting driver might drive a train to track 91 at 04.30, and walk to another train that must be moved from track 91 to track 102 at 05.00. He has approximately 5 min walking time and 25 min waiting time. But he could also be scheduled to walk to track 102 where a train must be moved at 04.55, which requires 20 min walking time and 5 min waiting time. Planners try to assign activities that are close together, which results in larger intervals in which no tasks are assigned. In these larger intervals, the shunting staff can wait in the canteen.

Note that during our task analysis, the planners did not use elaborate computer support in their work. The plans were created manually on paper, after which the plan is entered in the computer. The planners try to schedule the washing, internal cleaning, and refueling as much as possible immediately after a train arrives at the station because this reduces the number of train movements. The above mentioned planning activities must be performed regardless of the kind of event that caused them. However, the kind of event determines the task strategy. For example, if the arrival time of a train changes, the planner will try to minimize disrupting the existing schedule and absorbing the change by changing only the time of the shunting activities and thereby keeping the order of shunting activities intact. However, if an additional train has to be scheduled, the planner might have to start with matching the units again which could result in considerably more changes. Train unit matching, track assignment, routing, and shunting staff assignment conceptually are the natural level to describe the shunting scheduling task. However, if we need to design support, we can describe the task in more detail as well as on a more aggregate level. Although the purpose of this section is not to give a full description of the subtask structure of shunting planning, we will give two examples. A cognitive task analysis with one of the planners showed the following subtasks for finding an appropriate track or set of tracks for one train (Table 14.4):

Table 14.4 Subtasks for track assignment (Van Wezel and Jorna 2009; Van Wezel and Barten 2002)

1	Check if a track is free for the time interval that is needed. If not go to 2 or 5, otherwise go to 6
2	Check if a track is free for another time interval without changing the order of movements. If not go to 3 or 5, otherwise go to 6
3	Check if it is possible to internally clean the train at a non-cleaning track. Maybe it can be cleaned right after arrival at the arrival track. This might solve the problem. If not go to 4 or 5, otherwise go to 6
4	Check if the problem will be solved if the train will not be washed that night. If not go to 5, otherwise go to 6
5	Select a train that is in conflict with a chosen possible solution and go to 1 to find a track or set of tracks for the conflicting train
6	Apply the solution

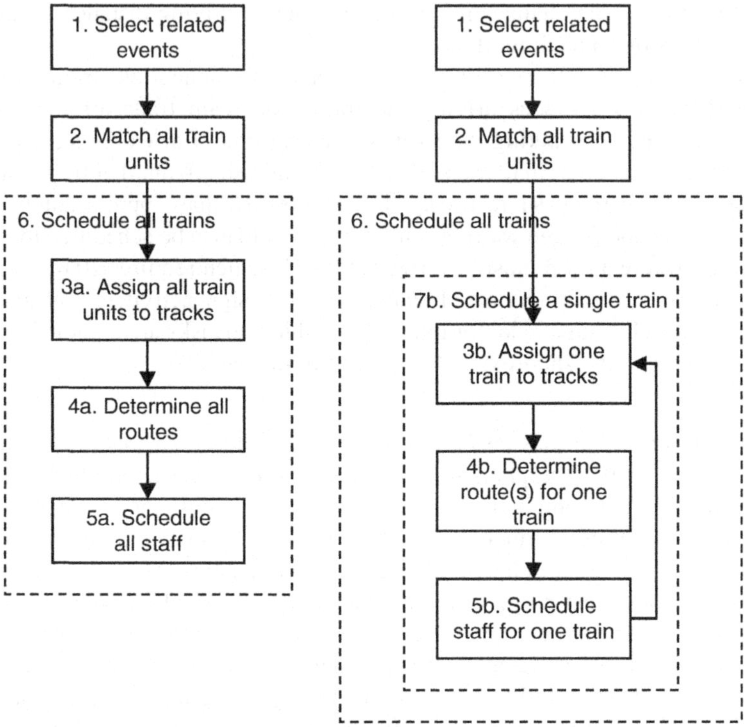

Fig. 14.6 Task strategies for shunting scheduling

In principle, each of these subtasks is a candidate for algorithmic support and, hence, for analyzing the underlying subproblems, the possible algorithms, the cognitive aspects of the human using the algorithm in the given subtask, and the resulting control mode.

Looking at a less detailed level is possible as well. There are multiple ways in which the lowest level subtasks are clustered in more or less independent packages. Our task analyses showed that events are processed in the following structure (the numbers between brackets refer to Fig. 14.6). Events are somehow sorted according to the relation between the events (1). Related events are handled as a set. For example, events can relate to the same unit type which makes it logical to analyze them in cohesion because units of the same type are exchangeable. Then, the arriving and departing units within the set are matched (2). We encountered two different strategies to perform the subtasks concerning scheduling of trains (6). In the first strategy, all trains are first assigned to the tracks (using the structure from Table 14.4), taking into account washing, cleaning, and refueling (3a). Then, the planner schedules the routes of all movements that are needed (4a), and he assigns the shunting staff to perform the connection, disconnection, and shunting activities (5a). In the second strategy, a train is assigned to a track, again using the structure in Table 14.4 (3b), immediately after which the planner creates the necessary route for this train (4b) and assigns the shunting staff for moving the train (5b). Whether the first or the second strategy is chosen mainly depends on the current situation on the shunting yard. If a lot of tracks are occupied already, it might be difficult to find routes in which case the planner decides to not postpone routing and staff scheduling. In this second strategy, there appears another subtask. The combination of 3b, 4b, and 5b can be seen as the higher level subtask of performing all necessary activities for a single train (7b). Next to all smaller subtasks, this subtask is a candidate for function allocation as well.

The descriptions of the various subtasks shows us that different task strategies are possible, and that different groupings of subtasks provide us with multiple candidates for analyzing possible computer support. In Fig. 14.6, one important aspect of applying the allocation model can be seen. Suppose, for example, that a combined algorithm we have for 3a, 4a, and 5a performs better than repeatedly applying algorithms we have for 3b, 4b, and 5b. Further suppose that we decide on the basis of an analysis of subtask 6 that situation awareness is not important. In this case, we would apply algorithms 3a, 4a, and 5a automatic or under supervisory control. However, if situation awareness is important, this would be a bad design choice because the planner would have no idea where all trains are on the station after running the algorithms, which would lead to the planner disregarding the algorithm and making schedules manually again. In the next subsection, we will analyze the characteristics of the separate subtasks that we identified for shunting scheduling.

14.5.3 Characteristics of the Subtasks

The first step in the function allocation model is to analyze the subtasks that are identified by the task analysis. What is the cost of suboptimal decisions, is there need for situation awareness, what is the maximum decision cycle time, and how complex is the problem? The values these characteristics have for the various

Table 14.5 Characteristics of subtasks

	Harm risk	Need for situation awareness	Maximum cycle duration	Complexity
1. Select related events	Medium	Low	Medium	Low
2. Match all train units	Medium	High	Short	Low
3a. Assign all train units to tracks	High	High	Long	High
3b. Assign one train to tracks	Medium	High	Short	Medium
4a. Determine all routes	Medium	Low	Medium	High
4b. Determine route(s) for one train	Low	Low	Short	Low
5a. Schedule all staff	High	Low	Medium	Medium
5b. Schedule staff for one train	Medium	Low	Short	Low
6. Schedule all trains	High	High	High	High
7b. Schedule a single train	Medium	High	Medium	Medium

subtasks can be found in Table 14.5; we will describe how we evaluated the various subtasks.

14.5.3.1 Selecting Related Events

Selecting related events (1) is not a complex task. The number of events that must be processed is somewhere between 5 and 30, and the ordering is done to find groups of planning activities that are somehow related. A poor ordering could, but not necessarily will, lead to extra shunting movements, therefore the harm risk is medium. Furthermore, because of the low complexity, situation awareness is easily attained by looking at the solution, which means that there is no need to involve the human planner in the process of ordering itself. Selecting related events is the start of solving the scheduling problem and is done only once. Therefore, it is not a problem if it takes some time.

14.5.3.2 Matching Train Units

Matching the train units that are involved in the chosen events (2) is of medium importance. Matching is an intermediate step that does not directly influence the amount of shunting operations. However, a poor matching could lead to complex shunting operations. Although the complexity is rather low because of the low number of train units per type, the need for situation awareness is high. The reason is that the matching is done with some general idea of how the track assignments (which are done in the next step) will lead to the correct matching. This general idea can not be found easily by looking at the outcome of the matching procedure, which means that the planner needs to know the reasoning followed in the process of matching.

14.5.3.3 Determine Tracks

Assigning trains to tracks (3a/3b) is the core of the shunting scheduling problem. It determines directly the number of activities that must be performed by the shunting staff, and it determines for the most important constrains (washing, cleaning, and refueling) whether they are being followed or not. Therefore, the harm risk is high. Because trains can block each other on tracks, track assignments are mutually dependent. As a consequence, the need for situation awareness is high, because the planners need to know the internal structure of the schedule, i.e., where the trains are on the shunting yard, in order to be able to make adaptations to the schedule at a later stage. Because track assignment is the main task of the planners, the decision cycle duration of the subtask to assign all trains (task 3a) might be long. However, it is very complex as well, because it is about *all* trains during the whole night. In contrast, determining the track assignments of one train (3b) is less complex. Because it is only a small part of the total task, it must be done fast, and for the same reason the harm risk is not high. However, poor track assignment can lead to more movements than necessary, which means the harm risk is neither low.

14.5.3.4 Determine Routes

Unlike the track assignments of train units, routes to drive a train from one track to another are mutually independent. And although the complexity to determine the optimal solution for all routes (4a) in a single step is high, the need for situation awareness is low because changing a single route is easy and does not influence other parts of the schedule. Furthermore, the train drivers are not the bottleneck during the night, so poor routes have no high costs in itself. However, if many routes are poor, tracks might become blocked more than necessary which impairs the possibilities for track assignment. Therefore, the harm risk is medium. Determining a single route (4b) is not complex and the harm risk and need for situation awareness are low. However, as it might have to be done often, the maximum decision cycle duration is short.

14.5.3.5 Assign Staff

Assigning shunting staff has largely the same characteristics as routing. The harm risk is somewhat higher for both assigning staff for all shunting activities in one step (5a) and for one shunting activity at the time (5b) because a poor staff schedule might lead to much waiting and walking time, which results in dissatisfied shunting staff. The need for situation awareness is low for both 5a and 5b. The activities the shunting staff must perform are only slightly interrelated (a change in track assignment might lead to activities that are too far apart which might mean that the train driver can not be there in time), but these relations are easy to track by looking at the plan itself. The complexity of scheduling all staff is lower than the routing problem, and scheduling staff for one activity is a simple problem.

14.5.3.6 Combined Subtasks

As can be deduced from the above, assigning tracks, routes, and shunting staff to all trains in one step (6) is very complex, has a high need for situation awareness, a high harm risk, and a long maximum cycle duration. Scheduling a single train (7b) has medium harm risk, medium maximum cycle duration, medium complexity, and a high need for situation awareness.

14.5.4 Matching Algorithms to Subtasks

As discussed in the introduction of the section, six algorithms were developed for solving parts of the described shunting problem. Note that more algorithms for shunting planning are described in literature (see Haijema et al. (2006) for a literature review), but using the algorithms that were developed for the Rintel project ensures that we only include algorithms that were actually developed to solve the described shunting problem. We will provide a concise description of the algorithms and their characteristics.

Abbink (2006) developed a constraint satisfaction approach for integrated matching and assigning tracks. The algorithm does not optimize, it stops once it finds a valid solution. Furthermore only two movements per train unit are determined: form arrival track to parking track, and from parking track to departure track. So, washing, cleaning, and refueling are not taken into account. The algorithm also determines the routes but does not check for conflicts in the routes.

Lentink et al. (2006) describe a mathematical approach for the subtasks matching, track assignment, and routing. In contrast with the approach of Abbink (2006), these three steps are performed sequentially instead of integrated, which means that the planner can adapt the outcome of a step before the next step is performed.

Haijema et al. (2006) describe a multi-period heuristics that applies a set of decision rules hierarchically. They first make a blueprint of arriving and departing trains. Based on this, they match train units. Given the matching, a heuristics that is inspired by dynamic programming assigns trains to tracks.

Riezebos and Van Wezel (2006) describe two heuristics. The first heuristics finds a set of tracks to assign a train to a given time interval. The second heuristics determines a route for moving a train. The heuristics are both based on the K-shortest path algorithm and can both provide a number of alternative solutions.

Lamberts (2007) describes an algorithm for assigning shunting staff. A shortest augmenting path algorithm is used that combines shunting activities in cycles in such a way that within a cycle the walking distance is minimized (Jonker and Volgenant 1987). Multiple cycles are assigned to a shift. Longer cycles means less overall walking distance but at the same time longer waiting times. The balance is found by calculating multiple solutions with different cycle lengths and combining the cycles to shifts such that waiting time is minimized without violating constraints.

The characteristics of the algorithms can be found in Table 14.6.

The aim of the function allocation model is to determine what kind of algorithmic support is acceptable from a cognitive perspective. According to function allocation literature, this is a prerequisite for successful application. In the next subsection, we will describe the algorithms and heuristics from the perspective of the user, taking into account the cognitive dimensions that were discussed in Sect. 14.4.

Table 14.6 Characteristics of algorithms

Algorithm	Subtasks	Speed	Error rate	Quality	Interaction
Riezebos and Van Wezel (2006)	3b. Assign one train to tracks (provides *n* alternatives)	<1 s	No errors	Near optimal	Planner can choose alternatives
Riezebos and Van Wezel (2006)	4b. Determine the *n* shortest routes for one train	<4 s	No errors	Near optimal	Planner can choose alternatives. Planner can block tracks or include tracks to be used in the route
Lamberts (2007)	5a. Schedule all staff	<1 min	No errors	Near optimal	Specify weights of goals (walking versus waiting time)
Lentink et al. (2006)	2+3a+4a. Matching, track assignment, and routing; performed sequentially	<1 h	Washing, cleaning, refueling not included. Routes might contain conflicts	High	Matching can be predetermined. Outcomes of the steps can be changed Planner can specify weights for objectives
Haijema et al. (2006)	2+3a. Matching, track assignment	<1 s	Washing, cleaning, refueling not included. Solution might contain conflicts	Heuristic; high quality	None
Abbink (2006)	2+3a+4a. Integrated matching, track assignment, and routing	<5 min	Washing, cleaning, refueling not included. Routes might contain conflicts	Conflict free; no optimization	None

14.5.5 Cognitive Aspects of the Joint Human/Computer System

Using an algorithm to determine parts of the schedule has several consequences for the task of human planners. As discussed in Sect. 14.4, the use of algorithms may lead to complacency, skill loss, and unacceptable remaining tasks. In this subsection, we will analyze for the various algorithms how they score on these aspects. Note that these characteristics should be evaluated in the context of the subtask they support.

The single-train algorithms of Riezebos and Van Wezel (2006) are not prone to skill loss or complacency. The subtasks they support are not complex, the outcome is easy to interpret, and the algorithms provide multiple alternative solutions. The situation awareness when using these algorithms can be high because each time the algorithm is applied, the outcome can be evaluated, which means that the planner deliberates on each individual track assignment. The overall quality of the outcome is not necessary high, because the possibilities of combinatorics are ignored; always choosing the best alternative that is offered by the "single-train" algorithms results in a greedy strategy.

Although the staff assignment problem is relatively simple, the algorithm of Lamberts (2007) results in little situation awareness. The slack in the staff schedule depends on the flexibility in the shunting schedule, and because the algorithm of Lamberts schedules all staff assignments simultaneously, the planner has no feeling for the relation between the staff schedule and the shunting schedule. For the same reason, there are no left-over tasks but the risk of skill loss and complacency is high. The overall quality in terms of waiting time and walking distance for the shunters is high.

The algorithm of Lentink et al. (2006) provides a complete solution for track assignment and routing. However, the algorithm does not schedule washing, cleaning, and refueling, and routes might contain errors. The left-over task is to add those activities and check the routes. Because the planner still has to do so much work manually, the risk of skill loss and complacency is limited. However, because of the black-box nature of the algorithm (the algorithm provides a complete solution without feedback during the process), the situation awareness when the algorithm is used is low. Due to the black-box nature, the algorithms of Haijema et al. (2006) and Abbink (2006) largely have the same properties as the algorithm of Lamberts et al. (2007), although there are small differences in the left-over tasks.

Table 14.7 shows the human aspects of the algorithms.

14.5.6 Determining the Control Mode

According to the model for function allocation proposed in Sect. 14.4, the characteristics of the sub-problems, algorithms, and human should lead to the choice of control mode. For this, we first have to determine whether respectively algorithm

Table 14.7 Characteristics of human aspects using the algorithm

Algorithm	Skill loss/ complacency	Situation awareness	Left-over Tasks	Overall quality
Riezebos and Van Wezel (2006), track assignment	Low	High	Applying the algorithm sequentially	Risk of greedy strategy
Riezebos and Van Wezel (2006), routing	Low	High	Applying the algorithm sequentially	Risk of greedy strategy
Lamberts (2007)	High	Low	None	High
Lentink et al. (2006)	Low	Low	Schedule washing, cleaning, refueling	Unknown; depends on possibility to perform left-over tasks
Haijema et al. (2006)	Low	Low	Schedule washing, cleaning, refueling; check for errors	Unknown; depends on possibility to perform left-over tasks
Abbink (2006)	Low	Low	Schedule washing, cleaning, refueling; check for errors in routes	Unknown; depends on possibility to perform left-over tasks

Table 14.8 Possible control modes for the subtasks

	Human	Algorithm	Possible control modes
1. Select related events	Optional	Optional	Dynamic allocation/supervisory control/automatic
2. Match all train units	Necessary	Optional	Dynamic allocation/advisory control/interaction
3a. Assign all train units to tracks	Necessary	Necessary	Interaction
3b. Assign one train to tracks	Necessary	Optional	Dynamic allocation/advisory control/interaction
4a. Determine all routes	Optional	Necessary	Interaction/supervisory control/ automatic
4b. Determine route(s) for one train	Optional	Optional	Supervisory control/automatic
5a. Schedule all staff	Optional	Optional	Supervisory control or dynamic control (medium harm risk precludes automatic scheduling)
5b. Schedule staff for one train	Optional	Optional	Supervisory control/automatic
6. Schedule all trains	Necessary	Necessary	Interaction
7b. Schedule a single train	Necessary	Optional	Dynamic allocation/advisory control/interaction

and human are impossible, necessary, or optional. Table 14.8 shows the control modes that are possible given the characteristics of the sub-problems. This is determined as follows. When a sub-problem has a high need for situation awareness, the human is required. When complexity is high and maximum cycle duration is low, the human is not able to perform the task. This latter situation is not encountered in the sub-problems. In all other situations, the role of the human is optional. For the algorithms we look at the complexity and harm risk. If one of these is high, we deem algorithmic support to be necessary; in other cases an algorithm is optional. On the basis of this classification we can determine the control modes that are possible. We will demonstrate this line of reasoning with subtask 2: match all train units. The need for situation awareness is high which indicates that the human is required. The complexity is low, hence an algorithm is optional. Thus, the control modes advisory control and interaction are possible (see Table 14.8). The harm risk is medium, so when an algorithm for this subtask would result in skill loss or complacency, dynamic control should be used to prevent these effects. The possible control modes for the other subtasks are determined likewise.

Given the optional control modes, we can analyze whether the available algorithms match the requirements. For this, it must first be determined whether the algorithm can work in one of the possible control modes. Furthermore, we have to look at the human characteristics. The black box nature of the comprehensive algorithms of Lentink et al. (2006); Haijema et al. (2006), and Abbink (2006) leads to low situation awareness. These three algorithms all perform subtask 3a, a task for which high situation awareness is needed. As a consequence, the algorithms can not be used. The track assignment algorithm of Riezebos and Van Wezel (2006) performs subtask 3b. The analysis shows that use of the algorithm would lead to high situation awareness but also a risk of skill loss and complacency. Hence, dynamic allocation is proposed, where advisory control and interactive control are alternated. The routing algorithm of Riezebos and Van Wezel (2006) performs task 4b. The algorithm itself can be used in full automatic mode. Furthermore, the algorithm could be useful for subtask 4a as well in automatic mode, interactive mode, or using supervisory control by running the algorithm repetitively for each train movement. Lastly, the staff scheduling algorithm of Lamberts (2007) can be used in supervisory control mode, letting the user know intermediate solutions.

14.6 Discussion

An important feature of the function allocation model can be seen in Fig. 14.4. The problem decomposition, appropriate algorithms, and role of the human planner interact. In this section, we showed a real world example, but as we only examined existing algorithms we did not demonstrate this feature. We would like to point out some aspects that are inherent in the function allocation model and that should be considered in a real world setting because it would strengthen the shunting case analysis considerably. First, the analysis shows several subtasks for which no algorithm was readily available. Formulating usability requirements for algorithms to support these subtasks is quite straightforward using the function allocation model. For example, none of the algorithms explicitly supports the task of selecting related events. Second, the algorithms for which we concluded that they were not usable given the characteristics of the sub problems, might be adaptable to meet the requirements. For example, the algorithms might allow more interaction if it would be possible to apply them to sub problems. Third, the analysis of sub problems, available algorithms, and role of the human planners might reveal the need for a more detailed problem decomposition (for example Table 14.4), and thereby show more possibilities for algorithmic support.

14.7 Conclusions

Making algorithms to solve scheduling problems is a challenging task. It involves mathematical as well as practical aspects in a continuously changing business environment. And precisely because of the close relationship between algorithms and business practice, it is surprising that there is so little theory about the role of the human in using scheduling algorithms, at least when compared to other fields such as car driving, control of nuclear power plants, air-traffic control, etc. Still, this human planner is both the user of the system and the one who is responsible for creating the plan and making the decisions.

In other empirical fields like the ones mentioned, the role distribution between the human and the control system is investigated by cognitive ergonomics. Several theories and methods are developed that can be used to determine the way in which functions should be allocated to the human and the computer. In this article we have investigated the applicability of function allocation theory for planning and scheduling tasks. Our analysis shows considerable differences between dynamic situations and scheduling. In particular, dynamic situations require high risk decisions under time pressure, whereas solving scheduling problems requires much information processing capacity. We propose a function allocation model in which the decision about functional allocation depends on the characteristics of the scheduling problem, the possibilities and requirements of the algorithm, and the cognitive limitations of the human user.

Currently, we are testing the model in a real-world scenario of a partially failed implementation of algorithms. In future research we will develop more detailed contents, and we will extend the model with findings from empirical research.

References

Abbink, E. (2006). Intelligent shunting: dealing with constraints (satisfaction). In W. Van Wezel, R. J. Jorna, & A. Meystel (Eds.), *Planning in intelligent systems: Aspects, motivation, and methods*. New York: Wiley.

Annett, J. (2000). Theoretical and pragmatic influences on task analysis methods. In J. M. Schraagen, S. F. Chipman, & V. L. Shalin (Eds.), *Cognitive task analysis* (pp. 25–37). Mahwah, NJ: Lawrence Erlbaum Associates.

Annett, J., & Duncan, K. D. (1967). Task analysis and training design. *Occupational Psychology, 41*, 211–221.

Bainbridge, L. (1983). Ironies of automation. *Automatica, 19*, 775–779.

Bernardo, J. J., & Lin, K. S. (1994). An interactive procedure for bi-criteria production scheduling. *Computers and Operations Research, 21*(6), 677–688.

Campbell, D. J. (1988). Task complexity: A review and analysis. *Academy of Management Review, 13*(1), 40–52.

Cegarra, J. (2008). A cognitive typology of scheduling situations: A contribution to laboratory and field studies. *Theoretical Issues in Ergonomics Science, 9*(3), 201–222.

Cegarra, J., & Hoc, J. M. (2008). The role of algorithm and results comprehensibility of automated scheduling on complacency. *Human Factors and Ergonomics in Manufacturing, 18*(6), 603–622.

Davis, F. E., & Kottemann, J. E. (1995). Determinants of decision rule use in a production planning task. *Organizational Behavior and Human Decision Processes, 63*(2), 145–157.

Dearden, A., Harrison, M., & Wright, P. (2000). Allocation of function: scenarios, context and the economics of effort. *International Journal of Human Computer Studies, 52*(2), 289–318.

Dockx, K., De Boeck, Y., & Meert, K. (1997). Interactive scheduling in the chemical process industry. *Computers and chemical engineering: an international journal, 21*(9), 925–946.

Dutoon, J. M. (1964). Production scheduling: A behavior model. *International Journal of Production Research, 4*, 21–41.

Endsley, M. R. (1995). Measurement of situation awareness in dynamic systems. *Human Factors, 37*, 65–84.

Endsley, M. R., & Kiris, E. O. (1995). The out-of-the-loop performance problem and level of control in automation. *Human Factors, 37*(1), 381–394.

Endsley, M. R., Kiris, E. O., Endsley, M. R., & Kaber, D. B. (1999). Level of automation effects on performance, situation awareness and workload in a dynamic control task. *Ergonomics, 42*(3), 462–492.

Fitts, P. M. (1951). *Human engineering for an effective air navigation and traffic control system.* Washington, DC: National Research Council.

Gabrel, V., & Vanderpooten, D. (2002). Enumeration and interactive selection of efficient paths in a multiple criteria graph for scheduling an earth observing satellite. *European Journal of Operational Research, 139*(3), 533–542.

Haijema, R., Duin, C. W., & van Dijk, N. M. (2006). Train shunting: A practical heuristic inspired by dynamic programming. In W. Van Wezel, R. J. Jorna, & A. Meystel (Eds.), *Planning in intelligent systems: Aspects, motivation, and methods.* New York: Wiley.

Hoc, J. M. (2000). From human-machine interaction to human-machine cooperation. *Ergonomics, 43*, 833–843.

Hoc, J. M. (2001). Towards a cognitive approach to human-machine cooperation in dynamic situations. *International Journal of Human Computer Studies, 54*, 509–540.

Hoc, J. M., & Amalberti, R. (2007). Cognitive control dynamics for reaching a satisficing performance in complex dynamic situations. *Journal of Cognitive Engineering and Decision Making, 1*(1), 22–55.

Hoc, J. M., & Debernard, S. (2002). Respective demands of task and function allocation on human-machine co-operation design: a psychological approach. *Connection Science, 14*, 283–295.

Hoc, J. M., Mebarki, N., & Cegarra, J. (2004). L'assistance à l'opérateur humain pour l'ordonnancement dans les ateliers manufacturiers. *Le Travail Humain, 67*, 181–208.

Jonker, R., & Volgenant, A. (1987). A shortest augmenting path algorithm for dense and sparse linear assignment problems. *Computing, 38*, 325–340.

Kaber, D. B., Riley, J. M., Tan, K.-W., & Endsley, M. R. (2001). On the design of adaptive automation for complex systems. *International Journal of Cognitive Ergonomics, 5*, 37–57.

Kantowitz, B. H., & Sorkin, R. D. (1987). Allocation of functions. In G. Salvendy (Ed.), *Handbook of human factors* (pp. 355–369). New York: Wiley.

Lamberts, D. (2007). Algorithmic support for worker assignment planning. Unpublished master thesis, University of Groningen, The Netherlands.

Lauer, J., Jacobs, L. W., Brusco, M. J., & Bechtold, S. E. (1994). An interactive, optimization-based decision support system for scheduling part-time, computer lab attendants. *Omega, 22* (6), 613–626.

Lee, J., & Moray, N. (1992). Trust, control strategies and allocation of function in human-machine systems. *Ergonomics, 35*(10), 1243–1270.

Lee, J., & Moray, N. (1994). Trust, self-confidence, and operators adaptation to automation. *International Journal of Human Computer Studies, 40*(1), 153–184.

Lee, J. D., & See, K. A. (2004). Trust in automation: Designing for appropriate reliance. *Human Factors, 46*, 50–80.

Lentink, R. M., Fioole, P. J., Kroon, L. G., & Van't Woudt, C. (2006). Applying operations research techniques to planning of train shunting. In W. Van Wezel, R. J. Jorna, & A. Meystel (Eds.), *Planning in intelligent systems: Aspects, motivation, and methods*. New York: Wiley.

Marsden, P., & Kirby, M. (2005). Allocation of functions. In N. Stanton (Ed.), *Handbook of human factors and ergonomics methods*. London: CRC.

McKay, K. N., & Buzacott, J. A. (2000). The application of computerized production control systems in job shop environments. *Computers in Industry, 42*, 79–97.

McKay, K. N., & Wiers, V. C. S. (2003). Integrated decision support for planning, scheduling, and dispatching tasks in a focused factory. *Computers in Industry, 50*, 5–14.

Mietus, D. M. (1994). *Understanding planning for effective decision support*. PhD Thesis, University of Groningen, The Netherlands.

Moray, N., Dessouky, M. I., Kijowski, B. A., & Adapathya, R. (1991). Strategic behavior, workload, and performance in task scheduling. *Human Factors, 33*, 607–629.

Mosier, K. L., & Skitka, L. J. (1996). Human decision-makers and automated decision aids: Made for each other? In R. Parasuraman & M. Mouloua (Eds.), *Automation and human performance: theory and applications*. Mahwah, NJ: Lawrence Erlbaum Associates.

Myers, K. L., Tyson, W. M., Wolverton, M. J., Jarvis, P. A., Lee, T. J., & desJardins, M. (2002). "PASSAT: A User-centric Planning Framework". In: *Proceedings of the 3rd International NASA Workshop on Planning and Scheduling for Space*, Houston, TX.

Neerincx, M. A., & Griffioen, E. (1996). Cognitive task analysis: harmonising tasks to human capacities. *Ergonomics, 39*(4), 543–561.

Oddi, A., & Cesta, A. (2000). Toward interactive scheduling systems for managing medical resources. *Artificial Intelligence in Medicine, 20*(2), 113–138.

Older, M. T., Waterson, P. E., & Clegg, C. W. (1997). A critical assessment of task allocation methods and their applicability. *Ergonomics, 40*(2), 151–171.

Parasuraman, R., Molloy, R., & Singh, I. L. (1993). Performance consequences of automation-induced "complacency". *The International Journal of Aviation Psychology, 3*, 1–23.

Parasuraman, R., & Riley, V. (1997). Humans and automation: Use, misuse, disuse, abuse. *Human Factors, 39*(2), 230–253.

Parasuraman, R., Sheridan, T. B., & Wickens, C. D. (2000). A model for types and levels of human interaction with automation. *Systems, Man and Cybernetics, Part A, 30*(3), 286–297.

Price, H. E., Maisano, R. E., & VanCott, H. P. (1982). *The allocation of function in man-machine systems: A perspective and literature review*. Oak Ridge, TN: Oak Ridge National Laboratory.

Pulliam, R., & Price, H. E. (1984). Allocating functions to man or machine in nuclear power plant control. *The Nuclear Engineer, 25*(3), 79–85.

Riedel, R., Fransoo, J., & Wiers, V. C. S. (2007). *Building decision support systems for acceptance – the role of trust. Proceedings of the 14th Annual Euroma Conference*, Ankara, Turkey.

Riezebos, J., & van Wezel, W. M. C. (2006). Planner-oriented design of algorithms for train shunting scheduling. In W. M. C. van Wezel, R. J. Jorna, & A. M. Meystel (Eds.), *Planning in intelligent systems: Aspects, motivations, and methods* (pp. 477–496). New Jersey: Wiley.

Riezebos, J., & van Wezel, W. M. C. (2009). k-Shortest routing of trains on shunting yards. *OR Spectrum, 31*(4), 745–758.

Rouse, W. B. (1976). Adaptive allocation of decision making responsibility between supervisor and computer. In T. B. Sheridan & G. Johannsen (Eds.), *Monitoring behavior and supervisory control* (pp. 295–306). New York: Plenum Press.

Sanderson, P. M. (1989). The human planning and scheduling role in advanced manufacturing systems: an emerging human factors domain. *Human Factors, 31*, 635–666.

Scerbo, M. W. (1996). Theoretical perspectives on adaptative automation. In R. Parasuraman & M. Mouloua (Eds.), *Automation and human performance* (pp. 37–64). Mahwah, NJ: Lawrence Erlbaum Associates.

Schakel, B. (1976). Process control: Simple and sophisticated display devices as decision aids. In T. B. Sheridan & G. Johanssen (Eds.), *Monitoring behavior and supervisory control* (pp. 429–444). New York: Plenum.

Schraagen, J. M., Chipman, S. F., & Shalin, V. L. (Eds.). (2000). *Cognitive task analysis*. Mahwah, NJ: Lawrence Erlbaum Associates.

Sharit, J. (1997). Allocation of functions. In G. Salvendy (Ed.), *Handbook of human factors and ergonomics* (2nd ed.). New York: Wiley.

Sheridan, T. B., & Verplanck, W. L. (1978). *Human and computer control of undersea teleoperators*. Cambridge, MA: MIT Man–Machine Systems Laboratory Report.

Smed, J., Johtela, T., Johnsson, M., Puranen, M., & Nevalainen, O. (2000). An interactive system for scheduling jobs in electronic assembly. *The international journal of advanced manufacturing technology, 16*(6), 450–459.

Smith, H. T., & Crabtree, R. G. (1975). Interactive planning: a study of computer aiding in the execution of a simulated scheduling task. *International Journal of Man-Machine Studies, 7*, 213–231.

Smith, P. J., McCoy, C. E., & Layton, C. (1997). Brittleness in the design of cooperative problem-solving systems: the effects on user performance. *Systems, Man and Cybernetics, Part A, 27*(3), 360–371.

Ulusoy, G., & Özdamar, L. (1996). A framework for an interactive project scheduling system under limited resources. *European Journal of Operations Research, 90*(2), 362–375.

van Wezel, W. M. C., & Barten, B. (2002). Hierarchical mixed-initiative planning support. In: T. Grant and C. Witteveen (Eds.), *Plansig 2002. Proceedings of the 21th workshop of the UK Planning and Scheduling Special Interest Group* (pp 7–17). Delft: Delft University of Technology.

Van Wezel, W. M. C., & Jorna, R. J. (2006). Introduction. In W. Van Wezel, R. J. Jorna, & A. Meystel (Eds.), *Planning in intelligent systems: Aspects, motivation, and methods*. New York: Wiley.

Van Wezel, W. M. C., & Jorna, R. J. (2009). Cognition, tasks, and planning; supporting the planning of shunting operations at Netherlands Railways. *Cognition, Technology & Work, 11* (2), 165–176.

Van Wezel, W., Jorna, R. J., & Meystel, A. (Eds.). (2006a). *Planning in intelligent systems: Aspects, motivation, and methods*. New York: Wiley.

Van Wezel, W. M. C., & Kiewiet, D. J. (2006). Perspectives on shunting planning: Research in planning support at the Netherlands railways. In W. Van Wezel, R. J. Jorna, & A. Meystel (Eds.), *Planning in intelligent systems: Aspects, motivation, and methods*. New York: Wiley.

Van Wezel, W. M. C., van Donk, D. P., & Gaalman, G. J. C. (2006b). The planning flexibility bottleneck in food processing industries. *Journal of Operations Management, 24*(3), 287–300.

Wäfler, T. (2001). Planning and scheduling in secondary work systems. In B. MacCarthy & J. Wilson (Eds.), *Human performance in planning and scheduling*. London, UK: Taylor and Francis.

Chapter 15
Design of Scheduling Algorithms: Applications

Jan Riezebos, Jean-Michel Hoc, Nasser Mebarki, Christos Dimopoulos, Wout Van Wezel, and Guillaume Pinot

Abstract This chapter discusses the insights developed for designing scheduling algorithms according to three design projects where algorithms have been developed. The choice of applications covers a broad spectrum. The methods used are from three different fields, namely combinatorial optimization, genetic (evolutionary) algorithms, and mathematical optimization. The application areas differ also in terms of the role of a human user of the algorithm. Some of these algorithms have been developed without detailed study of the competences of the perceived users. Others have examined humans when performing the scheduling tasks manually, but have not considered the change in cognitive load if the process of planning changes due to the new algorithm and computerized support. Although none of the design projects fulfils all criteria developed in the framework of Chap. 12, we show that the framework helps to assess the design projects and the resulting algorithms, and to identify the main weaknesses in these applications. Finally, we show how they can be addressed in future.

J. Riezebos (✉) and W. Van Wezel,
Faculty of Economics and Business, University of Groningen, Groningen, The Netherlands
e-mail: j.riezebos@rug.nl, w.m.c.van.wezel@rug.nl

J.-M. Hoc and G. Pinot
Centre National de la Recherche Scientifique, (CNRS: French National Research Centre), University of Technology of Compiègne, Compiègne, France
e-mail: Jean-Michel.Hoc@irccyn.ec-nantes.fr, guillaume.pinot@irccyn.ec-nantes.fr

N. Mebarki
Département Qualité, Logistique Industrielle et Organisation (QLIO), University of Nantes, Nantes, France
e-mail: nasser.mebarki@univ-nantes.fr

C. Dimopoulos
European University Cyprus, Nicosia, Cyprus
e-mail: c.dimopoulos@euc.ac.cy

J.C. Fransoo et al. (eds.), *Behavioral Operations in Planning and Scheduling*,
DOI 10.1007/978-3-642-13382-4_15, © Springer-Verlag Berlin Heidelberg 2011

The three application areas are:

1. Decision support for shunting yard scheduling using a network flow heuristic.
2. An evolutionary multi-objective decision tool for job-shop scheduling.
3. *Group sequencing.* A predictive-reactive scheduling method for job-shop scheduling.

15.1 Introduction

This chapter describes three design projects where algorithms have been developed. These applications stem from totally different origins in terms of methods (combinatorial optimization, genetic of evolutionary algorithms, and mathematical optimization). But even more important, they are developed for totally different application areas in terms of the role of a human user of the algorithm. Some of these algorithms have been developed without detailed study of the competences of the perceived users. Others have examined humans when performing the scheduling tasks manually, but have not considered the change in cognitive load if the process of planning changes due to the new algorithm and computerized support. Therefore, none of the design projects fulfils the criteria developed in the framework of Chap. 12. However, by assessing the various design projects and the resulting algorithms, we will be able to identify the main weaknesses in these applications and show how they can be addressed in future.

The structure of this chapter is as follows. Each application will be described and assessed in a separate section. The description and assessment is presented in brief in Figs. 15.1–15.3, which follow the structure of respectively Chaps. 12–14. The description will follow Mitroff's model of problem solving (Mitroff et al. 1974),

Describe the problem solving process (implicitly) proposed
Assess the problem structure
Characterize the algorithm
Describe an object model of the scheduling problem
Assess the planning problem
Human & organizational noise in the planning & scheduling environment

Fig. 15.1 Description of the applications according to Chap. 12

Describe the perspective:
    Descriptive, prescriptive, formative
Describe the analysis used:
    Cognitive/Hierarchical/Work domain,
Characterize the problem requirements:
    Device/Event dependence, psychological relevance

Fig. 15.2 Assessment of problem analysis in the applications according to Chap. 13

Problem characteristics
Human characteristics
Algorithm characteristics
Possible modes of automation

Fig. 15.3 Assessment of human–machine interaction according to Chap. 14

and ends with an evaluation where the other elements of Figs. 15.1–15.3 will be addressed.

The first step in describing the problem analyzes the problem solving process according to a process view. Mitroff discussed four stages and six steps in such a process. He showed that the starting point might differ between different problems and problem solvers. So we will describe what path through the network of Mitroff has been used in designing the application.

Next, the problem structure is characterized according to the distinction of Newell with respect to well-structured, semi-structured or ill-structured problems. Moreover, it is stated whether it concerns a transformation or design problem (Hoc 1988).

A lot of attention is given to the description and characterization of the algorithm that is the result of the design project. Several questions will be addressed, such as on the computational complexity of the problem (does it belong to P or NP), whether the algorithm produces a heuristic or optimal solution, what type of solution approach (evolutionary, greedy, genetic, constraint programming, MIP, single-pass, etc.) it applies, whether there is an (implicit) preference of risk of decision maker, how robust the algorithm is, what its speed of model solving is, and finally, where the project considers when the problem has been solved. If possible, a description of the object model of the scheduling problem is provided.

The planning problem will be assessed according to, e.g., the categories environment (i.e., predictability, triggering events, and goals/constraints); planned and planning entities, the plan that is being generated, and the methods applied. Finally, the human & organizational noise in the planning & scheduling environment will be addressed.

After this extensive description and characterization of the planning problem and the designed algorithm, we will continue with assessing the problem context from a task perspective. Figure 15.2 shows the three main questions that will be answered.

First, the perspective of the task analysis has to be characterized. If a normative perspective has been used, a prescription of how the task should be done has to be given. Typically, a list of sub-tasks the user must perform to complete the main task will be listed. In case a descriptive perspective has been used, a description of how the task is currently performed in practice will be presented. A descriptive perspective generally focuses on modeling the knowledge, thought processes and goals that underlie the task effectively done by the human.

Finally, if a formative perspective has been used, a specification of an exhaustive description of the work domain and its physical, functional interrelations will be given. A formative perspective enables a designed system to allow tasks that were not prescribed or analyzed during the design of the system. It allows to support humans in situations where events happen that did not occur during a task analysis and that were not anticipated by the designers.

Second, the analysis itself will be addressed, if appropriate. We discussed the three types of analysis, cognitive task analysis, hierarchical task analysis, and work domain analysis. If we can characterize the analysis that has taken place in designing the algorithm, we will list it accordingly. However, it might not always be the case that one of these analyses has been used. Sometimes, algorithms are designed without a proper analysis of the planning tasks. This holds true especially for more recent developments in the design of multi-objective generic solution methodologies, which will be the subject of Sect. 15.3.

Finally, the problem analysis will be characterized. We will characterize both the device and event dependence of the analysis and its psychological relevancy.

We end the assessment of the algorithms by focusing on the human–machine interaction that is required when using the algorithm. This is assessed according to the criteria of Chap. 14. First, we discuss issues related to the problem itself, i.e., harm risk, need for situation awareness for the humans that use the algorithm, maximum duration of a solution cycle, and complexity of the problem. Next, some characteristics of the human user of the algorithm are addressed. For example, we will address benefits and threats if the human involvement in solving the actual problem changes. We will do an analysis of the characteristics of the algorithm as well, i.e., technical feasibility, computation time, error rate, development costs, need for interaction, et cetera. We end with listing the possible modes of automation when using the algorithm.

15.2 Decision Support for Shunting Yard Scheduling Using a Network Flow Heuristic

15.2.1 Introduction

The first application that we describe is a decision support system for shunting yard scheduling. The algorithm that we designed for this support system is based on combinatorial optimization and denoted as a network flow heuristic.

15.2.2 Problem Context

This subsection addresses the problem context for which the decision support has been developed. An important task of a shunting planner is to plan the reconfiguration and relocation of all trains that arrive at a station at the end of the day such that they can depart in the correct configuration at the start of the next day. The planner also has to schedule the washing and cleaning activities and plan the movement of the trains to and from these facilities during the night. Finally, the planner has to modify already generated plans if additional information is released, for example if the maintenance schedule for the railway system is updated.

There are many planners involved in the process of shunting plan generation (Wezel and Jorna 2009). They are supported by information systems that contain information on the railway network, equipment, and all expected and planned events related to both trains and shunting teams over time. Literature on train routing and scheduling problems suggests that there is a high potential for the application of decision support in these information systems. However, a survey of 153 papers on train routing and scheduling problems concludes (Cordeau et al. 1998: 399): "even though most proposed models are tested on realistic data instances, very few are actually implemented and used in railway operations". We developed a decision support tool for shunting yard planning for the Netherlands Railways. This decision support tool includes various types of algorithmic support.

This subsection will describe the specific problem faced by the shunting yard planners. Trains that have to be cleaned during the night cannot be left at the track where they arrive, as these tracks are regularly used for trespassing trains. Planners often have to reconfigure the trains, as they consist of various coaches, and the departing train may need more or less coaches, depending on expected passenger load. "Storage" capacity of a shunting yard is limited. There is only one track with cleaning equipment. The task of the shunting planner is to plan the configurations and movements of the trains and to decide on what tracks trains stay during the night.

The station of Zwolle, The Netherlands, is used to illustrate the planning problem. This station has also been studied in Zwaneveld et al. (1996) and Freling et al. (2005). Figure 15.4 shows an example of a part of a shunting plan. The horizontal axis denotes the time. The vertical axis contains the tracks. The bars are trains that occupy a track during a certain amount of time. For example, the train ZN1 is located on track 3B from 05:52 until 07:02. At that time, it is moved to track 4B, where it stays until 08:14. In Fig. 15.4, the movement of train ZN1 from track 3B to 4B seems instantaneous. In practice, however, the movement takes a few minutes and the route has to be determined by the planner. An example of the routing of train ZN1 is shown in Fig. 15.5.

Figure 15.5 shows the route from track 3B (lower right) to track 4B (upper left) in bold. It consists of two saw-movements (one at track 13, the other four tracks before). These saw-movements are unavoidable. A shorter route via track 3A would have been possible if other trains had not been blocking that track. Sometimes it is impossible to find a feasible route between two tracks.

We studied the task of planners that are responsible for making short-term adjustments (1 week ahead) to already created plans (Wezel and Jorna 2009; Riezebos and Wezel 2009). Sometimes, they need to adjust a plan because of a

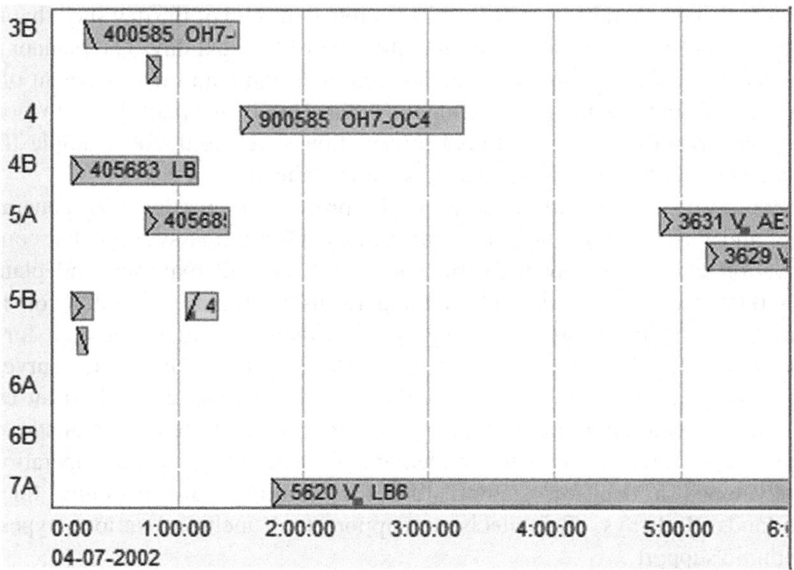

Fig. 15.4 Example of a shunting plan

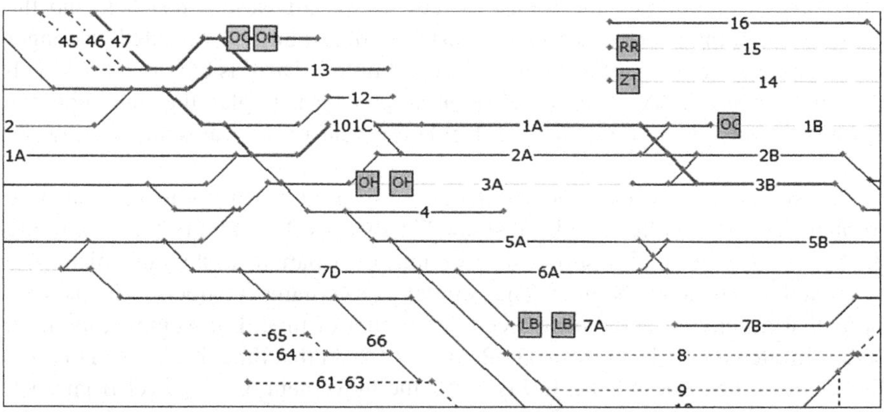

Fig. 15.5 Route found from track 3B to 4B

notification of a different configuration of an arriving train. Other reasons might be the scheduled maintenance of a track during a part of the night, prescribed changes in the arrival/departure track, et cetera. While updating the plan, a planner has to take several constraints and goals into account. In general, the planner aims at efficient schedules for both drivers and shunters, resulting in minimal walking and waiting times. The actions that the planner takes can be repetitive or non-repetitive. Some events, e.g., a scheduled maintenance of a track, generate a highly repetitive set of actions for the planner, as all trains that were scheduled on this track during a part of the night need to be relocated on other tracks. The routes of the trains to and from these new tracks will have to be planned as well. Other actions are more or less non-repetitive. A notification of a change in arrival time of a train will cause a single non-repetitive action.

The planning tasks are performed manually. Some computer programs are used to collect information, but the plan itself is made on paper before it is put in the computer. From the total group of 130 planners, about 60 are involved in planning these short-term adjustments. The short-term planners are geographically specialized, such that each planner performs these tasks for specific stations.

The Netherlands Railways has asked to develop a prototype of a decision support system for these planners. This system should make use of intelligent algorithmic support in order to improve the speed and quality of planning. A prototype of a support system for routing decisions has already been designed (Riezebos and Wezel 2009). This system is able to present various alternatives for routings between two tracks. It is used to support the planner when planning relocations, as the feasible routings are presented directly when rescheduling a train to another track. The routings are being determined for the train that is being relocated, assuming that all other things remain unchanged.

However, if the number of trains that need to be rerouted and relocated is high, two problems arise. First, the task of the planner is still highly repetitive and not attractive from a human perspective. Second, the assumption that all other things remain equal is not valid anymore, which results in sub-optimization. Therefore we decided to develop intelligent algorithmic support for relocating and rerouting a group of trains. As the task of the human planner for rescheduling a group of trains differs from that of rescheduling sequentially single trains, we will first analyze this task of the human planner. Next, we will design the algorithm for relocating and rerouting a group of trains.

First a remark on problem complexity. The task of planning the relocation and rerouting of a group of trains is more difficult if the number of available shunting tracks is constraining. If more than two trains are put on one track, the trains in the middle can get blocked. Furthermore, less tracks available means also that it becomes more difficult to find free routes.

We analyzed the strategy of a planner that has a long experience in solving this task. The planner was requested to solve actual problems and to think aloud during his work. The thinking-aloud protocols were analyzed, after which we held sessions with the planner in which we actively asked for, and discussed explanations of, decisions.

The process of shunting planning for a group of trains appears to be as follows. The planner receives a message from the information system that makes it necessary to relocate a group of trains. He generates this list of trains, which are ordered by the information system on arrival time at the track. The planner sorts this list according to his own criteria and generates a plan for the trains on this list, using a graph similar to Fig. 15.4, as well as pencil and fixer. Backtracking is almost never applied. The depth of searching is therefore limited. If a planner cannot find a solution in a few steps, he reverts to a constraint violation (for example: a train that will not be cleaned). The resulting modifications are put into the information system and made available for other planners and/or the shunting operators. The information system controls whether other planners have to modify their partial plans in order to make the total plan feasible.

In order to find possibilities for improvements, we first investigated performance gaps with respect to quality, speed, flexibility and costs. The management of the Netherlands Railways stressed the importance of responsibility in the process of plan creation. Planners should feel responsible for the quality of the plans they generate. Next, management would like to see an improvement in the quality of the plans, i.e., no constraint violations and accurately configured plans. Planners stressed the importance of robustness of the plan, which can be seen as a flexibility measure. They complained about the speed or responsiveness of the information system, which delayed the process of plan generation and made it less efficient.

15.2.3 Modeling

15.2.3.1 Decision Support System

The prototype shunt scheduling support system that we developed includes elaborate functionality for manual planning and a number of algorithms. The prototype contains a graphical user interface that resembles the current desktop of the planners. The schedule shows the location of train and coaches on the tracks over time. The path of individual coaches resembles detailed information for the planner. The list of attributes, constraints and violations uses hyperlinks to enable the planner to search for relevant information and to make modifications. Finally, the blackboard enables backtracking. The planner can easily modify the layout of the screen and open additional windows, such as the one shown in Fig. 15.5 that is used for supporting the routing decision. Figure 15.6 shows a screenshot of the system.

15.2.3.2 Algorithm

The problem of relocating and rerouting a group of trains can be modeled as a multi-commodity time–space network flow problem (Ahuja et al. 1993). Network flow problems aim to find efficient paths from a source node to a sink node in a network. Nodes are connected through arcs with associated costs and capacity. A path is more efficient if the sum of the associated costs of the arcs in the path is lower. A time–space network uses multiple nodes to represent both different locations on the shunting yard and the state of a location over time. A multi-commodity network model allows for modeling flows of different commodities (e.g., train types) in the same network. The capacity constraints are modeled independent of the commodity type.

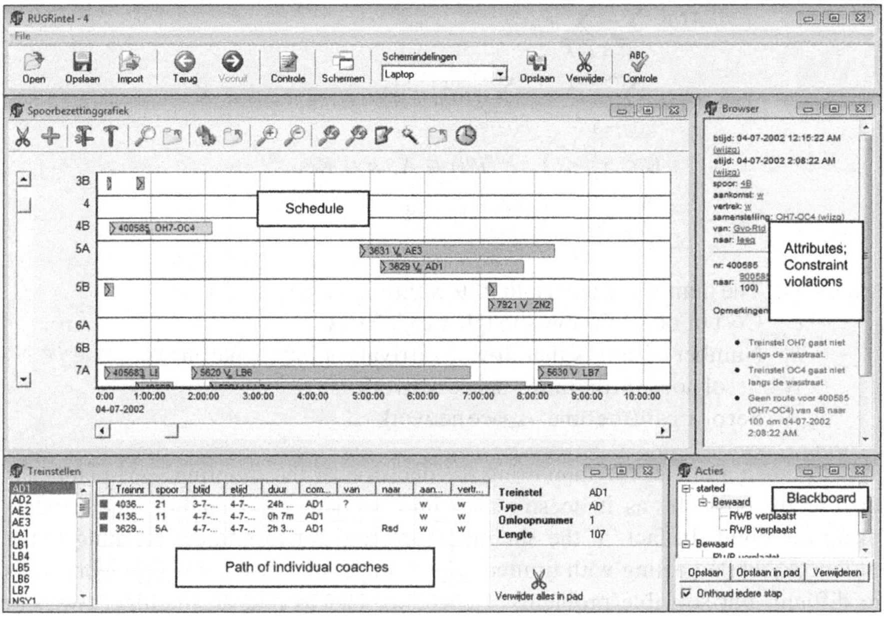

Fig. 15.6 Screenshot of decision support system

Network flow models are frequently used for large optimization problems, as the mathematical structure of the problem enables the usage of specific algorithms that make it possible to find optimal solutions. However, multi-commodity network flow problems belong to the class of NP-complete problems (discussed in Chap. 12), so no polynomial-efficient algorithm is available to solve such problems. Algorithms may still try to use the available structure in the problems to reduce the time needed for finding good solutions.

In our model, the state of the system over time is represented using discrete moments in time and an interval of 2 min. So the availability of tracks is modeled using small time intervals. The network flow model uses arcs to represent the possibility of a state change between two nodes. A node represents the existence of infrastructure (i.e., tracks) on some moment in time. An arc between two nodes may either indicate a change in time or in location. Two nodes *i* and *j* that represent the same track are connected using a directed arc *(i,j)* if the time represented by node *j* is exactly one time interval later than that represented by node *i*. Two nodes *i* and *j* that represent different tracks are connected using a directed arc *(i,j)* if the tracks are directly physically connected, i.e., no other intermediate tracks between them. Such an arc identifies a routing possibility. We use track positions with a maximum capacity of one train at the same time. Therefore, the maximum capacity of an arc equals one.

Mathematically, the multi-commodity time–space network flow model for this problem can be formulated as:

$$
\begin{aligned}
& Min \sum_{(i,j)\in A} \sum_{k\in K} c_{ij} x_{ij}^k \\
& s.t. \sum_{i:(i,j)\in A} x_{ij}^k - \sum_{i:(j,i)\in A} x_{ji}^k = b_j^k \quad \forall_j \in N,\ k \in K \\
& 0 \le x_{ij}^k \le 1 \quad \forall (i,j) \in A,\ k \in K \\
& \sum_{k\in K} x_{ij}^k \le 1 \quad \forall (i,j) \in A
\end{aligned}
$$

where x_{ij}^k The number of units k that flow through arc $(i,j) \in A$
c_{ij} Cost of one unit flow though arc $(i,j) \in A$
b_j^k Number of units k that flow in (arrival) or out (departure) of node $j \in N$
N Set of nodes in the time - space network
A Set of arcs in the time - space network

Using such a model for supporting the decision of the shunting yard planner is not directly possible, as it doesn't take into account the cleaning requirements during the night. In fact, at the shunting yard that we examined, cleaning is the bottleneck, while dealing with limited storage capacity and finding efficient routes are difficult but solvable problems. Hence, we had to find an intelligent way to include cleaning in a solution approach.

The heuristic solution approach that we developed is a three stage approach, as illustrated in Fig. 15.7. The three stage approach is as follows:

I. Generate a schedule for the bottleneck, the two internal cleaning tracks. Take a minimal time lag between the expected arrival time at the shunting yard and the earliest start of cleaning this train into account, in order to be able to route the train to the internal cleaning tracks. The same holds true for the minimal time between finishing cleaning and departure from the shunting yard. Details of the scheduling algorithm are provided in Fig. 15.8. The algorithm is based on a minimal slack time rule (see Demeulemeester and Herroelen 2002:237 for a further description of solution procedures for the similar resource-constrained project scheduling problem).

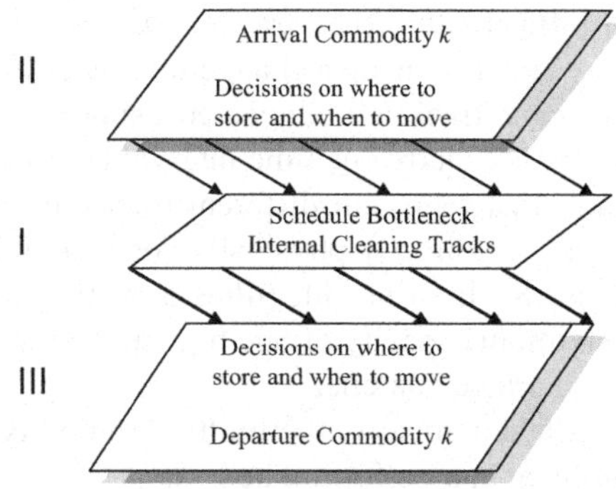

Fig. 15.7 Solution approach

II. Use the scheduled start moment of cleaning and the assigned cleaning track of each train as destination (sink) in the multi-commodity network flow model (i.e., $b_j^k = -1$ for all nodes j at which a train of type k is scheduled to arrive at a cleaning track). Solve the multi-commodity network flow problem directly or (if the expected calculation time exceeds 5 min) apply a sequential heuristic of single-commodity network flow models with modified arcs (based on the outcomes of preceding optimizations).

III. Use the scheduled finish moment of cleaning and the internal cleaning track as arrival (source) in the multi-commodity network flow model (i.e., $b_j^k = +1$ for all nodes j at which a train of type k is scheduled to depart from the cleaning track). Update the availability of tracks based on the routings and storage schedules of the preceding stage II. Solve the multi-commodity network flow problem optimally or heuristically.

15.2.4 Model Solving

As an illustration, the algorithm of Fig. 15.8 has been tested using data from railway station Zwolle. Details on the size of the problem that we used for this experiment can be found in Table 15.1. We compared the outcomes of the model with the plan that actually has been generated by a human planner. Table 15.2 shows the results of this comparison.

The results show that the model generates a schedule in almost 3 min. The planner used far more time. However, the schedule differs significantly in two respects. First, the number of trains located at a trespassing track for more than 1 h. Planners allowed half of the trains that need to be cleaned to stay for a rather long time at such a track. Their verbal description of the objectives and constraints of this problem pointed towards avoiding such behavior. When actually generating a schedule, they aim at a balance between decreasing the number of train movements and avoiding usage of these tracks. Second, the planners allowed for combined movement of trains, sometimes even of different types. This reduces the total number of movements as well as the number of saw-movements. The model only allows for single movements of trains. The possibility of combined movement of the same type of trains can easily be implemented in the model. We have done some experiments with such a model, which showed that the outcomes much better resemble the decision of the planner. However, the description of the extended model goes beyond the scope of this section. The current model provides a good starting point for planning the relocation and reconfiguration of the trains during the night. This includes a schedule for cleaning all trains.

We have shown that algorithmic support for planning decisions can be designed, taking into account several organizational and human aspects. The Netherlands Railways has organized the planning process in such a way that a planner sometimes needs to modify a set of plan entries at once. Planners are used to modifying plan entries sequentially (i.e., one by one). We designed a support tool that enables them to find a starting point solution for a set of plan entries that need to be modified.

P Set of trains that still need to be scheduled

K Set of train types

$g(p)$ Train type of train $p \in P : g(p) \in K$

t_x Time at track x

$CT(k)$ Cleaning time of train type $k \in K$

$EA(p)$ Earliest arrival time of train $p \in P$ at a cleaning track

$LD(p)$ Latest departure time of train $p \in P$ from a cleaning track

$S(p)$ $= [x,t]$: Scheduled arrival time t of train $p \in P$ at cleaning track x

Scheduling heuristic

Initialize $t_x := 0 \ \forall x$

Repeat $x^* := \arg\min_x t_x$

$$t := \max\left(t_{x^*}, \min_{p \in P}(EA(p))\right)$$

$$p^* := \arg\min_{p \in \{p \in P | EA(p) \le t\}} (LD(p) - CT(g(p)))$$

$$S(p^*) := [x^*, t]$$

$$t_{x^*} := t + CT(g(p^*))$$

$$P := P \setminus p^*$$

until $P = \varnothing$

Solution $\{S\}$ is the set of scheduled arrival times of trains at the cleaning tracks x

Fig. 15.8 Bottleneck scheduling heuristic for stage I

15.2.5 Evaluation

15.2.5.1 Assessment of Problem Description

In the description of the problem and of the algorithm design, we have already referred to the stages of Mitroff et al. (1974). The starting point for this problem solving process was clearly the finding that the task for planning a group of trains differs from the task of planning routes for individual trains, which would require another type of algorithmic support. However, it was not initiated by the planner, but by the researchers that had to develop several showcases of types of intelligent algorithmic support for the Netherlands Railways.

Table 15.1 Characteristics of the experiment at Railway station Zwolle

Number of tracks	92
Number of time periods (length 2 min)	240
Number of nodes (track-time combinations)	22.080
Number of arcs	240.960
Number of train types	6
Number of trains	22

Table 15.2 Results of the experiment

Schedule	Experiment	
	Model	Planner
Number of trains located at a trespassing track for more than 1 h	1	11
Number of movements	95	60
Single train	95	32
Connected trains (2 or 3)	0	29
Number of saw-movements	21	18
CPU Time (s)	187	–

The problem is a semi-structured problem. A description of the current state and definition of operators/actions are present. Criteria for optimization are even defined. But the weights of the various goals (the costs of either physical transfer to another or staying at the same track) are not easy to determine, which makes it difficult to verify that the achieved state corresponds to the goal. This is left to the human planner. The planning problem has characteristics of a design problem (Hoc 1988), although it is not an ill-structured problem.

The object model of this problem is identical to the model presented in Chap. 14.

The planning environment is relatively predictable, as some of the changes a planner has to take into account are due to predictable events, such as additional trains needed for concerts of well-known artists, planned maintenance of trains or tracks, et cetera. There is some human and organizational noise in the planning environment, as the planner depends on the input from other planners. For example, changes due to bad performance of other planners cause such noise.

15.2.5.2 Assessment of Problem Analysis

The algorithm focuses on a specific subtask of the human scheduler: groupwise relocation of a number of shunting activities. The perspective that has been used for the analysis of this task is descriptive since we used a cognitive task analysis. The analysis was device dependent as we only looked at the "devices" used to create the schedule: the human planners themselves, the paper, and the computer systems that they used. As a consequence, the event dependence was low but the psychological relevance was high.

15.2.5.3 Assessment of Human–Machine Interaction

The interface facilities for the human scheduler can be used for supporting this planning task as well. However, additional input possibilities are required for providing relevant problem information, i.e., selecting the group of trains that has to be relocated. The planner could also need to interact with the algorithm, as the determination of the sequence of the train types (commodities) may have to be changed manually. No special consideration is given to the analysis of the planning problems involved and the design of these facilities.

In the following sections a comparative analysis of the user interface and algorithm will be provided based on the human–machine interaction evaluation framework of Chap. 14, which consists of four main categories: problem characteristics, human characteristics, algorithm characteristics, and possible automation modes.

Problem Characteristics

Harm risk. The cost of sub-optimal decisions in case of sequential route determination is one of the main reasons for designing a new algorithm. However, we can characterize this cost as being low to medium, since a sub-optimal solution might result in profit cost, but it will not bring catastrophic consequences. When entering a proposed solution in the information system, feasibility and safety issues are checked automatically.

Need for situation awareness. The human planner needs to be aware of the details on the shunting yard in order to assess the solution provided by the algorithm. This is mainly due to the important left-over task of finalizing the plan before submitting it to the information system of all planners.

Decision cycle duration. The problem is part of a weekly schedule and the planner has generally sufficient time available for finding a solution. However, there exists a clear time/quality trade-off, as a human planner cannot afford to spend too much time on a specific routing problem. Improving the speed of problem solving of a partial problem will therefore provide more time for solving the remaining planning problems related to the event.

Complexity. The scheduling task is of the highest complexity possible, since as discussed earlier, the multi-commodity network flow problem is a well-known NP-hard problem.

Human Characteristics

Level of human involvement in the task. If full automation of the planning of a group of trains would be possible, the human planners would loose skills in routing trains. In the algorithm, explicit knowledge on the position of the bottleneck (the cleaning station) is included. Due to investments or changes in regulations, this might change over time. It is therefore important that the human planner keeps training his skills in solving the problem manually (using the GUI and the other type of algorithmic support for sequentially solving the routing problems). This will help him to assess the quality of the outcomes of the algorithm as well.

Performance level and performance variability. This characteristic refers to the algorithm's perceived usefulness and the variability of its performance, and is directly related to the trust of the human scheduler in its operation. Due to the fact that the algorithm is a heuristic, the planner might loose confidence in the performance of the algorithm. However, the task analysis revealed the bottleneck and the algorithm is designed such that it provides good support to the decision maker on the schedule at the bottleneck and the relocation of trains at the tracks. We therefore expect that the performance level will be sufficiently high.

Providing relevant feedback. The performance of the cooperation between the human planner and the support system will increase if the human scheduler is able to comprehend the operation of the algorithm and the results that it produces. The algorithm scores very high in this category, since the sequence of the heuristic (steps I, II, and III) is logically and the intermediate solutions are reported visually.

Flexibility with constraints. The human planner has to use his knowledge on the constraints in the model. Several assumptions depend on train characteristics, such as length, maintenance history, et cetera. This allows for flexibility which is not supported by the algorithm.

Algorithm Characteristics

Computation time. The computation time for realistic problem sizes was 3 min, which is much more than finding a single route (less than a second), but still acceptable when generating a large plan for a group of trains.

Error rate. The algorithm does not always produce valid solutions. The solutions are valid within the set of constraints that is implemented, but several other constraints might have to be considered as well.

Mode of automation. The algorithm has to interact with a human planner, as it is an input decision for which trains a solution has to be determined. The generation of the solutions does not require the presence of the human planner. We denote the mode of automation therefore as supervisory control.

Development cost. The high development cost for such intelligent algorithms is partly due to gathering knowledge of the constraints and goals at the shunting yard. This is more or less a one-off cost. For applying the algorithm in practice, special mathematical programming software has to be used. We tested the algorithm using CPLEX and AIMS. CPLEX has special solvers for network flow problems, which makes it very fast. However, licenses for the software are quite expensive. The results reported are obtained using mixed integer programming in AIMS.

15.2.6 Conclusion

In this section we described the development of a network flow algorithm for the solution of a multi-commodity routing problem in shunting yard scheduling. The design of this algorithm followed a typical technological approach, but took into account a hierarchical task analysis of the planner. This approach did not focus on the task environment after the algorithm would have been developed. Therefore, several human and organizational aspects have not been addressed when designing this algorithm.

We used the evaluation categories of Chaps. 12–14 to assess the problem and the design of the algorithm. Some qualitative characteristics of the system have been addressed. The main points of this evaluation that was presented in the previous section are the following:

The problem description revealed that the problem is semi-structured, and the planning environment rather predictable, which makes it suitable for algorithmic support at a higher level than for very small partial problems.

The problem analysis was based on hierarchical task analysis. However, the characteristics of the human tasks when using the designed algorithm were not addressed.

The analysis of the human–machine interaction revealed that in solving the problem involvement of a human planner is still desirable, although it will be helpful if an algorithm is used for important partial tasks.

15.3 An Evolutionary Multi-Objective Decision Tool for Job-Shop Scheduling

15.3.1 Introduction

As discussed in Chap. 12, the existence of trade-off considerations within a scheduling environment is a realistic situation frequently neglected by designers of scheduling algorithms.

This section introduces GP-MOS (Genetic Programming Multi-Objective Scheduler), a prototype framework that attempts to address scheduling considerations of this type. In its current form the framework provides the basis for the generation of a set of candidate schedules when multiple optimization objectives are considered. On a later stage the framework will be able to address additional cognitive and organizational considerations by explicitly involving the human operator in the multi-objective optimization process through a suitably designed graphical user interface.

The proposed framework bases its optimization process on the use of evolutionary computation methodologies. Evolutionary Algorithms (EAs) are population-based heuristic optimization techniques that contain stochastic elements in their search. Their operation is based on a loose interpretation of the Darwinian process of the survival of the fittest. The term "EAs" is used interchangeably to describe various optimization techniques that follow the above principles, but utilize different solution representations, and/or genetic operations in their optimization process (Eiben and Smith 2003).

15.3.2 Problem Context

The development of the GP-MOS algorithm is based on the traditional production research view of the scheduling environment, as this has already been described in Chap. 12.

Based on this view, GP-MOS plays the role of the scheduling algorithm that generates solutions for a job-shop scheduling problem. All necessary input data are readily available for the algorithm at the time they are required. The source of the input data is not specified. There is also no explicit consideration of the human scheduler's involvement in the overall process.

Implicitly, it is assumed that the human scheduler will initiate the problem solving process. In the context of the GP-MOS application, this includes the setting of algorithmic parameters such as the population size, the number of generations etc. It is also assumed that the human scheduler will be the recipient of generated schedules. In the case of the GP-MOS scheduling algorithm, the generation of multiple trade-off solutions also implies that a decision maker should choose one of the possible alternative solutions that will be applied on the problem considered. The decision maker does not necessarily have to be a human scheduler, since another algorithm, or even an artificial intelligence system can also play this role. However, the fact that the output of the algorithm cannot be directly applied to the problem environment, since a choice from the set of potential solutions has to be made by another entity, introduces some realistic considerations in the traditional production research view of the job-shop scheduling problem.

15.3.3 Modeling

15.3.3.1 Basic Multi-Objective Optimization Concepts

A general multi-objective optimization problem can be defined as follows (Zitzler and Thiele 1999):

$$
\begin{array}{rll}
\min / \max & y = f(x) = (f_1(x), f_2(x), \ldots, f_n(x)) & \\
\text{subject to} & x = (x_1, x_2, \ldots, x_n) \in X & \\
& y = (y_1, y_2, \ldots, y_n) \in Y & \\
\text{where} & x & \text{decision vector} \\
& y & \text{objective vector} \\
& X & \text{parameter space} \\
& Y & \text{objective space}
\end{array}
$$

A single solution (a decision vector) that results to an optimal objective vector does not generally exist for the general multi-objective optimization problem. Instead, there exists a set of Pareto-optimal decision vectors that form the Pareto set of solutions. A decision vector α is said to dominate decision vector b if and only if (assuming a minimization problem and without loss of generality):

$$\forall i \in \{1, 2, \ldots, n\} : f_i(a) \leq f_i(b) \quad \exists i \in \{1, 2, \ldots, n\} : f_i(a) < f_i(b).$$

The aim of multi-objective optimization methods is to find a set of non-dominated solutions that provide a reasonable approximation of the Pareto set of solutions. The non-dominated set of solutions produced by multi-objective optimization methods does not necessarily correspond to the actual Pareto set of solutions, which, in real-life cases of multi-objective optimization, is not known in advance.

15.3.3.2 The Multi-objective Job-Shop Scheduling Problem

In the job-shop scheduling problem n jobs have to be processed on m available machines. Each job comprises of a number of operations, which have to be processed in a predefined fixed sequence. Each operation can be processed on only one of the available machines. The following constraints have to be respected for the typical static job-shop scheduling problem:

- No job operation can start unless its predecessor job operation has been finished
- Each machine can process only one job operation at a time
- No pre-emption of job operations is allowed
- The processing times of job operations are integral, known in advance, and they include any transportation or set-up times
- All jobs are available at time zero
- It is assumed that there are no unexpected machine breakdowns

The single-objective job-shop scheduling problem has been the subject of extensive academic research over the last decades due to its computational complexity (Garey and Johnson 1979). While a number of efficient heuristic algorithms have been devised for its solution, there has been some considerable criticism on the relevance of this research to industrial practice. This criticism concerns both the static nature of the instances considered, as well as the use of the makespan as the principal optimization objective (Fang et al. 1996). In addition, T'kindt and Billaut

(2002) reported that there were very few algorithms that addressed the multi-objective version of the job-shop scheduling problem. In recent years there have been some attempts to produce non-dominated sets of schedules for multi-objective job-shop problem instances through the use of modern heuristic techniques. A short description of these algorithms is provided in the following sections.

Esquivel et al. (1996) designed an evolutionary algorithm for the solution of single and multi-objective job-shop scheduling problems. They employed a sub-population-based approach as a multi-objective optimization technique. A sub-population of solutions was generated and evolved for each optimization criterion concerned. Using this approach the authors were able to generate non-dominated sets of schedules for some single-objective test cases taken from the literature, suitably modified to allow the consideration of due-date based criteria.

Baykasoglu et al. (2002) employed a multi-objective tabu search technique that utilized the concept of standard dispatching rules within Giffler and Thompson's scheduling algorithm (1969) for the indirect evolution of job-shop schedules. The authors presented limited results of non-dominated sets of schedules on test problems taken from the literature.

Petrovic et al. (2004) developed a hybrid evolutionary algorithm for the solution of fuzzy multi-objective scheduling problems. Their algorithm allowed the interaction of the decision maker in the setting of aspiration levels. The application of the algorithm was illustrated on an industrial case study.

Recently, Chiang and Fu (2006) developed a dispatching-rule based evolutionary algorithm for the solution of the multi-objective job-shop scheduling problem. They also proposed an improved evolutionary multi-objective optimization technique, and illustrated its efficiency on some test problems taken from the literature.

15.3.4 Model Solving

The operation of the GP-MOS framework is based on the use of a genetic programming evolutionary machine responsible for the generation of potential schedules, combined with NSGA-II, (Non-dominated Sorting Genetic Algorithm – II) (Deb et al. 2002), a typical evolutionary multi-objective optimization technique which allows the generation of a set of non-dominated solutions for the scheduling problem considered. The scheduling application described in this chapter concerns the multi-objective job-shop scheduling problem, although the proposed framework can be easily adapted to consider various instances of scheduling problems. A basic description of the genetic programming algorithm is necessary for a proper understanding of the proposed framework.

15.3.4.1 Genetic Programming Fundamentals

Genetic programming (GP) (Koza 1992) is a variant of Evolutionary Algorithms that evolves solutions in the form of computer executable programs. Evolved programs are usually represented as parse tree expressions. Genetic programs are created from a set of functions and a set of terminals that act as arguments to these functions. A typical application of GP is symbolic expression: Given a set of input–output fitness cases, the user defines a set of appropriate functions and terminals for the problem considered. A population of candidate expressions is initially constructed randomly from these sets. The expressions propagate probabilistically to subsequent generations based on their ability to fit the cases provided. In addition, they continuously evolve with the help of sexual and asexual genetic operations in order to explore new regions of the search space.

A possible GP expression (using parse tree representation) that can be evolved from the function set of four basic arithmetic operations and the terminal set of two independent variables X and Z is shown in Fig. 15.9.

Parse trees are translated in a depth-first, left-to-right manner. The corresponding formula of the GP expression is indicated in the same figure. While GP differs from other evolutionary paradigms on the representation scheme employed for potential solutions, the evolutionary cycle employed is similar to the one used by all evolutionary computation algorithms.

15.3.4.2 Solution Generation Mechanism

The schedule generation mechanism is the central part of the overall GP-MOS framework. It provides a mechanism for generating and evaluating potential job-shop schedules. This mechanism is placed within a typical evolutionary process, which allows the progressive evolution of a set of non-dominated solutions for the scheduling problem considered.

The proposed framework generates schedules indirectly by evolving artificial dispatching rules as tie-breakers for the well-known Giffler and Thompson's job-shop heuristic (1969). Giffler and Thompson's heuristic is a methodology that progressively assigns job operations to machines until a complete schedule has been constructed. It comprises of the following steps:

1. Find the schedulable operation *o* (an operation is schedulable if all predecessor operations have already been scheduled) with the earliest finishing time *t*.

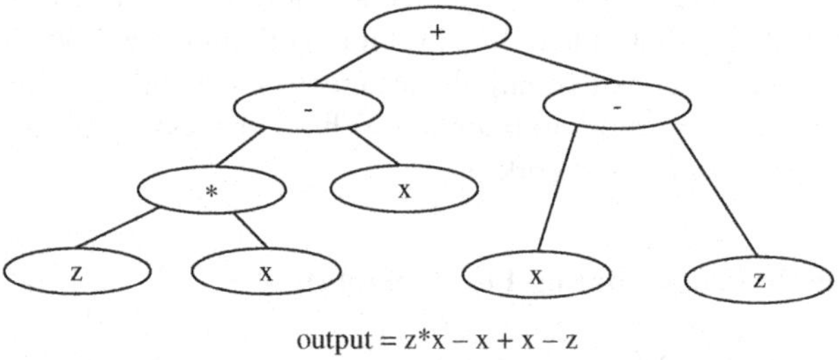

Fig. 15.9 An example genetic program for symbolic regression

2. Choose the machine M on which o is supposed to run.
3. Find the set of operations schedulable on machine M with starting times earlier than t.
4. Schedule one of the operations from this set.
5. Update the set of schedulable operations.
6. Iterate until all operations have been scheduled.

During the execution of Giffler's and Thompson's algorithm, step 4 requires the selection of an operation to be scheduled among all job operations that satisfy the requirements. Dispatching rules are commonly used as tie-breakers at this step. A dispatching rule uses problem parameters to calculate a priority value for all schedulable operations. The operation with the minimum (or maximum, depending on the rule used) priority value is selected for scheduling. Typical dispatching rules that have been used in the past for this purpose are the following:

SPT. Schedule the operation o of job i on machine M with the shortest processing time $p_{oi}M$

LPT. b i on machine M with the longest processing time $p_{oi}M$

EDD. Schedule the operation o of job i with the earliest due date d_i

MNS. Schedule the operation o of job i with the largest number of subsequent job operations MS_i

LNS. schedule the operation o of job i with the smallest number of subsequent job operations LS_i

MWR. Schedule the operation o of job i with the most work remaining MW_i

LWR. Schedule the operation o of job i with the least work remaining LW_i

The use of a dispatching rule provides a quick and dirty approach for the generation of a job-shop schedule. However, the performance of generated schedules depends heavily on the characteristics of the particular problem instance considered and the type of the optimization objective(s) used. GP-MOS adds a dimension of flexibility in the above process by making a genetic programming machine responsible for the generation and evolution of artificial candidate dispatching rules for the job-shop scheduling problem considered. Genetic programming generates these candidate rules using typical problem parameters, similar to the ones used as building blocks by the man-made dispatching rules described earlier in this section. These parameters form the terminal set of the genetic programming algorithm.

15.3.4.3 The Optimization Process

The operation of GP-MOS starts with the random generation of a population of candidate dispatching rules. Each of these rules is used for the indirect generation of a job-shop schedule, through the process described in the previous subsection. Once a job-shop schedule has been generated, the algorithm calculates the corresponding objective values, based on the objective functions that need to be optimized. These values are associated with the genetic program (i.e., dispatching rule) that generated the schedule.

Once the performance of the initial generation of dispatching rules has been evaluated, a new generation of dispatching rules is created through the processes of crossover and mutation. For a single-objective optimization problem, the selection of the rules that would participate in this process would be based on a single value associated with the objective function considered. However, in this case there are multiple conflicting objective values, and a single "best" solution does not generally exist, as explained in an earlier section. GP-MOS employs NSGA-II (Deb et al. 2002) as a ranking and parent selection mechanism for the evolutionary process. In short, this process ranks individual solutions according to their level of non-domination, based on the objective functions considered. Ranking values are used to determine probabilistically the solutions to participate in the genetic operations of crossover and mutation. The generated offspring are evaluated on the problem considered through the solution evaluation mechanism and a set of objective values is associated with them.

A new population of candidate dispatching rules is then created from the combined pool of old rules and all generated offspring rules, based on their ranking values. The evolutionary process is repeated for a predefined number of generations. The final generation contains a population of dispatching rules that corresponds to a set of non-dominated solutions for the problem considered. A detailed description of the NSGA-II process is beyond the scope of this chapter. Interested readers are referred to Deb et al. (2002) for a detailed explanation of the evolutionary multi-objective technique.

A block diagram of the solution generation and evaluation mechanism of GP-MOS is illustrated in Fig. 15.11. A summary of the GP-MOS operation in pseudo-code format is illustrated in Fig. 15.10.

15.3.5 Presentation of Results

15.3.6 Experimental Set

An extensive number of single-objective job-shop scheduling problems exist in the literature, which can be used for comparing experimental results. However, these problems do not contain information that can be used for the calculation of optimization of due-date related objectives such as the total tardiness or the total number of tardy jobs.

For this reason there are very few problems in the literature that can be readily used for experimentation that explicitly considers multiple objectives. In this chapter we illustrate the operation of the GP-MOS framework on a multi-objective job-shop scheduling problem that has been used in the experimental analysis of Baykasoglou et al. (2002). The authors have provided a limited number of

```
Procedure GP-MOS
Generate random population of dispatching rules P0 of size N
Run Giffler and Thompson's algorithm for each dispatching rule in P0
Sort P0 (non-dominated sorting)
Create new population of solutions Q0 using recombination operators and tournament selection
Run procedure Giffler and Thompson's algorithm for each dispatching rule in Q0
Loop
   Create combined population Rt=Pt+Qt of size 2N
   Sort Rt (non-dominated sorting)
   Create Pt+1 from first N solutions in Rt (according to their non-domination sorting)
   Create new population of solutions Qt+1 from Pt+1 using recombination operators and tournament
         selection
   Run procedure Giffler and Thompson's algorithm for each dispatching rule in Qt+1
Until termination criterion is true
```

Fig. 15.10 The GP-MOS algorithm in pseudo-code format

experimental results related to this problem, which can be used for comparison with the GP-MOS framework.

15.3.6.1 Operational Characteristics of the GP-MOS Framework

The application of the GP-MOS framework requires the specification of the operational characteristics for the experimental runs. In both experimental cases the GP-MOS runs were conducted using the following operational characteristics:

Function set. The four typical mathematical operations (addition, subtraction, multiplication, protected division) were used for the genetic evolution of the similarity coefficients. The protected division function is a special type of division that returns the value of "1" if the value of the divider is equal to "0".

Terminal set. GP-MOS employs the following parameters as members of its terminal set:

$p_{oi}M$ processing time of operation o of job i on machine M

d_i due date of job i

RO_i total number of remaining operations of job i

RW_i total time of work (in processing time units) left for job i

N total number of jobs for the problem considered

M total number of machines for the problem considered

TW_i total time of work (in processing time units) of job i

Note that these parameters can generate any of the dispatching rules described in the section "the multi-objective job-shop scheduling problem" when combined with the function set of the genetic programming machine algorithm. However, GP-MOS can also generate a large number of alternative artificial dispatching rules, based on the parameters provided in its terminal set.

Recombination operators. Typical subtree crossover and subtree mutation operators (Koza 1992) were used as mechanism for the generation of new similarity coefficients and the small-scale random introduction of new genetic material respectively.

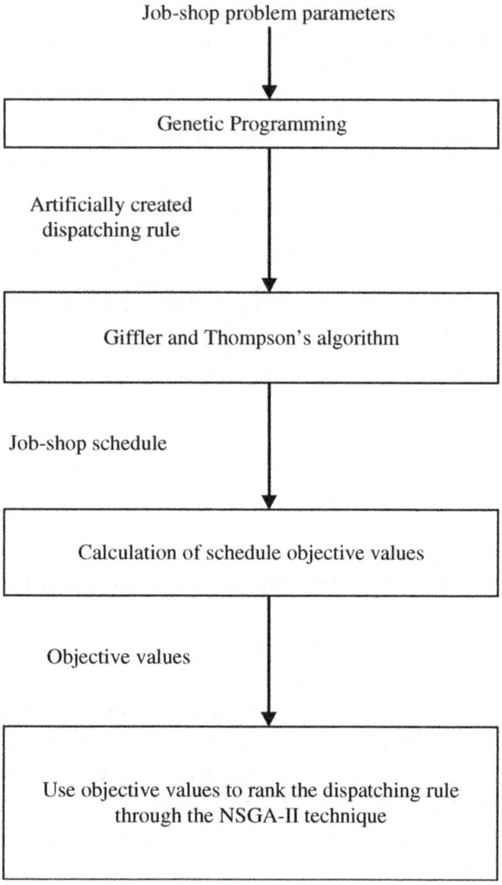

Fig. 15.11 Solution generation and evaluation mechanism of the GP-MOS framework

Objective functions. The optimization objectives that were used as the driving force of the evolutionary process in each of the experimental cases are the following:

- Minimization of makespan, defined as the latest completion time out of the completion times of all jobs for a generated schedule.
- Minimization of total tardiness, defined as the sum of tardiness of all jobs for a generated schedule.
- Minimization of average load balance, defined as the average processing load induced by jobs on the available machines for a generated schedule.

Parameters of the experimental runs. The values of additional parameters of the experimental runs are described in Table 15.3. Preliminary tests were conducted to determine the values of these parameters. Alternative settings for these parameters are certainly possible and might be able to further improve the results presented in the following subsection.

Table 15.3 Operational characteristics of the GP-MOS framework

Parameters	Values
Population size	500
Crossover probability	0.7
Mutation probability	0.3
Number of generations	1,000

15.3.7 Experimental Results

20 runs of the GP-MOS framework were conducted on the test problem. The non-dominated solutions that were found in at least 70% of the experimental runs are illustrated in Table 15.4. Table 15.5 contains the non-dominated solutions reported in the paper of Baykasoglu et al. (2002). It was reported that the algorithm found additional non-dominated solutions; however, it was not possible to obtain these solutions after personal communication with the authors.

An analysis of the results, presented in Table 15.4, indicates that the GP-MOS framework was able to automatically produce a wealth of non-dominated solutions for the given job-shop scheduling problem. This ability provided the decision maker with a range of possible choices that contain trade-off considerations. In addition, the GP-MOS framework was able to generate solutions that dominated the multi-objective optimization solutions no. 2 and 3 in Table 15.5, presented by Bayakasoglu et al. for the same test problem. The job-shop schedules that were generated by all GP-MOS non-dominated solutions can be made available to all interested researchers after communication with the author.

Additional experimentation is needed in order to fully evaluate the robustness of the proposed framework in relation to larger and harder instances of multi-objective job-shop scheduling problems. However, it should be noted that while the application of GP-MOS was illustrated using a multi-objective job-shop scheduling problem, any scheduling problem that utilizes the concept of dispatching rules could be considered in a similar way. It should also be noted that the GP-MOS framework could be easily modified to consider alternative optimization objectives, or to address problem characteristics and constraints that deviate from the typical job-shop scheduling case.

15.3.8 Evaluation of the GP-MOS Algorithm

15.3.8.1 Assessment of Problem Description

Problem solving process. The GP-MOS algorithm described in the previous sections was developed following the traditional production research approach as this has been discussed and reviewed in Chap. 12. In other words the process focused on the development of the solution mechanism rather than the consideration of the overall scheduling environment both from the cognitive and organizational points

Table 15.4 Objective function values for the set of non-dominated solutions evolved by the GP-MOS framework for the job-shop scheduling test problem

Solution no.	Makespan	Total tardiness	Load balance
1	55	108	0.1591070
2	57	100	0.0935836
3	57	101	0.0905143
4	58	83	0.0958563
5	60	83	0.0872210
6	60	129	0.0746299
7	62	95	0.0821613
8	62	104	0.0816928
9	62	140	0.0708914
10	63	97	0.0805155
11	63	103	0.0780516
12	64	68	0.1034970
13	64	138	0.0673989
14	65	71	0.1031280
15	65	87	0.0840788
16	65	93	0.0836248
17	65	99	0.0757534
18	65	110	0.0675695
19	66	102	0.0732881
20	67	76	0.0780410
21	67	96	0.0711364
22	67	106	0.0658023
23	67	108	0.0649602
24	67	109	0.0632978
25	67	132	0.0600818
26	68	97	0.0634283
27	68	111	0.0615259
28	68	112	0.0599111
29	68	118	0.0598302
30	69	108	0.0624869
31	70	90	0.0727812
32	70	140	0.0562771
33	71	75	0.0735542
34	72	114	0.0570475
35	72	133	0.0548364
36	72	136	0.0538334
37	73	126	0.0520311
38	73	129	0.0503502
39	74	102	0.0622048
40	76	94	0.0631200
41	76	99	0.0595547
42	78	113	0.0593685
43	79	124	0.0544342
44	80	120	0.0510115
45	85	136	0.0455035
46	87	145	0.0430516
47	89	123	0.0457901

Table 15.5 The non-dominated solutions in Baykasoglu et al. (2002)

Solution no.	Makespan	Total tardiness	Load balance
1	55	115	0.144
2	71	75	0.092
3	79	148	0.65

of view. The process started with the modeling stage (stage 2 in Mitroff's network), and proceeded with the development of the scientific model. The model solving process (stage 3) resulted in the generation of potential solutions. However, these solutions cannot be applied or validated in a realistic situation, since such a situation did not exist in the problem solving process.

The problem structure. The multi-objective job-shop scheduling problem is a well-structured problem (in the way it is modeled in this section).

Characterization of the algorithm. The multi-objective job-shop scheduling problem is an NP-hard problem in the strong sense. The algorithm proposed for its solution is an evolutionary heuristic approach. Its computational complexity is a complex interrelationship between the size and type of the problem and the setting of algorithmic parameters for the search process.

Object model description. An object model of the scheduling problem was not constructed during the development of the algorithm.

Assessment of the planning problem. No formal evaluation of the planning problem was conducted during the implementation of the framework.

Human & organizational noise in the planning & scheduling environment. The development of the framework did not explicitly consider the existence of a realistic scheduling environment.

15.3.8.2 Assessment of Problem Analysis

GP-MOS was applied on a version of the multi-objective job-shop scheduling problem for which alternative trade-off solutions were generated. This problem (subtask) was not formally generated through a plan decomposition process (see Chap. 13) from a larger planning and/or scheduling problem that needed to be solved, although it could have been the partial outcome of such a process. There was also neither a formal evaluation of the necessity/feasibility to develop an automated algorithmic solution, nor a formal placement of the scheduling subtask within the control categories of Chap. 14.

15.3.8.3 Assessment of Human–Machine Interaction

Problem Characteristics

Harm risk. This subcategory refers to the cost of a sub-optimal decision that might be proposed by the decision maker. For the case of the multi-objective job-shop scheduling problem we can characterize this cost as being low, since a sub-optimal solution might result in profit cost, but it will not bring catastrophic consequences.

Need for situation awareness. This subcategory refers to the need that exists for the scheduler to be aware about the characteristics of the solution and how this will affect the scheduling process. Since the proposed framework was designed for the theoretical version of the multi-objective scheduling problem, there can be no rigid evaluation for this category. In general, since the GP-MOS framework operates on an explicit full-automation mode until the step where one of the proposed solutions must be chosen, it decreases the situation awareness of a hypothetical human scheduler. The human scheduler has to refer to the same output files produced by the GP-MOS framework in order to understand the scheduling details of the solutions proposed.

Decision cycle duration. The GP-MOS framework does not explicitly consider the timing and duration of decisions that have to be made by the human scheduler.

Complexity. The scheduling task is of the highest complexity possible, since as discussed earlier, the job-shop scheduling problem is a well-known NP-hard problem even for the single-objective case.

Human Characteristics

Level of human involvement in the task. This characteristic refers to the potential loss of skills and the complacency phenomenon that might arise from the full automation of the scheduling task considered. In the case of the GP-MOS framework, which operates autonomously until the stage where a solution must be chosen, both phenomena might be observed.

Performance level and performance variability. This characteristic refers to the algorithm's perceived usefulness and the variability of its performance, and is directly related to the trust of the human scheduler in its operation. These characteristics are strongly related to the problem situation where the algorithm will be applied. Since the framework has only been applied to a small test problem taken from the literature, no definitive statements can be made about its robustness. However, the variability in performance is a typical characteristic of the framework. This is because the core evolutionary algorithm (Genetic Programming) is probabilistic in nature, thus evolved solutions will not –in general – be the same between different search runs. The variability of evolved solutions will increase with the size and difficulty of the problem.

Providing relevant feedback. The performance of the cooperation between the human scheduler and the support system will increase if the human scheduler is able to comprehend the operation of the algorithm and the results that it produces. The GP-MOS framework scores very low in this category, since it does not provide any feedback on the operation of the algorithm and the solutions are reported in the form of output text files that are difficult to read.

Flexibility with constraints. The human view of scheduling constraints, allows for an amount of flexibility depending on the case and the timing of the scheduling decision. This type of flexibility is not readily supported by the framework, which operates and searches under the strict guidance of the mathematical model of the situation. However, as already indicated, the GP-MOS framework, due to the nature of the evolutionary algorithm, can be modified to consider and operate under some user-defined constraints.

Algorithm Characteristics

Computation time. The computation time of any evolutionary algorithm like GP-MOS, is largely dependent on:

- The computation time needed for the calculation of the objective functions.
- The algorithm's search parameters values (population size, number of generations). Note that the core parameter values of the algorithm can only be changed by accessing the program code.

Error rate. The GP-MOS framework always produces valid scheduling solutions.

Mode of automation. The GP-MOS framework does not require (in principle) the existence of a human operator in order to operate (automatic control). In addition, the actual generation of the solutions does not require the presence of the human scheduler. The framework does not choose or propose a final solution out of the set of non-dominated solutions which are generated. This means that the framework implicitly assumes an interaction with a decision making system (human or other) at the final stage. However, the assistance of the framework to the decision making system is limited to a set of text-based output files.

Development cost. The bulk of the development cost is associated with the design and implementation of the core GP-MOS algorithm. This is because the GP-MOS is an experimental research-based search procedure. However, this is more or less a one-off cost. If the framework was to be applied to a realistic scheduling situation, the relative percentage of the cost would shift towards the design and implementation of interaction facilities, since the algorithm has already been implemented.

15.3.9 Conclusion

This section introduced GP-MOS, an evolutionary framework for the solution of multi-objective scheduling problems. The application of GP-MOS was illustrated on the multi-objective version of the job-shop scheduling problem, one of the hardest instances of scheduling problems that have been researched.

GP-MOS was developed through the traditional production research approach. It employs a genetic programming machine for the generation and evolution of dispatching rules for specific scheduling problems. These rules are used in conjunction with Giffler and Thompson's algorithm in order to provide job-shop schedules. The evolutionary process continuously optimizes the population of dispatching rules and the corresponding schedules. An evolutionary multi-objective optimization technique is responsible for the generation and progressive evolution of a non-dominated set of schedules for the problem considered. The proposed framework was tested on a typical test problem taken from the literature, producing promising results.

The framework was evaluated in terms of its technological and human (cognitive) characteristics. This evaluation illustrated that the use of the traditional production research approach for the design of scheduling decision support has certain limitations that constrain its use in realistic production environments.

15.4 Group Sequencing: A Predictive-Reactive Scheduling Method

Generally, the job-shop scheduling problem is solved using either discrete optimization or real-time control methods.

Discrete optimization uses a mathematical model of the job-shop, which tries to find the solution that fits constraints (i.e., a feasible solution) and that optimizes one or more objectives. GP-MOS, described in the previous section, is one of these methods. For the job-shop scheduling problem, which is a NP-hard optimization problem (Garey and Johnson 1979), these kinds of methods may give schedules which exhibit good performances, but it costs frequently a large amount of effort and time. These methods are called predictive methods because they deal with a prediction of the system. The problem is that in a real manufacturing system, there are a lot of uncertainties and randomness (e.g., the breakdown of a machine, late material, new orders to proceed immediately, et cetera). Moreover, the job-shop model can be more or less precise in regards with the real system (e.g., operating times considered as random, transfer times considered as negligible, et cetera).

To cope with these drawbacks, one can use real-time control methods. These kinds of methods do not make any plan and build incrementally the schedule in real-time. Generally, these methods use priority rules which determine for each resource of the shop the next operation to proceed among the waiting operations in the resource's queue. These methods consider the real-state of the shop, so the hazardous phenomenon can be effectively taken into account when they occur. As these methods deal with the real-state of the system, they are called reactive methods. The major drawback of the reactive methods is that their performance is generally poor.

Very few methods try to combine predictive and reactive methods (Esswein 2003). The idea of the application in this section is to combine the advantages of both predictive and reactive methods. A first phase is made using an optimization method to obtain a first prediction of the schedule, generally described as a set of predictive schedules. From this, during the execution of the schedule, there is a second phase which aims at proposing a schedule decision at each state of the scheduling process according to the prediction and the real-state of the shop. The first phase is called the predictive phase and the second phase is called the reactive phase. The group sequencing method, presented in Sect. 15.4.2, is one of the most studied predictive reactive methods.

15.4.1 Problem Context

To solve the scheduling problem under uncertainties, the context is:

- Before the execution of the schedule, the planner knows the quality of the future schedule in the worst case: he is able to know that all operations will be on time, and if it is not possible, can know the tardiness of each job in the worst case so that he can negotiate the new delivery date.
- Uncertainties are not modeled.
- During the execution of the schedule, the human operator monitors the quality of the schedule: he must know as soon as possible if a job will be late.
- The human operator needs to be involved in the generation of the schedule because in regards with the machine (i.e., the algorithm), he has a different knowledge of the shop: his knowledge can be used to avoid delay. So, the method has to be a decision support system.

15.4.2 Modeling

Group sequencing was first introduced in Erschler and Roubellat (1989). The goal of this method is to have a sequential flexibility during the execution of the schedule that might absorb uncertainties and to guarantee a minimal quality corresponding to the worst case. This method has been widely studied in the last 20 years, in particular in Erschler and Roubellat (1989), Billaut and Roubellat (1996), Wu et al. (1999) and Artigues et al. (2005). For a theoretical description of the method, see Artigues et al. (2005).

Group sequencing is a method to solve the job-shop problem, a problem already described in Sect. 15.3.3.2. A group of permutable operations is a set of operations to be performed on a given resource in an arbitrary order. It is named G_k. A group sequence is defined as an ordered list of groups (of permutable operations) on each machine, to be performed in this particular order.

A group sequence is feasible if every permutation among all the operations of the same group gives a feasible schedule (i.e., a schedule which satisfies all the constraints of the problem). As a matter of fact, a group sequence describes a set of valid schedules, without enumerating them.

The quality of a group sequence is expressed in the same way a classical schedule is. However, it is measured as the quality of the worst semi-active schedule found in the group sequence, as defined in Artigues et al. (2005).

To illustrate these definitions, let us study an example. Figure 15.12 left presents a job shop problem with three machines and three jobs, while Fig. 15.12 right presents a feasible group sequence solving this problem. This group sequence is made of seven groups: two groups of two operations and five groups of one operation. This group sequence describes four different semi-active schedules shown in Fig. 15.13. Note that these schedules do not always have the same makespan ($C_{\max}$): for this criterion, the best case quality is $C_{\max} = 10$and the worst case quality is $C_{\max} = 17$.

This method enables to describe a set of schedules in an implicit manner. It is used in two distinct phases. Firstly, using a discrete optimization method an effective schedule is generated. From this, a group sequence is created which brings flexibility in the execution of the schedule. This first phase is called the predictive phase. Each machine has a group sequence to realize, there is no order in a group

a

i	j	$M_{i,j}$	$p_{i,j}$
1	1	1	3
1	2	2	3
1	3	3	3
2	1	2	4
2	2	3	3
2	3	1	1
3	1	3	2
3	2	1	2
3	3	2	2

A job shop problem

b

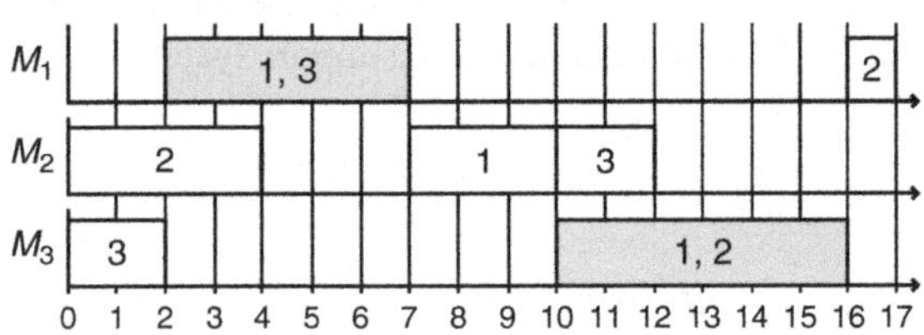

A group sequence solving the problem describe in Fig. 1a

Fig. 15.12 A job-shop problem solved by a group sequence (*left*: problem description; *right*: group schedule that solves this problem)

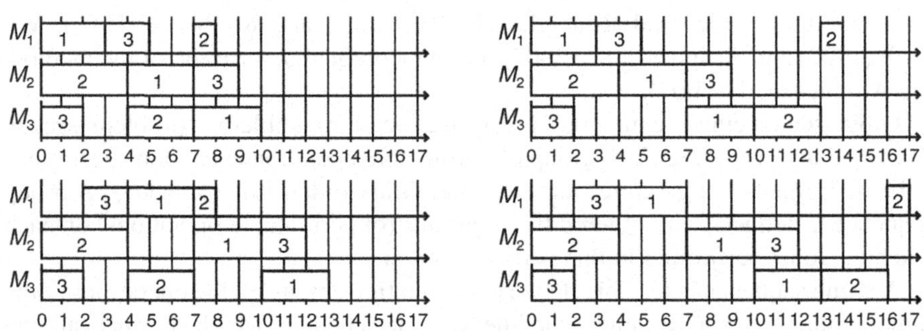

Fig. 15.13 Semi-active schedules described by Fig. 15.12 right

sequence, so the operations of a group sequence can be chosen in real time according to the real-state of the shop. This phase of choice is called the reactive phase.

15.4.3 Evaluation of a Group Sequence

Artigues et al (2005) proposes a polynomial algorithm that computes the exact value of C_i in the worst case. Because this algorithm takes insignificant time to run, it can be used in real time in the shop. It gives the exact evaluation of a group sequence with any minmax regular objective (as the makespan or the maximum lateness) in the worst case. For minsum regular objectives (like mean tardiness), due to the fact that every worst case C_i might not be in the same schedule, it is not possible to always have the worst case quality in a group sequence but only an upper bound. This tool also allows to determine which job can be late in a group sequence, and the amount of lateness in the worst case.

Thomas (1980) proposes a free margin adapted to group sequencing: sequential free margin. The sequential free margin of a given operation represents the maximum tardiness that can be tolerated to guarantee no late jobs in every schedule described by the group sequence. This margin gives the ability to monitor the group sequence in real time: if the sequential free margin becomes negative, then some schedules described by the group sequence have late jobs. Such a margin is a very useful tool to take proper decisions and monitor the quality of the group sequence.

Pinot and Mebarki (2008) and Pinot (2008) propose a complete study of the best case in a group schedule. First, Pinot and Mebarki (2008) proposes a polynomial algorithm to find a lower bound of the smallest C_i for each operation in a group schedule. This tool gives a lower bound for every regular objective (as mean tardiness, maximum lateness and makespan) in a group sequence. An improved lower bound for the makespan objective is also presented. It exhibits very good performances. Second, Pinot (2008) describes different heuristics for group sequencing. Heuristics give an upper bound of the quality of a group sequence, which complement lower bounds. In particular, the famous shifting bottleneck heuristic (Adams et al. 1988) is adapted to group sequencing. This heuristic gives very good performances: on average, the gap between the optimal and the schedule found is 1.5%. Third, Pinot (2008) proposes an exact method for finding the best case of any regular objective on a group sequence. Some experiments are done on the makespan. Most of the problems are solved in a limited time (the order of a minute), but some are not after several weeks of computation.

The best case evaluation of a group schedule can be used during the predictive phase as well as during the reactive phase. During the predictive phase, the best possible performance of the schedule along with its worst case evaluation give the range of all possible performances of the future schedule. During the reactive phase, it can be used to assess possible decisions: one can know the best quality obtained if an operation is executed first, thus it allows better performance for the execution of the overall schedule.

15.4.4 The Predictive Phase

Esswein (2003) formulates the predictive phase as a bi-objective problem: the first objective is the minimization of a regular objective (e.g., the makespan or the maximum lateness) in the worst case and the second is the maximization of the flexibility. The flexibility is expressed as a percentage according to the number of groups. A group sequence with a flexibility of 0% is a group sequence with one group per operation, and a group sequence with a flexibility of 100% is a group schedule with one group per machine. This flexibility represents the number of taken decisions divided by the maximum number of decisions.

Let us study the case of Fig. 15.12a. Because the problem has 3 operations on each machine, we can choose which operation will be executed (i.e., a decision) at maximum $2 \times 3 = 6$ (the last operation on a machine cannot be chosen because there is only one possibility left). There is no decision for groups having only one operation and only one decision for a group of two operations. So, there are 1+1 decisions on the group schedule presented in the example. The flexibility of this group schedule is $2/6 = 33\%$.

Esswein (2003) and Pinot (2008) propose an efficient algorithm named EBJG that generates a set of non-dominated solutions. It first begins with a group sequence with one operation per group (this first group sequence being obtained by another algorithm). Then, it merges two successive groups according to different criteria until no group merging is possible. During this process, it stores a set of non-dominated solution of the problem. Figure 15.14 illustrates a set of solutions for the job-shop instance la26 (Lawrence 1984) for the makespan objective.

To demonstrate the benefits of this algorithm, Esswein (2003) makes experiments with the makespan objective on a well known set of job-shop instances. He uses the exact algorithm described in Brucker et al. (1994) to get the initial group sequence with one operation per group.

With no degradation of the makespan, the flexibility obtained is 22% in average, which is not negligible. For example, the instance named la26, with no makespan degradation, has a flexibility of 17%. It corresponds to more that 6×109 different schedules, or 32 decisions on 190 possible decisions (for 20 operations on each machine and ten machines). Figure 15.14 presents the non-dominated solutions generated in terms of flexibility and degradation of the worst-case performance. It clearly shows that the more flexible the solution is, the more degraded is the worst-case performance.

15.4.5 Robustness of Group Sequencing

The flexibility added to the schedule should be able to absorb uncertainties. Three studies have tried to verify this property.

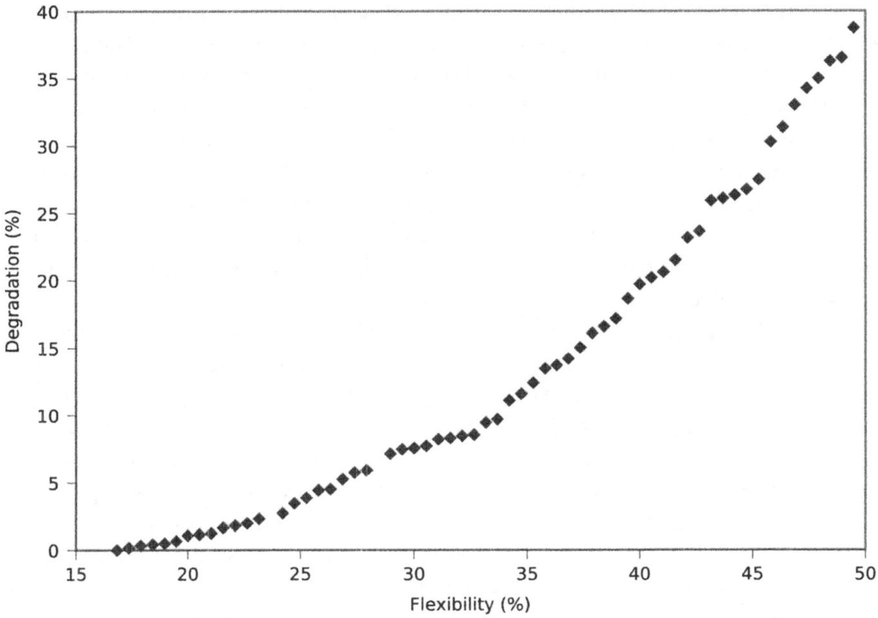

Fig. 15.14 Non-dominated solutions for instance la26

Wu et al. (1999) studies the impact of disturbed processing times on the objective of weighted sum of tardiness in comparison with static and dynamic heuristics. Whenever processing times are not so much disturbed, they observe that group sequencing obtain the best performances.

Esswein (2003) studies the impact of disturbed processing times, due dates and release dates on a one machine problem and compares its results with a static heuristic method. In average, performances are better with group sequencing than with the static method.

Pinot et al. (2007) studies the impact of non-modeled transportation time between two operations. The method exhibits good performances, even when transportation times and processing times are comparable.

These experiments show that group sequencing is very appropriate to cope with uncertainties. The flexibility provides robustness to group sequencing. But several studies have shown that cooperation between human and machine gives better performances than using only the machine or the human to solve the problem. Thus, using a decision support system to manage the flexibility provided by group sequencing might increase the performances of the method.

15.4.6 The Reactive Phase

ORDO software is a software tool that proposes a predictive-reactive method, based on the group sequence, to solve the scheduling problem in a manufacturing system.

The reactive phase implemented in ORDO software described in Billaut (1993) is based entirely on the group margin already presented in Sect. 15.4.3. The decision support system computes the margin of each operation, and thus can detect whenever the group sequence will eventually gives a schedule with tardy jobs. There is a human (an operator) on each workstation in charge of the schedule's execution on the machine. The decision support system provides the margins of each operation to the operator. He can choose which operation of the group to execute on the machine. This method has two important features: monitoring the group sequence can be done automatically and due to the margin the human operator has a large freedom in his decisions to achieve his goals.

Pinot (2008) points out that the main drawback of this reactive phase is the lack of assistance for the human operation: the only information given by the machine is the margin. The main risk is that the system will be underused: because without any more assistance another decision is very complex to evaluate for the human, the operator might just follows the proposition of the machine (the operation with the greatest margin) as suggested by Cegarra (2004). So, the flexibility proposed by the group sequence will be underused and the operator will almost never intervene.

Thus, Pinot (2008) proposes to use more indicators to describe each operation. These indicators can be the operation's margin, the worst case quality if the operation is executed next, and particularly the best case quality if the operation is executed next. As a matter of fact, the machine does not directly suggest the operation to execute, but transfers its knowledge of the situation to the operator.

To experiment these new indicators, Pinot (2008) intends to compare three different decision support systems:

- The machine does not give any information to the human.
- The machine gives the margin of each operation to the human.
- The machine gives the new indicators proposed by Pinot (2008) to the human.

Two kinds of indicators will be evaluated: performances of the schedule, and human factors (the trust in the system, the activity of the human in the decision process, etc.).

15.4.7 Evaluation

15.4.7.1 Assessment of Problem Description

Group sequencing is a scheduling method developed during the last 20 years, so it is difficult to analyze the complete modeling. First, we can say that the structure of groups of permutable operations was studied, which correspond to stage 2 and 3 of Mitroff et al. (1974). Then, the application to scheduling under uncertainties of this particular model is proposed (stage 1). Thanks to the validation (stage 6), special properties and algorithms are developed (stage 2 and 3). Then, group sequencing is implemented into scheduling software (stage 4). Several research works improved the model (element 3), the resolution (stage 3) and the implementation (stage 4).

The problem is well structured: it is based on a formal model (groups of permutable operations) and the objectives are well defined (flexibility and quality for the predictive phase, quality monitoring for the reactive phase). However, the reactive phase can be seen as a semi-structured problem: the objectives of the operator can be unknown formally, and he can merge the well-defined objectives (quality of the schedule) with his own preferences.

This scheduling method is made for small to medium manufacturing systems that made different products on demand. Different kind of uncertainties can be present in the shop: new important orders, machine breakdowns, late delivery, et cetera. The goal of this scheduling method is to be robust in regards with these uncertainties by providing flexibility.

15.4.7.2 Assessment of Problem Analysis

The perspective that has been used in this scheduling method is normative: the scheduling process is decomposed in several different subtasks and the resolution method is exposed.

The different subtasks are decomposed using a hierarchical task analysis: the global task (scheduling a shop under uncertainties) is decomposed in two subtasks: the predictive phase and the reactive phase. Then each subtask is decomposed in different objectives (flexibility and quality for the predictive phase) or in different subtasks (the different decisions to make).

Group sequencing is device and event dependent. The different devices are considered in the model: the planner, the operators, the machine and the shop. The reactive phase is highly dependent to the different events: each event will modify or add a new decision to make. The psychological relevance of group sequencing is globally low, except for two points: the structure used in group sequencing (groups of permutable operations) is comprehensible by the human, and the reactive phase proposed by Pinot (2008) takes into consideration the psychological aspect.

15.4.7.3 Assessment of Human–Machine Interaction

Problem Characteristics

Harm risk. A suboptimal solution in the predictive and the reactive phase might only result in profit cost. So, we can characterize this cost as being low to medium.

Need of situation awareness. During the predictive phase, the planner needs to know the context of the factory to choose the compromise between quality and flexibility. During the reactive phase, the operator needs to have a good knowledge of the state of the shop to make good decisions.

Decision cycle duration. The decision cycle duration for the predictive phase is the scheduling horizon. It is in practice a short-term period, usually a few weeks, from 1 day to 1 month. For this phase, the planner has the time to work on the schedule. The decision cycle duration for the reactive phase is the execution time of an operation. It is in practice between a minute to a few hours. For this phase, the decisions are done under time pressure.

Complexity. The problem of the predictive phase is a NP-hard problem (Artigues et al. 2005). The problem of the reactive phase is also NP-hard (Pinot 2008). Moreover, the possible uncertainties make the problem of the reactive phase not solvable optimally in practice.

Human Characteristics

Level of human involvement in the task. The level of human involvement in the predictive phase is low to medium. The planner has only to choose the compromise between quality and flexibility in a set of different group sequences. Loss of skills and complacency phenomena are possible. The reactive phase proposed by Pinot (2008) aims at involving the operator. This should limit loss of skills and complacency phenomena.

Performance level and performance variability. The predictive phase uses an efficient heuristic. The level of the performance should be seen as high by the human. Moreover, experimentations on different instances show that this heuristic has stable performances. The trust of the planner in the algorithm should be high. The performance level and the performance variability of the reactive phase are directly linked to the level of uncertainties. The higher is the level of uncertainties, the higher will be the performance variability and the lower will be the performance level. So, the trust of the operator in the algorithm should be correlated to the level of uncertainties.

Providing relevant feedback. Group sequencing is easily comprehensible to the human thanks to the notion of group of permutable operations. Moreover, Pinot (2008) proposes an improvement for the feedback of the algorithm to the human during the reactive phase.

Flexibility with constraints. The predictive phase does not integrate any flexibility with constraints. The reactive phase can manage flexibility with some constraints, as starting an operation before it is allowed by the model. These possibilities are presented in Billaut (1993).

Algorithm Characteristics

Computation time. EBJG, the algorithm used in the predictive phase to generate group sequences, runs less than 30 s for instances with ten machines and 30 jobs. EBJG needs an initial group sequence that has to be generated using a classical (heuristic or not) job-shop scheduling method. For example, with the algorithm described in Adams et al. (1988), the global computation time should be less than a minute for a quite good performance.

Error rate. The predictive phase always provides valid group sequences. During the reactive phase, errors in the different evaluations of the group sequence are directly related to the uncertainties.

Mode of automation. The predictive phase uses supervisory control: the algorithm proposes a set of group sequences, and the human chooses one. The reactive phase also uses the supervisory control: the algorithm evaluates and proposes the different decisions according to the group sequence, and the human choose the decision to execute.

Development cost. ORDO, the software which uses the group scheduling method is already in use in several manufacturing companies. The cost development of the method we propose to improve the interaction between the human and the software is a one-off cost. Nevertheless, we intend to conduct experiments on an experimental manufacturing system with students in production management to demonstrate the benefits of the development we propose.

15.4.8 Conclusion

This section presents group sequencing, a scheduling method adapted to the job shop problem. This scheduling method uses groups of permutable operations to provide a set of different schedules without enumerating them. Thanks to this set of schedules, the operator can choose the operation to execute during the execution of the schedule in order to manage uncertainties. This flexibility makes group sequencing a robust scheduling method.

This scheduling method is decomposed in two different phases: a predictive and a reactive phase. During the predictive phase, group sequences are generated according to two different objectives: flexibility and quality. The planner chooses the group sequence according to a compromise between these two objectives. During the reactive phase, the operator chooses which operation should be executed, according to the initial group sequence chosen with the help of the machine.

The scheduling method is evaluated according to the Chaps. 12–14. In this method some efforts have been done to take into consideration the human.

15.5 Conclusions

We have studied three different planning and scheduling applications focusing on the human–machine interaction. These applications are different in terms of methods, in terms of application areas as well as in terms of the human–machine interaction. Each application was described following Chaps. 12–14.

As stated in Chap. 12, the applications were described following Mitroff's model of problem solving (Mitroff et al. 1974); the problem structure was characterized according to Newell's distinction (well-structured, semi-structured, ill-structured); the algorithm was carefully examined in terms of computational complexity and in terms of the provided solutions (heuristic or optimal solution, solution approach, robustness, et cetera). Each planning problem was assessed according to various criteria and finally the human and organizational noise in the environment of reference was addressed for each application.

According to Chap. 13, problems context were analyzed from a task perspective (description of the perspective of the task analysis, description of the analysis itself, characterization of the problem requirements in terms of dependency and psychological relevance).

Using the criteria exposed in Chap. 14, the three algorithms were then assessed by focusing on the human–machine interaction (issued related to the problem itself, analysis of the interaction between human and the algorithm, description of the possible modes of automation).

The first application described the development of a network flow algorithm for the solution of a multi-commodity routing problem in shunting yard scheduling. This problem is semi-structured, and the planning environment rather predictable, which makes it suitable for algorithmic support.

The analysis of the human–machine interaction revealed that in solving the problem involvement of a human planner is still desirable, although it will be helpful if an algorithm is used for important partial tasks.

The second application described is named GP-MOS, an evolutionary framework for the solution of multi-objective scheduling problems used to solve the job-shop scheduling problem. GP-MOS uses a genetic programming machine for the generation and evolution of dispatching rules for specific scheduling problems.

The framework was evaluated in terms of its technological and human (cognitive) characteristics. This evaluation illustrated that the use of the traditional production research approach for the design of scheduling decision support has certain limitations that constrain its use in realistic production environments.

The third application described is the group sequencing method which aims at solving the job-shop problem by proposing not one schedule only but a set of different schedules. This set of schedules is presented through groups of permutable operations. This method needs the intervention of a human who chooses during the execution of the group schedule the operation to be executed in each group of permutable operations. Thus, this method brings flexibility and enables to choose in real-time the operation (among a group of permutable operations) that fits best to the real state of the system.

This method has been designed to provide a real interaction between the algorithm and the human for the job-shop scheduling problem. Yet, it still necessitates some developments to facilitate the activity of the human during the scheduling process in order to improve the benefits of this method.

In conclusion, these three different studies show the need of a framework to design planning and scheduling algorithms. Such a framework should take into account not only the structure of the algorithm but also the interaction with the human who is involved in the planning and scheduling process.

Acknowledgments We would like to thank Jan Banninga, master student at the University of Groningen, for his assistance in this research project.

References

Adams, J., Balas, E., & Zawack, D. (1988). The shifting bottleneck procedure for job shop scheduling. *Management Science, 34*(3), 391–401.

Ahuja, R. K., Magnanti, T. L., & Orlin, J. B. (1993). *Network flows: theory, algorithms, and applications*. Englewood Cliffs, NJ: Prentice Hall.

Artigues, C., Billaut, J.-C., & Esswein, C. (2005). Maximization of solution flexibility for robust shop scheduling. *European Journal of Operational Research, 165*(2), 314–328.

Baykasoglu, A., Özbakir, L., & Dereli, T. (2002). Multiple dispatching rule based heuristic for multi-objective scheduling of job shops using tabu search. *Proceedings of the 5th International Conference on Managing Innovations in Manufacturing* (pp 396–402). Milvaukee, WI.

Billaut, J.-C. (1993). *Prise en compte des ressources multiples et des temps de préparation dans les problèmes d'ordonnancement en temps réel*. Ph.D. Thesis, Université Paul Sabatier, Toulouse, France.

Billaut, J.-C., & Roubellat, F. (1996). A new method for workshop real-time scheduling. *International Journal of Production Research, 34*(6), 1555–1579.

Brucker, P., Jurisch, B., Sievers, B. (1994) A branch and bound algorithm for the job-shop scheduling problem, Discrete Applied Mathematics, *49*(1–3), 107–127.

Cegarra, J. (2004). *La gestion de la complexité dans la planification: le cas de l'ordonnancement*. Ph.D. Thesis, Université de Paris 8, France.

Chiang, T.-C., & Fu, L.-C. (2006). Multiobjective job shop scheduling using genetic algorithm with cyclic fitness assignment. *Proceedings of IEEE World Congress on Computational Intelligence* (pp 11035–11042).Vancouver, BC: IEEE.

Cordeau, J. F., Toth, P., & Vigo, D. (1998). A survey of optimization models for train routing and scheduling. *Transportation Science, 32*(4), 380–404.

Deb, K., Pratap, A., Agarwal, S., & Meyarivan, T. (2002). A fast and elitist multiobjective genetic algorithm: NSGA-II. *IEEE Transactions on Evolutionary Computation, 6*(2), 182–197.

Demeulemeester, E. L., & Herroelen, W. S. (2002). *Project scheduling, a research handbook*. Boston, MA: Kluwer Academic.

Eiben, A. E., & Smith, J. E. (2003). *Introduction to evolutionary computing*. Berlin: Springer.

Erschler, J., & Roubellat, F. (1989). *An approach for real time scheduling for activities with time and resource constraints, Advances in project scheduling*. Amsterdam: Elsevier.

Esquivel, S., Ferrero, S., Gallard, R., Salto, C., Alfonso, H., & Schütz, M. (1996). Enhanced evolutionary algorithms for single and multiobjective optimization in the job scheduling problem. *Knowledge-Based Systems, 15*(1–2), 13–25.

Esswein, C. (2003). *Un apport de flexibilité séquentielle pour l'ordonnancement robuste*. Ph.D. Thesis, Université François Rabelais, Tours, France.

Fang, H. L., Corne, D., & Ross, P. (1996). A Genetic Algorithm for job-shop problems with various schedule quality criteria. *Lecture Notes in Computer Science, 1143*, 39–49.

Freling, R., Lentink, R. M., Kroon, L. G., & Huisman, D. (2005). Shunting of passenger train units in a railway station. *Transportation Science, 39*(2), 261–272.

Garey, M., & Johnson, D. (1979). *Computers and intractability: A guide to the theory of NP-completeness*. San Francisco, CA: W.H. Freeman.

Giffler, B., & Thompson, G. L. (1969). Algorithms for solving production scheduling problems. *Operations Research, 8*, 487–503.

Hoc, J.-M. (1988). *Cognitive psychology of planning*. London: Academic.

Koza, J. R. (1992). *Genetic programming: On the programming of computers by means of natural selection*. Cambridge, MA: MIT.

Lawrence, S. (1984). *Resource constrained project scheduling: an experimental investigation of heuristic scheduling techniques*. Pittsburgh, PA: Carnegie-Mellon University.

Mitroff, I. I., Betz, F., Pondy, L. R., & Sagasti, F. (1974). On managing science in the systems age: two schemas for the study of science as a whole systems phenomenon. *Interfaces, 4*(3), 46–58.

Petrovic, D., Duenas, A., & Petrovic, S. (2004). A multi-objective job shop scheduling problem with linguistically quantified decision functions. In C. H. Antunes, & L. C. Dias (Eds.), *Proceedings of Mini-EURO Conference 2004, Managing Uncertainty in Decision Support Models* (pp. 1–6). Cuoimbra, Portugal: MUDSM04.

Pinot, G. (2008). *Coopération homme-machine pour l'ordonnancement sous incertitudes*. Ph.D. Thesis, Université de Nantes, France.

Pinot, G., Cardin, O., & Mebarki, M. (2007). A study on the group sequencing method in regards with transportation in an industrial FMS. *Proceedings of the IEEE SMC 2007 Conference* (pp. 151–156). Montréal, QC: SMC.

Pinot, G., & Mebarki, M. (2008). *Best-case lower bounds in a group sequence for the job shop problem: Proceedings of the 17th IFAC World Congress*. Seoul, Korea.

Riezebos, J., & van Wezel, W. M. C. (2009). K-shortest routing of trains on shunting yards. *OR Spectrum, 31*(4), 745–758.

T'kindt, V., & Billaut, J.-C. (2002). *Multicriteria scheduling*. Berlin: Springer.

Thomas, V. (1980). *Aide à la décision pour l'ordonnancement d'atelier en temps réel*. Ph.D. Thesis, Université Paul Sabatier, Toulouse, France.

van Wezel, W. M. C., & Jorna, R. (2009). Cognition, tasks and planning: supporting the planning of shunting operations at the Netherlands Railways. *Cognition, Technology & Work, 11*, 165–176.

Wu, S. D., Byeon, E.-S., & Storer, R. H. (1999). A graph-theoretic decomposition of the job shop scheduling problem to achieve scheduling robustness. *Operations Research, 47*(1), 113–124.

Zitzler, E., & Thiele, L. (1999). Multiobjective evolutionary algorithms: a comparative case study and the strength Pareto approach. *IEEE Transactions on Evolutionary Computation, 3*(4), 257–271.

Zwaneveld, P. J., Kroon, L. G., Romeijn, H. E., Salomon, M., Dauzère-Pérès, S., van Hoesel, S. P. M., et al. (1996). Routing trains through railway stations: model formulation and algorithms. *Transportation Science, 30*(3), 181–194.

Chapter 16
Case Study: Advanced Decision Support for Train Shunting Scheduling

Wout van Wezel and Jan Riezebos

16.1 Introduction to Advanced Support for Train Shunting Scheduling

Train shunting scheduling for the Netherlands Railways is a complex planning problem that is performed by about 130 full-time planners. It concerns the planning of day-to-day shunting operations at the large stations in the railway network. Input for this planning is based on the long term schedule of arrivals and departures at a station, rolling stock and track maintenance schedules, and specific circumstances that have to be taken into account when preparing the plan for a particular day or night. Currently, the planners mainly work manually. The plans are made and revised on paper first, after which the outcome is put in the computer. In our study, the central question is: how can these planners be supported in their task with an advanced planning system?

Advanced planning systems offer many advantages. Graphical manipulation of the plan, real time constraint checking, plan evaluation, real time data exchange, and automatic plan generation are functions that are commonly offered by advanced planning systems (Stadtler and Kilger 2005). Unfortunately, train shunting planning gets very little attention in literature. From a somewhat abstract perspective, shunting can be depicted as a combination of inventory location planning, routing, and staff scheduling. This, however, does not do justice to the specific aspects of train shunting planning. As a consequence, it is not trivial how these functionalities could be applied for shunting planning. To overcome the lack of knowledge about how to support shunting planning, we apply a task oriented approach to resolve our research question. In a task oriented approach, an analysis of the task as it is performed by the planners is input for the design of the system. In the paper, we describe the results of the task analysis and the developed prototype.

W. van Wezel (✉) and J. Riezebos
Faculty of Economics and Business, University of Groningen, Groningen, The Netherlands
e-mail: w.m.c.van.wezel@rug.nl

J.C. Fransoo et al. (eds.), *Behavioral Operations in Planning and Scheduling*, DOI 10.1007/978-3-642-13382-4_16,

In Sect. 16.2, we shortly elaborate upon the approach we applied. Section 16.3 describes domain and task models of the case. In Sect. 16.4, we discuss the technical background of the prototype, and the prototype itself is described in Sect. 16.5 (user interface) and Sect. 16.6 (algorithms). Section 16.7 provides conclusions and directions for further research.

16.2 Modeling Approach

In the analyses and prototype design we combined object oriented modeling with a task oriented approach, i.e., a mixture of an IT perspective and a cognitive perspective. In Chap. 12 (Design of Scheduling Algorithms) we described how planning problems can be modeled. Planning always somehow deals with arranging scarce resources. Examples of resources are products, machines, vehicles, staff, etc. More precisely, it always concerns multiple tokens of different object types. Production schedules, for example, arrange products, machines, and time intervals. In Chap. 12, we proposed the following definition: *A planning problem consists of groups of entities, whereby the entities from different groups must be assigned to each other. The assignments are subject to constraints, and alternatives can be compared on their level of goal realization.* In order to model, analyze, and design a decision support system for railway shunting planning, we use object oriented techniques. This adds (a) the distinction between classes or types and objects or tokens, and (b) hierarchic modeling of classes to the proposed definition. This approach is used throughout the project. First, we use the object oriented technique to describe the planning domain. Second, a planning task is described as manipulating and assigning objects in the domain model. Third, the user interface is modeled as the objects that are shown on the various windows, and fourth, the algorithms use the object model to automatically schedule certain parts of the plan. The prototype system uses a reusable library based on the object oriented modeling paradigm specifically created for planning systems.

16.3 Domain and Task Description of Shunting Planning

16.3.1 Methodology

We performed a two phase analysis following a descriptive perspective. First we analyzed the domain knowledge of the planners using a knowledge extraction tool. Three of the five planning departments participated with in total 27 planners. Multi Dimensional Scaling techniques were used to determine coherence in the mental models of the planning structure. The individual differences were small, and we did not find statistically significant structural differences when the mental models of different groups were compared (i.e., grouping planners according to the department where the planner works or the shunting yard that is planned). The resulting model is described in Sect. 16.3.3. Second, we performed extensive task analyses with two planners. This included observation, analysis of thinking-aloud protocols, and probing of decisions that were made. The resulting task structure was then verified with several other planners during meetings. The task model is discussed in Sect. 16.3.4. In Sect. 16.4, we describe the resulting system.

16.3.2 Generic Overview of Shunting Planning

Trains that arrive at a station do not necessarily leave in the same configuration. A train can be split in its individual coaches and these coaches can be combined again in a train. For example, the train from the city Groningen to the city Zwolle consists of two coaches and the train from the city Leeuwarden to Zwolle consists of one coach. In Zwolle the trains are connected, after which they leave as one train to the city Den Haag. See Fig. 16.1.

During the night the passenger trains stay at the station. The "storage" capacity, however, is limited. In addition to changing the configurations during the night, all trains must be washed at a track that contains the washing equipment. The task of the planner is to plan the movements of the trains and coaches and to decide on what track trains can stay. To plan the movements, the planner must assign train drivers, train shunters (the ones who connect and disconnect coaches), and routes to move trains from one track to another.

The major bottleneck in the shunting yard that we investigated is the washing track. Since the time it takes to wash a train is largely independent of the size of the train, it is advantageous to combine carriages in trains before they are washed. Therefore, the nightly transition is difficult to solve, since trains are shunted during the night for washing as well.

Because the main time table is stable, planning mainly means adjusting existing plans. Changes are made on the basis of events, for example a track that needs maintenance, an extra train that will pass through the station because of a pop concert, etc. The plans are made manually on paper, after which the result is entered in the computer. Figure 16.2 shows an example of a part of a shunting plan. The computer can check the validity of the plan once it has been entered and send the data to traffic control.

Approximately 130 planners from five major stations in The Netherlands plan the shunting operations for all stations. In our research project, the emphasis was on a task analysis of the shunting planners. In addition to building a graphical user

Fig. 16.1 Self-propelling coaches connected to trains

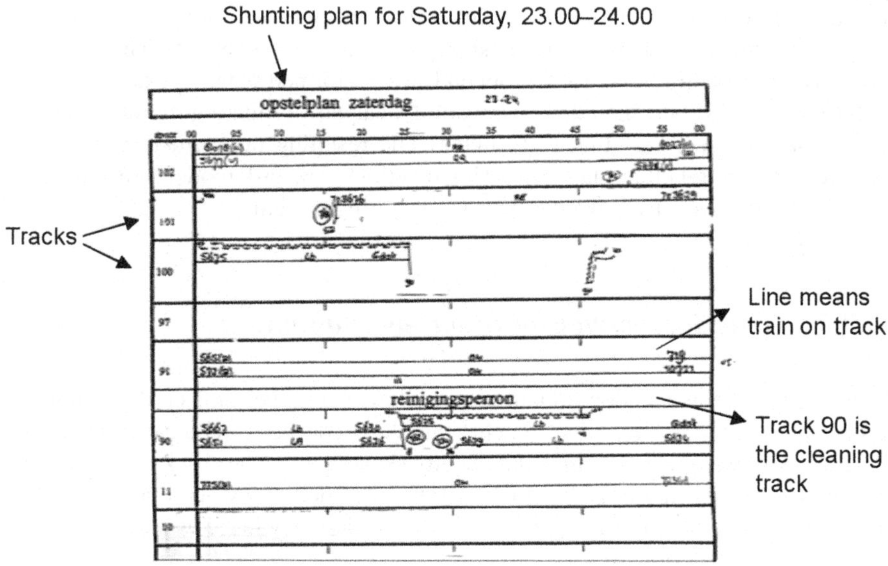

Fig. 16.2 Example of a train shunting plan as made by a planner using pencil and paper

interface for the planners, we have also combined the mathematical and task oriented approaches by designing algorithms in a task oriented fashion.

16.3.3 Domain Model

The domain model depicts the structure of the planning problem using an object oriented modeling approach (Van Wezel and Jorna 1999). Basically, *train coaches* are located on *track segments* on the shunting yard during a given *period*. Each train location is the result of a *shunting activity*. This shunting activity has to be performed by *shunting staff*, and uses a route on the shunting yard. A route consists of a number of *track segments* and a *time period*. Figure 16.3 shows the general structure.

There is a 2:2 relation between track occupation and shunting activity. Each track occupation is the result of a shunting activity, and each shunting activity transforms one track occupation in another track occupation.

The domain model is used as the basis to model the tasks and prototype system components. In the next section, we describe the task analysis.

16.3.4 Task Model

As indicated before, the task of the shunting planner is to plan the movements of the trains and carriages and to decide on what track trains stay during the night. To plan

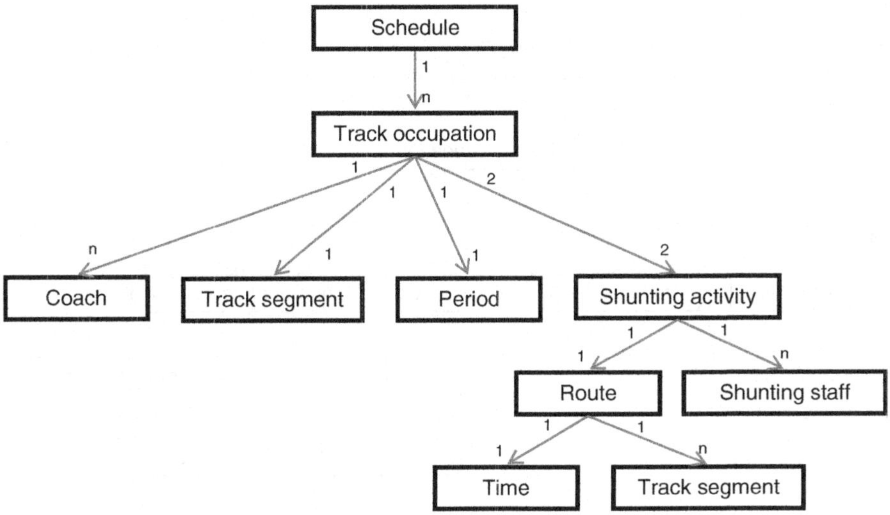

Fig. 16.3 Domain model of the train shunting plan

the movements, the planner must also assign engine drivers, train shunters (the ones who connect and disconnect carriages), and routes of the trains on the station. In the research project, we looked at two planners whose task it is to adjust already created plans. The task analysis revealed a hierarchical task structure. The activities of the planners contain the following high level subtasks:

1. Determine nightly transitions. Several trains of the same type stay overnight at the station. Coaches of the same type are interchangeable, so the planner must decide what incoming coach will become what outgoing coach.
2. Determine tracks. From the national timetable, the planner knows the amount, arrival/departure times, and tracks of both incoming and outgoing trains. After subtask 1, the planner can search for tracks that the coaches can stay on during the night. Because there are much more coaches than shunting tracks, multiple coaches are put on one track. Therefore, the dominant constraint is that a coach should not be blocked by trains to the left and right of it at the time it must be moved to the departure track.
3. Determine routes. When subtask 2 is finished, the planner must make sure that a relocation is indeed possible by determining the route the train should use.
4. Assign shunting staff. The route that is used determines the amount of time that is needed to move the train and the time that is needed for shunting staff to walk from one task to another. Consequentially, assigning tasks to shunting staff is the last planning subtask.

Figure 16.4 shows how the subtasks are related to the domain model.

The structure is hierarchical because each decision constrains the lower decisions. For example, the possible routes are determined by the tracks that the train

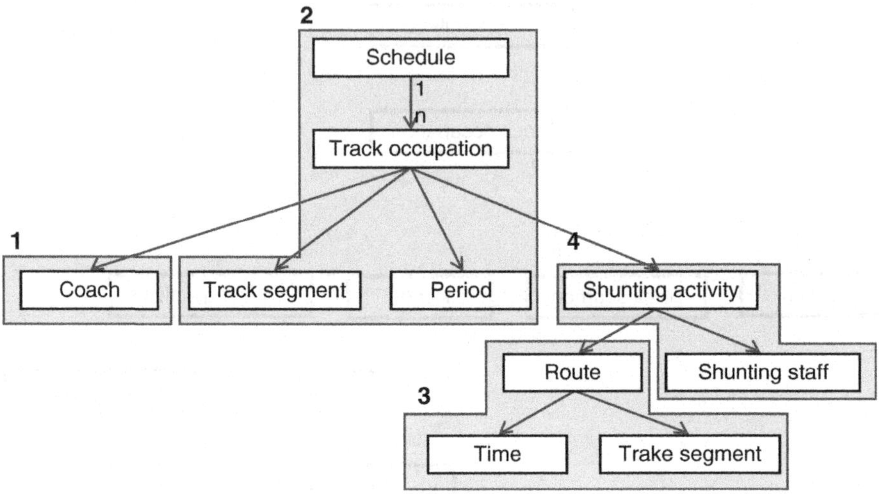

Fig. 16.4 Subtasks in the domain model

will be on. These tasks are performed in this order no matter the kind of event that triggered the task, but always for a few coaches at the time. In other words, the planner does not perform subtasks sequentially for all trains, but only for a limited set of related trains at a time.

Although the task order is fixed, the kind of event determines what task is started with. For example, when a track is scheduled for maintenance, the planner will start with subtask 2 and try to find new tracks for the trains that are scheduled for the track on which the maintenance will take place. Only if he can not find tracks (subtask 2 is overconstrained), he will change the nightly transitions and from there perform subtasks 2, 3 and 4. The consequence of changing a nightly transition will be that a part of the existing schedule becomes obsolete and that subtasks 2, 3, and 4 might have to be done again for several trains. This means much work and therefore the planner tries to avoid this.

In the prototype, we provide support for subtasks 2, 3, and 4. The reason for this is that in the task analyses and experiments we have focused on the day-planners, who reschedule for the short term and for whom subtask 1 is not a frequently performed task. Some of the activities of day planners are:

1. *Event*. One of the tracks on the station needs maintenance.
   Activity. Reschedule all trains that are on that track during the time of maintenance to other tracks.
2. *Event*. A train that is planned to arrive will not arrive due to maintenance. A few hours later, a similar extra train is put on.
   Activity. Replace the original train with the latter train in the plan.
3. *Event*. The time of departure of a train changes.
   Activity. Find out the consequences and fix the plan where appropriate.

4. *Event.* No driver for routing a train assigned.
   Activity. Find an engine driver available and allowed to move the train.
5. *Event.* No route assigned to a train relocation.
   Activity. Find a route to move from the initial track to the destination.
6. *Event.* Errors are encountered in the plan.
   Activity. Check the plan for errors.

Currently, the planning tasks are performed manually. Planners use some computer programs to collect information, but the plan itself is made on paper before it is put in the computer.

We will now focus on the first subtask that is mentioned in the list above: a number of tracks need maintenance for a couple of hours during the night, and all trains that are on those tracks during that time must be repositioned. In some aspects, this task is an easy one. The configuration of trains stays the same, so the planner only has to move trains. Unfortunately, the number of shunting tracks is limited already, and when there are even less to use, it becomes a difficult puzzle.

An important aspect in the task is weighing the depth of search against the quality of the plan. We will clarify this with an example.

Example. A train is on a track that is out of order. The planner searches for an alternative track, but there is none. The planner can now follow two strategies. First, he can violate some constraints so it does not have to be at the track, for example, by skipping washing the train or by letting it depart from another track in the morning. Second, he can put the train on one of the tracks, and search for another solution for the train that is already there. The planner will now (recursively) follow the same procedure with the chosen train: search for a solution that falls within the constraints or be satisfied with a constraint violation. Of course, because the planner makes his plan manually on paper, backtracking is difficult so the depth of searching is limited. The steps or subtasks are shown in the flowchart in Fig. 16.5.

The choice between searching depth and plan quality is apparent in step 2, 3, and 4 in the flowchart. There, the choice must be made between adding another search layer or violating a constraint (by choosing a solution in step 5, 6, or 7).

16.4 Reusable Library to Develop Scheduling Systems

The basic functionality needed to support the shunting planning task is no different from other planning tasks. Advanced Planning and Scheduling systems generally contain a graphical user interface, constraint and goal evaluator, algorithms, and links to external systems such as ERP systems. However, train shunting planning has many specific characteristics which makes that standard scheduling software can not be used, and hence, a system must be created from scratch. The object oriented modeling paradigm, however, enables reuse of system components. The basis for this is in the domain models, and a strong analogy in another problem domain comes to mind.

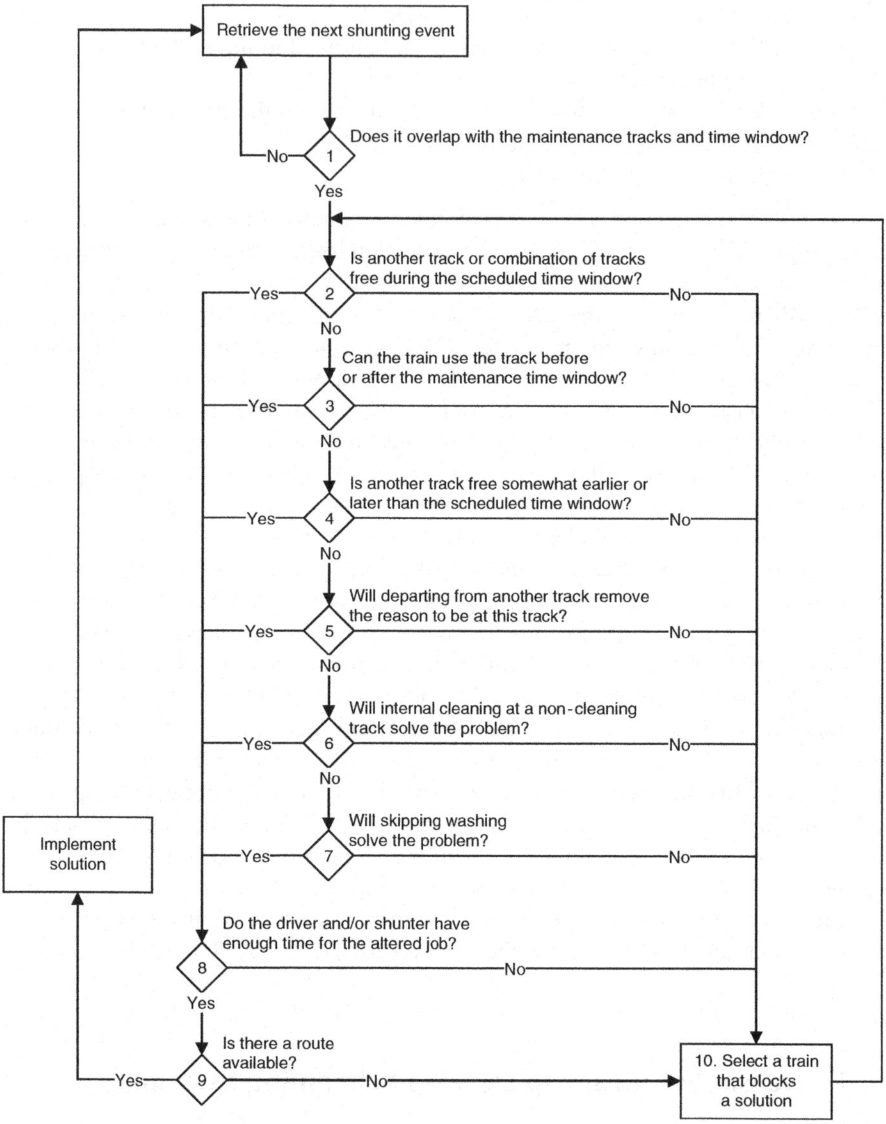

Fig. 16.5 Flowchart of task structure

In the database world, the introduction of relational databases, a standardized language (SQL), and standardized communication protocols (for example, ODBC and ADO) have almost eradicated the efforts that software developers must put in storing and retrieving data. As a result, developers can focus on the mapping of business model to data model and on the front-end of systems rather than on the back-end. Systems can easily be extended because data is separated from function.

Additionally, systems can easily use each others data by using the same tables in the database.

Unfortunately, the relational structure is not well equipped to store planning models for several reasons. First, the mapping that must be made from a hierarchical planning model to a relational table structure increases the complexity of queries that are used to manipulate the data. Second, queries for storage and retrieval of data are hardly reusable, because of this complex mapping problem. Third, some characteristics that are common to all planning situations (e.g., constraint checking, goal evaluation, reasoning at multiple levels of resolution) are neither embedded in the data query language nor in the relational structure. This complicates reuse of developed systems.

For that reason, we have extended the object oriented paradigm to implement a kind of Database Management System for planning and scheduling systems, thereby creating a domain specific modeling language (DSL) (Van Wezel 2006). In general the process of using a DSL would be as follows. Domain modeling is facilitated by a preconfigured set of classes. The base classes can be fixed, which means that the application scope is predetermined (e.g., machines and jobs in a manufacturing scheduling problem). Most approaches, however, allow extension of existing classes. The idea of most object oriented approaches is that software components are available for the different classes in the domain models. Specifying the domain therefore means putting together a number of software components. In some approaches these components together are a system, in other approaches the components provide a starting point and require further programming. France and Rumpe (2005, p. 1) describe the following advantages of domain specific modeling languages compared to generic modeling languages such as UML that have no domain specific content in their structure:

- "Domain specific constructs are better suited for communication with users in the domain. The users can better understand the models if they are presented in domain-specific terms. This leads to the claim that domain specific models are better suited for requirements engineering than UML.
- DSLs have restricted semantic scope (i.e., the number of semantic variations they have to deal with is small compared with general-purpose languages), thus developing a semantic framework is less challenging.
- Restricting semantic scope can lead to better support for generating implementations from models. Given an appropriate component-framework that implements individual constructs of the DSL, the composition of these components can lead to powerful implementations.
- DSLs increase domain specific reuse of components, which can lead to improved quality of systems and order-of-magnitude improvements in time-to-market and developer productivity. Examples from the telecommunications sector indicate that a speedup factor of 10 is possible."

Analogous to the separation between data, data structure, and functionality in systems that access data using a database management system, we have designed and implemented a generic structure that can be used to specify the hierarchical planning object model and object characteristics. This provides developers with a "domain model management system", which reduces the efforts that are needed to create code for manipulation of planning models.

Figure 16.6 shows the place of the DSL engine for planning systems. Generic components in such systems are the user interface, algorithms, constraint checker, goal evaluator, and links to external systems. The DSL specific Class and Object Administration component handles all data related tasks. It is used by the module Domain Model Specification to specify the domain structure.

The purpose of the Domain Model Specification is to make the transfer of the model of a planning sub-problem to the system transparent. The functioning can be compared to making a database model: when the developer provides a conceptual data model, he gets functionality to manipulate the data. Currently, the following classes are implemented in an object oriented programming language (Table 16.1):

The Context object contains functionality to load and save domain models; it functions as the blackboard. Referring to the analogue of relational database management systems, the architecture can be used for the following purposes:

- Create a domain model builder much like a graphical tool to specify a database structure.
- Build generic components for scheduling systems, e.g., a report generator.
- Build customizable components, for example a Gantt-chart of which the axes can be assigned to object types.
- Make specific components, for example a planning board for shunting scheduling

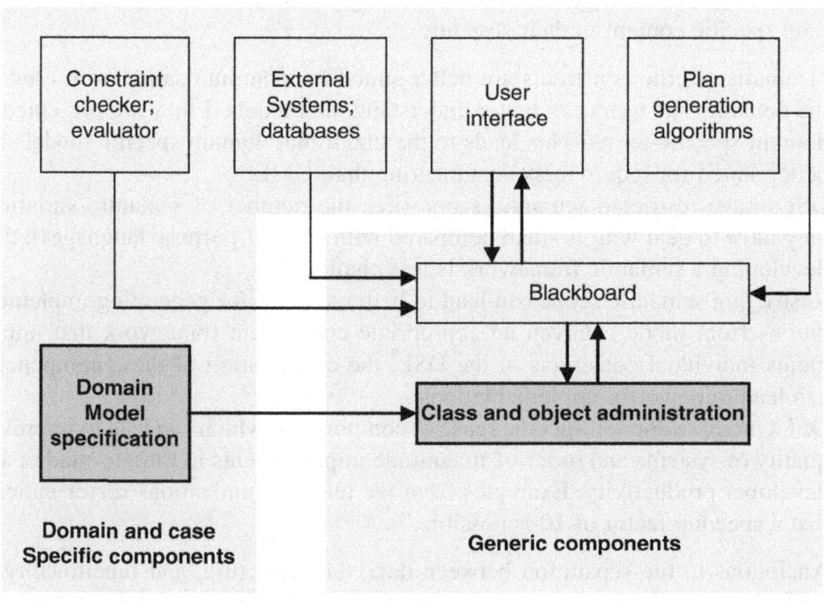

Fig. 16.6 Scheduling system architecture

Table 16.1 Classes of the data model (Van Wezel 2001)

	Concept	Function of class	Analogue in relational database	Examples
1	Class	Specifies a scheduling class	Table	Order, machine
2	Characteristic	Specifies a characteristic of a scheduling class	Field	Due date, capacity
3	Object	Specifies a scheduling object instance	Record	Order 1, production line 1
4	Characteristic value	Specifies a value of a characteristic of an object	Data value	1-2-2001, 100 km/h
5	Relation type	Relates two classes	Relation	Orders run on machines
6	Relation	Relates two objects	Two data values in a linking field have the same value	Order 1 runs on production line 1
7	Schedule	A container for the 6 classes above	Database	
8	Context	Contains a list with schedules; it functions as the blackboard	Database Management System	

- Make multiple systems that share the same domain model, and use event-based triggers to show the dependency between systems.

The taxonomy decreases the dependency on a complex data structure, since the domain models are implemented directly without a complex mapping. As a consequence, components can be made more independent, which means that changes are made easier. Furthermore, the link between tasks and sub-problems remains transparent throughout the system.

16.5 GUI Design

The Graphical User Interface (GUI) of the system is the communication link to the human planner. In the applied modeling paradigm, each representation of the plan must be expressible in the classes and objects that are shown and can be manipulated on the screen. The implementation of the DSL simplifies the development of views of the plan on the screen because of the inherent link between the GUI and the domain model. In other words, in the development of the GUI elements for the shunting planning systems we did not need to handle intricate data transactions as these were handled by the class and object administration module.

Several views on the plan were developed. Table 16.2 enumerates the various views and the objects that are being shown.

Table 16.2 Available views in the shunting planning prototype

Window	Objects in domain model	Objects
Track occupation details	Track occupation: coach, track segment, period, shunting activity: route	Fig. 16.7
List view	Schedule: track occupation: coach, track segment, period	Fig. 16.8
Matrix view	Schedule: track occupation: coach; track segment; period	Fig. 16.9
Coach view	Track occupation: coach, track segment, period	Fig. 16.10
Shunting yard view	Track occupation; track segment, coach, shunting activity: route	Fig. 16.11
Shunting activity overview	Schedule: track occupation: shunting activity, period, track segment, coach	Fig. 16.12

In each of the windows, several characteristics of the objects are shown, as are constraint violations. The objects that are shown can be manipulated, which means that the plan will be changed. Furthermore, views contain hyperlinks to other objects, which might be shown in a different window. Because all windows and all actions in a window are linked to the class and object administration module, the windows are always automatically kept up-to-date real-time.

An additional advantage of the class and object administration module is that plans can be backed up easily. This allows for trying several solutions with the ability to go back to previous partial solutions in case of a dead-end solution path.

16.6 Algorithms and Human/Computer Interaction

Looking at the subtasks as described in Sect. 16.3.4, we can discern four basic assignment tasks: match incoming to outgoing coaches, find a free track or combination of tracks, route a train, and assign tasks to shunting staff. Algorithms have been implemented for each of the basic assignment tasks:

1. *Train unit matching algorithm.* This algorithm determines how the coaches that enter the station are matched to the coaches that leave. This is performed by a mixed integer programming algorithm that matches arriving to departing train units. Initially, this algorithm was based on Freling et al. (2002). Recently, we designed and implemented a network flow algorithm that both matches arriving to departing trains and schedules the bottleneck operation at the station. The planner can decide whether this algorithm is applied to generate a totally new plan or to improve part of an existing plan. The network flow algorithm is being described in Sect. 15.2.
2. *Track-finding algorithm.* Find a track for a train that is available during a specific time window by varying the time window and constraints. Criteria that will affect the decision to what track the train should be moved are amongst others: the length of the time interval it can stay at this track, the routing distance (i.e., number of direction changes and total mileage) to this track, the previous

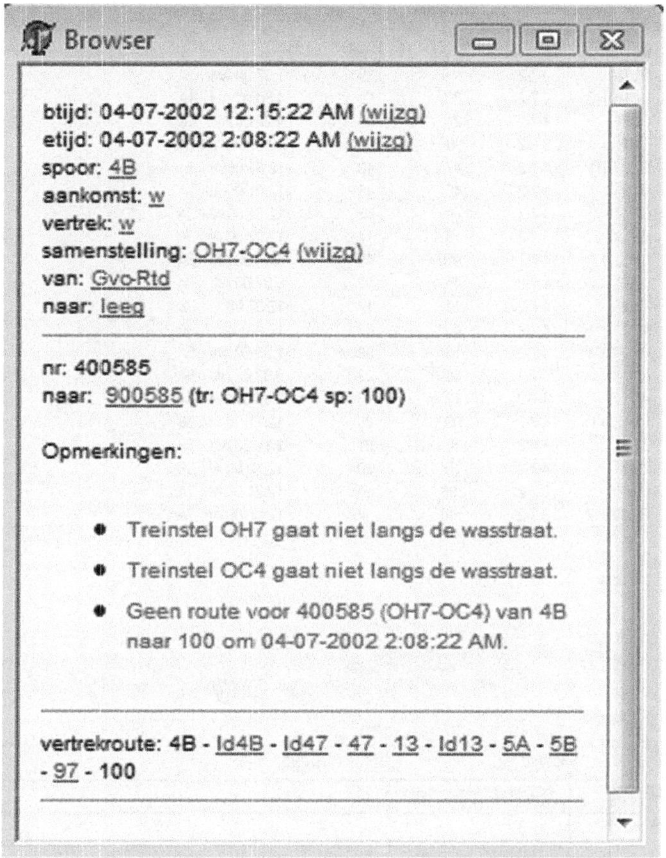

Fig. 16.7 Track occupation details

activities of driver and/or shunter, and the consequences for future actions with this train (i.e., internal cleaning, external cleaning, routing to the track from which it has to leave in the morning, etcetera). The problem the algorithm has to solve is defined as finding a sequence of partially overlapping time intervals from the moment of the actual move to the moment of departure. Sometimes, an additional feature of the sequence is that the cleaning track must be visited somewhere over time. Therefore, the algorithm includes the possibility of stating a set of intermittent nodes (i.e., intervals on the cleaning track) from which at least one has to be included in the final sequence before the departure track is reached. Finally, it has to be possible for the planner to block several tracks that may not be included at all in the final sequence, as they have to be reserved for other purposes such as maintenance or trains running through the station. The track finding problem is solved using a K-shortest path algorithm (Riezebos and van Wezel 2009).

Planvenster

trnnr	trst	stdatum	sttijd	etijd	duur	spoor	vspoor	nspoor
400786	OC1	3-7-2	:48	25	1:37:00 AM	1A		
524 V	OC1-OH2	4-7-2	28	18	4:50:00 AM	1A		
409188	OC2	4-7-2	24	30	4:06:00 AM	1B		
400783	OH2	3-7-2	:46	:53	12:07:00 AM	3A		
20787	OC3-OH1	4-7-2	47	48	12:01:00 AM	3A		
787 V	OC3	4-7-2	48	49	12:01:00 AM	3A		
10787 V	OH1	4-7-2	48	52	12:04:00 AM	3A		
408590	ZN1	4-7-2	15	17	12:02:00 AM	3A		
719 V	OH5	4-7-2	45	49	6:04:00 AM	3A		
10721 V	OH6	4-7-2	45	52	6:07:00 AM	3A		
407988	ZN2	4-7-2	12	17	12:05:00 AM	3B		
73682 V	AE2	4-7-2	45	53	12:08:00 AM	3B		
400585	OH7-OC4	4-7-2	15	08	1:53:00 AM	5A		
3631 V	AE3	4-7-2	50	23	3:33:00 AM	5A		
3629 V	AD1	4-7-2	12	49	2:37:00 AM	5A		
403680	P1-NSY1	4-7-2	09	20	12:11:00 AM	5B		
7921 V	ZN2	4-7-2	10	21	1:11:00 AM	5B		
428521	ZN1	4-7-2	10	20	12:10:00 AM	5B		
405679	LB5	3-7-2	:09	:30	12:21:00 AM	7A		
205681	RR1-RR2	3-7-2	:40	:45	12:05:00 AM	7A		

Fig. 16.8 List view

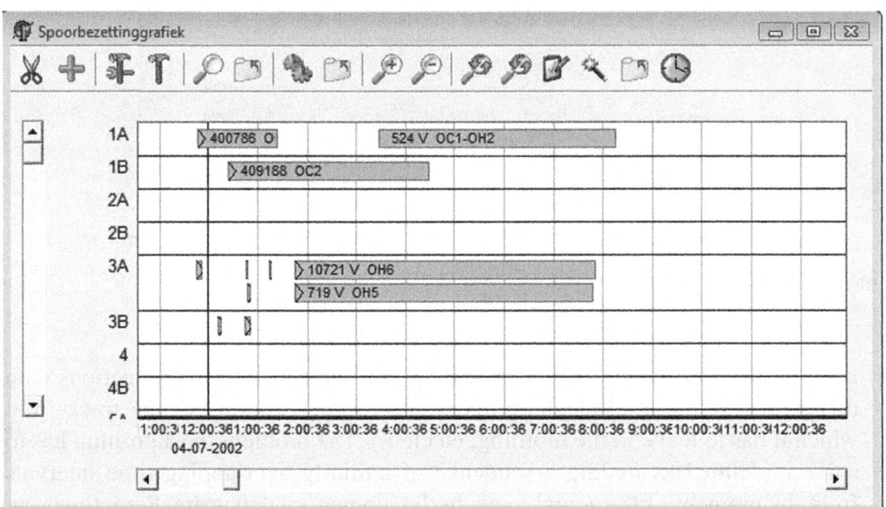

Fig. 16.9 Matrix view

3. *Routing algorithm.* Given the current plan and infrastructure information, the inputs for this subtask are the source and destination tracks for a train. The output of the subtask is a list with the shortest feasible routes for shunting the train to the proposed track. A modified version of the undirected K-shortest path algorithm of Shier (1976) is used to determine the K shortest paths from

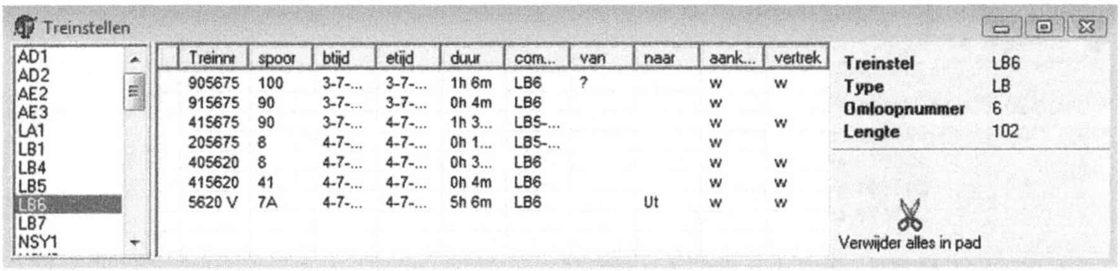

Fig. 16.10 Coach view

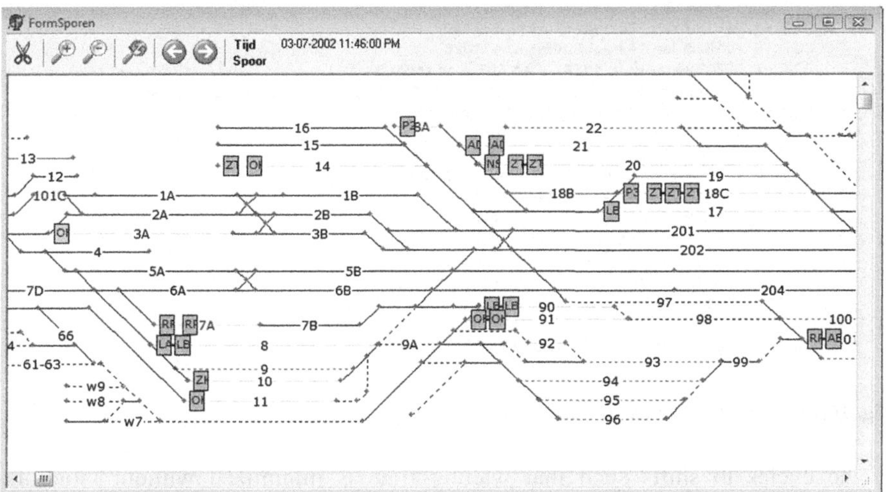

Fig. 16.11 Shunting yard

the source track to the destination track. Modification of the algorithm of Shier was necessary in order to determine the occurrence of direction changes in a route, which is the primary optimization criterion in the weighing of routing alternatives.

4. *Driver/shunter assignment*. The input for this task is the moment of train movements. In Zwolle, there are at night six train drivers available. After a movement, the train driver must walk to the track where he must move the next train. The main criterion is minimizing the overall walking distance, but a trade-off must be made with the time that drivers must wait at a track (which means a walking distance of zero) for the next train. Waiting too long means they would go to the canteen, which increases the walking distance. A shortest augmenting path algorithm is used that combines shunting activities in cycles in such a way that within a cycle the walking distance is minimized (Jonker and Volgenant 1987). Multiple cycles are assigned to a shift. Longer cycles means less overall walking distance but at the same time longer waiting times. The balance is found by calculating multiple solutions with different cycle lengths and combining

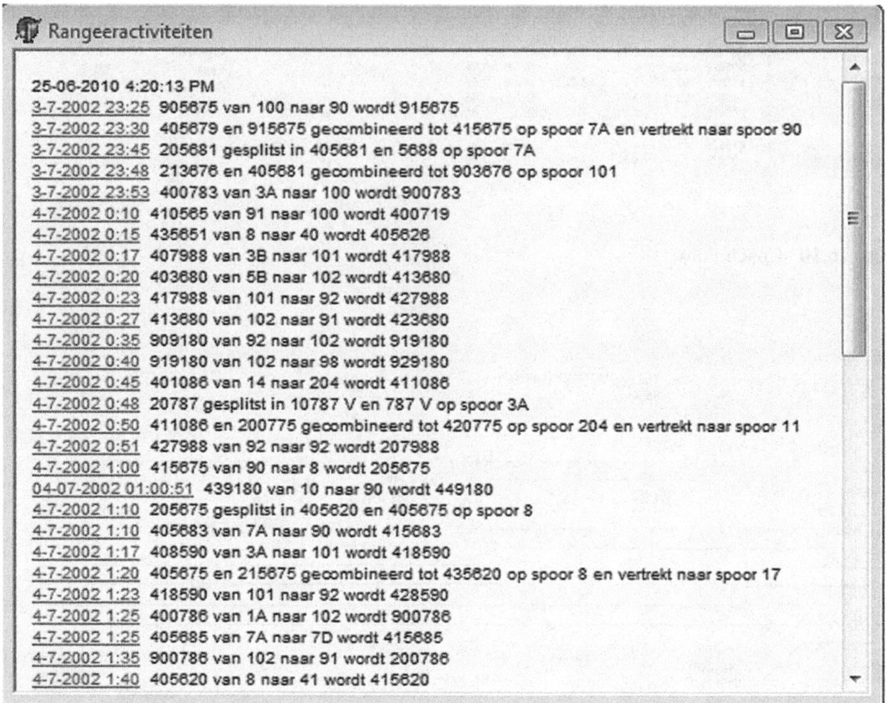

Fig. 16.12 Shunting activity overview

the cycles to shifts such that waiting time is minimized without violating constraints.

The task analysis that was performed showed a multitude of small subtasks. The execution of each of the subtasks showed much variety. The way in which a subtask was performed depended on the status of the schedule and on expectations of the planner regarding the remaining tasks to be performed. If the planners constrain themselves too much, they must backtrack which takes much time. Consequentially, the planners do not object to scheduling algorithms as such, but need to be in control so they can overrule the outcome of an algorithm. As mentioned, the algorithms are made for the basic assignment subtasks. This means that there is not an algorithm that creates the whole plan. Rather, the plan is created by making a number of consecutive planning decisions where each decision is either taken by the human or by the algorithm. This allows the planning process to be highly interactive, because the human planner and the algorithm can both participate in each step of the problem solving process. For this, however, the algorithms must be integrated in the user interface. We explain how this was accomplished for the routing algorithm.

Figure 16.13 shows how the interactive route planner was implemented. The algorithm proposes the optimal route and a number of alternative routes in the "route planner" window. This window shows what tracks are occupied during the time frame

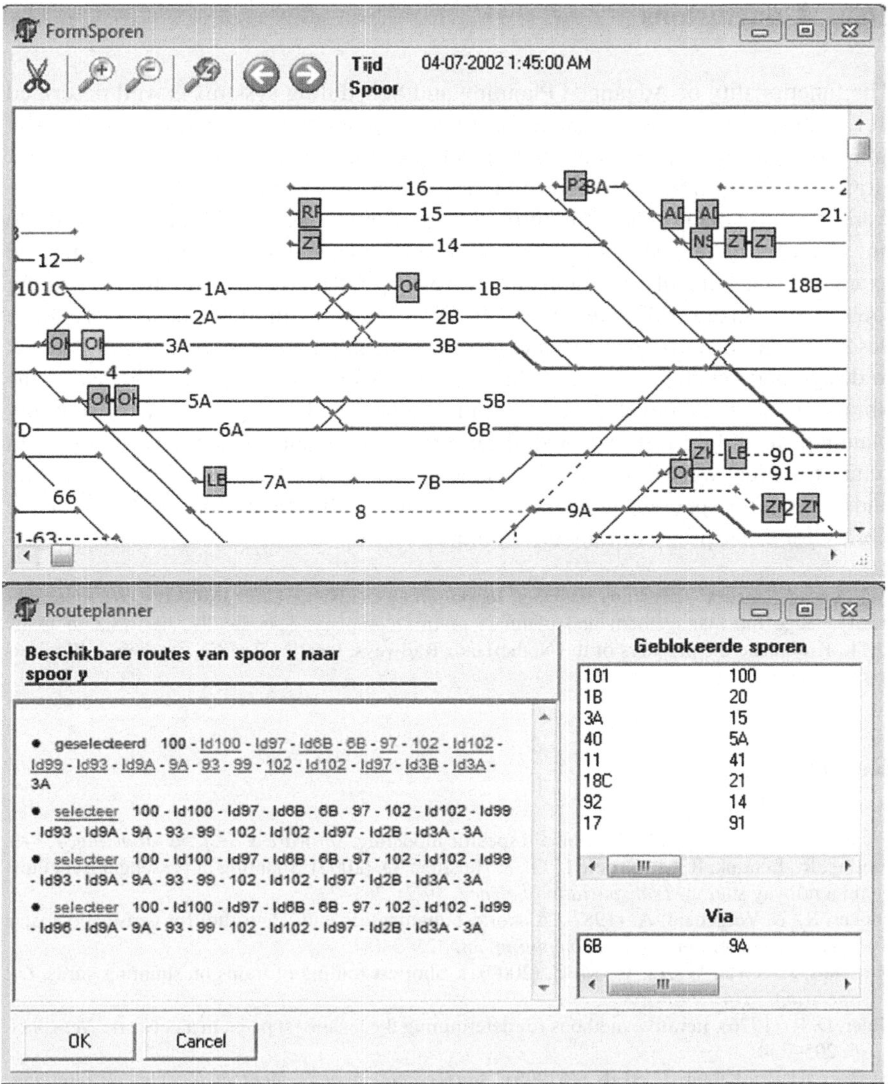

Fig. 16.13 Interactive routing; train moving from Track 11 to Track 5A

of the routing, so the planner understands why certain better routes are not proposed by the algorithm. Furthermore, the planner can add blocked tracks himself, for example because the optimal route of a train might use a track that a planner intends to use for another train. Additionally the planner can steer the algorithm by selecting tracks that must be used in the route. In the example, the optimal route uses the left side of the tracks, but by clicking on the 6B track on the track window, the planner has specified that this track should be used in the route. After clicking on the track, the algorithm proposes a new optimal route taking this new information into account.

16.7 Conclusions

The functionality of Advanced Planning and Scheduling systems is well described in literature. Such systems generally contain a graphical user interface, constraint and goal evaluator, algorithms, and links to external systems such as ERP systems. However, determining the actual functionality that is needed is not so straightforward. Because many scheduling problems are highly case specific, generic tools do not provide all the necessary functionality. Train shunting planning is a good example of a kind of planning that is not supported by mainstream scheduling tools. Customized software, however, is expensive. In this chapter, we have described an object oriented framework and accompanying library of components to design and develop Advanced Planning and Scheduling systems. The use of the library enabled fast development of a prototype for planning support for shunting planning, including task oriented algorithms and advanced interactive control. In future research, we will extend the framework with a service oriented architecture and a planning specific query language, which will increase accessibility of the library for other programming environments.

Acknowledgements This study has been supported by the Netherlands Railways. We gratefully acknowledge the management and planners of this company. Specifically, we want to thank Dr. L. Kroon and the planners of the Netherlands Railways, location Zwolle, for their willingness to co-operate.

References

France, R., & Rumpe, B. (2005). Domain specific modeling. *Software & Systems Modeling, 4*, 1–3.

Freling, R., Lentink, R. M., Kroon, L. G., & Huisman, D. (2002). Shunting of passenger train units in a railway station. *Transportation Science, 39*(2), 261–272.

Jonker, R., & Volgenant, A. (1987). A shortest augmenting path algorithm for dense and sparse linear assignment problems. *Computing, 38*, 325–340.

Riezebos, J., & van Wezel, W. M. C. (2009). k-Shortest routing of trains on shunting yards. *OR Spectrum, 31*(4), 745–758.

Shier, D. R. (1976). Iterative methods for determining the k shortest paths in a network. *Networks, 6*, 205–230.

Stadtler, H., & Kilger, C. (Eds.). (2005). *Supply chain management and advanced planning: Concepts, models, software and case studies* (3rd ed.). Berlin: Springer.

Van Wezel, W. M. C. (2001). *Tasks, hierarchies, and flexibility; planning in food processing industries*. Capelle a/d IJssel: Labyrint Publication.

Van Wezel, W. M. C. (2006). Interactive scheduling systems. In W. M. C. Van Wezel, R. J. Jorna, & A. M. Meystel (Eds.), *Planning in intelligent systems: Aspects, motivations, and methods* (pp. 205–242). Hoboken, NJ: Wiley.

Van Wezel, W. M. C., & Jorna, R. J. (1999). The SEC-system: reuse support for scheduling system development. *Decision Support Systems, 26*(1), 67–87.

Part IV
HOPSopedia

Chapter 17
An Open Source Encyclopedia and Debating Instrument for Planning Terms: The Hopsopedia

Cees De Snoo, Wout Van Wezel, and Jan Riezebos

Abstract This chapter describes the set-up of an online reference tool for planning terms. The tool, called Hopsopedia, enables the easy sharing of planning-related term definitions and descriptions (http://www.hops-research.org). The advanced search engine and linking features support researchers, students, and practitioners, for instance to find key references for planning terms. The tool is developed to enhance mutual understanding between people from different scientific disciplines by providing possibilities to share and discuss term descriptions. The Hopsopedia serves as an online glossary complementing this book as well as a permanent reference instrument for the further interdisciplinary shaping of the planning and scheduling sciences.

17.1 Introduction

Planning and scheduling are topics that are studied by a variety of academic disciplines, like management studies, operations research, ergonomics, industrial engineering, psychology, cognitive sciences, computer sciences and artificial intelligence. Within these disciplines, the same planning-related terms are used, but having sometimes different meanings. The operation researcher modeling a planning process mathematically has another interpretation and view of the "planning process" compared to the cognitive scientist investigating a planning process empirically. The former might focus on characteristics of plan objectives and algorithms, whereas the second's focus might be on human activities and behavior. A similar "language confusion" was once observed between a psychologist and a supply chain expert investigating collaboration in planning. The arguments for the necessity and impact

C. De Snoo (✉), W. Van Wezel, and J. Riezebos
Faculty of Economics and Business, University of Groningen, Groningen, The Netherlands
e-mail: c.de.snoo@rug.nl, w.m.c.van.wezel@rug.nl, j.riezebos@rug.nl

J.C. Fransoo et al. (eds.), *Behavioral Operations in Planning and Scheduling*,
DOI 10.1007/978-3-642-13382-4_17, © Springer-Verlag Berlin Heidelberg 2011

of collaborative planning appeared to be quite different between them. While the psychologist was emphasizing trust and mutual commitment, the supply chain expert emphasized issues like information sharing and event-handling procedures. The provision of definitions of "collaboration" and "collaborative planning" from both perspectives clarified and stimulated the discussion considerably.

Such different perspectives on similar key planning concepts can be found throughout this book. Almost all chapters in this book are a co-production of scientists using either an operation research and engineering perspective or a human behavioral perspective on planning. To facilitate the discussion between the authors, a project team started in 2006 to develop a tool to enhance the sharing of key term descriptions in our field of interest: human and organisational factors of planning and scheduling. As a part of the HOPS-website (Fig. 17.1), an online encyclopedia tool has been developed. The tool, called "Hopsopedia", enables the easy submission of, searching for and discussion of planning terms. This chapter describes the background and setup of the project and the functionality of the web-based planning encyclopedia. Subsequently, it discussed possible opportunities for further use and expansion of the tool.

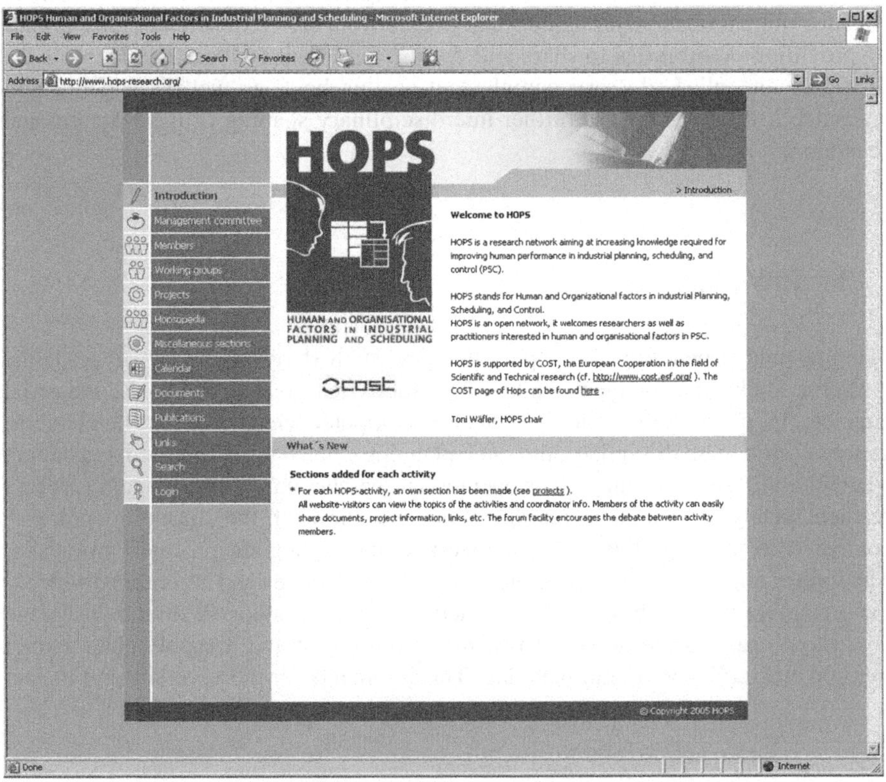

Fig. 17.1 Introduction page of the HOPS-website

17.2 The Hopsopedia and Other Encyclopedia

Since the start of the HOPS-project, the lack of a good reference book with planning-related terms has been recognized. Existing encyclopedias that contain planning-related terms are often oriented towards a single, particular domain or discipline, like the Encyclopedia of Operations Management (Hill 2007) and the MIT Encyclopedia of the Cognitive Sciences (Wilson and Keil 2001). Table 17.1 provides an overview of encyclopedia and their main characteristics that contain planning-related term descriptions. Although providing extensive term descriptions and references to key articles and books, these encyclopedia contain little *cross-discipline* descriptions or discussions. Moreover, each of these reference works covers a very broad field, often limiting the depth of the individual term descriptions.

As far as we know, there is no single work that adequately represents the full range of concepts, theories, and methods deployed by researchers in planning and scheduling. Because planning and scheduling are studied by different scientific disciplines, different views on similar concepts juice up but also confuse the scientific debate. Information about the precise meaning of concepts and constructs, including easy

Table 17.1 Overview of several encyclopedia containing planning-related term descriptions

Title	Edition and editors	Field/Domain	Content
Encyclopedia of operations management	2nd edition: Hill (2007)	Operations management	Over 1,000 entries; 288 pages
The Blackwell encyclopedia of management, volume x: Operations management	2nd edition: Slack and Lewis (2006) Also available online	Operations management	Over 250 entries; 376 pages
Encyclopedia of production and manufacturing management	1st edition: Swamidass (2000) Also available online	Operations management	Over 1,000 entries; 1,048 pages
Encyclopedia of operations research and management science	2nd edition: Gass and Harris (2001) Also available online	Operations research	Over 200 entries; 960 pages
Encyclopedia of optimization	2nd editionFloudas and Pardalos (2009) Also available online	Operations research computer sciences	Over 700 entries; 4,626 pages
Encyclopedia of algorithms	1st edition: Kao (2008) Also available online	Operations research computer sciences	Nearly 400 entries; 1,166 pages
MIT encyclopedia of the cognitive Sciences	1st edition: Wilson and Keil (2001) Also available online	Cognitive sciences	Nearly 500 entries; 1,104 pages
Encyclopedia of cognitive science	1st edition: Nadel (2003) Also available online	Cognitive sciences	Nearly 700 entries; 4,456 pages
The corsini encyclopedia of psychology and behavioral science	3rd edition: Craighead and Nemeroff (2001)	Psychology/ Behavioral science	Over 1,200 entries; 1,836 pages

reference to the basic articles or books in which these are introduced and explained, will help scientists and students during their research projects. In this way, the availability of a multidisciplinary dictionary of planning and scheduling related terms might improve the quality of the work of the scientific planning community. Furthermore, the encyclopedia aims to provide a basic reference work for any manager, planner, practitioner, student, lecturer, or scientist interested or involved in planning and scheduling.

17.3 Setup of the Hopsopedia: A Short History

During the first years of the Hopsopedia, terms have been submitted, defined, and discussed by the members of HOPS, i.e., the authors of this book. In this way, the HOPS-members shared, as experts in a subfield of planning their vision and knowledge about concepts, theories, and methods they are studying. Next to the collection of term descriptions, several technological functions were developed. For example, each term that was defined in the Hopsopedia appeared in a special color and with a hyperlink to the term description on all HOPS web pages. When clicking on such a colored word, the accompanying term description(s) were easily found. Besides, based on a content analysis of all available documents on the site (e.g., working papers, presentations, etc.), a list of documents that contained a particular term was given below the term description. In this way, for each term all documents in which a term was used could easily be found.

Currently, the Hopsopedia is ready for use for the wide planning and scheduling community. After registration, you are able to submit new term descriptions or to discuss existing ones. Registration is easy but needed to maintain and guarantee a high quality of the Hopsopedia.

17.4 Searching Term Descriptions

When opening the Hopsopedia (http://www.hops-research.org), you are asked to enter a term (Fig. 17.2). For example, entering the general term "planning" results into a long list of terms containing the word "planning" (Fig. 17.3).

After clicking on one of the terms, one or more definitions appear. Figure 17.4 shows the result for the term "vertical bullwhip". First, the context of and possible synonyms for the term are indicated. After that, a description follows, including several references to books, articles and working reports. Below this description and the references, comments on this description are shown. Comments are arranged by date of submission, and the name of the author of the comment is always shown.

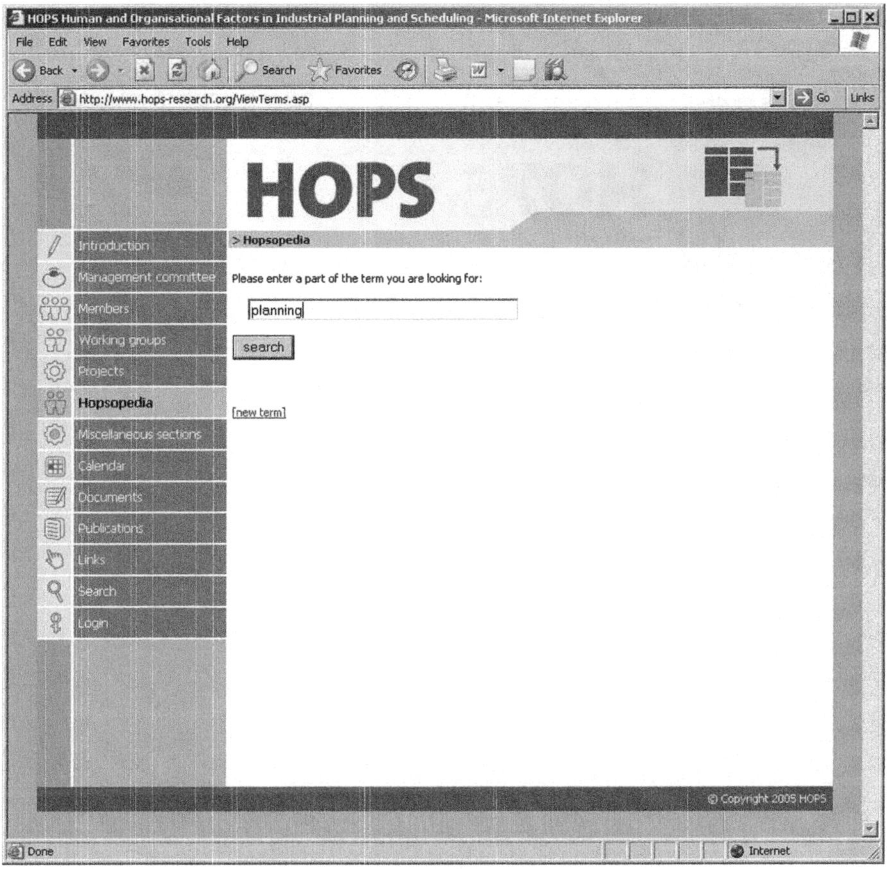

Fig. 17.2 The search function of the HOPS-website

At the bottom of the page, it is shown how often this term is found in documents that are available on the website. In Fig. 17.4, it can be seen that the term "vertical bullwhip" is used in seven documents: the term appears in three other Hopsopedia definitions, in three documents and in one project description. By clicking on one of these categories, a list with these documents is shown. In this way, all content of the website is cross-linked.

17.5 Contributing to the Hopsopedia

There are three ways to contribute to the Hopsopedia: (1) by proposing a new term for the encyclopedia; (2) by submitting a new description for an existing term; (3) by commenting an existing description. In Fig. 17.2, the link [new term] is shown, immediately below the box that is used to type a searching term. By clicking on this link, a new screen appears (Fig. 17.5) in which the web visitor can define

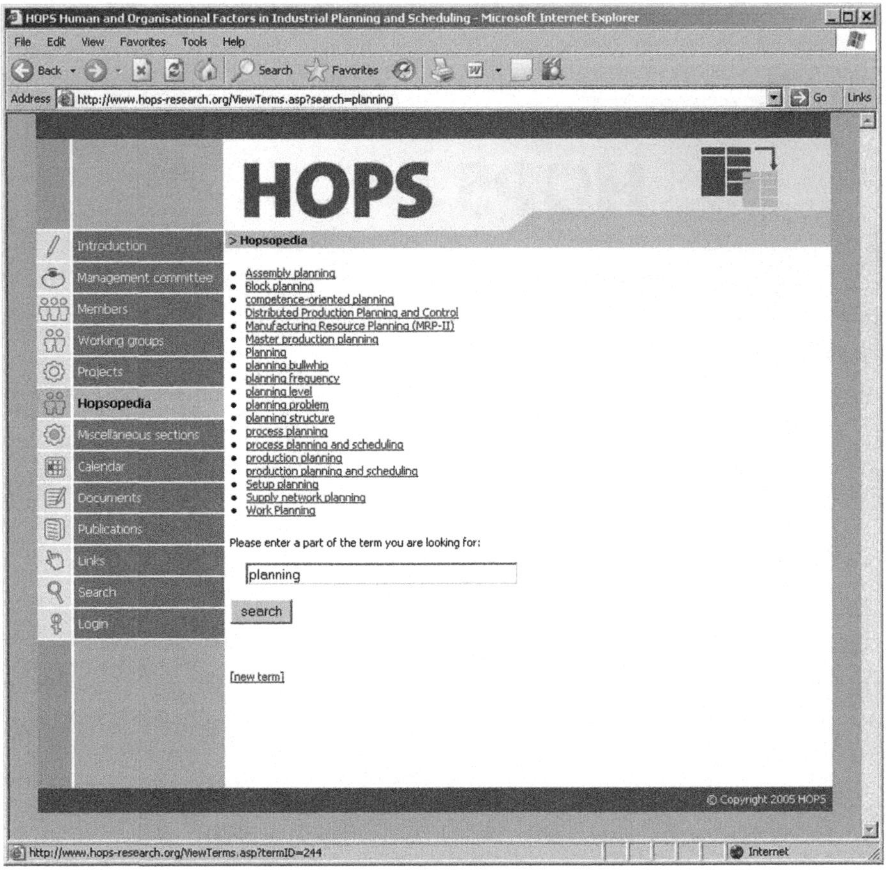

Fig. 17.3 List of terms containing the word "planning"

the new term. By clicking on the button "Add term", the new term is added to the Hopsopedia database. Subsequently, the user can submit one or more term descriptions for this term.

A new term description can be submitted by clicking on the link [new definition] in the term specific screen (Fig. 17.4). A screen with several fields to be filled in appears (Fig. 17.6). First, the (scientific or practical) context of the term definition has to be indicated. For example, the term "planning process" is conceptualized differently in a manufacturing context than in an artificial intelligence context: the process of aggregated production planning with MRP-II is largely different from the process of self-regulating agents in a computer simulation. Second, synonyms and abbreviations for the term can be provided. Third, the term definition is entered into the large white box, possibly including tables,

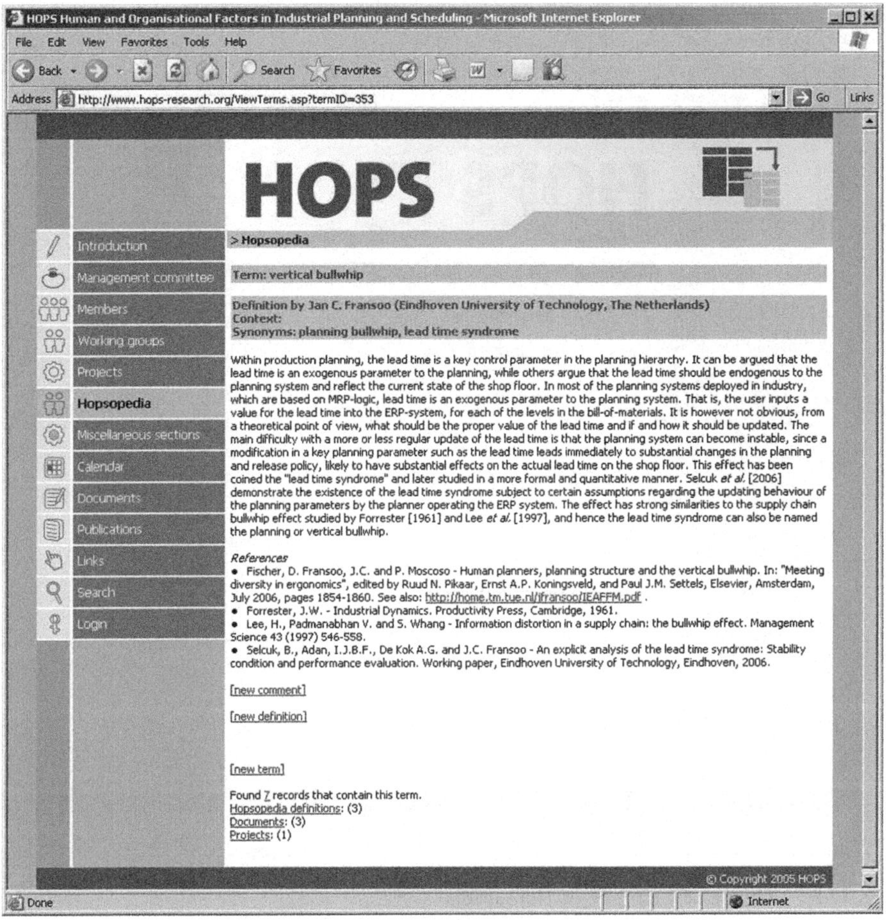

Fig. 17.4 Description, references, and cross-links for the term "vertical bullwhip"

figures or other specific lay out. All definitions and descriptions are provided with clear references to books, articles, or websites to enhance the users to refer to these sources.

Existing term definitions can be commented by registered users. Below a term definition and its references, a link [new comment] is available (Fig. 17.4). By clicking on this link, a screen with a large white box appears. In this box, the comment can be entered. The term description that is commented is shown on top of the screen, so quoting is easy possible.

Finally, term descriptions and comments can be edited easily. Editing is only possible by the authors, so authorship and "copyrights" for each contribution remain clear.

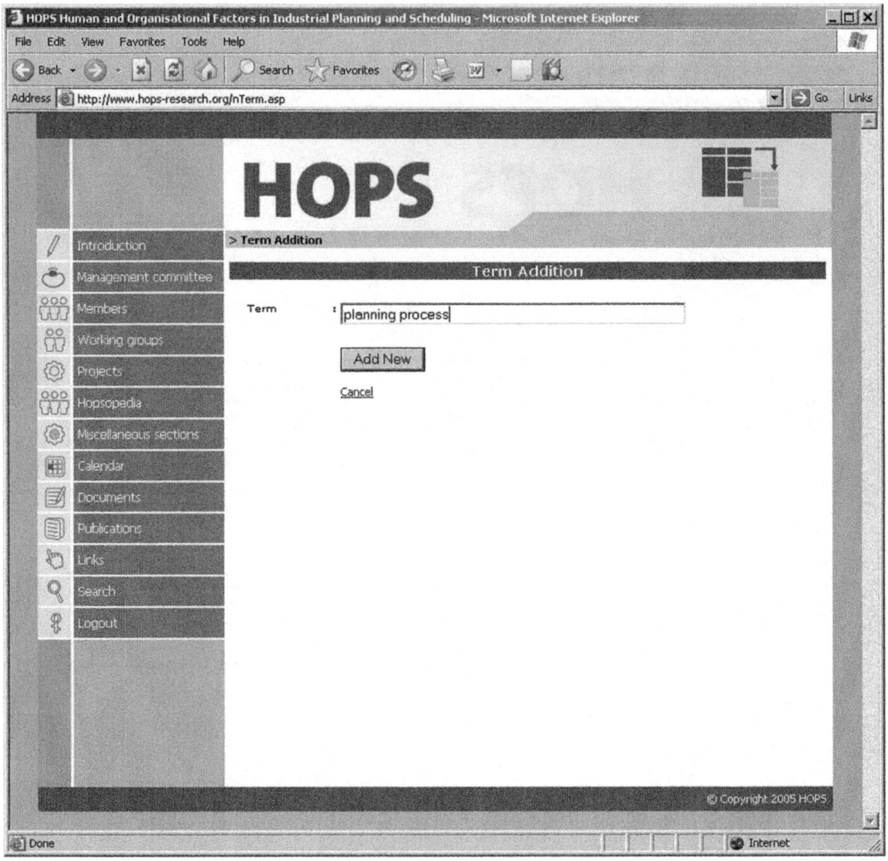

Fig. 17.5 Adding a new term to the Hopsopedia

17.6 Further Steps and Opportunities

The Hopsopedia project aims at becoming a standard reference book for the planning community. The many advantages and features of the web-based platform provide a strong basis for a lively debate about planning terms and concepts. Instead of providing an overview of monodisciplinary definitions per term, Hopsopedia wants to stimulate different viewpoints and the fundamental discussion about these differences.

As indicated, contributions from scientists, students and practitioners from all over the world and from a diversity of backgrounds, cultures, and disciplines are welcome to extend the Hopsopedia. In this way, the debate about human factors in planning and scheduling will be continued. It is our hope that the Hopsopedia will serve many scientists, students, and practitioners when investigating, managing, and practicing planning and scheduling within many contexts and situations.

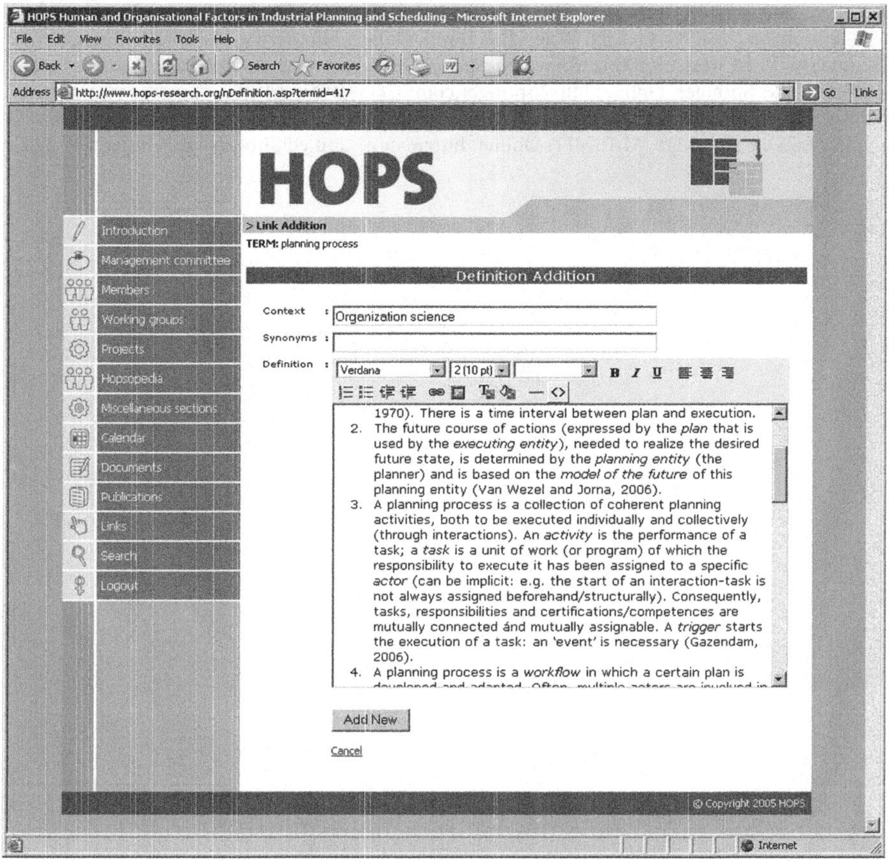

Fig. 17.6 Adding a new description to the Hopsopedia

References

Craighead, W. E., & Nemeroff, C. B. (Eds.). (2001). *The corsini encyclopedia of psychology and behavioral science*. New York: Wiley.

Floudas, C. A., & Pardalos, P. M. (Eds.). (2009). *Encyclopedia of optimization*. New York: Springer. Online: http://dx.doi.org/10.1007/978-0-387-74759-0.

Gass, S. I., & Harris, C. M. (Eds.). (2001). *Encyclopedia of operations research and management science*. Boston, MA: Kluwer Academic. Online: http://springer.com/978-1-4020-0611-1.

Hill, A. V. (2007). *The encyclopedia of operations management*. Eden Prairie, MN: Clamshell Beach Press.

Kao, M.-Y. (Ed.). (2008). *Encyclopedia of algorithms*. New York: Springer. http://springer.com/978-0-387-30770-1.

Nadel, L. (Ed.). (2003). *Encyclopedia of cognitive science*. London: Nature Publishing Group. http://www3.interscience.wiley.com/emrw/9780470018866/home/.

Slack, N., & Lewis, M. (Eds.). (2006). *The Blackwell encyclopedia of management: Operations management* (Vol. X). Oxford: Wiley-Blackwell. Online: http://www.blackwellreference.com.

Swamidass, P. M. (Ed.). (2000). *Encyclopedia of production and manufacturing management*. New York: Springer. Online: http://springer.com/978-1-4020-0612-8.

Wilson, R. A., & Keil, F. C. (Eds.). (2001). *The MIT encyclopedia of the cognitive sciences (MITECS)*. Cambridge, MA: MIT. Online: http://cognet.mit.edu/library/erefs/mitecs/.

Chapter 18
A Sample of Hopsopedia Term Descriptions

Cees De Snoo, Wout Van Wezel, and Jan Riezebos

Abstract This chapter presents a sample of term descriptions from the online Hopsopedia (www.hops-research.org) that is introduced in Chap. 17.

18.1 Introduction

This chapter presents a sample of term descriptions from the online Hopsopedia. The Hopsopedia project is introduced in the previous chapter. The term descriptions have been provided by the authors of the chapters in this book. The online Hopsopedia (www.hops-research.org) consists of a larger sample of term descriptions. For inclusion in this chapter, several criteria are used (1) the term has to be used in one or more chapters in the book, (2) strongly related terms are explained only once, and (3) the term description contains clear and relevant references to published work.

This chapter serves the reader in two ways. First, the term descriptions provide a glossary of key terms used in the various chapters of the book. Therefore, the descriptions are ordered alphabetically. Second, the references mentioned in the term descriptions refer the reader to the literature sources that were used for the term description. Comments and additions to the term descriptions can be submitted online, as described in the previous chapter.

C. De Snoo (✉), W. Van Wezel, and J. Riezebos
Faculty of Economics and Business, University of Groningen, Groningen, The Netherlands
e-mail: c.de.snoo@rug.nl, w.m.c.van.wezel@rug.nl, j.riezebos@rug.nl

J.C. Fransoo et al. (eds.), *Behavioral Operations in Planning and Scheduling*, DOI 10.1007/978-3-642-13382-4_18,

18.2 Term Descriptions and References

Advanced Planning and Scheduling/Advanced Planning System/APS – An Advanced Planning System (APS) is a decision support system to support decision making in planning and scheduling. Planning determines what and how many to manufacture and to purchase in order to satisfy future demand for end products. Scheduling takes place at the execution level of the plans and covers a step-by-step work or activity list, specifications of time at which every activity should start and end as well as the sequencing and re-sequencing of job orders; it is internally focused.

Advanced planning and scheduling (APS) covers simultaneous coordination of material and capacity constraints at an operational level to best meet market demand. Advanced planning is the synchronization of constrained material and resources to interdependent demand. Its purpose is to create a plan that is feasible with respect to all resources with sufficient operational slack to permit re-sequencing of work orders. Advanced scheduling covers the detailed sequencing of operations and material. Its purpose is to provide properly sequenced work orders under a lot of restrictions.

- Moon, C., Seo, Y., Yun, Y., & Gen, M. (2006). Adaptive genetic algorithm for advanced planning in manufacturing supply chain. *Journal of Intelligent Manufacturing, 17*(4), 509–522.
- Musselman, K., & Uzsoy, R. (2001). Advanced planning and scheduling for manufacturing. In G. Salvendy (Ed.), *Handbook of industrial engineering* (3rd ed.). New York: Wiley.
- Stadler, H., & Kilger, C. (2000). *Supply chain management and advanced planning – concepts, models, software and case studies*. Berlin: Springer.

Boundary Object – The term boundary object describes objects that are shared and shareable across different problem solving contexts. It establishes a shared syntax or language for individuals to represent their knowledge. A boundary object functions at syntactic, semantic, and pragmatic levels. It provides a means for individuals to specify and learn about their differences and dependencies across a given boundary, and facilitates a process where individuals can jointly transform their knowledge.

- Star, S. L. (1989). The structure of ill-structured solutions: Boundary objects and heterogeneous distributed problem solving. In M. Huhns (Ed.), *Distributed artificial intelligence* (Vol. 2). San Francisco: Morgan Kaufman Publishers.
- Carlile, P. R. (2002). A pragmatic view of knowledge and boundaries: Boundary objects in new product development. *Organization Science, 13* (4), 442–455.

Cognitive Task Analysis – See Task analysis.

Cognitive Typology – A cognitive typology makes it possible to describe situations according to factors that will directly determine the operators' strategies. For example, in the case of a dimension associated with the structural complexity of the process, Cowling (2001) presents an illustrative situation: when the order list is full (what could be defined, from a cognitive point of view, as high structural complexity), the schedulers' strategies are directed towards the maintenance of production flow by avoiding delays caused by changing produced components. On the other hand, when the order list is not full (i.e., low structural complexity), the schedulers focus on production quality. So, the use of a cognitive typology makes it possible to characterize situations (as a list of dimensions) on the basis of their implications for operators' strategies.

- Cegarra, J. (2008). A cognitive typology of scheduling situations: A contribution to laboratory and field studies. *Theoretical Issues in Ergonomics Science, 9*(3), 201–222.
- Cowling, P. (2001). Design and implementation of an effective decision support system: A case study in steel hot rolling mill scheduling. In B. L. MacCarthy, & J. R. Wilson (Eds.), *Human performance in planning and scheduling*. London: Taylor & Francis.

Cognitive Work Analysis – See Task analysis.

Collaborative Planning/CPFR – Regarding inter-organizational collaborative planning, it has been stated that collaborative planning focuses on eliminating supply and demand uncertainty through improved communication and collaboration between supply chain partners. Skjøtt-Larsen et al. (2003) describe collaborative planning as: "collaboration where two or more parties in the supply chain jointly plan a number of promotional activities and work out synchronized forecasts, on the basis of which the production and replenishment processes are determined" (p. 532). Sometimes the term supply chain planning is used to denote any collaborative planning activities in the supply chain.

Collaborative Planning, Forecasting and Replenishment (CPFR) is a business practice that combines the intelligence of multiple trading partners in the planning and fulfillment of customer demand. The CPFR framework includes planning activities on the strategic level, tactical level, and operational level. It also includes the operational control of the production and distribution of products and the monitoring of planning and execution activities. CPFR has been mainly used in the food and retail industry, but applications in other industries are present as well.

From a more abstract point of view, collaborative planning could be conceptualized as a form of joint decision making involving mutual understanding and mutual adjustment between the interacting planning actors. This description also applies for intra-organizational collaborative planning, e.g., between planners from different departments.

In the applied psychology literature, collaborative planning has been described as a mode of coordination that consists of deliberate and opportunistic processes through which actors jointly device and enact a plan. Collaborative planning, according to this view, necessitates that multiple actors orient their plans towards each other in order to reach a joint optimization of their planning.

- Akkermans, H., Bogerd, P., & Van Doremalen, J. (2004). Travail, transparency and trust: A case study of computer-supported collaborative supply chain planning in high-tech electronics. *European Journal of Operational Research, 153*(2), 445–456.
- Buehler, R., Messervey, D., & Griffin, D. (2005). Collaborative planning and prediction: Does group discussion affect optimistic biases in time estimation? *Organizational Behavior & Human Decision Processes, 97*(1), 47–63.
- Danese, P. (2007). Designing CPFR collaborations: insights from seven case studies. *International Journal of Operations & Production Management, 27* (2), 181–204.
- Günter, H. (2007). *Collaborative planning in heterarchic supply networks.* PhD-Thesis, ETH Zurich, Switzerland.
- Seifert, D. (2003). *Collaborative planning, forecasting, and replenishment: how to create a supply chain advantage.* New York: Amacom.
- Skjøtt-Larsen, T., Thernøe, C., & Andresen, C. (2003). Supply chain collaboration: Theoretical perspectives and empirical evidence. *International Journal of Physical Distribution & Logistics Management, 33*(6), 531–549.
- VICS. (2004). *Collaborative planning, forecasting and replenishment: An overview.* Available on: www.vics.org/committees/cpfr.
- Windischer, A., & Grote, G. (2003). Success factors for collaborative planning. In S. Seuring, M. Goldbach, & U. Schneidewind (Eds.) *Strategy and organization in supply chains.* Heidelberg: Physica-Verlag.

Complacency – In the design of human–machine cooperation, several authors have noted that expert operators, even aware of a machine's limits, could adopt its proposals without questioning them; this failure has been termed complacency. Complacency is traditionally associated with a decrease in machine supervision, implying a low cognitive workload. Cegarra and Hoc (2008) noted that complacency could also result from the high cognitive cost of interacting with the (scheduling) machine.

- Cegarra, J., & Hoc, J. M. (2008). The role of algorithm and result comprehensibility of automated scheduling on complacency. *Human Factors and Ergonomics in Manufacturing, 28*(6), 603–620.
- Parasuraman, R., Molloy, R., & Singh, I. L. (1993). Performance consequences of automation-induced complacency. *The International Journal of Aviation Psychology, 3*(1), 1–23.

Complexity – Complexity has always been a part of our environment, and therefore many scientific fields have dealt with complex systems and complex phenomena. While this has led some fields to come up with specific definitions of complexity, there is a more recent movement to regroup observations from different fields to study complexity in itself, whether it appears in anthills, human brains or stock markets.

One definition mentions that complexity is implicit in phenomena that are neither completely ordered nor completely random. This lack of certitude can be traced to the fact that the phenomenon has multiple scales, all of which are coupled to one another (West 2004). In this way, "complexity refers to the condition of the universe which is integrated and yet too rich and varied for us to understand in simple common mechanistic or linear ways. We can understand many parts of the universe in these ways, but the larger and more intricately related phenomena can only be understood by principles and patterns – not in detail. Complexity deals with the nature of emergence, innovation, learning, and adaptation" (Sherman and Shultz 1998 cited by McCarthy et al. 2000).

Complexity has also been defined as a measure of uncertainty in achieving a set of specific functions or functional requirements (Suh 1999).

- McCarthy, I. P., Rakotobe-Joel, T., & Fizelle, G. (2000). Complex systems theory: implications and promises for manufacturing organizations. *International Journal of Manufacturing Technology and Management, 2*(1–7), 559–579.
- Rojdestvenski, I., Cottam, M. G., Oquist, G., & Huner, N. (2003). Thermodynamics of complexity. *Physica A: Statistical Mechanics and its Applications, 320*, 318–329.
- Suh, N. P. (1999). A theory of complexity, periodicity and the design axioms. *Research in Engineering Design, 11*(2), 116–133.
- West, B. J. (2004). Comments on the renormalization group, scaling and measures of complexity. *Chaos, Solitons and Fractals, 20*(1), 33–45.

Constraint – A constraint is a predicate that describes a relationship among plan variables. It can be interpreted in three ways (a) in terms of object selection, a constraint is an elimination rule; (b) in terms of plan refinement, a constraint is a partial description that leaves options open; (c) in terms of processing interaction between sub-problems, a constraint is a communication medium.

- Stefik, M. (1981). Planning with constraints (MOLGEN: Part 1). *Artificial Intelligence, 16*(2), 111–139.

Control Capacity – Control capacity is the ability of a work system to cope with its actual and future tasks. It can be subdivided in anticipative, operational, and structural control capacity.

- Von der Weth, R. (2001). *Management der Komplexität (Management of complexity)*. Bern: Huber.

Cooperation – Two agents are in a cooperative situation if they meet two minimal conditions (1) each one strives towards goals and can interfere with the other on goals, resources, procedures, etc.; (2) each one tries to manage the interference to facilitate the individual activities and/or the common task. The symmetric nature of this definition can be only partly satisfied.

- Hoc, J. M. (2001). Towards a cognitive approach to human-machine co-operation in dynamic situations. *International Journal of Human-Computer Studies, 54*(4), 509–540.

Coordination Stock – Coordination stock is excess inventory kept specifically in order to reduce order amplification (cf., bullwhip effect) by providing a buffer against the endogenous risk of coordination failure. Such coordination stock differs from traditional safety stock, which buffers against exogenous demand uncertainty.

- Croson, R., Donohue, K., Katok, E., Sterman, J. (2005). *Order stability in supply chains: Coordination risk and the role of coordination stock.* Working paper series MIT, ESD-WP-2004-04.

Decision Support System/DSS – A Decision Support System (DSS) is an interactive, flexible, and adaptable computer-based information system, especially developed for supporting the solution of non-structured management problems for improved decision-making. It utilizes data, provides an easy-to-use interface, and allows for the decision makers own insights. The DSS supports technological and managerial decision making by assisting in the organization of knowledge about ill-structured, semi-structured or unstructured issues. A DSS supports humans in formal steps of problem solving: formulation of alternatives, analysis of their impacts, interpretation and selection of appropriate options. The primary purpose of a DSS is to support cognitive activities that involve human information processing and associated judgment and choice. A DSS usually consists of three components: a database management system, a model-based management system, and a dialogue generating and management system. The following areas contribute to the development of DSS: computer science, management science and operations research, organizational behavior, behavioral and cognitive science, and systems engineering.

- Bennet, J. L. (Ed.). (1983). *Building decision support systems*. Reading, MA: Addison Wesley.
- Druzdzel, M. J., & Flynn, R. R. (1999). Decision support systems. In A. Kent, (Ed.), *Encyclopedia of library and information science*. New York: Marcel Dekker.
- Parsai, H. R., Kolli, S., & Handley, T. R. (Eds.). (1997). *Manufacturing decision support systems*. New York: Chapman & Hall.
- Sage, A. P. (1991). *Decision support systems engineering*. New York: Wiley.
- Sage, A. P. (2001). Decision support systems. In G. Salvendy (Ed.), *Handbook of industrial engineering* (3rd ed.). New York: Wiley.
- Turban, E. (1995). *Decision support and expert systems: Management support systems*. Englewood Cliffs, NJ: Prentice Hall.

Deliberate Planning – Deliberate planning processes are often described in difference to flexible, ad-hoc planning processes: "Deliberate planning processes lead to the development of formal plans, which prescribe principal courses of action and provide an overall framework within which faster cycling episodic plans can be formulated and executed" (Mathieu and Schulze 2006). Deliberate plans are thus seen to pre-structure or even control the execution of tasks.

- Mathieu, J. E., & Schulze, W. (2006). The influence of team knowledge and formal plans on episodic team processes-performance relationships. *Academy of Management Journal, 49*(3), 605–619.
- Miller, G. A., Galanter, E., & Pribram, K. (1960). *Plans and the structure of behavior*. New York: Holt, Rinehart, and Winston.

Function Allocation – A function is defined as a unit coming from task decomposition that the integrated human–machine is required to be capable of performing. Each function might be allocated to the actor that would achieve the supposed best performance. Originally, function allocation focused on a static allocation of functions between human and machine on the basis of qualifications with respect to the task at hand. Nowadays, in dynamic function allocation, the allocation is not determined beforehand but during system functioning taking into account situational factors.

- Dearden, A., Harrison, M., & Wright, P. (2000). Allocation of function: Scenarios, context and the economics of effort. *International Journal of Human-Computer Studies, 52*(2), 289–318.
- Rouse, W. B. (1976). Adaptive allocation of decision-making responsibility between supervisor and computer. In T. B. Sheridan, & G. Johannsen (Eds.), *Monitoring behavior and supervisory control*. New York: Plenum Press.
- Scerbo, M. W. (1996). Theoretical perspectives on adaptive automation. In R. Parasuraman, & M. Mouloua (Eds.), *Automation and human performance: Theory and applications*. Mahwah: Lawrence Erlbaum Associates.

Hierarchical Planning – In industrial practice, production planning is typically done following a hierarchical approach. A hierarchical planning approach divides a large, complex planning problem into simpler sub-problems. The set of problems is then solved in a sequential fashion, increasing the level of detail and decreasing the planning horizon, whereby the higher planning level imposes constraints on the lower planning level.

- Bitran, G., & Hax, A. C. (1977). On the design of hierarchical production planning systems, *Decision Sciences, 8*(1), 28–55.
- Bowers, M., & Jarvis, J. (1992). A hierarchical production planning and scheduling model, *Decision Sciences, 23*(1), 144–157.
- McKay, K. N., Safayeni, F. R., & Buzacott, J. A. (1995). A review of hierarchical production planning and its applicability for modern manufacturing. *Production Planning & Control, 6*(5), 384–395.

Hierarchical Task Analysis – See Task analysis.

Human Computer Interaction (HCI) – A discipline concerned with the design, evaluation, and implementation of interactive computing systems for human use and with the study of major phenomena surrounding them.

- Hewett, T., Baecker, R., Card, S., Carey, T., Gasen, J., Mantei, M., et al. (1992). *ACM SIGCHI Curricula for Human-Computer Interaction*. Report of the ACM SIGCHI Curriculum Development Group.

Interdependence – Interdependence between humans has been defined as the extent to which individuals are dependent on each other in fulfilling their tasks and reaching set goals. Within the context of teamwork, two types are generally distinguished: task interdependence (i.e., the extent humans must share materials, information, or expertise in order to achieve the desired output or performance) and outcome interdependence (i.e., the extent to which significant consequences of work are contingent on the collective performance of goals). The latter type could be further decomposed into three sub-types: reward, goal, and feedback interdependence.

Interdependence between activities has been conceptualized differently. Three types are distinguished: pooled, sequential and reciprocal dependences.

- Kiggundu, M. (1983). Task interdependence and the theory of job design. *Academy of Management Review, 6*(3), 499–508.
- Thompson, J. D. (1967). *Organizations in action*. New York: McGraw-Hill.
- Van de Ven, A. H., Delbecq, A. L., & Koenig, R. (1976). Determinants of coordination modes within organizations. *American Sociological Review, 41*(2), 322–338.
- Wageman, R. (2001). The meaning of interdependence. In M. E. Turner (Ed.), *Groups at work: Theory and research*. Mahwah: Lawrence Erlbaum Associates.

Knowledge – Knowledge is a high value form of information that is ready to be applied to decisions and actions. Knowledge is a combination of information, ideas, procedures, and perceptions that guide actions and decisions. Knowledge differentiates from information by being relevant to context and being formed by experience.

The term tacit knowledge was originally defined by Polanyi (1966) who differentiated tacit from explicit knowledge. Tacit knowledge resides in the individual's experience and action, whereas explicit knowledge is codified and communicated in symbolic form or language. From an organizational perspective, tacit knowledge may be viewed as "sticky" if it is difficult to separate know-how and practice and so to transfer it to any other context. Similarly, Von Hippel and Katz (2002) defined "sticky information" in design innovation as information that is costly to transfer and that is generally highly context specific.

- Bolisani, E., & Scarso, E. (1999). Information technology management: A knowledge-based perspective. *Technovation, 19*(4), 209–217.
- Brown, J.S., & Duguid, P. (1998). Organizing knowledge. *California Management Review, 40*(3), 90–111.
- Choo, C. W. (1996). The knowing organization: How organizations use information to construct meaning, create knowledge and make decisions. *International Journal of Information Management, 16*(5), 329–340.

- Polanyi, M. (1966). *The tacit dimension*. London: Routledge & Kegan Paul.
- Shin, M., Holden, T., & Schmidt, N. (2001). From knowledge theory to management practice: Towards an integrated approach. *Information Processing and Management, 37*(2), 335–355.
- Von Hippel, E., & Katz, R. (2002). Shifting innovation to users via toolkits. *Management Science, 48*(7), 821–833.

Lead Time Syndrome – See Vertical bullwhip.

Multi-Objective Scheduling – The allocation of limited resources to tasks over time that has as a goal the simultaneous optimization of more than one objective functions.

- Pinedo, M. (1995). *Scheduling: Theory, algorithms and systems*. Upper Saddle River, NJ: Prentice Hall.

Perceived Ease of Use and Perceived Usefulness (of Information Systems) – Perceived ease of use is the degree to which an individual believes that using a particular system would be free of effort.

Perceived usefulness is the degree to which an individual believes that using a particular system would enhance his or her job performance.

- Davis, F. D. (1989). Perceived usefulness, perceived ease of use, and user acceptance of information technology. *MIS Quarterly, 13*(3), 318–340.
- Davis, F. D. (1993). User acceptance of information technology: System characteristics, user perceptions and behavioral impacts. *International Journal of Man-Machine Studies, 38*(3), 475–487.
- Davis, F. D., Bagozzi, R. P. & Warshaw, P. R. (1989). User acceptance of computer technology: A comparison of two theoretical models. *Management Science, 35*(8), 982–1003.
- Venkatesh, V., & Davis, F. D. (1996). A model of the antecedents of perceived ease of use: Development and test. *Decision Sciences, 27*(3), 451–481.

Performance Indicator – A performance indicator or performance criterion is the specific characteristic to be measured for estimating performance. Logistic performance indicators analyze the effect of logistics on company objectives in the four target areas of quality, cost, delivery, and flexibility.

Based upon Neely et al. (1997), we define supply chain performance as the effectiveness and efficiency of supply chain action. Effectiveness is the extent to which customer requirements are met and efficiency measures how economically a supply chain's resources are utilized when providing a pre-specified level of customer satisfaction.

- Neely, A., Richards, H., Mills, J., Platts, K., & Bourne, M. (1997). Designing performance measures: A structured approach. *International Journal of Operations & Production Management, 17*(11), 1131–1152.
- Schönsleben, P. (2007). *Integral logistics management. Operations and supply chain management in comprehensive value-added networks* (3rd ed.). Boca Raton, FL: Auerbach Publications.

Plan – A plan is a schematic and/or hierarchical representation to guide activity. Declarative plans represent static relational structures, whereas procedural plans represent procedures. The two main properties of plans are anticipation and schematization. A plan is a resource for action and is not always applied as it is.

- Hoc, J. M. (2001). Towards a cognitive approach to human-machine co-operation in dynamic situations. *International Journal of Human-Computer Studies, 54*(4), 509–540.

Planning – Description of Van Wezel and Jorna (2006): "Planning means that a planning entity determines a future course of actions for an executing entity. These actions should lead to a desired future state. The belief that the actions lead to the state is based on the model of the future of the planning entity. The future course of actions or the desired future state is expressed by the plan. Planning is a complex activity and often involves reasoning with incomplete information. Plans are usually made hierarchically."

Description of Tate (2001): "Planning is the process of generating (possibly partial) representations of future behavior prior to the use of such plans to constrain or control that behavior. The outcome is usually a set of actions, with temporal and other constraints on them, for execution by some agent or agents."

- Tate, A. (2001). Planning. In R. A. Wilson, & F. C. Keil (Ed.), *The MIT encyclopedia of the cognitive sciences* (*MITECS*). Cambridge: The MIT.
- Van Wezel, W., & Jorna, R. J. (2006). Introduction. In W. Van Wezel, R. J. Jorna, & A.M. Meystel (Eds.), *Planning in intelligent systems: Aspects, motivations, and methods.* Hoboken, NJ: Wiley.

Planning Bullwhip – See Vertical bullwhip.

Planning Problem – A planning problem consists of groups of entities, whereby the entities from different groups must be assigned to each other. The assignments are subject to constraints, and alternatives can be compared on their level of goal realization.

- Van Wezel, W., & Jorna, R. J. (2006). Introduction. In W. Van Wezel, R. J. Jorna, & A. M. Meystel (Eds.), *Planning in intelligent systems: Aspects, motivations, and methods.* Hoboken, NJ: Wiley.

Production Capability – Production capability is the temporal ability of production to fulfill orders. According to Slack et al. (1998), production planning and control activities are those that reconcile supply and demand, "in effect they co-ordinate and match production capability to customer requests for product and service."

At a strategic level, production capability is a set of capabilities consisting of "the resources and processes within the production operation which can be harnessed and exploited by the market place" (Slack and Lewis 2002).

- Slack, N., & Lewis, M. (2002). *Operations strategy*. London: Prentice Hall.
- Slack, N., Chambers, S., & Johnston, R. (1998). *Operations management* (4th ed.). London: Pitman Publishing.

Reciprocal Relationship – A reciprocal relationship exists between two variables A and B if A influences B and B influences A, that is, the two variables are interdependently related. "When any two events are related interdependently, designating one of those two "cause" and the other "effect" is an arbitrary designation" (Weick 1979, p. 77). Thus, in difference to unilateral relationships, no one-way cause-effect relationship can be established if variables are correlated in a reciprocal manner.

- Weick, K. E. (1979). *The social psychology of organizing* (2nd ed.). New York: Random House.

Relationship Quality – Relationship quality is often viewed as a higher-order construct composed of several dimensions: trust, commitment, coordination, adaptation, and satisfaction. Within a business-to-business context, relationship quality can be considered as both an input to the success of a joint venture, and an output of the interactions between the partners.

- Ariño, A., & De la Torre, J. (1998). Learning from failure: towards an evolutionary model of collaborative ventures. *Organization Science, 9*(3), 307–325.
- Crosby, L. A., Evans, K. R., & Cowles, D. (1990). Relationship quality in services selling: An interpersonal influence perspective. *Journal of Marketing, 54*(3), 68–81.
- Fynes, B., Voss, C. A., & De Burca, S. (2005). The impact of supply chain relationship quality on quality performance. *International Journal of Production Economics, 96*(3), 339–354.
- Settoon, R., & Mossholder, K. (2002). Relationship quality and relationship context as antecedents of person- and task-focused interpersonal citizenship behavior. *Journal of Applied Psychology, 87*(2), 255–267.

Role – The notion of role is related to patterns of human conduct, to expectations, identities, social positions, context, social structure as well as individual response. A common notion in role theory is that roles are related to social positions, and that roles are induced through the sharing of expectations for role behavior. The role is then learned through role-playing by practicing the roles performed by others and role taking by internalizing expectations from others.

- Biddle, B. J. (1979). *Role theory: Expectations, identities, and behaviors.* New York: Academic.
- Katz, D., & Kahn, R. (1978). *The social psychology of organizations.* New York: Wiley.

Skill Loss – Having an algorithm autonomously performing the main scheduling tasks leads the human scheduler to be in a monitoring role. This role might lead the human to become passive, which can have severe consequences when the human must perform actions in case of exceptions or automation failures. Being in a monitoring role reduces the opportunity for the human to learn from experience and to maintain his skills. Bainbridge (1983) pointed this progressive loss of skill as one "irony of automation".

- Bainbridge, L. (1983). Ironies of automation. *Automatica, 19*(6), 775–779.
- Hoc, J. M. (2000). From human–machine interaction to human–machine cooperation. *Ergonomics, 43*(7), 833–843.

Situation Awareness – Situation awareness is the perception of environmental elements within a volume of time and space, the comprehension of their meaning, and the projection of their status in the near future.

- Endsley, M. R. (1995). Toward a theory of situation awareness in dynamic systems. *Human Factors, 37*(1), 32–64.

Supply Chain Performance – See Performance indicator.

Supply Chain Relationship – A supply chain relationship describes any upstream and downstream relationship a firm entertains with other supply chain members. It describes relationships between customers and suppliers in a business-to-business context. Other terms used in the literature are supply relationship and buyer–supplier relationship.

- Harland, C. M., Zheng, J., Johnsen, T. E., & Lamming, R. C. (2004). A conceptual model for researching the creation and operation of supply networks. *British Journal of Management, 15*(1), 1–21.
- Johnsen, T. E., Lamming, R. C., & Harland, C. M. (2008). Interorganizational relationships, chains, and networks. A supply perspective. In S. Cropper, M. Ebers, C. Huxam, & P. S. Ring (Eds.), *The Oxford handbook of interorganizational relationships*. Oxford: Oxford University Press.

System Trust – See Trust in automation.

Task – Description 1: A set of actions leading to a set of goals taking into account a set of constraints (Kiewiet et al. 2006).

Description 2: Specific demands on an anticipative, operational and structural level for a work system coming from a mandate or an order for this work system. It can be described as a relation between actual situation, goals and means (Von der Weth 2001).

- Kiewiet, D. J., Jorna, R. J., & Van Wezel, W. (2006). Task analysis for problems of shunting planning. In W. Van Wezel, R.J. Jorna, & A. M. Meystel (Eds.), *Planning in intelligent systems: Aspects, motivations, and methods*. Hoboken, NJ: Wiley.
- Von der Weth, R. (2001). *Management der Komplexität (Management of complexity)*. Bern: Huber.

Task Analysis – The principle of a task analysis is to break down and study the elements of a task. The usual goal is to improve the task execution by asking questions such as "why is the work performed as it is," "what is needed to perform it in this way," and "how can the work method be improved."

In 1960, Miller, Galanter, and Pribram suggested that human behavior is fundamentally goal directed and can be understood in terms of a hierarchical goal structure; the attainment of primary goals is served by the attainment of its subgoals, which could, at their turn, be decomposed into sub-sub-goals. This idea has been expanded to the analysis of process control tasks, which lead to the method Hierarchical Task Analysis (HTA). The designer's goal is to describe the hierarchical goal-structure. Each level (goal or sub-goal) is recursively decomposed in 3–10 immediate sub-goals. For this decomposition to happen not indefinitely, it is common to consider the purpose of the analysis as a stop point. For example, when studying teamwork, Stanton (2006) suggested stopping the hierarchical decomposition when the subtasks deal with the individual exchange of information.

During the 1960s and 1970s, there was a shift from the description of work analysis methods from purely physical tasks towards more cognitive-oriented tasks such as air traffic control or troubleshooting diagnosis. Traditional task analysis only has an implicit formulation from the cognitive point of view. Cognitive Task Analysis (CTA) appeared as an extension of traditional task analysis to yield information about knowledge, thought processes and goals that underlie task performance. Different authors applied a CTA in planning and scheduling, but they do not always refer to this method.

Taking into account the importance of novelty in socio-technical systems, Rasmussen and others have developed a framework that is called Cognitive Work Analysis (CWA). This method has its root in ecological psychology that implies to adopt the human–environment system as the fundamental unit of analysis and to focus the analysis on the examination of constraints that the environment imposes on behavior. So, in contrast with the two previous methods, the CWA does not focus on the human's hypothesized or real practice, as was stressed by Vicente (2002, p.63): "A task can be defined as the set of actions that can or should be performed by one or more actors to achieve a particular goal. In contrast, a work domain is the system being controlled, independent of any particular worker, automation, event, task, goal, or interface." The decomposition in CWA is based on an abstraction hierarchy: the higher levels of the hierarchy describe functional information, whereas lower levels describe physical information. Besides this physical to functional decomposition, there is a part– whole dimension, which takes into account several levels of detail (e.g., system, subsystem, components). Five different levels of abstraction are considered in CWA: functional purposes, abstract functions, generalized functions, physical functions and the physical form.

• Annett, J. (2000). Theoretical and pragmatic influences on task analysis methods. In J. M. Schraagen, S. F. Chipman, & V. L. Shalin (Eds.), *Cognitive task analysis*. Mahwah: Lawrence Erlbaum Associates.

- Annett, J. (2003). Hierarchical task analysis. In E. Hollnagel (Ed.), *Handbook of cognitive task design*. Mahwah: Lawrence Erlbaum Associates.
- Chipman, S. F., Schraagen, J. M., & Shalin, V. L. (2000). Introduction to cognitive task analysis. In J. M. Schraagen, S. F. Chipman, & V. L. Shalin (Eds.), *Cognitive task analysis*. Mahwah: Lawrence Erlbaum Associates.
- Crawford, S., MacCarthy, B. L., Wilson, J. R., & Vernon, C. (1999). Investigating the work of industrial schedulers through field study. *Cognition, Technology & Work, 1*(2), 63–77.
- Gibson, J. J. (1986). *The ecological approach to visual perception*. Hillsdale, NJ: Lawrence Erlbaum Associates.
- Higgins, P. G. (1999). *Job shop scheduling: Hybrid intelligent human-computer paradigm*. PhD-Thesis, University of Melbourne, Australia.
- Higgins, P. G. (2001). Architecture and interface aspects of scheduling decision support. In B. L. MacCarthy & J. R. Wilson (Eds.), *Human performance in planning and scheduling*. London: Taylor & Francis.
- Hollnagel, E., & Woods, D. D. (1983). Cognitive systems engineering: New wine in new bottles. *International Journal of Man-Machine Studies, 18*(6), 583–600.
- Miller, G. A., Galanter, E., & Pribram, K. (1960). *Plans and the structure of behavior*. New York: Holt, Rinehart and Winston.
- Rasmussen, J. (1986). *Information processing and human machine interaction: An approach to cognitive engineering*. New York: Elsevier Science.
- Rasmussen, J., Pejtersen, A. M., & Goodstein, L. P. (1994). *Cognitive systems engineering*. New York: Wiley.
- Stanton, N. A. (2006). Hierarchical task analysis: Developments, applications, and extensions. *Applied Ergonomics, 37*(1), 55–79.
- Usher, J. M., & Kaber, D. B. (2000). Establishing information requirements for supervisory controllers in a flexible manufacturing system using GTA. *Human Factors and Ergonomics in Manufacturing, 10*(4), 431–452.
- Van Wezel, W., Jorna, R. J., & Mietus, D. (1996). Scheduling in a generic perspective. *International Journal of Expert Systems, 3*(9), 357–381.
- Vicente, K. J. (1999). *Cognitive work analysis: Toward safe, productive, and healthy computer-based work*. Mahwah: Lawrence Erlbaum Associates.
- Vicente, K. J. (2002). Ecological interface design: Progress and challenges. *Human Factors, 44*(1), 62–78.

Technology Acceptance Model – The Technology Acceptance Model (TAM) specifies the causal relationships between system design features, perceived usefulness, perceived ease of use, attitude toward using, and actual usage behavior. Overall, the TAM provides an informative representation of the mechanisms by which design choices influence user acceptance, and should therefore be helpful in applied contexts for forecasting and evaluating user acceptance of information technology.

- Davis, F. D., Bagozzi, R. P., & Warshaw, P. R. (1989). User acceptance of computer technology: A comparison of two theoretical models. *Management Science, 35*(8), 982–1003.

- Venkatesh, V., & Davis, F. D. (2000). A theoretical extension of the technology acceptance model: Four longitudinal field studies. *Management Science, 46*(2), 186–204.

Trust in Automation – According to Luhmann (1979), trust is a mechanism to reduce internal complexity of a system of interaction through the adoption of specific expectations about the future behavior of the other by selecting amongst a range of possibilities. Trust absorbs complexity insofar as someone who trusts acts as if the trustee's acts are – at least to some degree – predictable.

Trust in automation guides reliance when the complexity of the automation makes a complete understanding impractical and when the situation demands adaptive behavior that procedures cannot guide.

System trust plays a role when a "trustor", the user, interacts with a trustee, the system. The necessary precondition is that this user-system interaction takes place voluntarily. In addition, the incentive for the user to engage in interaction with the system is the expected benefit from the outcomes of the task. The actual outcome, however, is uncertain, due to a lack of sufficient evidence. The degree to which the actual amount of evidence available can be called sufficient depends on the risk involved: compared to situations of low risk, high-risk situations may require more evidence for uncertainty to be reduced.

- De Vries, P. (2004). *Trust in systems: Effects of direct and indirect information*. PhD-Thesis, Eindhoven University of Technology, Eindhoven, The Netherlands.
- De Vries, P., Midden, C., & Bouwhuis, D. (2003). The effects of errors on system trust, self-confidence, and the allocation of control in route planning. *International Journal of Human-Computer Studies, 58*(6), 719–735.
- Dzindolet, M. T., Peterson, S. A., Pomranky, R. A., Pierce, L. G., & Beck, H. P. (2003). The role of trust in automation reliance. *International Journal of Human-Computer Studies, 58*(6), 697–718.
- Lee, J. D., & See, K. A. (2004). Trust in automation: Designing for appropriate reliance. *Human Factors, 46*(1), 50–80.
- Lee, J. D., & Moray, N. (1994). Trust, self-confidence, and operators' adaptation to automation. *International Journal of Human-Computer Studies, 40*(1), 153–184.
- Luhmann, N. (1979). *Trust and power: Two works by Niklas Luhmann*. Chichester: Wiley.
- Muir, B. M. (1987). Trust between humans and machines, and the design of decision aids. *International Journal of Man-Machine Studies, 27*(5–6), 527–539.
- Riley, V. (1996). Operator reliance on automation: Theory and data. In R. Parasuraman, & M. Mouloua (Eds.), *Automation and human performance: Theory and applications*. Mahwah: Lawrence Erlbaum Associates.

User Participation/User Involvement – User involvement or user participation is the participation in the system development process by representatives of the target user group. There are two facets of user involvement: the type and the degree of involvement. The type of involvement can range from indirect (user representatives) to direct (participates in development process). The degree of involvement refers to the influence that users have over the final product. User involvement is measured as a set of behaviors or activities that such individuals perform.

In information systems theory, involvement is seen as user influence during system development, where involved users perform activities that are influential in leading and managing particular aspects of the system development process. There are four areas of influence (1) explaining and clarifying information needs; (2) detailing input and output requirements; (3) stating system needs and objectives; (4) asking questions and providing answers.

In psychology, user involvement is considered a psychological state triggered by the importance and personal relevance that users attach either to a particular system or to an information system in general, depending on the users' focus. User participation is an observable behavior respectively a behavioral engagement. User participation is considered to be critical for system quality and system use.

In the literature, there is often no clear distinction between user influence, engagement, participation in the process, and user involvement. This influences also the measurability of these constructs and makes the verification of the effects of user participation or involvement more difficult.

- Barki, H., & Hartwick, J. (1989). Rethinking the concept of user involvement. *MIS Quarterly, 13*(1), 53–63.
- Barki, H., & Hartwick, J. (1994). Measuring user participation, user involvement, and user attitude. *MIS Quarterly, 18*(1), 59–82.
- Baroudi, J. J., Olson, M. H., & Ives, B. (1986). An empirical study of the impact of user involvement on system usage and information satisfaction. *Communications of the ACM, 29*(3), 232–238.
- Hartwick, J., & Barki, H. (1994). Explaining the role of user participation in information system use. *Management Science, 40*(4), 440–465.
- Hunton, J. E., & Beeler, J. D. (1997). Effects of user participation in systems development: A longitudinal field experiment. *MIS Quarterly, 21*(4), 359–388.
- Ives, B., & Olson, M.H. (1984). User involvement and MIS success: A review of research. *Management Science, 30*(5), 586–603.
- Kappelman, L. A. (1995). Measuring user involvement: A diffusion of innovation perspective. *ACM SIGMIS Database, 26*(2–3), 65–86.
- Kujala, S. (2003). User involvement: A review of the benefits and challenges. *Behavior and Information Technology, 22*(1), 1–16.
- Palanisamy, R. (2001). User involvement and flexibility in strategic MIS planning: A path analytic study. *Global Journal of Flexible Systems Management, 2*(4), 15–32.

Within a production planning system, the lead time is a key control parameter in the planning hierarchy. While some authors argue that the lead time is an exogenous parameter to planning, other state that the lead time should be endogenous to the planning system and reflect the current state of the shop floor. Most of the planning systems deployed in industry are based on some kind of MRP-logic, and then the lead time is an exogenous parameter to the planning system. That is, the user inputs a value for the lead time into the ERP-system for each of the levels in the bill-of-materials. However, it is theoretically not fully clear, what the proper value and updating frequency of the lead time should be. These decisions, however, are critical for the planning performance, as in dependence of the updating policy of the lead time the planning system may lose its stability. This effect has been coined the "lead time syndrome" (Mather and Plossl 1978) and only much later studied in a more formal and quantitative manner by Selçuk et al. (2009).

The effect has strong similarities to the supply chain bullwhip effect studied by Forrester (1961) and Lee et al. (1997), and hence the lead time syndrome can also be named the planning or vertical bullwhip. The planning bullwhip is a complex dynamic phenomenon where a planning system ends up with an erratic ordering and updating behavior in response to changing workload levels, resulting in uncontrolled order release patterns, generating eventually a larger variability in the workload levels and lead (flow) times.

- De Kok, T. G., & Fransoo, J. C. (2003). Planning supply chain operations: Definition and comparison of planning concepts. In A. G. De Kok & S. C. Graves (Eds.), *Supply chain management: Design, coordination and operation*. Amsterdam: Elsevier.
- Forrester, J. W. (1961). *Industrial Dynamics*. Cambridge: Productivity Press.
- Lee, H., Padmanabhan, V., & Whang, S. (1997). Information distortion in a supply chain: The bullwhip effect. *Management Science, 43*(4) 546–558.
- Mather, H., & Plossl, G. W. (1978). Priority fixation versus throughput planning. *Production and Inventory Management, 19*, 27–51.
- Moscoso, P. G., Fransoo, J. C., & Fischer, D. (2010). An empirical study on reducing planning instability in hierarchical planning systems. *Production Planning & Control, 19*(8), 781–787
- Selçuk, B., Fransoo, J. C., & De Kok, A. G. (2006). The effect of updating lead times on the performance of hierarchical planning systems. *International Journal of Production Economics, 104*(2), 427–440.
- Selçuk, B., Adan, I. J. B. F., De Kok, A. G., & Fransoo, J. C. (2009). An explicit analysis of the lead time syndrome: Stability condition and performance evaluation. *International Journal of Production Research, 47*(9), 2507–2529.
- Zijm, W. H. M., & Buitenhek, R. (1996). Capacity planning and lead time management. *International Journal of Production Economics, 46/47*, 165–179.

Visualization – Visualization can be defined as a graphical representation of data or concepts, which is either an internal construct of the mind or an external artifact supporting decision making.

- Ware, C. (2000). *Information visualization: Perception for design.* San Francisco, CA: Morgan Kaufmann.
- Tory, M., & Möller, T. (2004). Human factors in visualization research. *IEEE Transactions on Visualization and Computer Graphics, 10*(1), 72–84.

About the Editors

Jan Fransoo
Eindhoven University of Technology, School of Industrial Engineering, J.C. Fransoo@tue.nl

Jan C. Fransoo is a full professor of Operations Planning & Control at Eindhoven University of Technology. He specializes in Operations Planning and Supply Chain Management in multiple industries including process, food, pharmaceutical and retail. He also serves as Research Director of the European Supply Chain Forum, and Vice-President of the Dutch Institute of Advanced Logistics Dinalog. Jan held visiting appointments at Clemson University, Stanford University, and the University of California at Los Angeles. He serves as Senior Editor of *Production and Operations Management*, and on the editorial board of five other journals. He has published in academic journals such as *Management Science*, *Journal of Operations Management*, *IIE Transactions*, *Production and Operations Management*, *Manufacturing and Service Operations Management* and *European Journal of Operational Research*.

Toni Wäfler
School of Applied Psychology, University of Applied Sciences Northwestern Switzerland, toni.waefler@fhnw.ch

Toni Wäfler is a professor for work and organizational psychology at the University of Applied Sciences Northwestern Switzerland (FHNW), School of Applied Psychology (APS), where he established the Institute "Humans in Complex Systems (MikS)". The Institute conducts research projects in the domain of human factors, sociotechnical system design, collaborative planning processes, occupational health, security, and safety. In 1998 Toni Wäfler was also co-founder of iafob GmbH (www.iafob.ch) a private consulting company in Zurich, Switzerland, where he still is a senior consultant. As a consultant his main topics include organizational and job design in highly automated work systems; with a main focus on process efficiency and reliability as well as on system safety.

John R. Wilson
Professor of Occupational Ergonomics, University of Nottingham, john.wilson@nottingham.ac.uk
John R. Wilson is Professor of Human Factors in the School of Mechanical, Materials and Manufacturing Engineering. He was founder and first Director of the University's Institute for Occupational Ergonomics and the Virtual Reality Applications Research Team (VIRART), and co-founder of the University of Nottingham Mixed Reality Laboratory. Earlier appointments were at Loughborough University, Birmingham University, UC Berkeley and University of New South Wales. He is a Chartered Psychologist and a Chartered Engineer, and Fellow of the Ergonomics Society, and of the British Psychological Society, the Human Factors Society and the Institution of Engineering and Technology. Professor Wilson researches and consults in all areas of human factors, with particular interests in socio-technical systems, human-centred design, collaborative work, virtual environments, participative ergonomics and rail human factors. He has authored 500 publications with over 250 in refereed books, journals or conferences.

About the Authors

Martina Berglund
Quality Technology and Management, Department of Management and Engineering, Linköping University, martina.berglund@liu.se
Martina Berglund is a guest lecturer at Linköping University, Sweden and engaged as a consultant in production, organisation and workplace development. She received her MSc in Industrial Engineering and Management from Linköping Institute of Technology and her PhD in Technology and Health from KTH, the Royal Institute of Technology, Sweden. Her research interests include human factors in production planning and scheduling, workplace learning and the integration of the human, technology and organization in various domains. She is also the director of HFN, the Swedish Network for Human Factors. Papers on her work have been published in *International Journal of Production Economics* and *Human Factors and Ergonomics in Manufacturing*.

Jessica Bruch
Department of Industrial Engineering and Management, Jönköping University, jessica.bruch@jth.hj.se
Jessica Bruch is a PhD candidate in production systems at the Department of Industrial Engineering and Management at the School of Engineering, Jönköping University and is also connected to the School of Innovation, Design and Engineering at Mälardalens University. Her major research interests include requirement specification, production system development and information management.

Julien Cegarra

Université de Toulouse, julien.cegarra@univ-jfc.fr

Julien Cegarra is assistant professor at the University of Toulouse - Champollion University Centre (FR). His research interests focus on the design of human-machine cooperation in optimization-related problems (such as planning, scheduling, dispatching and routing problems).

Naoufel Cheikhrouhou

Ecole Polytechnique Fédérale de Lausanne, naoufel.cheikhrouhou@ epfl.ch

Naoufel Cheikhrouhou is the head of the Operations Management group at the Swiss Federal Institute of Technology at Lausanne. He received his industrial engineer diploma from the Ecole Nationale d'Ingenieurs de Tunis and his PhD in Industrial Engineering from the Institut National Polytechnique de Grenoble in France. His main research interests are in the areas of modelling, simulation and optimisation of supply networks, the identification and integration of human factors in production management and the reduction of management complexity in logistics and services, for which Dr. Cheikhrouhou received the Burbidge Award in 2003. Leading different projects with the collaboration of industrial partners, Dr. Cheikhrouhou has also published several papers in various journals and international conferences.

Cees De Snoo

Faculty of Economics and Business, University of Groningen, c.de.snoo@rug.nl

Cees De Snoo is a researcher and lecturer at the University of Groningen. From 2005 till 2009, he has been working on a PhD-project about organizational aspects of planning and scheduling. A variety of research methods like case studies, surveys, and laboratory experiments has been used to investigate planning and scheduling process-related issues. He intends to complete and defend his PhD-thesis in 2011. His research interests are planning process design, planning performance measurement, event handling and rescheduling, collaborative planning, behavioral operations, and healthcare logistics.

Christos Dimopoulos

European University Cyprus, c.dimopoulos@euc.ac.cy

Christos Dimopoulos is Assistant Professor of Computer Science & Engineering and Vice Chair of the Department of Computer Science & Engineering in European University Cyprus. Christos' current research interests involve the use of systems analysis techniques in the development of industrial IT systems and educational software. His scientific research has been published in journals such as *International Journal of Production Research*, *Engineering Optimization* and *IEEE Transactions in Production Research*. He is an active reviewer of journals such as *International Journal of Production Research*, *Computers & Industrial Engineering*, *Computers & Operational Research*, and many others.

Gael Farine
Ecole Polytechnique Fédérale de Lausanne
Gael Farine was a Master's student at EPFL at the time of writing the first version of the book chapter. He is about to start his industry career.

Dieter Fischer
Institute for Business Engineering, University of Applied Sciences Northwestern Switzerland, dieter.fischer@fhnw.ch
Dieter Fischer is a professor for Systems Engineering at the University of Applied Sciences Northwestern Switzerland. His main areas of research are finding conditions for the target-oriented and sustainable organisation of innovation and change in the field of intra-plant an inter-plant logistics, ERP and SCM.

Katrin Fischer
School of Applied Psychology, University of Applied Sciences Northwestern Switzerland, katrin.fischer@fhnw.ch
Katrin Fischer is a professor for work and organizational psychology at the University of Applied Sciences Northwestern Switzerland (FHNW), School of Applied Psychology (APS), where she works at the Institute Humans in Complex Systems (MikS). The Institute conducts research projects in the domain of human factors, sociotechnical system design, collaborative planning processes, occupational health, security, and safety. Her current research topics include decision processes and decision support in complex business environments, human factors in safety management and system reliability.

Kary Främling
Aalto University, Kary.Framling@tkk.fi
Kary Främling received his MSc in Computer Science from Helsinki University of Technology in 1990 and his PhD from Ecole Nationale Supérieure des Mines de Saint-Etienne, France, in 1996. He is currently Adjunct Professor at Helsinki University of Technology, which is part of the new Aalto University. His current research topics are on information management practices and applications for product lifecycle management and ubiquitous computing. His main areas of competence are distributed systems, middleware, multi-agent systems, autonomously learning agents, neural networks and decision support systems.

Roland Gasser
Department of Mechanical and Industrial Engineering, University of Toronto, roland.gasser@utoronto.ca
Roland Gasser is a visiting researcher at the University of Toronto, funded through a fellowship of the Swiss National Science Foundation. He is associated with the Cognitive Engineering Lab (CEL) at the Department of Mechanical and Industrial Engineering. Previously he has been working as a project manager in research and development at MEDGATE, a Swiss e-health company, and as a research assistant at the University of Applied Sciences Northwestern Switzerland. His research interests are rooted within human factors and engineering psychology. More recently, he has been working on a cognitive work analysis approach to planning and scheduling in collaboration with the CEL.

Kathrin Gärtner
School of Applied Psychology, University of Applied Sciences Northwestern Switzerland, kathrin.gaertner@fhnw.ch
Kathrin Gärtner is an industrial and organizational psychologist at the University of Applied Sciences Northwestern Switzerland (FHNW). As a research assistant she conducts research in several projects concerning system safety and human factors engineering in the sectors of transport, energy and engine building. She focuses on decision making under high uncertainty in complex business environments such as planning and scheduling or top management. Currently she does her PhD at the University of Innsbruck on decision support in safety management processes.

Bernard Grabot
Université de Toulouse – ENIT, bernard@enit.fr
Bernard Grabot is Professor in the National Engineering School of Tarbes, France (ENIT). His research activities are oriented on supply chain management, scheduling, competence management and decision support systems based on artificial intelligence tools. Pr Grabot is member of the IFAC working groups 3.2 "Computational Intelligence in Control" and 5.1 "Manufacturing Plant Control", and of the IFIP working group 5.7 "Advances in Production Management Systems". He was member of the Network of Excellence I*PROMS of FP6 and participated to several European projects (CRAFT, INTERREG, etc.). Professor Grabot is Editor in Chief of the IFAC journal *Engineering Applications of Artificial Intelligence*. He also belongs to the editorial boards of *International Journal of Production Research*, *Journal of Mechanical Engineering Science* and *International Journal of Computational Intelligence Research*.

Jane Guinery
Nottingham University Business School, jane.guinery@nottingham.ac.uk
Dr Jane Guinery is a Lecturer in Operations Management at Nottingham University Business School. Her current research is addressing the need to improve order fulfillment processes through analysis and transfer of good practice. Prior to academia she worked in industry where she undertook a variety of roles from Systems Engineer to Operations Manager. During this time she project managed CADCAM, MRPII, ISO 9000, Quality program and TOC implementations. Her PhD is on knowledge integration in production planning and control, and was gained whilst undertaking an industrial consortium based EPSRC funded project that established ways of designing the organisation and processes to improve the responsiveness of order fulfillment processes. Her primary research interests are in knowledge management, human decision making, lean thinking and organisation design in industrial and Healthcare contexts.

Hannes Günter
Department of Organization and Strategy, Maastricht University, h.guenter@maastrichtuniversity.nl
Hannes Günter is an assistant professor at Maastricht University in the Department of Organization and Strategy (The Netherlands). Dr. Günter studied Psychology

with focus on Work and Organizational Psychology at University of Freiburg (Germany), University of Melbourne (Australia), and Dresden University of Technology (Germany). In October 2007, he obtained a Ph.D. in Work Psychology from ETH Zürich (Switzerland) with research in the area of collaborative planning. His research interests include team and multi-team planning, time in organizations with a special focus on delays, and organizational behavior in supply chains. At Maastricht University, he teaches courses on organizational change, consultancy, and organizational learning.

Jean-Michel Hoc

Centre National de la Recherche Scientifique (CNRS: French National Research Centre), Jean-Michel.Hoc@irccyn.ec-nantes.fr

Jean-Michel Hoc is cognitive psychologist in ergonomics, Research Director at CNRS (the French National Research Centre), Head of the CNRS French National Network in Cognitive Ergonomics (GDR 3169), Head of the PyCoTec research team (Psychology, Cognition, Technology) within IRCCyN (Research Institute for Communications and Cybernetics in Nantes), Research unit associated to CNRS (UMR 6597), Associate Editor of the multidisciplinary and international journal *Le Travail Humain.* He is member of *Cognition, Technolog & Work, Cognitive Science Quarterly, Information-Interaction-Intelligence,* and *Journal of Cognitive Engineering and Decision Making* Editorial Boards. He is member of boards within ANR (French National Research Agency) and AERES (French National Research Evaluation Agency). His research team is studying cognitive activities brought into play within dynamic environment supervision and control (flexible manufacturing systems, maritime navigation, and car-driving), more specifically stressing the integration of human operators into automated systems and production of results directly usable within cognitive modelling or human-machine cooperation design.

Jan Holmström

School of Science and Technology, Aalto University, jan.holmstrom@tkk.fi

Jan Holmström is Professor of Industrial Services and Maintenance and a team member of Logistics research Group of Aalto University. He is an expert in supply chain management. His research is practice oriented and focuses on problem solving and solution spotting. He has published extensively on the improvement of industrial, project, and retail supply chains. Currently his focus is on integrating service supply chains and asset management. He holds an M.Sc. in Computer Sciences (1990) and Dr.Tech in Industrial Engineering and Management from Helsinki University of Technology (1995). He is a member of the Editorial Advisory Board of the *Journal of Operations Management* since 1998. In 2006 he received a Citation of Excellence Award by the Emerald Management Review Board as co-author with Matthias Holweg, Steven Disney, and Johanna Småros of one of the top fifty management articles for "Supply chain collaboration: making sense of the strategy continuum".

David Jentsch

Department of Factory Planning and Factory Management, Chemnitz University of Technology (Germany), david.jentsch@mb.tu-chemnitz.de

David Jentsch has been with the Department of Factory Planning and Factory Management as a research assistant and PhD student since 2009. He obtained a Master of Science degree in Systems Engineering from Chemnitz University of Technology in 2009. His research projects are in the area of innovation processes in production management and logistics.

Johan Karltun

Department of Industrial Engineering and Management, Jönköping University, johan.karltun@jth.hj.se

Johan Karltun is a senior lecturer at School of Engineering, Jönköping University, Sweden in work organisation and production management. He received his MSc in Mechanical Engineering from Chalmers Institute of Technology. His PhD is on change management in woodworking SMEs from Linköping Institute of Technology, Sweden. Prior to academia he worked in industry in various positions and as a research engineer within production management and ergonomics at the Swedish Institute for Wood Technology Research. His research interests include human factors in production planning and scheduling, change management and the integration of humans, technology and organization in various domains. Papers on his work have been published in *International Journal of Production Economics, Human Factors and Ergonomics in Manufacturing* and *Applied Ergonomics.*

Stefan Marsina

University of Economics in Bratislava, marsina@euba.sk

Stefan Marsina got a Master degree in Industrial Engineering, University of Zilina, Slovakia, then attended a Management Development Program, in Banff, Canada. He obtained a Ph.D. in Project Management from University of Economics in Bratislava, Slovakia. His was successively employed in the Machine industry, Geological survey industry and University of Economics where he teaches Management, Corporate Planning, Project Management (Bc., MSc., MBA, Ph.D. levels; Continuing education courses). His research interests are on practical application of the project management theory and methodology. He also has consulting activities on Project management, Strategic change, Business planning. Dr Marsina also performed various expertises (feasibility study for UNIDO, assessment of development projects for SMEs, assessment of the Research Networking Program application within the European Science Foundation).

Anne Mayère

University of Toulouse, anne.mayere@iut-tlse3.fr

Anne Mayère is Professor in Information and Communication Sciences at the University Technology Institute of Toulouse. She is a member of CERTOP CNRS Research Laboratory (UMR 5044), and more specifically of a team dealing with Communication and Risk in Health and Environment. She is interested in Information and Communication Technologies, the way they are implemented, and their contribution to organizing, with a focus on the related evolution of coordination between activities, professional groups and firms. In such coordination, planning plays a crucial role, and she questions the way different types of organization succeed, or not, in this continuous process. Regarding the different types of organizations, she had a specific interest in SMEs; nowadays, she studies healthcare delivery organizations, inquiring their risk management system and the computerization of patient medical information.

Nasser Mebarki
Département Qualité, Logistique Industrielle et Organisation (QLIO), University of Nantes, nasser.mebarki@univ-nantes.fr
Nasser Mebarki is currently assistant professor in Technology Management at the Technology Institute of Nantes (University of Nantes). He teaches production management, manufacturing scheduling, discrete event simulation and database management. His research interests involve discrete event simulation, dynamic scheduling, cooperation between human and machine for manufacturing scheduling. Dr Mebarki was the head of the Quality and Industrial Logistic department of the Technology Institute of Nantes (of the University of Nantes) between 2003 and 2006. His scientific research has been published in journals such as *International Journal of Production Research*, *International Journal of Production Economics*, *Journal of Intelligent Manufacturing*, and *Simulation Practice and Theory*.

Philip Moscoso
IESE Business School, Universidad de Navarra, pmoscoso@iese.edu
Philip G. Moscoso is a professor in the Operations and Technology Management department at IESE Business School, University of Navarra. He teaches courses for both executives and master students. Prior to joining IESE faculty, Philip Moscoso was a manager at the management consulting firm Bain & Co. Philip has completed senior executive programs at Harvard Business School and IESE, and obtained his M.Sc. and Doctorate at the Swiss Federal Institute of Technology (ETH) in Zurich in the department of Industrial Engineering and Management. His primary research interest is the development of strategies and systems that help firms achieve excellence in their service and supply chains. He has published his work in academic journals as *Ergonomics* and the *Journal of Engineering Design*, as well as in managerial journals. Additionally, he regularly writes opinion articles and is quoted in the business press.

Guillaume Pinot
Centre National de la Recherche Scientifique (CNRS: French National Research Centre), University of Technology of Compiègne, guillaume.pinot@irccyn.ec-nantes.fr
Guillaume Pinot is an engineer in computer science from the University of Technology of Compiègne and has a MSc degree in computer science from the University of Nantes. He made his Ph.D at the IRCCyN laboratory in Nantes. His subject was "cooperation between human and machine for scheduling under uncertainties". He used group sequencing to provide flexibility in the schedule. Thanks to the

understandable structure of group sequencing, the human can evaluate the different solutions allowed. Combinatorial optimisation is used to help the human to choose the operation to execute according to the group schedule. During his Ph.D, he was an active member of the HOPS research group. He defended his Ph.D in 2009. He is now a teaching assistant at the University of Nantes. He is particularly interested in combinatorial optimisation and especially scheduling using exact and hybrid methods.

Johann Riedel
Nottingham University Business School, johann.riedel@nottingham.ac.uk
Johann Riedel is a Senior Research Fellow in the Centre for Concurrent Enterprise. His main interest is in design, innovation and the management thereof. Dr Riedel holds a PhD from the University of Wolverhampton and Masters from Imperial College. He is active in a variety of European Union funded research projects. He has held the positions of coordinator, technical manager and workpackage manager. His current projects encompass the design of accessible interfaces, experience design and emotional design. He has also devised innovative educational tools, such as the Cosiga new product development simulation game. He has published in journals such as the International Journal of Operations and Production Management, Technovation, and R&D Management.

Ralph Riedel
Department of Factory Planning and Factory Management, Chemnitz University of Technology, ralph.riedel@mb.tu-chemnitz.de
Ralph Riedel has been working with the department of Factory Planning and Factory Management Chemnitz University of Technology since 2003. He has a diploma in Industrial Engineering and Management and in Mechanical Engineering. At the department he works a scientific assistant and he is responsible for the research group “Factory Organization and Factory Management”. His teaching subjects comprise Production Planning and Control, Industrial Engineering and Industrial Management, Logistics and Project Management. His research interests cover the fields of Project Management, Supply Chain Management and Supply Networks - especially in a globalised work – Innovation Processes and Innovation Management as well as Planning Processes. He has supervised numerous Diploma candidates and several PhD candidates. He is also regularly teaching at the National University of Ireland Galway.

Jan Riezebos
Faculty of Economics and Business, University of Groningen, j.riezebos@rug.nl
Jan Riezebos is director of the bachelor and master programs in Technology Management of the University of Groningen. He is Associate Professor in the Department of Operations of the Faculty of Economics and Business. Jan’s current research interests involve the application of operations research techniques to planning problems in and between firms.
His scientific research has been published in several books and in journals such as *Computers in Industry, European Journal of Operational Research, International Journal of Production Research, International Journal of Production Economics,*

International Journal of Project Management, Journal of Intelligent Manufacturing, and *Production Planning & Control.* He is a member of the editorial board of the *European Journal of Industrial Engineering* and the *International Journal of Information Systems and Supply Chain Management.* He has co-organized the third GTCM world conference (GTCM2006) on Group Technology and Cellular Manufacturing.

Craig Shepherd
Nottingham University Business School, craig.shepherd@nottingham.ac.uk
Craig Shepherd is a lecturer in Organisation Studies at Nottingham University Business School. Previously, he spent time working as an information technology consultant for Shell on Enterprise Resource Planning projects. Craig is secretary of the British Computer Society Sociotechnical Group. His research interests include new technology, the management of change, research methods and supply chain management.

Ulrike Starker
University of Bamberg, ulrike.starker@uni-bamberg.de
Ulrike Starker works as coach, trainer and psychotherapist. In 1987 she got a grant of the "Studienstiftung des Deutschen Volkes". She received a Master degree equivalent (Diplom) in psychology from the Otto-Friedrich-Universiy of Bamberg in 1991. Her PhD thesis was about the psychology of aesthetics in West- and East Germany (1996). Since 1991 she works at the Otto-Friedrich-University of Bamberg as research assistant, professor and lecturer in the fields of Cognitive Science and Educational Research. Her academic teachers were Dietrich Dörner and Lothar Laux. Domains of research: Individual problem solving strategies of managers. Role of motivational and emotional aspects in technical and organizational innovation processes. Cognitive science esp. computer models about the interaction of cognitive, emotional and motivational processes in problem solving. She has published a book and 17 papers in international and German journals and books.

Wout van Wezel
Faculty of Economics and Business, University of Groningen, w.m.c.van.wezel@rug.nl
Wout van Wezel is Assistant Professor of Production System Design, Faculty of Economics and Business, University of Groningen, where he also obtained his PhD. His research interests include planning and scheduling, human-centered decision support systems, and production control.

Rüdiger von der Weth
HTW, University of Applied Sciences, weth@wiwi.htw-dresden.de
Rüdiger von der Weth received his Master degree equivalent (Diplom) in psychology from Technical University in Berlin in 1985. His PhD thesis was on goal conflicts in complex management tasks (1989). He became a full professor at the "Hochschule für Technik Stuttgart" in 2000 (Applied Psychology) and changed to

the HTW Dresden in 2003 (Human Resource Management and Ergonomics). His research focus is on the following: the role of motivational and emotional aspects in technical and organizational innovation processes. Planning and decision making in professional contexts (industry, public service, agroforestry). Development of methods for participation in industrial, public and political decision making. Training for project managers and evaluation of the effects. Rüdiger von der Weth works as consultant and management trainer in these fields. He has published three scientific books and 56 papers in international and German journals and books. He works as scientific reviewer for several international research journals (e.g. *Research in Engineering Design*, *Learning and Education*) and was member of the management committee of the research network HOPS.

Vincent Wiers
School of Industrial Engineering, Eindhoven University of Technology, v.c.s.wiers @tue.nl
Vincent Wiers is an Industrial Fellow in the area of Human Factors in Planning and Scheduling at Eindhoven University of Technology, and a self-employed consultant in the area of Advanced Planning Systems implementation. He obtained a PhD from Eindhoven University of Technology in 1997, and has since then worked for most of his time in parallel in industry and academia. Together with Kenneth McKay, he is the author of the book "Practical Production Control: A survival guide for planners and schedulers". He has published extensively in journals such as Journal of Operations Management, Production Planning and Control, and Computers in Industry.
Furthermore, I noted that in some of the resumes, the titles of journals were formatted in italics, while in others they were not. My suggestion is to have them all in italics. I have underlined the journal titles in the document.

Peter Williams
University of Limerick, peter.williams@ul.ie
Peter Williams lectures in operations management at the Department of Manufacturing and Operations Engineering, UL, and is coordinator of the Health Systems research group in the Enterprise Research Centre. He has bachelor and PhD degrees in engineering (TCD 1977; UCD 1984), and was Course Director for the BSc Production Management degree course 1997–2009. His is interested in modelling and management skills in supply-chains, and more recently in the healthcare delivery operations. He is particularly interested in questions related to the scope of operational models and relevance at the coalface, what skills people apply in real decision-making, and how work and peoples' preparation for it might be better designed and supported, especially from the perspective of those charged with operating in large systems. He gratefully acknowledges the encouragement of Ken McKay, Bart McCarthy and Peter Higgins in pre-HOPS days to engage with the challenges posed in this important facet of work.

Index

The manufacturer's authorised representative in the EU is Springer Nature Customer Service Centre GmbH, Europaplatz 3, 69115 Heidelberg, Germany. If you have any concerns regarding our products, please contact ProductSafety@springernature.com

Printed and bound by CPI Group (UK) Ltd, Croydon, CR0 4YY

07/01/2026

02030921-0015